THE
ROBOT BUILDER'S
BONANZA

GORDON McCOMB

SECOND EDITION

McGraw-Hill

New York   San Francisco   Washington, D.C.   Auckland   Bogotá
Caracas   Lisbon   London   Madrid   Mexico City   Milan
Montreal   New Delhi   San Juan   Singapore
Sydney   Tokyo   Toronto

Library of Congress Cataloging-in-Publication Data

McComb, Gordon.
    The robot builder's bonanza / Gordon McComb.—2nd ed.
        p.   cm.
    Includes bibliographical references and index.
    ISBN 0-07-136296-7
    1. Robotics.   I. Title.

TJ211.M418 2000
629.8'92— dc21   00-058426

McGraw-Hill

*A Division of The **McGraw·Hill** Companies*

  2 3 4 5 6 7 8 9 0 AGM/AGM   0 5 4 3 2 1 0

ISBN 0-07-136296-7

The sponsoring editor for this book was Scott Grillo. The editing supervisor was Caroline Levine, and the production supervisor was Sherri Souffrance. It was set in Times New Roman PS per the TAB4 Design by Deirdre Sheean of McGraw-Hill's Professional Book Group composition unit, Hightstown, New Jersey.

Printed and bound by Quebecor/Martinsburg.

 This book is printed on recycled, acid-free paper containing a minimum of 50% recycled, de-inked fiber.

CONTENTS

This time for my son, Max,
who dreams of robots.

ACKNOWLEDGMENTS

Only until you've climbed the mountain can you look behind you and see the vast distance that you've covered, and remember those you've met along the way who made your trek a little easier.

Now that this book is finally finished, after the many miles of weary travel, I look back to those who helped me turn it into a reality and offer my heartfelt thanks: To the gang on comp.robotics.misc, for the great ideas, wisdom, and support; to Scott Savage, designer of the OOPic; to Frank Manning and Jack Schoof of NetMedia for their help with the BasicX; to Tony Ellis, a real-life "Q" if I ever met one; to Scott Grillo and the editors at McGraw-Hill; to my agents Matt Wagner and Bill Gladstone; and last and certainly not least, to my wife Jennifer.

INTRODUCTION

The word *robot* is commonly defined as a mechanical device capable of performing human tasks, or behaving in a human-like manner. No argument here. The description certainly fits.

But to the robotics experimenter, "robot" has a completely different meaning. A robot is a special brew of motors, solenoids, wires, and assorted electronic odds and ends, a marriage of mechanical and electronic gizmos. Taken together, the parts make a half-living but wholly personable creature that can vacuum the floor, serve drinks, protect the family against intruders and fire, entertain, educate, and lots more. In fact, there's almost *no limit* to what a well-designed robot can do.

In just about any science, it is the independent experimenter who first establishes the pioneering ideas and technologies. Robert Goddard experimented with liquid-fuel rockets during World War I; his discoveries paved the way for modern-day space-flight. In the mid-1920s, John Logie Baird experimented with sending pictures of objects over the airwaves. His original prototypes, which transmitted nothing more than shadows of images, were a precursor to television and video.

Robotics—like rocketry, television, and countless other technology-based endeavors—started small. But progress in the field of robots has been painfully slow. Robotics is still a cottage industry, even considering the special-purpose automatons now in wide use in automotive manufacturing. What does this mean for the robotics experimenter? There is plenty of room for growth, with a lot of discoveries yet to be made—perhaps more so than in any other high-tech discipline.

Inside Robot Builder's Bonanza

Robot Builder's Bonanza, Second Edition takes an educational but fun approach to designing working robots. Its modular projects take you from building basic motorized platforms to giving the machine a brain—and teaching it to walk and talk and obey commands.

If you are interested in mechanics, electronics, or robotics, you'll find this book a treasure chest of information and ideas on making thinking machines. The projects in *Robot Builder's Bonanza* include all the necessary information on how to construct the essential building blocks of a personal robot. Suggested alternative approaches, parts lists, and sources of electronic and mechanical components are also provided where appropriate.

Several good books have been written on how to design and build your own robot. But most have been aimed at making just one or two fairly sophisticated automatons, and at a fairly high price. Because of the complexity of the robots detailed in these other books, they require a fairly high level of expertise and pocket money on your part.

Robot Builder's Bonanza is different. Its modular "cookbook" approach offers a mountain of practical, easy to follow, and inexpensive robot experiments. Taken together, the modular projects in *Robot Builder's Bonanza* can be combined to create several different types of highly intelligent and workable robots of all shapes and sizes—rolling robots, walking robots, talking robots, you name it. You can mix and match projects as desired.

About the Second Edition

This book is a completely revised edition of *Robot Builder's Bonanza*, first published in 1987. The first edition of this book has been a perennial bestseller, and is one of the most widely read books ever published on hobby robotics.

This new edition provides timely updates on the latest technology and adds many new projects. In the following pages you'll find updated coverage on exciting technologies such as robotic sensors, robot construction kits, and advanced stepper and DC motor control. Plus, you'll find new information on microcontrollers such as the Basic Stamp, digital compasses, open- and closed-loop feedback mechanisms, new and unique forms of "soft touch" sensors including those using lasers and fiber optics, radio-controlled servo motors, and much, much more.

Book Updates

Periodic updates to *Robot Builder's Bonanza* can be found at *http://www.robot-oid.com*. You'll find new and updated links to Web sites and manufacturer addresses, a robot product and parts finder, tutorials on robot construction, a robot builder's discussion board, and more.

What You Will Learn

In the more than three dozen chapters in this book you will learn about a sweeping variety of technologies, all aimed at helping you learn robot design, construction, and application. You'll learn about:

- *Robot-building fundamentals.* How a robot is put together using commonly available parts such as plastic, wood, and aluminum.
- *Locomotion engineering.* How motors, gears, wheels, and legs are used to propel your robot over the ground.
- *Constructing robotic arms and hands.* How to use mechanical linkages to grasp and pick up objects.

- *Sensor design.* How sensors are used to detect objects, measure distance, and navigate open space.
- *Adding sound capabilities.* Giving your robot creation the power of voice and sound effects so that it can talk to you, and you can talk back.
- *Remote control.* How to operate and "train" your robot using wired and wireless remote control.
- *Computer control.* How to use and program a computer or microcontroller for operating a robot.

…plus much more.

How to Use This Book

Robot Builder's Bonanza is divided into six main sections. Each section covers a major component of the common personal or hobby (as opposed to commercial or industrial) robot. The sections are as follows:

- *Robot Basics.* What you need to get started; setting up shop; how and where to buy robot parts.
- *Robot Construction.* Robots made of plastic, wood, and metal; working with common metal stock; converting toys into robots; using LEGO parts to create robots; using the LEGO Mindstorms Robotics Invention System.
- *Power, Motors, and Locomotion.* Using batteries; powering the robot; working with DC, stepper, and servo motors; gear trains; walking robot systems; special robot locomotion systems.
- *Practical Robotics Projects.* Over a half-dozen step-by-step projects for building wheels and legged robot platforms; arm systems; gripper design.
- *Computers and Electronic Control.* "Smart" electronics; robot control via a computer or microcontroller; infrared remote control; radio links.
- *Sensors and Navigation.* Speech synthesis and recognition; sound detection; robot eyes; smoke, flame, and heat detection; collision detection and avoidance; ultrasonic and infrared ranging; infrared beacon systems; track guidance navigation.

Many chapters present one or more projects that you can duplicate for your own robot creations. Whenever practical, I designed the components as discrete building blocks, so that you can combine the blocks in just about any configuration you desire. The robot you create will be uniquely yours, and yours alone.

I prefer to think of *Robot Builder's Bonanza* not as a textbook on how to build robots but as a *treasure map*. The trails and paths provided between these covers lead you on your way to building one or more complete and fully functional robots. You decide how you want your robots to appear and what you want your robots to do.

Expertise You Need

Robot Builder's Bonanza doesn't contain a lot of hard-to-decipher formulas, unrealistic assumptions about your level of electronic or mechanical expertise, or complex designs that only a seasoned professional can tackle. This book was written so that just about anyone can enjoy the thrill and excitement of building a robot. Most of the projects can be duplicated without expensive lab equipment, precision tools, or specialized materials, and at a cost that won't contribute to the national debt!

If you have some experience in electronics, mechanics, or robot building in general, you can skip around and read only those chapters that provide the information you're looking for. Like the robot designs presented, the chapters are very much stand-alone modules. This allows you to pick and choose, using your time to its best advantage.

However, if you're new to robot building, and the varied disciplines that go into it, you should take a more pedestrian approach and read as much of the book as possible. In this way, you'll get a thorough understanding of how robots tick. When you finish with the book, you'll know the kind of robot(s) you'll want to make, and how you'll make them.

Conventions Used in This Book

You need little advance information before you can jump head first into this book, but you should take note of a few conventions I've used in the description of electronic parts, and the schematic diagrams for the electronic circuits.

TTL integrated circuits are referenced by their standard 74XX number. The common "LS" or "HC" identifier is assumed. I built most of the circuits using LS or HC TTL chips, but unless otherwise indicated, the projects should work with the other TTL families. However, if you use a type of TTL chip other than LS or HC, you should consider current consumption, fanout, and other design criteria. These may affect the operation or performance of the circuit.

The chart in Fig. I-1 details the conventions used in the schematic diagrams. Note that nonconnected wires are shown by a direct cross or lines, or a broken line. Connected wires are shown by the connecting dot.

Details on the specific parts used in the circuits are provided in the parts list tables that accompany the schematic. Refer to the parts list for information on resistor and capacitor type, tolerance, and wattage or voltage rating.

In all full circuit schematics, the parts are referenced by component type and number.

- *IC#* means an integrated circuit (IC).
- *R#* means a resistor or potentiometer (variable resistor).
- *C#* means a capacitor.

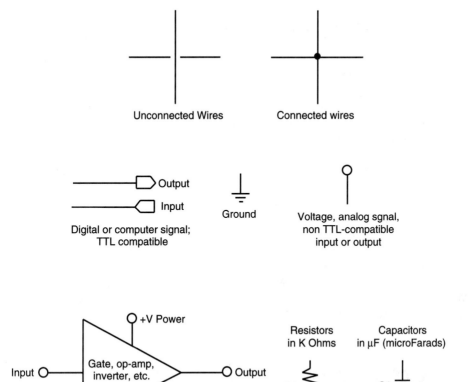

FIGURE I.1 Schematic diagram conventions used in this book.

- *D#* means a diode, a zener diode, and, sometimes a light-sensitive photodiode.
- *Q#* means a transistor and, sometimes, a light-sensitive phototransistor.
- *LED#* means a light-emitting diode (most any visible LED will do unless the parts list specifically calls for an infrared or other special purpose LED).
- *XTAL#* means a crystal or ceramic resonator.
- Finally, *S#* means a switch; RL# means a relay; SPKR#, a speaker; TR#, a transducer (usually ultrasonic); and MIC#, a microphone.

Enough talk. Turn the page and begin the journey. The treasure awaits you.

PART 1

ROBOT BASICS

THE

ROBOT

EXPERIMENTER

There he sits, as he's done countless long nights before, alone and deserted in a dank and musty basement. With each strike of his ball-peen hammer comes an ear-shattering bong and an echo that seems to ring forever. Slowly, his creation takes shape and form—it first started as an unrecognizable blob of metal and plastic, then it was transformed into an eerie silhouette, then...

Brilliant and talented, but perhaps a bit crazed, he is before his time: a social outcast, a misfit who belongs neither to science nor fiction. He is the robot experimenter, and all he wants to do is make a mechanical creature that serves drinks at parties and wakes him up in the morning.

Okay, maybe this is a rather dark view of the present-day hobby robotics experimenter. But though you may find a dash of the melodramatic in it, the picture is not entirely unrealistic. It's a view held by many outsiders to the robot-building craft. It's a view that's over 100 years old, from the time when the prospects of building a human-like machine first came within technology's grasp. It's a view that will continue for another 100 years, perhaps beyond.

Like it or not, if you're a robot experimenter, you are an oddball, an egghead, and—yes, let's get it all out—a little on the *weird* side!

As a robot experimenter, you're not unlike Victor Frankenstein, the old-world doctor from Mary Wollstonecraft Shelley's immortal 1818 horror-thriller. Instead of robbing graves in the still of night, you "rob" electronic stores, flea markets, surplus outlets, and other specialty shops in your unrelenting quest—your *thirst*—for all kinds and sizes of motors, batteries, gears, wires, switches, and other odds and ends. Like Dr. Frankenstein, you galvanize life from these "dead" parts.

If you have yet to build your first robot, you're in for a wonderful experience. Watching your creation scoot around the floor or table can be exhilarating. Those around you may not immediately share your excitement, but you know that you've built something—however humble—with your own hands and ingenuity.

If you're one of the lucky few who has already assembled a working robot, then you know of the excitement I refer to. You know how thrilling it is to see your robot obey your commands, as if it were a trusted dog. You know the time and effort that went into constructing your mechanical marvel, and although others may not always appreciate it (especially when it marks up the kitchen floor with its rubber tires) you are satisfied with the accomplishment and look forward to the next challenge.

And yet if you have built a robot, you also know of the heartache and frustration inherent in the process. You know that not every design works and that even a simple engineering flaw can cost weeks of work, not to mention ruined parts. This book will help you—beginner and experienced robot maker alike—get the most out of your robotics hobby.

The Building-block Approach

One of the best ways to experiment with—and learn about—hobby robots is to construct individual robot components, then combine the completed modules to make a finished, fully functional machine. For maximum flexibility, these modules should be interchangeable whenever possible. You should be able to choose locomotion system "A" to work with appendage system "B," and operate the mixture with control system "C"—or any variation thereof.

The robots you create are made from building blocks, so making changes and updates is relatively simple and straightforward. When designed and constructed properly, the building blocks, as shown in diagram form in Fig. 1.1, may be shared among a variety of robots. It's not unusual to reuse parts as you experiment with new robot designs.

Most of the building-block designs presented in the following chapters are complete, working subsystems. Some operate without ever being attached to a robot or control computer. The way you interface the modules is up to you and will require some forethought and attention on your part (I'm not doing *all* the work, you know!). Feel free to experiment with each subsystem, altering it and improving upon it as you see fit. When it works the way you want, incorporate it into your robot, or save it for a future project.

Basic Skills

What skills do you need as a robot experimenter? Certainly, if you are already well versed in electronics and mechanical design, you are on your way to becoming a robot experimenter *extraordinaire*. But an intimate knowledge of electronics and mechanical design is not absolutely necessary.

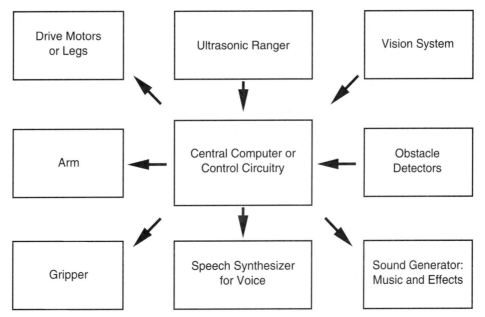

FIGURE 1.1 The basic building blocks of a fully functional robot, including central processor (brain), locomotion (motors), and sensors (switches, sonar, etc.).

All you really need to start yourself in the right direction as a robot experimenter is a basic familiarity with electronic theory and mechanics (or time and interest to study the craft). The rest you can learn as you go. If you feel that you're lacking in either beginning electronics or mechanics, pick up a book or two on these subjects at the bookstore or library. See Appendix A, "Further Reading," for a selected list of suggested books and magazines. In addition, you may wish to read through the seven chapters in Part 1 of this book to learn more about the fundamentals of electronics and computer programming.

ELECTRONICS BACKGROUND

Study analog and digital electronic theory, and learn the function of resistors, capacitors, transistors, and other common electronic components. Your mastery of the subject need not be extensive, just enough so that you can build and troubleshoot electronic circuits for your robot. You'll start out with simple circuits with a minimum of parts, and go from there. As your skills increase, you'll be able to design your own circuits from scratch, or at the very least, customize existing circuits to match your needs.

Schematic diagrams are a kind of recipe for electronic circuits. The designs in this book, as well as those in most any book that deals with electronics, are in schematic form. You owe it to yourself to learn how to read a schematic. There are really only a few dozen common schematic symbols, and memorizing them takes just one evening of concentrated study. Several books have been written on how to read schematic diagrams, and the basics are also covered in Chapter 5, "Common Electronic Components." See also Appendix A for a list of other suggested books on robotics.

Sophisticated robots use a computer or microcontroller to manage their actions. In this book you'll find plenty of projects, plans, and solutions for connecting the hardware of your robot to any of several kinds of robot "brains." Like all computers, the ones for robot control need to be programmed. If you are new or relatively new to computers and programming, start with a beginners' computer book, then move up to more advanced texts. Chapter 7, "Programming Concepts—The Fundamentals," covers programming basics.

MECHANICAL BACKGROUND

Some robot builders are more comfortable with the mechanical side of robot building than the electronic side—they can *see* gears meshing and pulleys moving. Regardless of your comfort level with mechanical design, you do not need to possess an extensive knowledge of mechanical and engineering theory to build robots. This book provides some mechanical theory as it pertains to robot building, but you may want to supplement your learning with books or study aids.

There is a wealth of books, articles, and online reading materials on mechanical design equations, and engineering formulas, so this book will not repeat the information. This means we will have more room to describe more robotics projects you can experiment with. Appendix A, "Further Reading," and Appendix C, "Robot Information on the Internet," include a multitude of sources that provide good, solid design equations and formulas.

THE WORKSHOP APTITUDE

To be a successful robot builder, you must be comfortable working with your hands and thinking problems through from start to finish. You should know how to use common shop tools, including all safety procedures, and have some basic familiarity with working with wood, lightweight metals (mostly aluminum), and plastic. Once more, if you feel your skills aren't up to par, read up on the subject and try your hand at a simple project or two first.

You'll find construction tips and techniques throughout this book, but nothing beats hands-on shop experience. With experience comes confidence, and with both comes more professional results. Work at it long enough, and the robots you build may be indistinguishable from store-bought models (in appearance, not capability; yours will undoubtedly be far more sophisticated!).

THE TWO MOST IMPORTANT SKILLS

So far, I've talked about basic skills that are desirable for the hobby robotics field. There are others. Two important skills that you *can't* develop from reading books are *patience* and the *willingness to learn*. Both are absolutely essential if you want to build your own working robots. Give yourself time to experiment with your projects. Don't rush into things because you are bound to make mistakes if you do. If a problem continues to nag at you, put the project aside and let it sit for a few days. Keep a small notebook handy and jot down your ideas so you won't forget them.

If trouble persists, perhaps you need to bone up on the subject before you can adequately tackle the problem. Take the time to study, to learn more about the various sciences and disciplines involved. While you are looking for ways to combat your current dilemma, you are increasing your general robot-building knowledge. *Research is never in vain.*

Ready-Made, Kits, or Do It Yourself?

This is a wonderful time to be an amateur robot builder. Not only can you construct robots "from scratch," you can buy any of several dozen robot kits and assemble them using a screwdriver and other common tools. If you don't particularly like the construction aspects of robotics, you can even purchase ready-made robots—no assembly required. With a ready-made robot you can spend all your time connecting sensors and other apparatuses to it and figuring out new and better ways to program it.

Whether you choose to buy a robot in ready-made or kit form, or build your own from the ground up, it's important that you match your skills to the project. This is especially true if you are just starting out. While you may seek the challenge of a complex project, if it's beyond your present skills and knowledge level you'll likely become frustrated and abandon robotics before you've given it a fair chance. If you want to build your own robot, start with a simple design—a small rover, like those in Chapters 8 through 12. For now, stay away from the more complex walking and heavy-duty robots.

The Mind of the Robot Experimenter

Robot experimenters have a unique way of looking at things. They take nothing for granted:

- At a restaurant, it's the robot experimenter who collects the carcasses of lobster and crabs to learn how these ocean creatures use articulated joints, in which the muscles and tendons are *inside* the bone. Perhaps the articulation and structure of a lobster leg can be duplicated in the design of a robotic arm . . .
- At a county fair, it's the robot experimenter who studies the way the "egg-beater" ride works, watching the various gears spin in perfect unison. Perhaps the gear train can be duplicated in an unusual robot locomotion system . . .
- At a phone booth, it's the robot experimenter who listens to the tones emitted when the buttons are pressed. These tones, the experimenter knows, trigger circuitry at the phone company office to call a specific telephone out of all the millions in the world. Perhaps these or similar tones can be used to remotely control a robot . . .
- At work on the computer, it's the robot experimenter who rightly assumes that if a computer can control a printer or plotter through an interface port, the same computer and interface can be used to control a robot . . .
- When taking a snapshot at a family gathering, it's the robot experimenter who studies the inner workings of the automatic focus system of the camera. The camera uses ultrasonic sound waves to measure distance and automatically adjusts its lens to keep things in focus. The same system should be adaptable to a robot, enabling it to judge distances and "see" with sound . . .

The list could go on and on. The point? All around us, from nature's designs to the latest electronic gadgets, are an infinite number of ways to make better and more sophisticated robots. Uncovering these solutions requires *extrapolation*—figuring out how to apply one design and make it work in another application, then experimenting with the contraption until everything works.

From Here

To learn more about . . .	*Read*
Fundamentals of electronics	Chapter 5, *Common Electronic Components*
Basics on how to read a schematic	
Electronics construction techniques	Chapter 6, *Electronic Construction Techniques*
Computer programming fundamentals	Chapter 7, *Programming Concepts—The Fundamentals*
Robot construction using wood, plastic, and metal	Chapters 8-10
Making robots from old toys found in your closet	Chapter 11, *Constructing High-tech Robots from Toys*
What your robot should do	Chapter 42, *Tips, Tricks, and Tidbits for the Robot Experimenter*
Other sources about electronics and mechanics	Appendix A, *Further Reading* Appendix C, *Robot Information on the Internet*

2

ANATOMY

OF A ROBOT

We humans are fortunate. The human body is, all things considered, a nearly perfect machine: it is (usually) intelligent, it can lift heavy loads, it can move itself around, and it has built-in protective mechanisms to feed itself when hungry or to run away when threatened. Other living creatures on this earth possess similar functions, though not always in the same form.

Robots are often modeled after humans, if not in form then at least in function. For decades, scientists and experimenters have tried to duplicate the human body, to create machines with intelligence, strength, mobility, and auto-sensory mechanisms. That goal has not yet been realized, but perhaps some day it will.

Nature provides a striking model for robot experimenters to mimic, and it is up to us to take the challenge. Some, but by no means all, of nature's mechanisms—human or otherwise—can be duplicated to some extent in the robot shop. Robots can be built with eyes to see, ears to hear, a mouth to speak, and appendages and locomotion systems of one kind or another to manipulate the environment and explore surroundings.

This is fine theory; what about real life? Exactly what constitutes a real hobby robot? What basic parts must a machine have before it can be given the title "robot"? Let's take a close look in this chapter at the anatomy of robots and the kinds of materials hobbyists use to construct them. For the sake of simplicity, not every robot subsystem in existence will be covered, just the components that are most often found in amateur and hobby robots.

Tethered versus Self-Contained

People like to debate what makes a machine a "real" robot. One side says that a robot is a completely *self-contained, autonomous* (self-governed) machine that needs only occasional instructions from its master to set it about its various tasks. A self-contained robot has its own power system, brain, wheels (or legs or tracks), and manipulating devices such as claws or hands. This robot does not depend on any other mechanism or system to perform its tasks. It's complete in and of itself.

The other side says that a robot is anything that moves under its own motor power *for the purpose of performing near-human tasks* (this is, in fact, the definition of the word *robot* in most dictionaries). The mechanism that does the actual task is the robot itself; the support electronics or components may be separate. The link between the robot and its control components might be a wire, a beam of infrared light, or a radio signal.

In the experimental robot from 1969 shown in Fig. 2.1, for example, a man sat inside the mechanism and operated it, almost as if driving a car. The purpose of the four-legged "lorry" was not to create a self-contained robot but to further the development of *cybernetic anthropomorphous machines*. These were otherwise known as *cyborgs*, a concept further popularized by writer Martin Caidin in his 1973 novel *Cyborg* (which served as the inspiration for the 1970s television series, *The Six Million Dollar Man*).

We won't argue the semantics of robot design here (this book is a *treasure map* after all, not a textbook on theory), but it's still necessary to establish some of the basic characteristics of robots. What makes a robot a robot and just not another machine? For the purposes of this book, let's consider a robot as *any device that—in one way or another—mimics human or animal functions.* The way the robot does this is of no concern; the fact that it does it at all is enough.

The functions that are of interest to the robot builder run a wide gamut: from listening to sounds and acting on them, to talking and walking or moving across the floor, to picking up objects and sensing special conditions such as heat, flames, or light. Therefore, when we talk about a robot it could very well be a self-contained automaton that takes care of itself, perhaps even programming its own brain and learning from its surroundings and environment. Or it could be a small motorized cart operated by a strict set of predetermined instructions that repeats the same task over and over again until its batteries wear out. Or it could be a radio-controlled arm that you operate manually from a control panel. Each is no less a robot than the others, though some are more useful and flexible. As you'll discover in this chapter and those that follow, how complex your robot creations are is completely up to you.

Mobile versus Stationary

Not all robots are meant to scoot around the floor. Some are designed to stay put and manipulate some object placed before them. In fact, outside of the research lab and hobbyist garage, the most common types of robots, those used in manufacturing, are *stationary*. Such robots assist in making cars, appliances, and even *other* robots!

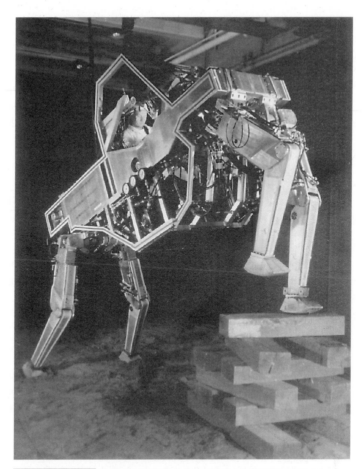

FIGURE 2.1 This quadruped from General Electric was controlled by a human operator who sat inside it. The robot was developed in the late 1960s under a contract with the U.S. government. Photo courtesy of General Electric.

Other common kinds of stationary robots act as shields between a human operator or supervisor and some dangerous material, such as radioactive isotopes or caustic chemicals. Stationary robots are armlike contraptions equipped with grippers or special tools. For example, a robot designed for welding the parts of a car is equipped with a welding torch on the end of its "arm." The arm itself moves into position for the weld, while the car slowly passes in front of the robot on a conveyor belt.

Conversely, *mobile* robots are designed to move from one place to another. Wheels, tracks, or legs allow the robot to traverse a terrain. Mobile robots may also feature an armlike appendage that allows them to manipulate objects around them. Of the two—stationary or mobile—the mobile robot is probably the more popular project for hobbyists to build. There's something endearing about a robot that scampers across the floor, either chasing or being chased by the cat.

As a serious robot experimenter, you should not overlook the challenge and education you can gain from building both types of robots. Stationary robots typically require greater precision, power, and balance, since they are designed to grasp and lift objects—hopefully not destroying the objects they handle in the process. Likewise, mobile robots present their own difficulties, such as maneuverability, adequate power supply, and avoiding collisions.

Autonomous versus Teleoperated

Among the first robots ever demonstrated for a live audience were fake "robots" that were actually machines remotely controlled by a person off stage. No matter. People thrilled at the concept of the robot, which many anticipated would be an integral part of their near futures (like flying to work in your own helicopter and colonies on Mars by 1975…yeah, right!).

These days, the classic view of the robot is a fully autonomous machine, like Robby from *Forbidden Planet*, Robot B-9 from *Lost in Space*, or that R2-D2 thingie from *Star Wars*. With these robots (or at least the make-believe fictional versions), there's no human operator, no remote control, no "man behind the curtain." While many actual robots are indeed fully autonomous, many of the most important robots of the past few decades have been *teleoperated*. A teleoperated robot is one that is commanded by a human and operated by remote control. The typical "tele-robot" uses a video camera that serves as the eyes for the human operator. From some distance—perhaps as near as a few feet to as distant as several million miles—the operator views the scene before the robot and commands it accordingly.

The teleoperated robot of today is a far cry from the radio-controlled robots of the world's fairs of the 1930s and 1940s. Many tele-robots, like the world-famous Mars Rover Sojourner, the first interplanetary dune buggy, are actually half remote controlled and half autonomous. The low-level functions of the robot are handled by a microprocessor on the machine. The human intervenes to give general-purpose commands, such as "go forward 10 feet" or "hide, here comes a Martian!" The robot is able to carry out basic instructions on its own, freeing the human operator from the need to control every small aspect of the machine's behavior.

The notion of tele-robotics is certainly not new—it goes back to at least the 1940s and the short story "Waldo" by noted science fiction author Robert Heinlein. It was a fantastic idea at the time, but today modern science makes it eminently possible. Stereo video cameras give a human operator 3-D depth perception. Sensors on motors and robotic arms provide feedback to the human operator, who can actually "feel" the motion of the machine or the strain caused by some obstacle. Virtual reality helmets, gloves, and motion platforms literally put the operator "in the driver's seat."

This book doesn't discuss tele-robotics in any extended way, but if the concept interests you, read more about it and perhaps construct a simple tele-robot using a radio or infrared link and a video camera. See Appendix A, "Further Reading," for more information.

The Body of the Robot

Like the human body, the body of a robot—at least a self-contained one—holds all its vital parts. The body is the superstructure that prevents its electronic and electromechanical "guts" from spilling out. Robot bodies go by many names, including *frame* and *chassis*, but the idea is the same.

SKELETAL STRUCTURES

In nature and in robotics, there are two general types of support frames: endoskeleton and exoskeleton. Which is better? Both: In nature, the living conditions of the animal and its eating and survival tactics determine which skeleton is best. The same is true of robots.

- *Endoskeleton* support frames are the kind found in many critters—including humans, mammals, reptiles, and most fish. The skeletal structure is on the inside; the organs, muscles, body tissues, and skin are on the outside of the bones. The endoskeleton is a characteristic of vertebrates.
- *Exoskeleton* support frames have the "bones" on the outside of the organs and muscles. Common exoskeletal creatures are spiders, all shellfish such as lobsters and crabs, and an endless variety of insects.

FRAME CONSTRUCTION

The main structure of the robot is generally a wood, plastic, or metal frame, which is constructed a little like the frame of a house—with a bottom, top, and sides. This gives the automaton a boxy or cylindrical shape, though any shape is possible. It could even emulate the human form, like the "robot" in Fig. 2.2. For a machine, however, the body shape of men and women is a terribly inefficient one.

Onto the frame of the robot are attached motors, batteries, electronic circuit boards, and other necessary components. In this design, the main support structure of the robot can be considered an exoskeleton because it is outside the "major organs." Further, this design lacks a central "spine," a characteristic of endoskeletal systems and one of the first things most of us think about when we try to model robots after humans. In many cases, a shell is sometimes placed over these robots, but the "skin" is for looks only (and sometimes the protection of the internal components), not support. Of course, some robots are designed with endoskeletal structures, but most such creatures are reserved for high-tech research and development projects and science fiction films. For the most part, the main bodies of your robots will have an exoskeleton support structure because they are cheaper to build, stronger, and less prone to problems.

SIZE AND SHAPE

The size and shape of the robot can vary greatly, and size alone does not determine the intelligence of the machine nor its capabilities. Homebrew robots are generally the size of a small dog, although some are as compact as an aquarium turtle and a few as large as

FIGURE 2.2 The *android* design of robots is the most difficult to achieve, not only because of its bipedal (two-leg) structure, but because it distributes the weight toward the mid and top sections of the body. In reality, this "android" is science fiction writer J. Steven York modeling the latest in casual robot ready-to-wear.

Arnold Schwarzenegger (if one of these asks you "Are you Sarah Conner?" answer "No!"). The overall shape of the robot is generally dictated by the internal components that make up the machine, but most designs fall into one of the following "categories":

- *Turtle*. Turtle robots are simple and compact, designed primarily for "tabletop robotics." Turtlebots get their name from the fact that their bodies somewhat resemble the shell of a turtle and also from early programming language, turtle graphics, which was adapted for robotics use in the 1970s.
- *Vehicle*. These scooter-type robots are small automatons with wheels. In hobby robotics, they are often built using odds and ends like used compact discs, extra LEGO parts, or the chassis of a radio-controlled car. The small vehicular robot is also used in science and industry: the Rover Sojourner, built by NASA, explored the surface of Mars in July 1997.
- *Rover*. Greatly resembling the famous R2-D2 of *Star Wars* fame, rovers tend to be short and stout and are typically built with at least some humanlike capabilities, such as fire-fighting or intruder detection. Some closely resemble a garbage can—in fact, not a few hobby robots are actually built from metal and plastic trash cans! Despite the euphemistic title, "garbage can" robots represent an extremely workable design approach.
- *Walker*. A walking robot uses legs, not wheels or tracks, to move about. Most walker 'bots have six legs, like an insect, because they provide excellent support and balance. However, robots with as few as one leg ("hoppers") and as many as 8 to 10 legs have been successfully built and demonstrated.
- *Appendage*. Appendage designs are used specifically with robotic arms, whether the arm is attached to a robot or is a stand-alone mechanism.
- *Android*. Android robots are specifically modeled after the human form and are the type most people picture when talk turns to robots. Realistically, android designs are the most restrictive and least workable, inside or outside the robot lab.

This book provides designs and construction details for at least one robot in every one of the preceding types except *Android*. I'll leave that to another book.

FLESH AND BONE

In the 1926 movie classic *Metropolis*, an evil scientist, Dr. Rotwang, transforms a cold and calculating robot into the body of a beautiful woman. This film, generally considered to be the first science fiction cinema epic, also set the psychological stage for later movies, particularly those of the 1950s and 1960s. The shallow and stereotypical character of Dr. Rotwang, shown in the movie still in Fig. 2.3, proved to be a common theme in countless movies. The shapely robotrix changed form for these other films, but not its evil character. Robots have often been depicted as metal creatures with hearts as cold as their steel bodies.

Which brings us to an interesting question: Are all "real" robots made of heavy-gauge steel, stuff so thick that bullets, disinto-ray guns, even atomic bombs can't penetrate? Indeed, while metal of one kind or another is a major component of robot bodies, the list of materials you can use is much larger and diverse. Hobby robots can be easily constructed from aluminum, steel, tin, wood, plastic, paper, foam, or a combination of them all:

FIGURE 2.3
The evil Dr. Rotwang
and the robot, from the
classic motion picture
Metropolis.

- *Aluminum*. Aluminum is the best all-around robot-building material for medium and large machines because it is exceptionally strong for its weight. Aluminum is easy to cut and bend using ordinary shop tools. It is commonly available in long lengths of various shapes, but it is somewhat expensive.
- *Steel*. Although sometimes used in the structural frame of a robot because of its strength, steel is difficult to cut and shape without special tools. Stainless steel is sometimes used for precision components, like arms and hands, and also for parts that require more strength than a lightweight metal (such as aluminum) can provide. Expensive.
- *Tin, iron, and brass*. Tin and iron are common hardware metals that are often used to make angle brackets, sheet metal (various thickness from $\frac{1}{32}$ inch on up), and (when galvanized) nail plates for house framing. Brass is often found in decorative trim for home construction projects and as raw construction material for hobby models. All three metals are stronger and heavier than aluminum. Cost: fairly cheap.
- *Wood*. Surprise! Wood is an excellent material for robot bodies, although you may not want to use it in all your designs. Wood is easy to work with, can be sanded and sawed to any shape, doesn't conduct electricity (avoids short circuits), and is available everywhere. Disadvantage: ordinary construction plywood is rather weak for its weight, so you need fairly large pieces to provide stability. Better yet, use the more dense (and expensive) multi-ply hardwoods for model airplane and sailboat construction. Common thicknesses are $\frac{1}{4}$- to $\frac{1}{2}$-inch—perfect for most robot projects.
- *Plastic*. Everything is going plastic these days, including robots. Pound for pound, plastic has more strength than many metals, yet is easier to work with. You can cut it, shape it, drill it, and even glue it. To use plastic effectively you must have some special tools, and extruded pieces may be hard to find unless you live near a well-stocked plastic specialty store. Mail order is an alternative.
- *Foamboard*. Art supply stores stock what's known as "foamboard" (or "Foam Core"), a special construction material typically used for building models. Foamboard is a sandwich of paper or plastic glued to both sides of a layer of densely compressed foam. The material comes in sizes from $\frac{1}{8}$ inch to over $\frac{1}{2}$ inch, with $\frac{1}{4}$ to $\frac{1}{3}$ inch being fairly common. The board can be readily cut with a small hobby saw (paper-laminated foamboard

can be cut with a sharp knife; plastic-laminated foamboard should be cut with a saw). Foamboard is especially well suited for small robots where light weight is of extreme importance.

■ *Rigid expanded plastic sheet.* Expanded sheet plastics are often constructed like a sandwich, with thin outer sheets on the top and bottom and a thicker expanded (air-filled) center section. When cut, the expanded center section often has a kind of foam-like appearance, but the plastic itself is stiff. Rigid expanded plastic sheets are remarkably lightweight for their thickness, making them ideal for small robots. These sheets are known by various trade names such as *Sintra* and are available at industrial plastics supply outlets.

Power Systems

We eat food that is processed by the stomach and intestines to make fuel for our muscles, bones, skin, and the rest of our body. While you could probably design a digestive system for a robot and feed it hamburgers, french fries, and other semi-radioactive foods, an easier way to generate the power to make your robot go is to take a trip to the store and buy a set of dry-cell batteries. Connect the batteries to the robot's motors, circuits, and other parts, and you're all set.

TYPES OF BATTERIES

There are several different types of batteries, and Chapter 15, "All about Batteries and Robot Power Supplies," goes into more detail about them. Here are a few quick details to start you off.

Batteries generate DC current and come in two distinct categories: rechargeable and nonrechargeable (for now, let's forget the nondescriptive terms like *storage*, *primary*, and *secondary*). *Nonrechargeable* batteries include the standard zinc and alkaline cells you buy at the supermarket, as well as special-purpose lithium and mercury cells for calculators, smoke detectors, watches, and hearing aids. A few of these (namely, lithium) have practical uses in hobby robotics.

Rechargeable batteries include nickel-cadmium (Ni-Cad), gelled electrolyte, sealed lead-acid cells, and special alkaline. Ni-Cad batteries are a popular choice because they are relatively easy to find, come in popular household sizes ("D," "C," etc.) and can be recharged many hundreds of times using an inexpensive recharger. Gelled electrolyte ("Gel-cell") and lead-acid batteries provide longer-lasting power, but they are heavy and bulky.

ALTERNATIVE POWER SOURCES

Batteries are required in most fully self-contained mobile robots because the automaton cannot be connected by power cord to an electrical socket. That doesn't mean other power sources, including AC or even solar, can't be used in some of your robot designs. On the contrary, stationary robot arms don't have to be capable of moving around the room; they are designed to be placed about the perimeter of the workplace and perform within this predefined area. The motors and control circuits may very well run off AC power, thus freeing you from replacing batteries and worrying about operating times and recharging periods.

This doesn't mean that AC power is the preferred method. High-voltage AC poses greater shock hazards. Should you ever decide to make your robot independent, you must also exchange all the AC motors for DC ones. Electronic circuits ultimately run off DC power, even when the equipment is plugged into an AC outlet.

One alternative to batteries in an all-DC robot system is to construct an AC-operated power station that provides your robot with regulated DC juice. The power station converts the AC to DC and provides a number of different voltage levels for the various components in your robot, including the motors. This saves you from having to buy new batteries or recharge the robot's batteries all the time.

Small robots can be powered by solar energy when they are equipped with suitable solar cells. Solar-powered robots can tap their motive energy directly from the cells, or the cells can charge up a battery over time. Solar-powered 'bots are a favorite of those in the "BEAM" camp—a type of robot design that stresses simplicity, including the power supply of the machine.

PRESSURE SYSTEMS

Two other forms of robotic power, which will not be discussed in depth in this book, are hydraulic and pneumatic. *Hydraulic* power uses oil or fluid pressure to move linkages. You've seen hydraulic power at work if you've ever watched a bulldozer move dirt from pile to pile. And while you drive you use it every day when you press down on the brake pedal. Similarly, *pneumatic* power uses air pressure to move linkages. Pneumatic systems are cleaner than hydraulic systems, but all things considered they aren't as powerful.

Both hydraulic and pneumatic systems must be pressurized to work, and this pressurization is most often performed by a pump. The pump is driven by an electric motor, so in a way robots that use hydraulics or pneumatics are fundamentally electrical. The exception to this is when a pressurized tank, like a scuba tank, is used to provide air pressure in a pneumatic robot system. Eventually, the tank becomes depleted and must either be recharged using some pump on the robot or removed and filled back up using a compressor.

Hydraulic and pneumatic systems are rather difficult to implement effectively, but they provide an extra measure of power in comparison to DC and AC motors. With a few hundred dollars in surplus pneumatic cylinders, hoses, fittings, solenoid valves, and a pressure supply (battery-powered pump, air tank, regulator), you could conceivably build a hobby robot that picks up chairs, bicycles, even people!

Locomotion Systems

As mentioned earlier, some robots aren't designed to move around. These include robotic arms, which manipulate objects placed within a work area. But these are exceptions rather than the rule for hobby robots, which are typically designed to get around in this world. They do so in a variety of ways, from using wheels to legs to tank tracks. In each case, the locomotion system is driven by a motor, which turns a shaft, cam, or lever. This motive force affects forward or backward movement.

WHEELS

Wheels are the most popular method for providing robots with mobility. There may be no animals on this earth that use wheels to get around, but for us robot builders it's the simple and foolproof choice. Robot wheels can be just about any size, limited only by the dimensions of the robot and your outlandish imagination. Turtle robots usually have small wheels, less than two or three inches in diameter. Medium-sized rover-type robots use wheels with diameters up to seven or eight inches. A few unusual designs call for bicycle wheels, which despite their size are lightweight but very sturdy.

Robots can have just about any number of wheels, although two is the most common. The robot is balanced on the two wheels by one or two free-rolling casters, or perhaps even a third swivel wheel. Four- and six-wheel robots are also around. You can read more about wheel designs in Part 3.

LEGS

A small percentage of robots—particularly the hobby kind—are designed with legs, and such robots can be conversation pieces all their own. You must overcome many difficulties to design and construct a legged robot. First, there is the question of the number of legs and how the legs provide stability when the robot is in motion or when it's standing still. Then there is the question of how the legs propel the robot forward or backward—and more difficult still!—the question of how to turn the robot so it can navigate a corner.

Tough questions, yes, but not insurmountable. Legged robots are a challenge to design and build, but they provide you with an extra level of mobility that wheeled robots do not. Wheel-based robots may have a difficult time navigating through rough terrain, but leg-based robots can easily walk right over small ditches and obstacles.

A few daring robot experimenters have come out with two-legged robots, but the challenges in assuring balance and control render these designs largely impractical for most robot hobbyists. Four-legged robots (*quadrapods*) are easier to balance, but good locomotion and steering can be difficult to achieve. I've found that robots with six legs (called *hexapods*) are able to walk at brisk speeds without falling and are more than capable of turning corners, bounding over uneven terrain, and making the neighborhood dogs and cats run for cover. Leg-based robots are discussed more fully in Chapter 22, "Build a Heavy-duty, Six-legged Walking Robot," where you can learn more about the Walkerbot, a brutish, insectlike 'bot strong enough to carry a bag of groceries.

TRACKS

The basic design of *track-driven* robots is pretty simple. Two tracks, one on each side of the robot, act as giant wheels. The tracks turn, like wheels, and the robot lurches forward or backward. For maximum traction, each track is about as long as the robot itself.

Track drive is practical for many reasons, including the fact that it makes it possible to mow through all sorts of obstacles, like rocks, ditches, and potholes. Given the right track material, track drive provides excellent traction, even on slippery surfaces like snow, wet concrete, or a clean kitchen floor. Alas, all is not rosy when it comes to track-based robots. Unless you plan on using your robot exclusively outdoors, you should probably stay away from track drive. Making the drive work can be harder than implementing wheels or even legs.

Arms and Hands

The ability to manipulate objects is a trait that has enabled humans, as well as a few other creatures in the animal kingdom, to manipulate the environment. Without our arms and hands, we wouldn't be able to use tools, and without tools we wouldn't be able to build houses, cars, and—hmmm, robots. It makes sense, then, to provide arms and hands to our robot creations so they can manipulate objects and use tools. A commercial industrial robot "arm" is shown in Fig. 2.4. Chaps. 24 through 27 in Part 4 of this book are devoted entirely to robot arms and hands.

You can duplicate human arms in a robot with just a couple of motors, some metal rods, and a few ball bearings. Add a gripper to the end of the robot arm and you've created a complete arm-hand module. Of course, not all robot arms are modeled after the human appendage. Some look more like forklifts than arms, and a few use retractable push rods to move a hand or gripper toward or away from the robot. See Chapter 24, "An Overview of Arm Systems," for a more complete discussion of robot arm design. Chaps. 25 and 26 concentrate on how to build several popular types of robot arms using a variety of construction techniques.

STAND-ALONE OR BUILT-ON MANIPULATORS

Some arms are complete robots in themselves. Car manufacturing robots are really arms that can reach in just about every possible direction with incredible speed and accuracy. You can build a stand-alone robotic arm trainer, which can be used to manipulate objects within a defined workspace. Or you can build an arm and attach it to your robot. Some arm-robot designs concentrate on the arm part much more than the robot part. They are, in fact, little more than arms on wheels.

GRIPPERS

Robot hands are commonly referred to as *grippers* or *end effectors*. We'll stick with the simpler sounding "hands" and "grippers" in this book. Robot grippers come in a variety of styles; few are designed to emulate the human counterpart. A functional robot claw can be built that has just two fingers. The fingers close like a vise and can exert, if desired, a surprising amount of pressure. See Chapter 27, "Experimenting with Gripper Designs" for more information.

FIGURE 2.4.
A robotic arm from
General Electric is
designed for precision
manufacturing. Photo
courtesy General Electric.

Sensory Devices

Imagine a world without sight, sound, touch, smell, or taste. Without these senses, we'd be nothing more than an inanimate machine, like the family car, the living room television, or that guy who hosts the Channel 5 late-night movie. Our senses are an integral part of our lives—if not life itself.

It makes good sense (pardon the pun) to build at least one of these senses into your robot designs. The more senses a robot has, the more it can interact with its environment. That capacity for interaction will make the robot better able to go about its business on its own, which makes possible more sophisticated tasks. Sensitivity to *sound* is a sensory system commonly given to robots. The reason: Sound is easy to detect, and unless you're trying to listen for a specific kind of sound, circuits for sound detection are simple and straightforward.

Sensitivity to *light* is also common, but the kind of light is usually restricted to a slender band of infrared for the purpose of sensing the heat of a fire or navigating through a room using an invisible infrared light beam.

Robot eyesight is a completely different matter. The visual scene surrounding the robot must be electronically rendered into a form the circuits on the robot can accept, and the machine must be programmed to understand and act on the shapes it sees. A great deal of experimental work is underway to allow robots to distinguish objects, but true robot vision is limited to well-funded research teams. Chapter 37, "Robotic Eyes," provides the basics on how to give crude sight to a robot.

In robotics, the sense of *touch* is most often confined to collision switches mounted around the periphery of the machine. On more sophisticated robots, pressure sensors may be attached to the tips of fingers in the robot's hand. The more the fingers of the hand close in around the object, the greater the pressure detected by the sensors. This pressure information is relayed to the robot's brain, which then decides if the correct amount of pressure is being exerted. There are a number of commercial products available that register pressure of one kind or another, but most are expensive. Simple pressure sensors can be constructed cheaply and quickly, however, and though they aren't as accurate as commercially manufactured pressure sensors, they are more than adequate for hobby robotics. See Chapter 35, "Adding the Sense of Touch," and Chapter 36, "Collision Avoidance and Detection," for details.

The senses of smell and taste aren't generally implemented in robot systems, though some security robots designed for industrial use are outfitted with a gas sensor that, in effect, smells the presence of dangerous toxic gas.

Output Devices

Output devices are components that relay information from the robot to the outside world. A common output device in computer-controlled robots (discussed in the next section) is the video screen or (liquid crystal display) panel. As with a personal computer, the robot communicates with its master by flashing messages on a screen or panel. A more common output device for hobby robots is the ordinary light-emitting diode, or a seven-segment numeric display.

Another popular robotic output device is the speech synthesizer. In the 1968 movie *2001: A Space Odyssey*, Hal the computer talks to its shipmates in a soothing but electronic voice. The idea of a talking computer was a rather novel concept at the time of the movie, but today voice synthesis is commonplace.

Many hobbyists build robots that contain sound and music generators. These generators are commonly used as warning signals, but by far the most frequent application of speech, music, and sound is for entertainment purposes. Somehow, a robot that wakes you up to an electronic rendition of Bach seems a little more human. Projects in robot sound-making circuits are provided in Chapter 40, "Sound Output and Input."

Smart versus "Dumb" Robots

There are smart robots and there are dumb robots, but the difference really has nothing to do with intelligence. Even taking into consideration the science of *artificial intelligence*, all self-contained autonomous robots are fairly unintelligent, no matter how sophisticated the electronic brain that controls it. Intelligence is not a measurement of computing capacity but the ability to reason, to figure out how to do something by examining all the variables and choosing the best course of action, perhaps even coming up with a course that is entirely new.

In this book, the difference between dumb and smart is defined as the ability to take two or more pieces of data and decide on a preprogrammed course of action. Usually, a *smart* robot is one that is controlled by a computer. However, some amazingly sophisticated actions can be built into an automaton that contains no computer; instead it relies on simple electronics to provide the robot with some known "behavior" (such is the concept of BEAM robotics). A *dumb* robot is one that blindly goes about its task, never taking the time to analyze its actions and what impact they may have.

Using a computer as the brains of a robot will provide you with a great deal of operating flexibility. Unlike a control circuit, which is wired according to a schematic plan and performs a specified task, a computer can be electronically "rewired" using software instructions—that is, programs. To be effective, the electronics must be connected to all the *control* and *feedback* components of the robot. This includes the drive motors, the motors that control the arm, the speech synthesizer, the pressure sensors, and so forth. Connecting a computer to a robot is a demanding task that requires many hours of careful work. This book presents several computer-based control projects in later chapters.

Note that this book does not tell you how to construct a computer. Rather than tell you how to build a specially designed computer for your robot, the projects in this book use readily available and inexpensive *microcontrollers* and *single-board computers* as well as ready-built personal computers based on the ubiquitous IBM PC design. You can permanently integrate some computers, particularly the portable variety, with your larger robot projects.

The Concept of Robot "Work"

The term *robota*, from which the common word *robot* is derived, was first coined by Czech novelist and playwright Karel Capek in his 1917 short story "Opilec." The word *robota* was used by Capek again in his now-classic play *R.U.R.* (which stands for "Rossum's Universal Robots"), first produced on stage in 1921. *R.U.R.* is one of many plays written by Capek that have a utopian theme. And like most fictional utopias, the basic premise of the play's "perfect society" is fatally flawed. In *R.U.R.* the robots are created by humans to take over all labor, including working on farms and in factories. When a scientist attempts to endow the robot workforce with human emotions—including pain—the automatons conspire against their flesh-and-bone masters and kill them.

In Czech, the term *robota* means "compulsory worker," a kind of machine slave. In many other Baltic languages the term simply means "work." It is the *work* aspect of robotics that is often forgotten, but it defines a "robot" more than anything else. A robot that is not meant to do something—for example, one that simply patrols the living room looking for signs of warm-blooded creatures—is not a robot at all but merely a complicated toy.

That said, designing and building lightweight "demonstrator" robots provides a perfectly valid way to learn about the robot-building craft. Still, it should not be the end-all of your robot studies. Never lose sight of the fact that *a robot is meant to do something*—the more, the better! Once you perfect the little tabletop robot you've been working on the past several months, think of ways to apply your improved robot skills to building a more substantial robot that actually performs some job. The job does not need to be labor saving. We'd all like to have a robot maid like Rosie the Robot on the *Jetsons* cartoon series, but, realistically, it's a pretty sophisticated robot that knows the difference between a clean and dirty pair of socks left on the floor.

From Here

3

TOOLS AND SUPPLIES

Take a long look at the tools in your garage or workshop. You probably already have all the implements you will need to build your own robots. Unless your robot designs require a great deal of precision (and most hobby robots don't), a common assortment of hand tools is all that's really required to construct robot bodies, arms, drive systems, and more. Most of the hardware, parts, and supplies you need are also things you probably already have, left over from old projects around the house. You can readily purchase the pieces you don't have at a hardware store, a few specialty stores around town, or through the mail.

This chapter discusses the basic tools and supplies needed for hobby robot building and how you might use them. You should consider this chapter only as a guide; suggestions for tools and supplies are just that—*suggestions*. By no means should you feel that you must own each tool or have on hand all the parts and supplies mentioned in this chapter. Once again, the concept behind this book is to provide you with the know-how to build robots from discrete modules. In keeping with that open-ended design, you are free to exchange parts in the modules as you see fit. Some supplies and parts may not be readily available to you, so it's up to you to consider alternatives and how to work them into your design. Ultimately, it will be your task to take a trip to the hardware store, collect the items you need, and hammer out a unique creation that's all your own.

Construction Tools

Construction tools are the things you use to fashion the frame and other mechanical parts of the robot. These include a hammer, a screwdriver, and a saw. We will look at the tools needed to assemble the electronics later in this chapter.

BASIC TOOLS

No robot workshop is complete without the following:

- *Claw hammer*. These can be used for just about any purpose you can think of.
- *Rubber mallet*. For gently bashing together pieces that resist being joined nothing beats a rubber mallet; it is also useful for forming sheet metal.
- *Screwdriver assortment*. Have several sizes of flat-head and Philips-head screwdrivers. It's also handy to have a few long-blade screwdrivers, as well as a ratchet driver. Get a screwdriver magnetizer/demagnetizer; it lets you magnetize the blade so it attracts and holds screws for easier assembly.
- *Hacksaw*. To cut anything, the hacksaw is the staple of the robot builder. Buy an assortment of blades. Coarse-tooth blades are good for wood and PVC pipe plastic; fine-tooth blades are good for copper, aluminum, and light-gauge steel.
- *Miter box*. To cut straight lines, buy a good miter box and attach it to your work table (avoid wood miter boxes; they don't last). You'll also use the box to cut stock at near-perfect 45° angles, which is helpful when building robot frames.
- *Wrenches, all types*. Adjustable wrenches are helpful additions to the shop but careless use can strip nuts. The same goes for long-nosed pliers, which are useful for getting at hard-to-reach places. One or two pairs of Vise-Grips will help you hold pieces for cutting and sanding. A set of nut drivers will make it easy to attach nuts to bolts.
- *Measuring tape*. A six- or eight-foot steel measuring tape is a good length to choose. Also get a cloth tape at a fabric store so you can measure things like chain and cable lengths.
- *Square*. You'll need one to make sure that pieces you cut and assemble from wood, plastic, and metal are square.
- *File assortment*. Files will enable you to smooth the rough edges of cut wood, metal, and plastic (particularly important when you are working with metal because the sharp, unfinished edges can cut you).
- *Drill motor*. Get one that has a variable speed control (reversing is nice but not absolutely necessary). If the drill you have isn't variable speed, buy a variable speed control for it. You need to slow the drill when working with metal and plastic. A fast drill motor is good for wood only. The size of the chuck is not important since most of the drill bits you'll be using will fit a standard 1/4-inch chuck.
- *Drill bit assortment*. Use good sharp ones only. If yours are dull, have them sharpened (or do it yourself with a drill bit sharpening device), or buy a new set.
- *Vise*. A vise is essential for holding parts while you drill, nail, and otherwise torment them. An extra large vise isn't required, but you should get one that's big enough to handle the size of the pieces you'll be working with. A rule of thumb: A vice that can't close around a two-inch block of metal or wood is too small.

■ *Safety goggles*. Wear them when hammering, cutting, and drilling as well as any other time when flying debris could get in your eyes. Be sure you use the goggles. A shred of aluminum sprayed from a drill bit while drilling a hole can rip through your eye, permanently blinding you. No robot project is worth that.

If you plan to build your robots from wood, you may want to consider adding rasps, wood files, coping saws, and other woodworking tools to your toolbox. Working with plastic requires a few extra tools as well, including a burnishing wheel to smooth the edges of the cut plastic (the flame from a cigarette lighter also works but is harder to control), a strip-heater for bending, and special plastic drill bits. These bits have a modified tip that isn't as likely to rip through the plastic material. Small plastic parts can be cut and scored using a sharp razor knife or razor saw, both of which are available at hobby stores.

OPTIONAL TOOLS

There are a number of other tools you can use to make your time in the robot shop more productive and less time consuming. A *drill press* helps you drill better holes because you have more control over the angle and depth of each hole. Be sure to use a drill press vise to hold the pieces. *Never* use your hands! A *table saw* or *circular saw* makes it easier to cut through large pieces of wood and plastic. To ensure a straight cut, use a guide fence or fashion one out of wood and clamps. Be sure to use a fine-tooth saw blade if you are cutting through plastic. Using a saw designed for general woodcutting will cause the plastic to shatter.

A motorized *hobby tool*, such as the model shown in Fig. 3.1, is much like a handheld router. The bit spins very fast (25,000 rpm and up), and you can attach a variety of wood, plastic, and metal working bits to it. The better hobby tools, such as those made by Dremel and Weller, have adjustable speed controls. Use the right bit for the job. For example, don't use a wood rasp bit with metal or plastic because the flutes of the rasp will too easily fill with metal and plastic debris.

A *RotoZip* tool (that's its trade name) is a larger, more powerful version of a hobby tool. It spins at 30,000 rpm and uses a special cutting bit—it looks like a drill bit, but it works like a saw. The RotoZip is commonly used by drywall installers, but it can be used to cut through most any material you'd use for a robot (exception: heavy-gauge steel).

Hot-melt glue guns are available at most hardware and hobby stores and come in a variety of sizes. The gun heats up glue from a stick; press the trigger and the glue oozes out the tip. The benefit of hot-melt glue is that it sets very fast—usually under a minute. You can buy glue sticks for normal- or low-temperature guns. I prefer the normal-temperature sticks and guns as the glue seems to hold better. Exercise caution when using a hot-melt glue gun: the glue is hot, after all! You'll know what I'm talking about when a glob of glue falls on your leg. Use a gun with an appropriate stand; this keeps the melting glue near the tip and helps protect you from wayward streams of hot glue.

A *nibbling tool* is a fairly inexpensive accessory (under $20) that lets you "nibble" small chunks from metal and plastic pieces. The maximum thickness depends on the bite of the tool, but it's generally about 1/16 inch. Use the tool to cut channels and enlarge holes. A *tap and die set* lets you thread holes and shafts to accept standard-sized nuts and bolts. Buy a good set. A cheap assortment is more trouble than it's worth.

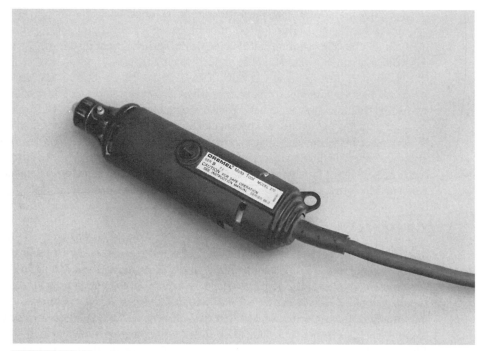

FIGURE 3.1 A motorized hobby tool is ideal for drilling, sanding, and shaping small parts.

A *thread size gauge,* made of stainless steel, may be expensive, but it helps you determine the size of any standard SAE or metric bolt. It's a great accessory for tapping and dieing. Most gauges can be used when you are chopping threads off bolts with a hacksaw. They will provide a cleaner cut.

A *brazing tool* or *small welder* lets you spot-weld two metal pieces together. These tools are designed for small pieces only. They don't provide enough heat to adequately weld pieces larger than a few inches in size. Be sure that extra fuel and oxygen cylinders or pellets are readily available for the brazer or welder you buy. There's nothing worse than spending $30 to $40 for a home welding set, only to discover that supplies are not available for it. Be sure to read the instructions that accompany the welder and observe all precautions.

Electronic Tools

Constructing electronic circuit boards or wiring the power system of your robot requires only a few standard tools. A *soldering iron* leads the list. For maximum flexibility, invest in a modular soldering pencil, the kind that lets you change the heating element. For routine electronic work, you should get a 25- to 30-watt heating element. Anything higher may damage electronic components. You can use a 40- or 50-watt element for wiring switches, relays, and power transistors. *Stay away from "instant-on" soldering irons.* For any application other than soldering large-gauge wires they put out far too much heat.

Supplement your soldering iron with these accessories:

- *Soldering stand.* This is useful for keeping the soldering pencil in a safe, upright position.
- *Soldering tip assortment.* Get one or two small tips for intricate printed circuit board work and a few larger sizes for routine soldering chores.
- *Solder.* Don't buy just any kind of solder; get the resin or flux core type. Acid core and silver solder should never be used on electronic components.
- *Sponge.* Sponges are useful for cleaning the soldering tip as you use it. Keep the sponge damp, and wipe the tip clean every few joints.
- *Heat sink.* Attach the heat sink to sensitive electronic components during soldering. It draws the excess heat away from the component, so it isn't damaged. (See Chapter 6.)
- *Desoldering vacuum tool.* This is useful for soaking up molten solder. Use it to get rid of excess solder, remove components, or redo a wiring job.
- *Dental picks.* These are ideal for scraping, cutting, forming, and gouging into the work.
- *Resin cleaner.* Apply the cleaner after soldering is complete to remove excess resin.
- *Solder vise.* This vise serves as a "third hand," holding together pieces to be soldered so you are free to work the iron and feed the solder.

Read Chapter 6, "Electronic Construction Techniques," for more information on soldering.

Volt-Ohm Meter

A *volt-ohm meter*, or *multitester*, is used to test voltage levels and the resistance of circuits. This moderately priced tool is the basic prerequisite for working with electronic circuits of any kind. If you don't already own a volt-ohm meter you should seriously consider buying one. The cost is small considering the usefulness of the device.

There are many volt-ohm meters on the market today. For robotics work, you don't want a cheap model, but you don't need an expensive one. A meter of intermediate quality is sufficient and does the job admirably at a price of between $30 and $75 (it tends to be on the low side of this range). Meters are available at Radio Shack and most electronics outlets. Shop around and compare features and prices.

DIGITAL OR ANALOG

There are two general types of volt-ohm meters available today: digital and analog. The difference is not that one meter is used on digital circuits and the other on analog circuits. Rather, digital meters employ a numeric display not unlike a digital clock or watch. Analog meters use the older-fashioned—but still useful—mechanical movement with a needle that points to a set of graduated scales. Digital meters used to cost a great deal more than the analog variety, but the price difference has evened out recently. Digital meters, such as the one shown in Fig. 3.2, are fast becoming the standard. In fact, it's hard to find a decent analog meter these days.

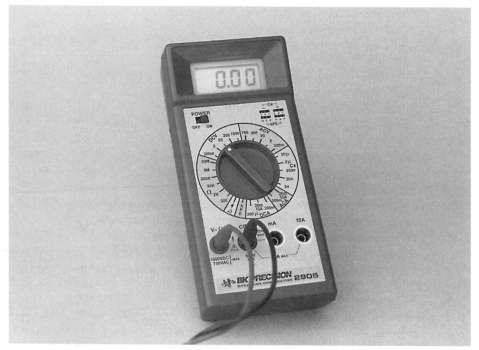

FIGURE 3.2 A volt-ohm meter (or multitester) checks resistance, voltage, and current. This model is digital and has a 3 1/2-digit liquid crystal display (LCD) readout.

AUTOMATIC RANGING

Some volt-ohm meters, analog or digital, require you to select the range before it can make an accurate measurement. For example, if you are measuring the voltage of a 9-volt transistor battery, you set the range to the setting closest to, but above, 9 volts (with most meters it is the 20- or 50-volt range). Auto-ranging meters don't require you to do this, so they are inherently easier to use. When you want to measure voltage, you set the meter to volts (either AC or DC) and take the measurement. The meter displays the results in the readout panel.

ACCURACY

Little of the work you'll do with robot circuits will require a volt-ohm meter that's superaccurate. A meter with average accuracy is more than enough. The accuracy of a meter is the minimum amount of error that can occur when taking a specific measurement. For example, the meter may be accurate to 2000 volts, plus or minus 0.8 percent. A 0.8 percent error at the kinds of voltages used in robots—typically, 5 to 12 volts DC—is only 0.096 volts.

Digital meters have another kind of accuracy. The number of digits in the display determines the maximum resolution of the measurements. Most digital meters have three and a half digits, so they can display a value as small as .001 (the half digit is a "1" on the left side of the display). Anything less than that is not accurately represented; then again, there's little cause for accuracy higher than this.

FUNCTIONS

Digital volt-ohm meters vary greatly in the number and type of functions they provide. At the very least, all standard meters let you measure AC volts, DC volts, milliamps, and ohms. Some also test capacitance and opens or shorts in discrete components like diodes and transistors. These additional functions are not absolutely necessary for building general-purpose robot circuits, but they are handy to have when troubleshooting a circuit that refuses to work.

The maximum ratings of the meter when measuring volts, milliamps, and resistance also vary. For most applications, the following maximum ratings are more than adequate:

DC volts	1000 volts
AC volts	500 volts
DC current	200 milliamps
Resistance	2 megohms

One exception to this is when you are testing current draw for the entire robot versus just for motors. Many DC motors draw in excess of 200 milliamps, and the entire robot is likely to draw 2 or more amps. Obviously, this is far out of the range of most digital meters. You need to get a good assessment of current draw to anticipate the type and capacity of batteries, but to do so you'll need either a meter with a higher DC current rating (digital or analog) or a special-purpose AC/DC current meter. You can also use a resistor in series with the motor and apply Ohm's law to calculate the current draw. The technique is detailed in Chapter 17, "Choosing the Right Motor for the Job."

METER SUPPLIES

Volt-ohm meters come with a pair of test leads, one black and one red. Each is equipped with a needlelike metal probe. The quality of the test leads is usually minimal, so you may want to purchase a better set. The coiled kind are handy; they stretch out to several feet yet recoil to a manageable length when not in use.

Standard leads are fine for most routine testing, but some measurements may require that you use a clip lead. These attach to the end of the regular test leads and have a spring-loaded clip on the end. You can clip the lead in place so your hands are free to do other things. The clips are insulated to prevent short circuits.

METER SAFETY AND USE

Most applications of the volt-ohm meter involve testing low voltages and resistance, both of which are relatively harmless to humans. Sometimes, however, you may need to test high voltages—like the input to a power supply—and careless use of the meter can cause serious bodily harm. Even when you're not actively testing a high-voltage circuit, danger-ous currents can still be exposed.

The proper procedure for using meters is to set it beside the unit under test, making sure it is close enough so the leads reach the circuit. Plug in the leads, and test the meter oper-ation by first selecting the resistance function setting (use the smallest scale if the meter is not auto-ranging). Touch the leads together: the meter should read 0 ohms. If the meter does not respond, check the leads and internal battery and try again. If the display does not read 0 ohms, double-check the range and function settings, and adjust the meter to read 0 ohms (not all digital meters have a 0 adjust, but most analog meters do).

Once the meter has checked out, select the desired function and range and apply the leads to the circuit under test. Usually, the black lead will be connected to ground, and the red lead will be connected to the various test points in the circuit.

Logic Probe

Meters are typically used for measuring analog signals. *Logic probes* test for the presence or absence of low-voltage DC signals, which represent digital data. The 0s and 1s are usually electrically defined as 0 and 5 volts, respectively, with TTL integrated circuits (ICs). In practice, the actual voltages of the 0 and 1 bits depend entirely on the circuit. You can use a meter to test a logic circuit, but the results aren't always predictable. Further, many logic circuits change states (pulse) quickly, and meters cannot track the voltage switches quickly enough.

Logic probes, such as the model in Fig. 3.3, are designed to give a visual and (usually) aural signal of the logic state of a particular circuit line. One LED (light emitting diode) on the probe lights up if the logic is 0 (or LOW); another LED lights up if the logic is 1 (or HIGH). Most probes have a built-in buzzer that has a different tone for the two logic levels. This prevents you from having to keep glancing at the probe to see the logic level.

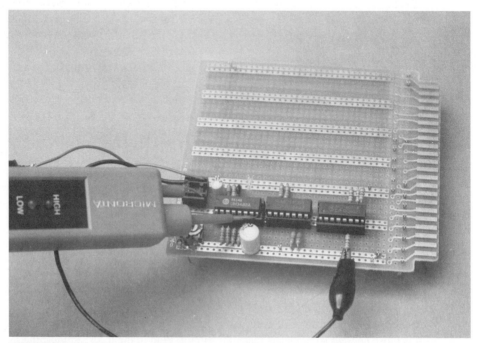

FIGURE 3.3 The logic probe in use. The probe derives its power from the circuit under test.

A third LED or tone may indicate a pulsing signal. A good logic probe can detect that a circuit line is pulsing at speeds of up to 10 MHz, which is more than fast enough for robotic applications, even when using computer control. The minimum detectable pulse width (the time the pulse remains at one level) is 50 nanoseconds, which again is more than sufficient.

Although logic probes may sound complex, they are really simple devices, and their cost reflects this. You can buy a reasonably good logic probe for under $20. Most probes are not battery operated; rather, they obtain operating voltage from the circuit under test. You can also make a logic probe on your own if you wish. A number of project books provide plans.

USING A LOGIC PROBE

The same safety precautions for using a meter apply when you are using a logic probe. Be cautious when working close to high voltages. Cover them to prevent accidental shock (for obvious reasons, logic probes are not meant for anything but digital circuits, so never apply the leads of the probe to an AC line). Logic probes cannot operate with voltages exceeding about 15 volts DC, so if you are unsure of the voltage level of a particular circuit test it with a meter first.

To use the logic probe successfully you really must have a circuit schematic to refer to. Keep it handy when troubleshooting your projects. It's nearly impossible to blindly use the logic probe on a circuit without knowing what you are testing. And since the probe receives its power from the circuit under test, you need to know where to pick off suitable power. To use the probe, connect the probe's power leads to a voltage source on the board, clip the black ground wire to circuit ground, and touch the tip of the probe against a pin on an integrated circuit or the lead of some other component. For more information on using your probe, consult the manufacturer's instruction sheet.

Logic Pulser

A handy troubleshooting accessory to have when you are working with digital circuits is the *logic pulser*. This device puts out a timed pulse, making it possible for you to see the effect of the pulse on a digital circuit. Normally, you'd use the pulser with a logic probe or an oscilloscope (discussed in the next section). The pulser can be switched between one pulse and continuous pulsing.

Most pulsers obtain their power from the circuit under test. It's important that you remember this. With digital circuits, it's generally a bad idea to present an input signal to a device when it's greater than the supply voltage for that device. In other words, if a chip is powered by 5 volts, and you give it a 12-volt pulse, you'll probably ruin the chip. Some circuits work with split (+, −, and ground) power supplies (especially circuits with op amps), so be sure you connect the leads of the pulser to the correct power points.

Also be sure that you do not pulse a line that has an output but no input. Some integrated circuits are sensitive to unloaded pulses at their output stages, and if the pulse is applied inappropriately it can destroy the chip.

Oscilloscope

An *oscilloscope* is a pricey tool—good ones start at about $350. For really serious work, however, an oscilloscope is an invaluable tool that will save you hours of time and frustration. Other test equipment will do some of the things you can do with a scope, but oscilloscopes do it all in one box and generally with greater precision. Among the many applications of an oscilloscope, you can do the following:

■ Test DC or AC voltage levels
■ Analyze the waveforms of digital and analog circuits
■ Determine the operating frequency of digital, analog, and RF circuits
■ Test logic levels
■ Visually check the timing of a circuit to see if things are happening in the correct order and at the prescribed time intervals

The designs provided in this book don't absolutely require that you use an oscilloscope, but you'll probably want one if you design your own circuits or want to develop your electronic skills. A basic, no-nonsense model is enough, but don't settle for the cheap, single-trace units. A dual-trace (two-channel) scope with a 20- to 25-MHz maximum input frequency should do the job nicely. The two channels let you monitor two lines at once, so you can easily compare the input and output signals at the same time. You do not need a scope with storage or delayed sweep, although if your model has these features you're sure to find a use for them sooner or later.

Scopes are not particularly easy to use; they have lots of dials and controls for setting operation. Thoroughly familiarize yourself with the operation of your oscilloscope before using it for any construction project or for troubleshooting. Knowing how to set the time-per-division knob is as important as knowing how to turn the scope on. As usual, exercise caution when using the scope with or near high voltages.

OF OSCILLOSCOPE BANDWIDTH AND RESOLUTION

One of the most important specifications of an oscilloscope is its bandwidth. If 20 MHz is too low for your application, you should invest in a more expensive oscilloscope with a bandwidth of 35, 60, or even 100 MHz. Prices go up considerably as the bandwidth increases.

The resolution of the scope reveals its sensitivity and accuracy. On an oscilloscope, the X (horizontal) axis displays time, and the Y (vertical) axis displays voltage. The sweep time indicates the X-axis resolution, which is generally 0.5 microseconds or faster. The sweep time is adjustable so you can test signal events that occur over a longer time period, usually as long as a half a second to a second. Note that signal events faster than 0.5 microseconds can be displayed on the screen, but the signal may appear as a fleeting glitch or voltage spike.

The sensitivity indicates the Y-axis resolution. The low-voltage sensitivity of most average-priced scopes is about 5 mV to 5 volts. You turn a dial to set the sensitivity you want. When you set the dial to 5 mV, each tick mark on the face of the scope tube represents a difference of 5 mV. Voltage levels lower than 5 mV may appear, but they cannot be accurately measured. Most scopes will show very low-level voltages (in the microvolt range) as a slight ripple.

ENHANCED FEATURES

Over the years, oscilloscopes have improved dramatically, with many added features and capabilities. Among the most useful features are the following:

- *Delayed sweep*. This is helpful when you are analyzing a small portion of a long, complex signal.
- *Digital storage*. This feature records signals in computerized memory for later recall. Once signals are in the memory you can expand and analyze specific portions of them. Digital storage also lets you compare signals, even if you take the measurements at different times.
- *Selectable triggering*. This feature lets you choose how the scope will trigger on the input signal. When checking DC signals, no triggering is necessary, but for AC and digital signals you must select a specific part of the signal so the scope can properly display the waveform. At the very least, a scope will provide automatic triggering, which will lock onto most stable AC and digital signals.

USE GOOD SCOPE PROBES

The probes used with oscilloscopes are not just wires with clips on the end of them. To be effective, the better scope probes use low-capacitance/low-resistance shielded wire and a capacitive-compensated tip. These ensure better accuracy.

Most scope probes are passive, meaning they employ a simple circuit of capacitors and resistors to compensate for the effects of capacitive and resistive loading. Many passive probes can be switched between 1X and 10X. At the 1X setting, the probe passes the signal without attenuation (weakening). At the 10X setting, the probe reduces the signal strength by 10 times. This allows you to test a signal that might otherwise overload the scope's circuits.

Active probes use operational amplifiers or other powered circuitry to correct for the effects of capacitive and resistive loading as well as to vary the attenuation of the signal. Table 3.1 shows the typical specifications of passive and active oscilloscope probes.

Table 3.1 SPECIFICATIONS FOR TYPICAL OSCILLOSCOPE PROBE

PROBE TYPE	FREQUENCY RANGE	RESISTIVE LOAD	CAPACITIVE LOAD
Passive 1X	DC - 5 MHz	1 megohm	30 pF
Passive 10X	DC - 50 MHz	10 megohms	5 pF
Active	DC - 500 MHz	10 megohms	2 pF

USING A PC-BASED OSCILLOSCOPE

As an alternative to a stand-alone oscilloscope you may wish to consider a PC-based oscilloscope solution. Such oscilloscopes not only cost less but may provide additional features, such as long-term data storage. A PC-based oscilloscope uses your computer and the software running on it as the active testing component.

Most PC-based oscilloscopes are comprised of an interface card or adapter. The card-adapter connects to your PC via an expansion board or a serial, parallel, or USB port (different models connect to the PC in different ways). A test probe then connects to the interface. Software running on your PC interprets the data coming through the interface and displays the results on the monitor.

Prices for low-end PC-based oscilloscopes start at about $100. The price goes up the more features and bandwidth you seek. For most robotics work, you don't need the most fancy-dancy model. PC-based oscilloscopes that connect to the parallel, serial, or USB port—rather than internally through an expansion card—can be readily used with a portable computer. This allows you to take your oscilloscope anywhere you happen to be working on your robot.

Frequency Counter

A *frequency counter* (or frequency meter) tests the operating frequency of a circuit. Most models can be used on digital, analog, and RF circuits for a variety of testing chores—from making sure the crystal in the robot's computer is working properly to determining the radio frequency of a transmitter. You need only a basic frequency counter, which represents a $100 to $200 investment. You can save some money by building a frequency counter kit. Frequency counters have an upward operating limit, but it's generally well within the region applicable to robotics experiments. A frequency counter with a maximum range of up to 50 MHz is enough.

Breadboard

You should test each of the circuits you want to use in your robot (including the ones in this book) on a *solderless breadboard* before you commit it to a permanent circuit. Solderless breadboards consist of a series of holes with internal contacts spaced one-tenth of an inch apart, which is just the right spacing for ICs. To create your circuit, you plug in ICs, resistors, capacitors, transistors, and 20- or 22-gauge wire in the proper contact holes.

Solderless breadboards come in many sizes. For the most flexibility, get a double-width board that can accommodate at least 10 ICs. A typical double-width model is shown in Fig. 3.4. You can use smaller boards for simple projects. Circuits with a high number of components require bigger boards. While you're buying a breadboard, purchase a set of prestripped wires. These wires come in a variety of lengths and are already stripped and bent for use in breadboards. The set costs $5 to $7, but you can bet they are well worth the price.

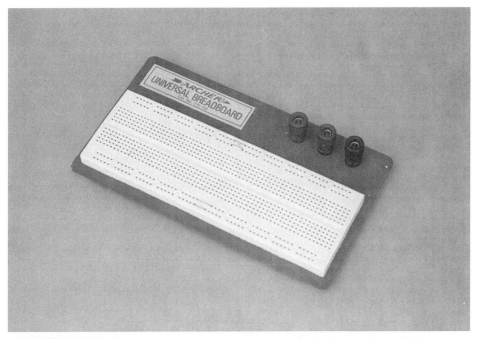

FIGURE 3.4 Solderless breadboards are used to "try out" a circuit before soldering. Some robot makers even use them in their final prototypes.

Wire-Wrapping Tools

Making a printed circuit board for a one-shot application is time consuming, though it can be done with the proper kits and supplies. Conventional point-to-point solder wiring is not an acceptable approach when you are constructing digital circuits, which represent the lion's share of electronics you'll be building for your robots.

The preferred construction method is to use *wire-wrapping*. Wire-wrapping is a point-to-point wiring system that uses a special tool and extra-fine 28- or 30-gauge wrapping wire. When done properly, wire-wrapped circuits are as sturdy as soldered circuits, and you have the added benefit of being able to go back and make modifications and corrections without the hassle of desoldering and resoldering.

A manual wire-wrapping tool is shown in Fig. 3.5. You insert one end of the stripped wire into a slot in the tool, and place the tool over a square-shaped wrapping post. Give the tool five to ten twirls, and the connection is complete. The edges of the post keep the wire anchored in place. To remove the wire, you use the other end of the tool and undo the wrapping.

Several different wire-wrapping tools are available. Some are motorized, and some automatically strip the wire for you, which frees you of this task and of the need to purchase the more expensive prestripped wire. I recommend that you use the basic manual tool initially. You can graduate to other tools as you become proficient in wire-wrapping. Wrapping wire comes in many forms, lengths, and colors, and you need to use special wire-wrapping sockets and posts. See the next section on electronics supplies and components for more details.

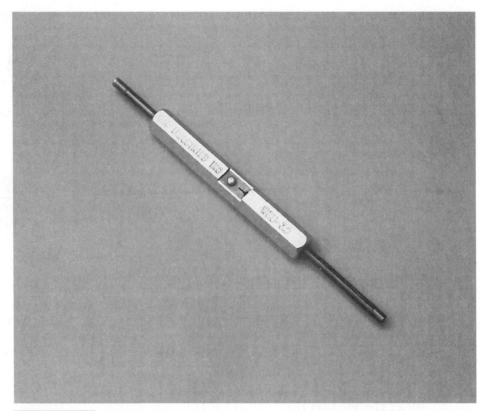

FIGURE 3.5 A wire-wrapping tool. The long end is for wrapping the wire around the post; the short end for unwrapping (should it be necessary). The blade in the middle is for stripping the insulation off the wire.

A number of companies offer proprietary alternatives for wire-wrapping; some are better than others. Which system you choose will depend on your own personal preferences as well as the availability of parts and supplies in your area. Visit a well-stocked electronics parts store or catalog Web site and examine their offerings.

Hardware Supplies

A robot is about 75 percent hardware and 25 percent electronic and electromechanical components. Most of your trips for parts for your robots will be to the local hardware store. The following sections describe some common items you'll want to have around your shop.

NUTS AND BOLTS

Number 6, 8, and 10 nuts and pan-head stove bolts (6/32, 8/32, and 10/24, respectively) are good for all-around construction. Get a variety of bolts in 1/2-, 3/4-, 1-, 1 1/4-, and 1 1/2-inch lengths. You may also want to get some 2- and 3-inch-long bolts for special applications.

Motor shafts and other heavy-duty applications require 1/4-inch 20 or 5/16-inch hardware. Pan-head stove bolts are the best choice; you don't need hex-head carriage bolts unless you have a specific requirement for them. You can use number 6 (6/32) nuts and bolts for small, lightweight applications. Or for even smaller work, use the miniature hardware available at hobby stores, where you can get screws in standard 5/56, 4/40, and 2/20 sizes.

WASHERS

While you're at the store, stock up on flat washers, fender washers (large washers with small holes), tooth lock washers, and split lock washers. Get an assortment so you have a variety of nut and bolt sizes. Split lock washers are good for heavy-duty applications because they provide more compression locking power. You usually use them with bolt sizes of 1/4 inch and above.

ALL-THREAD ROD

All-thread is two- to three-foot lengths of threaded rod stock. It comes in standard thread sizes and pitches. All-thread is good for shafts and linear motion actuators. Get one of each in 8/32, 10/24, and 1/4-inch 20 threads to start. If you need small sizes, hobby stores provide all-thread rod (typically used for push/pull rods in model airplanes) in a variety of diameters and threads.

SPECIAL NUTS

Coupling nuts are just like regular nuts except that they have been stretched out. They are designed to couple two bolts or pieces of all-thread together, end to end. In robotics, you might use them for everything from linear motion actuators to grippers. *Locking nuts* have a piece of nylon built into them that provides a locking bite when they are threaded onto a bolt. It is preferable to use locking nuts over two nuts tightened together.

Extruded Aluminum

If you're making metal robots, you can take advantage of a rather common hardware item: extruded aluminum stock. This stuff, which is designed for such things as building bathtub enclosures, picture frames, and other handyman applications, comes in various sizes, thicknesses, and configurations. The standard length is usually 12 feet, but if you need less most hardware stores will cut to order (you save when you buy it in full lengths). The stock is available in plain (dull silver) anodized aluminum and gold anodized aluminum. Get the plain stuff: it's 10 to 25 percent cheaper.

Two particularly handy stocks are 41/64-by-1/2-by-1/16-inch channel and 57/64-by-9/16-by-1/16-inch channel. I use these extensively to make the frames, arms, legs, and other parts of my robots. Angle stock measuring 1-by-1-by-1/16-inches stock is another often-used item, which is usually employed to attach crossbars and other structural components. No matter what size you eventually settle on for your own robots, keep several feet of extruded aluminum handy at all times. You'll always use it.

If extruded aluminum is not available, you can use shelving standards, the barlike channel stock used for wall shelving. It's most often available in steel, but some hardware stores carry it in aluminum (silver, gold, and black anodized). Use this as a last resort, however, because shelving standards can add a considerable amount of weight to your robot.

Angle Brackets

Also ideal for metal robots (and some plastic and wood ones) is an assortment of 3/8-inch and 1/2-inch galvanized iron brackets. These are used to join the extruded stock or other parts together. Use 1 1/2-inch by 3/8-inch flat corner irons when joining pieces cut at 45° angles to make a frame. Angle irons measuring 1 inch by 3/8 inch and 1 1/2 inch by 3/8 inch are helpful when attaching the stock to baseplates and when securing various components to the robot.

Beware of weight! Angle brackets are heavy when you use a lot of them in your robot. If you find the body of your robot is a few pounds more than it should be, consider substituting the angle brackets with other mounting techniques, including gluing, brazing (for metal), or nails (for wood robots).

Electronic Supplies and Components

Most of the electronic projects you'll assemble for this book, and for other books involving digital and analog circuits, depend on you having a regular stable of common electronic components. If you do any amount of electronic circuit building, you'll want to stock up on the following standard components. If you keep spares handy you won't have to make repeat trips to the electronics store. If you're new to electronics, see Chapter 5, "Common Electronic Components" and refer to Appendix A, "Further Reading," for more information.

RESISTORS

Get a good assortment of 1/8- and 1/4-watt resistors. Make sure the assortment includes a variety of common values and that there are several of each value. Supplement the assortment with individual purchases of the following resistor values: 270 ohm, 330 ohm, 1K ohm, 3.3K ohm, 10K ohm, and 100K ohm. The 270- and 330-ohm values are often used with light-emitting diodes, and the other values are common to TTL and CMOS digital circuits.

VARIABLE RESISTORS

Variable resistors, or potentiometers (pots) as they are also called, are relatively cheap and are a boon to anyone designing and troubleshooting circuits. Buy an assortment of the small PC-mount pots (about 80 cents each retail) in the 1-megohm and 2.5K-, 5K-, 10K-, 50K-, 100K-, 500K-, and 250K-ohm values. You'll find that 500K-ohm and 1-megohm pots are often used in op amp circuits, so buy a few extra of these.

CAPACITORS

Like resistors, you'll find yourself returning to the same standard capacitor values project after project. For a well-stocked shop, get a dozen or so each of the following inexpensive ceramic capacitors: 0.1, 0.01, and 0.001 μF (microfarad).

Many circuits use in-between values of 0.47, 0.047, and 0.022 μF. You may want to get a couple of these as well. Power supply, timing, and audio circuits often use larger polarized electrolytic or tantalum capacitors. Buy a few each of the 1.0-, 2.2-, 4.7-, 10-, and 100-μF values. Some projects call for other values (in the picofarad range and the thousandth of a microfarad range). You can buy these as you need them unless you find yourself returning to standard values repeatedly.

TRANSISTORS

There are thousands of transistors available, and each one has slightly different characteristics. Most applications need nothing more than "generic" transistors for simple switching and amplifying. Common NPN signal transistors are the 2N2222 and the 2N3904 (some transistors are marked with an "MPS" prefix instead of the "2N" prefix; nevertheless, they are the same). Both kinds are available in bulk packages of 10 for about $1. Common PNP signal transistors are the 2N3906 and the 2N2907. (See Chapter 5 for the difference between NPN and PNP.)

If the circuit you're building specifies a transistor other than the generic kind, you may still be able to use one of the generic ones if you first look up the specifications for the transistor called for in the schematic. A number of cross-reference guides provide the specifications and replacement-equivalents for popular transistors.

There are common power transistors as well, and these are often used to provide current to larger loads, such as motors. The NPN TIP 31 and TIP 41 are familiar to most anyone who has dealt with power switching or amplification of up to 1 amp or so. PNP counterparts are the TIP 32 and TIP 42. These transistors come in the TO-220 style package (see Fig. 3.6). A common larger-capacity NPN transistor that can switch 10 amps or more is the 2N3055. It comes in the TO-3 style package and is available everywhere. It costs between 50 cents and $2, depending on the source.

DIODES

Common diodes are the 1N914, for light-duty signal-switching applications, and the 1N4000 series (1N4001, 1N4002, 1N4003, and 1N4004). Get several of each, and use the proper size to handle the current in the circuit. Refer to a data book on the voltage- and power-handling capabilities of these diodes. A special kind of diode is the zener, which is typically used to regulate voltage (see Chapter 5). Zener diodes are available in a variety of voltages and wattages.

LEDs

All semiconductors emit light, but light-emitting diodes (LEDs) are designed especially for the task. LEDs last longer than regular filament lamps and require less operating current. They are available in a variety of sizes, shapes, and colors. For general applications, the medium-sized red LED is perfect. Buy a few dozen and use them as needed.

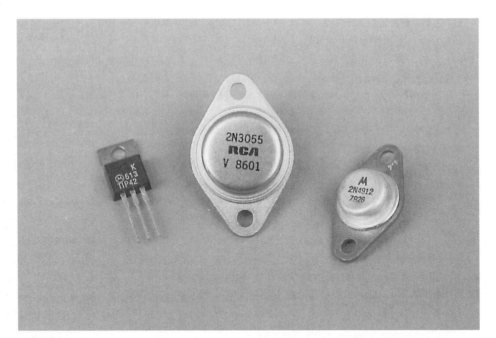

FIGURE 3.6 An assortment of common power transistors. From left to right, the transistors are packaged in TO-220, TO-3, and TO-66 cases. All three can be mounted on a heat sink for enhanced operation.

Many of the projects in this book call for infrared LEDs. These emit no visible light and are used in conjunction with an infrared-sensitive phototransistor or photodiode. Use care when separating your LEDs so you don't mistake an infrared LED for a regular LED. If you do, you may try to use the infrared LED, see that it outputs no visible light, and throw it away. Keep visible-light and infrared LEDs in separate drawers, containers, or plastic bags. Get into the habit of marking the bags so you don't forget what's inside them!

Likewise, if you use light-sensitive diodes in any of your robotics projects keep these separate as well. These diodes look a lot like LEDs but instead of emitting light they are designed to sense light falling on them.

INTEGRATED CIRCUITS

Integrated circuits will enable you to construct fairly complex circuits from just a couple of components. Although there are literally thousands of different ICs, some with exotic applications, a small handful crops up again and again in hobby projects. You should keep the following common ICs in ready stock:

■ *555 timer*. This is, by far, the most popular integrated circuit for hobby electronics. With just a couple of resistors and capacitors, the 555 can be made to act as a pulser, a timer, a time delay, a missing pulse detector, and dozens of other useful things. The chip is usually used as a pulse source for digital circuits. It is available in dual versions as the 556 and in quadruple versions as the 558.

- *741 op amp.* The 741 comes in second in popularity to the 555. The 741 can be used for signal amplification, differentiation, integration, sample-and-hold, and a host of other useful applications. The 741 is available in a dual version, the 1458. The chip comes in different package configurations. The schematics in this book and in other sources specify the pins for the common 8-pin DIP package. Note that there are numerous op amps available and that some have design advantages over the 741.
- *TTL logic chips.* TTL ICs are common in computer circuits and other digital applications. There are many types of TTL packages, but you won't use more than 10 to 15 of them unless you're heavily into electronics experimentation. Specifically, the most common and most useful TTL ICs are the 7400, 7401, 7402, 7404, 7407, 7408, 7414, 7430, 7432, 7473, 7474, 74154, 74193, and 74244.
- *CMOS logic chips.* Because CMOS ICs require less power to operate than the TTL variety, you'll often find them specified for use with low-power robotic and remote control applications. Like TTL, there is a relatively small number of common packages: 4001, 4011, 4013, 4016, 4017, 4027, 4040, 4041, 4049, 4060, 4066, 4069, 4071, and 4081.
- *339 quad comparator.* Comparators are used to compare two voltages. The output of the comparator changes depending on the voltage levels at its two inputs. The comparator is similar to the op amp, except that it does not use an external feedback resistor. You can use an op amp as a comparator, but a better approach is to use something like the 339 chip, which contains four comparators in one package.

WIRE

Solid-conductor, insulated 22-gauge hookup wire can be used in your finished projects as well as to connect wires in breadboards. Buy a few spools in different colors. Solid-conductor wire can be crimped sharply, and when excessively twisted and flexed it can break. If you expect that the wiring in your project may be flexed repeatedly, use stranded wire instead. You must use heavier 12- to 18-gauge hookup wire for connections to heavy-duty batteries, drive motors, and circuit board power supply lines. Appendix E, "Reference," lists the power-handling capabilities of standard wire gauges.

Wire-wrap wire is available in spool or precut and prestripped packages. For ease of use, buy the more expensive precut stuff unless you have a tool that does it for you. Get several of each length. The wire-wrapping tool has its own stripper built in (which you must use instead of a regular wire stripper), so you can always shorten the precut wires as needed. Some special wire-wrapping tools can only be used with their own wrapping wire. Check the instructions that came with the tool for details.

Circuit Boards

Simple projects can be built on *solder breadboards*. These are modeled after the solderless breadboard (discussed earlier in the chapter). You simply transfer the tested circuit from the solderless breadboard to the solder board. You can cut the board with a hacksaw or razor saw if you don't need all of it.

Larger projects require *perforated boards*. Get the kind that have solder tabs or solder traces on them. You'll be able to secure the components onto the boards with solder. Many "perf" boards are designed for wire-wrapping.

SOCKETS

You should use sockets for ICs whenever possible. Sockets come in sizes ranging from 8-pin to 40-pin. You should use the sockets with extra-long square leads for wire-wrapping and when assembling a wire-wrapped project.

You can also use IC sockets to hold discrete components like resistors, capacitors, diodes, LEDs, and transistors. You can, if you wish, wire-wrap the leads of these components. However, because the leads are not square the small wire doesn't have anything to bite into, and therefore the connection isn't very strong. After assembly and testing, when you are sure the circuit works, apply a dab of solder to the leads to hold the wires in place.

Setting up Shop

You'll need a worktable to construct the mechanisms and electronic circuits of your robots. The garage is an ideal location because it affords you the freedom to cut and drill wood, metal, and plastic without worrying about getting the pieces in the carpet. Electronic assembly can be done indoors or out, but I've found that when working in a carpeted room, it's best to spread another carpet or some protective cover over the floor. When the throw rug gets filled up with solder bits and little pieces of wire and component leads, I can take it outside, beat it with a broom handle, and it's as good as new.

Whatever space you choose to set up your robot lab, make sure all your tools are within easy reach. Keep special tools and supplies in an inexpensive fishing tackle box. The tackle box provides lots of small compartments for screws and other parts. For the best results, your work space should be an area where the robot-in-progress will not be disturbed if you have to leave it for several hours or several days, as will usually be the case. The worktable should also be off limits or inaccessible to young children.

Good lighting is a must. Both mechanical and electronic assembly require detail work, and you will need good lighting to see everything properly. Supplement overhead lights with a 60-watt desk lamp. You'll be crouched over the worktable for hours at a time, so a comfortable chair or stool is a must. Be sure you adjust the seat for the height of the worktable.

From Here

4

BUYING PARTS

Building a robot from scratch can be hard or easy. It's up to you. Personally, I go for the easy route; life is too demanding as it is. From experience, I've found that the best way to simplify the construction of a robot is to use standard, off-the-shelf parts, things you can get at the neighborhood hardware, auto parts, and electronics store.

Exactly where can you find robot parts? The neighborhood robot store would be the logical place to start—if only such a store existed! Not yet, anyway. Fortunately, other local retail stores are available to fill in the gaps. Moreover, there's a veritable world of places that sell robot junk, thanks to mail order and the Internet.

Hobby and Model Stores

Hobby and model stores are *the* ideal sources for small parts, including lightweight plastic, brass rod, servo motors for radio control (R/C) cars and airplanes, gears, and construction hardware. Most of the products available at hobby stores are designed for building specific kinds of models and toys. But that shouldn't stop you from raiding the place with an eye to converting the parts for robot use.

In my experience many hobby store owners and sales people have little knowledge about how to use their line of products for anything but their intended purpose. So you'll likely receive little substantive help in solving your robot construction problem. Your best bet is to browse the store and look for parts that you can put together to build a robot. Some of the parts, particularly those for R/C models, will be behind a counter, but they should still be visible enough for you to conceptualize how you might use them. If you don't have a well-stocked hobby and model store in your area, there's always mail order and the Internet.

Craft Stores

Craft stores sell supplies for home crafts and arts. As a robot builder, you'll be interested in just a few of the aisles at most craft stores, but what's in those aisles will be a veritable gold mine! Look for these useful items:

- *Foam rubber sheets*. These come in various colors and thicknesses and can be used for pads, bumpers, nonslip surfaces, tank treads, and lots more. The foam is very dense; use a sharp scissors or knife to cut it (I like to use a rotary paper cutter to get a nice, straight cut).
- *Foamboard*. Constructed of foam sandwiched between two heavy sheets of paper, foamboard can be used for small, lightweight robots. Foamboard can be cut with a hobby knife and glued with paper glue or hot-melt glue. Look for it in different colors and thicknesses.
- *Parts from dolls and teddy bears*. These can often be used in robots. Fancier dolls use "articulations"—movable and adjustable joints—that can be used in your robot creations. Look also for linkages, bendable posing wire, and eyes (great for building robots with personality!).
- *Electronic light and sound buttons*. These are designed to make Christmas ornaments and custom greeting cards, but they work just as well in robots. The electric light kits come with low-voltage LEDs or incandescent lights, often in several bright colors. Some flash at random, some sequentially. Sound buttons have a built-in song that plays when you depress a switch. Don't expect high sound quality with these devices. You could use these buttons as touch sensors, for example, or as a "tummy switch" in an animal-like robot.
- *Plastic crafts construction material*. This can be used in lieu of more expensive building kits, such as LEGO or Erector Set. For example, many stores carry the plastic equivalent of that old favorite, the wooden Popsicle sticks (the wooden variety is also available, but these aren't as strong). The plastic sticks have notches in them so they can be assembled to create frames and structures.
- *Model building supplies*. Many craft stores have these, sometimes at lower prices than the average hobby-model store. Look for assortments of wood and metal pieces, adhesives, and construction tools.

There are, of course, many other interesting products of interest at craft stores. Visit one and take a stroll down its aisles.

Hardware Stores

Hardware stores and builder's supply outlets (usually open to the public) are the best source for the wide variety of tools and parts you will need for robot experimentation. Items like nuts and bolts are generally available in bulk, so you can save money. As you tour the hardware stores, keep a notebook handy and jot down the lines each outlet carries. Then, when you find yourself needing a specific item, you have only to refer to your notes. Take an idle stroll through your regular hardware store haunts on a regular basis. You'll always find something new and laughably useful for robot design each time you visit.

Electronic Stores

Ten or twenty years ago electronic parts stores were plentiful. Even some automotive outlets carried a full range of tubes, specialty transistors, and other electronic gadgets. Now, Radio Shack remains as the only national electronics store chain. In many towns across the country, it's the only thing going.

Radio Shack continues to support electronics experimenters, but they stock only the very common components. If your needs extend beyond resistors, capacitors, and a few integrated circuits, you must turn to other sources. Check the local yellow pages under *Electronics-Retail* for a list of electronic parts shops near you. Note that Radio Shack offers a more complete line of electronic components online, which you can purchase via the Internet or mail order.

Electronics Wholesalers and Distributors

Most electronic stores carry a limited selection, especially if they serve the consumer or hobby market. Most larger cities across the United States—and in other countries throughout the world, for that matter—host one or more electronics wholesalers or distributors. These companies specialize in providing parts for industry.

Wholesalers and distributors are two different kinds of businesses, and it's worthwhile to know how they differ so you can approach them effectively. Wholesalers are accustomed to providing parts in quantity; they offer attractive discounts because they can make up for them with higher volume. Unless you are planning to buy components in the hundreds or thousands, a wholesaler is likely not your best choice.

Distributors may also sell in bulk, but many of them are also set up to sell parts in "onesies and twosies." Cost per item is understandably higher, and not all distributors are willing to sell to the general public. Rather, they prefer to establish relationships with companies and organizations that may purchase thousands of dollars worth of parts over the course of a year. Still, some electronics parts distributors, particularly those with catalogs on the Internet (see "Finding Parts on the Internet," later in this chapter) are more than happy to work with individuals, though minimum-order requirements may apply. Check with the companies near you and ask for their "terms of service."

When buying through a distributor, keep in mind that you are seldom able to browse their warehouse to look for goodies. Most distributors provide a listing of the parts they carry. Some only list the "lines" they offer. You are required to know the make, model, and part number of what you want to order. Fortunately, these days most of the electronics manufacturers provide free information about their products on the Internet. Many such Internet sites offer a search tool that allows you to look up parts by function. Once you find a part you want, jot down its number, and use it to order from the local distributor.

If you belong to a local robotics club or user's group, you may find it advantageous to establish a relationship with a local electronics parts distributor through the club. Assuming the club has enough members to justify the quantities of each part you'll need to buy, the same approach can work with electronics wholesalers. You may find that the buying power of the group gets you better service and lower prices.

Samples from Electronics Manufacturers

Some electronics manufacturers are willing to send samples—some free, some at a small surcharge—of their products. Most manufacturers prefer to deal with engineers in the industry, but a few will respond to requests from individuals. Keep in mind that most manufacturers are not set up to deal directly with the public, and, in fact, many have standing relationships with wholesalers and distributors that *prohibit* them from dealing directly with individuals.

Still, over the years electronics manufacturers have found that the best way to promote their products is to provide free and low-cost engineering samples. Make sure you stay on the manufacturer's mailing list, so you can obtain updates on new technologies that might become available.

If you belong to a school or a robotics club, send your request (preferably on professional-looking letterhead) under the auspices of your school or club. Most electronics manufacturers realize that today's robotics students and hobbyists are tomorrow's engineers, and they want to actively foster a good working relationship with them.

Specialty Stores

Specialty stores are outlets open to the general public that sell items you won't find in a regular hardware or electronic parts store. They don't include surplus outlets, which are discussed in the next section. What specialty stores are of use to robot builders? Consider these:

- *Sewing machine repair shops*. Ideal for finding small gears, cams, levers, and other precision parts. Some shops will sell broken machines to you. Tear the machine to shreds and use the parts for your robot.
- *Auto parts stores*. The independent stores tend to stock more goodies than the national chains, but both kinds offer surprises on every aisle. Keep an eye out for things like hoses, pumps, and automotive gadgets.
- *Used battery depots*. These are usually a business run out of the home of someone who buys old car and motorcycle batteries and refurbishes them. Selling prices are usually between $15 and $25, or 50 to 75 percent less than a new battery.
- *Junkyards*. Old cars are good sources for powerful DC motors, which are used to drive windshield wipers, electric windows, and automatic adjustable seats (though take note: such motors tend to be terribly inefficient for battery-based 'bots). Or how about the hydraulic brake system on a junked 1969 Ford Falcon? Bring tools to salvage the parts you want. And maybe bring the Falcon home with you too.
- *Lawn mower sales-service shops*. Lawn mowers use all sorts of nifty control cables, wheel bearings, and assorted odds and ends. Pick up new or used parts for a current project or for your own stock at these shops.
- *Bicycle sales-service shops*. Not the department store that sells bikes, but a *real* professional bicycle shop. Items of interest: control cables, chains, brake calipers, wheels, sprockets, brake linings, and more.
- *Industrial parts outlets*. Some places sell gears, bearings, shafts, motors, and other industrial hardware on a one-piece-at-a-time basis. The penalty: fairly high prices and often the requirement that you buy a higher quantity of an item than you really need.

Shopping the Surplus Store

Surplus is a wonderful thing, but most people shy away from it. Why? If it's surplus, the reasoning goes, it must be worthless junk. That's simply not true. Surplus is exactly what its name implies: extra stock. Because the stock is extra, it's generally priced accordingly—to move it out the door.

Surplus stores that specialize in new and used mechanical and electronic parts (not to be confused with surplus clothing, camping, and government equipment stores) are a pleasure to find. Most urban areas have at least one such surplus store; some as many as three or four. Get to know each and compare their prices. Bear in mind that surplus stores don't have mass market appeal, so finding them is not always easy. Start by looking in the phone company's yellow pages under *Electronics* and also under *Surplus*.

MAIL ORDER SURPLUS

Surplus parts are also available through the mail. There a limited number of mail order surplus outfits that cater to the hobbyist, but you can usually find everything you need if you look carefully enough and are patient. See Appendix B, "Sources," for more information.

While surplus is a great way to stock up on DC motors, gears, roller chain, sprockets, and other odds and ends, you must shop wisely. Just because the company calls the stuff surplus doesn't mean that it's cheap or even reasonably priced. A popular item in a catalog may sell for top dollar. Always compare the prices of similar items offered by several surplus outlets before buying. Consider all the variables, such as the added cost of insurance, postage and handling, and COD fees. Be sure that the mail order firm has a lenient return policy. You should always be able to return the goods if they are not satisfactory to you.

WHAT YOU CAN GET SURPLUS

Shopping surplus—seither mail order or at a store near you—can be a tough proposition because it's hard to know what you'll need before you need it. And when you need it, there's only a slight chance that the store will have what you want. Still, certain items are almost always in demand by the robotics experimenter. If the price is right (especially on assortments or sets), stock up on the following:

- *Gears*. Small gears between 1/2 inch and 3 inches in size are extremely useful. Stick with standard tooth pitches of 24, 32, and 48. Try to get an assortment of sizes with similar pitches. Avoid "grab bag" collections of gears because you'll find no mates. Plastic and nylon gears are fine for most jobs, but you should use larger metal gears for the main drive systems of your robots.
- *Roller chain and sprockets*. Robotics applications generally call for 1/4-inch (#25) roller chain, which is smaller and lighter than bicycle chain. When you see this stuff, snatch it up, but make sure you have the master links if the chain isn't permanently riveted together. Sprockets come in various sizes, which are expressed as the number of teeth on the outside of the sprocket. Buy a selection. Plastic and nylon roller chain and sprockets are fine for general use; steel is preferred for main drives.
- *Bushings*. You can use bushings as a kind of ball bearing or to reduce the hub size of gears and sprockets so they fit smaller shafts. Common motor shaft sizes are 1/8 inch for small motors and 1/4 inch for larger motors. Gears and sprockets generally have 3/8-inch, 1/2-inch, and 5/8-inch hubs. Oil-impregnated Oilite bushings are among the best, but they cost more than regular bushings.
- *Spacers*. These are made of aluminum, brass, or stainless steel. The best kind to get have an inside diameter that accepts 10/32 and 1/4-inch shafts.
- *Motors*. Particularly useful are the 6-volt and 12-volt DC variety. Most motors turn too fast for robotics applications. Save yourself some hassle by ordering geared motors. Final speeds of 20 to 100 rpm at the output of the gear reduction train are ideal. If gear motors aren't available, be on the lookout for gearboxes that you can attach to your motors. Stepper motors are handy too, but make sure you know what you are buying. Chapters 17 through 20 discuss motors in detail.
- *Rechargeable batteries*. The sealed lead-acid and gel-cell varieties are common in surplus outlets. Test the battery immediately to make sure it takes a charge and delivers its full capacity (test it under a load, like a heavy-duty motor). These batteries come in 6-volt and 12-volt capacities, both of which are ideal for robotics. Surplus nickel-cadmium and nickel-metal hydride batteries are available too, in either single 1.2-volt cells or in combination battery packs. Be sure to check these batteries thoroughly.

Finding Parts on the Internet

The Internet has given a tremendous boost to the art and science of robot building. Through the Internet—and more specifically the World Wide Web—you can now search for and find the most elusive part for your robot. Most of the major surplus and electronics mail order companies provide online electronic catalogs. You can visit the retailer at their Web site and either browse their offerings by category or use a search feature to quickly locate exactly what you want.

Moreover, with the help of Web search engines—such as Google (*www.google.com*), Altavista (*www.altavista.com*), and HotBot (*www.hotbot.com*)—you can search for items of interest from among the millions of Web sites throughout the world. Search engines provide you with a list of Web pages that may match your search query. You can then visit the Web pages to see if they offer what you're looking for.

Of course, don't limit your use of the Internet and the World Wide Web to just finding parts. You can also use them to find a plethora of useful information on robot building. See Appendix B, "Sources," and Appendix C, "Robot Information on the Internet," for categorized lists of useful robotics destinations on the Internet. These lists are periodically updated at *www.robotoid.com*.

A number of Web sites offer individuals the ability to buy and sell merchandise. Most of these sites are set up as auctions: someone posts an item to sell and then waits for people to make bids on it. Robotics toys, books, kits, and other products are common finds on Web auction sites like eBay (*www.ebay.com*) and Amazon (*www.amazon.com*). If your design requires you to pull the guts out of a certain toy that's no longer made, try finding a used one at a Web auction site. The price should be reasonable as long as the toy is not a collector's item.

Keep in mind that the World Wide Web is indeed *worldwide*. Some of the sites you find may not be located in your country. Though many Web businesses ship internationally, not all will. Check the Web site's fine print to determine if the company will ship to your country, and note any specific payment requirements. If they accept checks or money orders, the denomination of each must be in the company's native currency.

Scavenging: Making Do with What You Already Have

You don't always need to buy new (or used or surplus) to get worthwhile robot parts. In fact, some of the best parts for hobby robots may already be in your garage or attic. Consider the typical used VCR, for example. It'll contain at least one motor (and possibly as many as five), numerous gears, and other electronic and mechanical odds and ends. Depending on the brand and when it was made, it could also contain belts and pulleys, infrared receiver modules, miniature push buttons, infrared light-emitting diodes and detectors, and even wire harnesses with multipin connectors. Any and all of these can be salvaged to help build your robot. All told, the typical VCR may have over $50 worth of parts in it.

Never throw away small appliances or mechanical devices without taking them apart and scavenging the good stuff. If you don't have time to disassemble that CD player that's skipping on all of your compact discs, throw it into a pile for a rainy day when you do have a free moment. Ask friends and neighbors to save their discards for you. You'd be amazed how many people simply toss old VCRs, clock radios, and other items into the trash when they no longer work.

Likewise, make a point of visiting garage sales and thrift stores from time to time, and look for parts bonanzas in used—and perhaps nonfunctioning—goods. I regularly scout the local thrift stores (Goodwill, Disabled American Veterans, Salvation Army, Amvets, etc.) and for very little money come away with a trunk full of valuable items I can salvage for parts. Goods that are still in functioning order tend to cost more than the broken stuff, but for robot building the broken stuff is just as good. Be sure to ask the store personnel if they have any nonworking items they will sell you at a reasonable cost.

Here is just a short list of the electronic and mechanical items you'll want to be on the lookout for and the primary robot-building components they have inside:

- *VCRs* are perhaps the best single-source for parts, and they are in plentiful supply (hundreds of millions of them have been built since the mid 1970s). As discussed above, you'll find motors, switches, LEDs, cable harnesses, and IR receiver modules on many models.
- *CD players* have optical systems you can gut out if your robot uses a specialty vision system. Apart from the laser diode, CD players have focusing lenses, miniature multi-cell photodiode arrays, diffraction gratings, and beam splitters, as well as micro-miniature motors and a precision lead-screw positioning device (used by the laser system to read the surface of the CD).
- *Fax machines* contain numerous motors, gears, miniature leaf switches, and other mechanical parts. These machines also contain an imaging array (it reads the page to fax it) that you might be able to adapt for use as robotic sensors.
- *Mice, printers, old scanners, disk drives, and other discarded computer peripherals* contain valuable optical and mechanical parts. Mice contain optical encoders that you can use to count the rotations of your robot's wheels, printers contain motors and gears, disk drives contain stepper motors, and scanners contain optics you can use for vision systems and other sensors on your robot. The old "handheld" scanners popular a few years ago are ideal for use as image sensors.
- *Mechanical toys,* especially the motorized variety, can be used either for parts or as a robot base. When looking at motorized vehicles, favor those that use separate motors for each drive wheel (as opposed to a single motor for both wheels).
- *Smoke alarms* have a useful life of about seven years. After that, the active element in the smoke sensor is depleted to the point where, for safety reasons, the detector should be replaced. Many homeowners replace their smoke alarms more frequently, and the sensor within them is still good. You can gut the parts and use them in your "Smokey the Robot" project.

From Here

5

COMMON ELECTRONIC COMPONENTS

Components are the things that make your electronic projects tick. Any given hobby robot project might contain a dozen or more components of varying types, including resistors, capacitors, integrated circuits, and light-emitting diodes. In this chapter, you'll read about the components commonly found in hobby electronic projects and their many specific varieties.

Fixed Resistors

A fixed resistor supplies a predetermined resistance to a circuit. The standard unit of value of a resistor is the ohm, represented by the symbol Ω. The higher the ohm value, the more resistance the component provides to the circuit. The value on most fixed resistors is identified by color coding, as shown in Fig. 5.1. The color coding starts near the edge of the resistor and comprises four, five, and sometimes six bands of different colors. Most off-the-shelf resistors for hobby projects use standard four-band color coding.

If you are not sure what the resistance is for a particular resistor, use a volt-ohm meter to check it. Position the test leads on either end of the resistor. If the meter is not auto-ranging, start at a high range and work down. Be sure you don't touch the test leads or the leads of the resistor; if you do, you'll add the natural resistance of your own body to the reading.

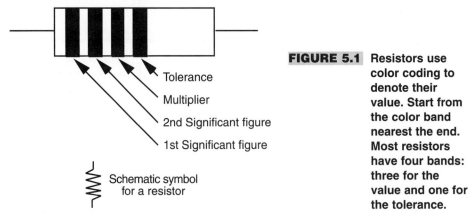

Tolerance

Multiplier

2nd Significant figure

1st Significant figure

Schematic symbol
for a resistor

FIGURE 5.1 Resistors use color coding to denote their value. Start from the color band nearest the end. Most resistors have four bands: three for the value and one for the tolerance.

Resistors are also rated by their wattage. The wattage of a resistor indicates the amount of power it can safely dissipate. Resistors used in high-load applications, like motor control, require higher wattages than those used in low-current applications. The majority of resistors you'll use for hobby electronics will be rated at 1/4 or even 1/8 of a watt. The wattage of a resistor is not marked on the body of the component; instead, you must infer it from the size of the resistor.

Variable Resistors

Variable resistors, first introduced in Chapter 3, are more commonly known as potentiometers, let you "dial in" a specific resistance. The actual range of resistance is determined by the upward value of the potentiometer. Potentiometers are thus marked with this upward value, such as 10K, 50K, 100K, 1M, and so forth. For example, a 50K potentiometer will let you dial in any resistance from 0 ohms to 50,000 ohms. Note that the range is approximate only.

Potentiometers are of either the dial or slide type, as shown in Fig. 5.2. The dial type is the most familiar and is used in such applications as television volume controls and electric blanket thermostat controls. The rotation of the dial is nearly 360°, depending on which potentiometer you use. In one extreme, the resistance through the potentiometer (or "pot") is zero; in the other extreme, the resistance is the maximum value of the component.

Some projects require precision potentiometers. These are referred to as multiturn pots or trimmers. Instead of turning the dial one complete rotation to change the resistance from, say, 0 to 10,000 ohms, a multiturn pot requires you to rotate the knob three, five, ten, even fifteen times to span the same range. Most are designed to be mounted directly on the printed circuit board. If you have to adjust them you will need a screwdriver or plastic tool.

Fixed Capacitors

After resistors, capacitors are the second most common component found in the average electronic project. Capacitors serve many purposes. They can be used to remove traces of

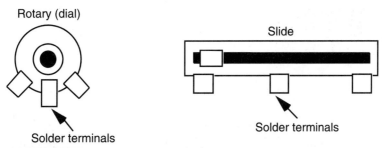

Rotary (dial)

Slide

Solder terminals

Solder terminals

FIGURE 5.2 **Potentiometers are variable resistors. You'll find them in rotary or slide versions; rotary potentiometers are the easiest to use in hobby circuits.**

alternating current ripple in a power supply, for example, to delay the action of some portion of the circuit, or to remove harmful glitches. All these applications depend on the ability of the capacitor to hold an electrical charge for a predetermined time.

Capacitors come in many more sizes, shapes, and varieties than resistors, though only a small handful are truly common. However, most all capacitors are made of the same basic stuff: a pair of conductive elements separated by an insulating dielectric (see Fig. 5.3). This dielectric can be composed of many materials, including air (in the case of a variable capacitor, as detailed in the next section), paper, epoxy, plastic, and even oil. Most capacitors actually have many layers of conducting elements and dielectric. When you select a capacitor for a particular job, you must generally also indicate the type, such as ceramic, mica, or Mylar.

Capacitors are rated by their capacitance, in farads, and by the breakdown voltage of their dielectric. The farad is a rather large unit of measurement, so the bulk of capacitors available today are rated in microfarads, or a millionth of a farad. An even smaller rating is the picofarad, or a millionth of a millionth of a farad. The "micro-" in the term *microfarad* is most often represented by the Greek "mu" (μ) character, as in 10 μF. The picofarad is simply shortened to pF. The voltage rating is the highest voltage the capacitor can withstand before the dielectric layers in the component are damaged.

For the most part, capacitors are classified by the dielectric material they use. The most common dielectric materials are aluminum electrolytic, tantalum electrolytic, ceramic, mica, polypropylene, polyester (or Mylar), paper, and polystyrene. The dialectic material used in a capacitor partly determines which applications it should be used for. The larger electrolytic capacitors, which use an aluminum electrolyte, are suited for such chores as power supply filtering, where large values are needed. The values for many capacitors are printed directly on the component. This is especially true with the larger aluminum electrolytic, where the large size of the capacitor provides ample room for printing the capacitance and voltage. Smaller capacitors, such as 0.1 or 0.01 μF mica disc capacitors, use a common three-digit marking system to denote capacitance and tolerance. The numbering system is easy to use, if you remember it's based on picofarads, not microfarads. A number such as 104 means 10, followed by four zeros, as in

100,000

or 100,000 picofarads. Values over 1000 picofarads are most often stated in microfarads. To make the conversion, move the decimal point to the left six spaces: 0.1 μF. Note that

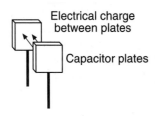

Electrical charge between plates

Capacitor plates

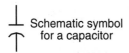

Schematic symbol for a capacitor

FIGURE 5.3 Capacitors store an electrical charge for a limited time. Along with the resistor, they are critical to the proper functioning of many electronic circuits.

values under 1000 picofarads do not use this numbering system. Instead, the actual value, in picofarads, is listed, such as 10 (for 10 pF).

The tolerance of the capacitor is most often indicated by a single letter code, which is sometimes placed by itself on the body of the capacitor or after the three-digit mark, such as

104Z

The letter Z donates a tolerance of +80 percent and −20 percent. That means the capacitor, which is rated at 0.1 µF, might be as much as 80 percent higher or 20 percent lower. More and more capacitor manufacturers are adopting the EIA (Electronic Industries Association) marking system for temperature tolerance. The three characters in the mark indicate the temperate tolerance and maximum variation within the stated temperature range. For example, a capacitor marked Y5P has the following characteristics:

- ■ −30°C low temperature requirement
- ■ +85°C high temperature requirement
- ■ +/− 10.0 percent variance in capacitance over the -30 to +85°C range

The maximum dielectric breakdown voltage is not always stated on the body of a capacitor, but if it is it is almost always indicated by the actual voltage, such as "35" or "35V." Sometimes, the letters *WV* are used after the voltage rating. This indicates the working voltage (really the maximum dielectric breakdown voltage) of the capacitor. You should not use the capacitor with voltages that exceed this rating.

One final mark you will find almost exclusively on larger tantalum and aluminum electrolytic is a polarity symbol, typically a minus (−) sign. The polarity symbol indicates the positive and/or negative lead of a capacitor. If a capacitor is polarized, it is extremely important that you follow the proper orientation when you install the capacitor in the circuit. If you reverse the leads to the capacitor—connecting the + side to the ground rail, for example—the capacitor may be ruined. Other components in the circuit could also be damaged.

Variable Capacitors

Variable capacitors are similar to variable resistors in that they allow you to adjust capacitance to suit your needs. Unlike potentiometers, however, variable capacitors operate on a drastically reduced range of values, and seldom do they provide "zero" capacitance.

The most common type of variable capacitor you will encounter is the air dielectric type, as found in the tuning control of an AM radio. As you dial the tuning knob, you move one set of plates within another. Air separates the plates so they don't touch. Smaller variable capacitors are sometimes used as "trimmers" to adjust the capacitance within a narrow band. You will often find trimmers in radio receivers and transmitters as well as in circuits that use quartz crystals to gain an accurate reference signal. The value of such trimmers is typically in the 5–30 pF range.

Diodes

The diode is the simplest form of semiconductor. They are available in two basic flavors, germanium and silicon, which indicate the material used to manufacture the active junction within the diode. Diodes are used in a variety of applications, and there are numerous subtypes. Here is a list of the most common:

- *Rectifier*. The "average" diode, it rectifies AC current to provide DC only.
- *Zener*. It limits voltage to a predetermined level. Zeners are used for low-cost voltage regulation.
- *Light-emitting*. These diodes emit infrared of visible light when current is applied.
- *Silicon controlled rectifier (SCR)*. This is a type of high-power switch used to control AC or DC currents.
- *Bridge rectifier*. This is a collection of four diodes strung together in sequence; it is used to rectify an incoming AC current.

Other types of diodes include the diac, triac, bilateral switch, light-activated SCR, and several other variations. Diodes carry two important ratings: *peak inverse voltage* (PIV) and *current*. The PIV rating roughly indicates the maximum working voltage for the diode. Similarly, the current rating is the maximum amount of current the diode can withstand. Assuming a diode is rated for 3 amps, it cannot safely conduct more than 3 amps without overheating and failing.

All diodes have positive and negative terminals (polarity). The positive terminal is the *anode*, and the negative terminal is the *cathode*. You can readily identify the cathode end of a diode by looking for a colored stripe near one of the leads. Fig. 5.4 shows a diode that has a stripe at the cathode end. Note how the stripe corresponds with the heavy line in the schematic symbol for the diode.

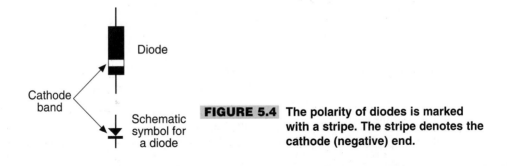

Diode

Cathode band

Schematic symbol for a diode

FIGURE 5.4 The polarity of diodes is marked with a stripe. The stripe denotes the cathode (negative) end.

All semiconductors emit light when an electric current is applied to them. This light is generally very dim and only in the infrared region of the electromagnetic spectrum. The light-emitting diode (LED) is a special type of semiconductor that is expressly designed to emit copious amounts of light. Most LEDs are engineered to produce red, yellow, or green light, but special-purpose types are designed to emit infrared and blue light.

LEDs carry the same specifications as any other diode. The LED has a PIV rating of about 100 to 150 volts, with a maximum current rating of under 40 milliamps. Most LEDs are used in low-power DC circuits and are powered with 12 volts or less. Even though this voltage is far below the PIV rating of the LED, the component can still be ruthlessly damaged if you expose it to currents exceeding 40 or 50 mA. A resistor is used to limit the current to the LED.

Transistors

Transistors were designed as an alternative to the old vacuum tube, and they are used in similar applications, either to amplify a signal or to switch a signal on and off. At last count there were *several thousand* different transistors available. Besides amplifying or switching a current, transistors are divided into two broad categories:

- *Signal*. These transistors are used with relatively low current circuits, like radios, telephones, and most other hobby electronics projects.
- *Power*. These transistors are used with high-current circuits, like motor drivers and power supplies.

You can usually tell the difference between the two merely by size. The signal transistor is rarely larger than a pea and uses slender wire leads. The power transistor uses a large metal case to help dissipate heat and heavy spokelike leads.

Transistors are identified by a unique code, such as 2N2222 or MPS6519. Refer to a data book to ascertain the characteristics and ratings of the particular transistor you are interested in. Transistors are rated by a number of criteria, which are far too extensive for the scope of this book. These ratings include collector-to-base voltage, collector-to-emitter voltage, maximum collector current, maximum device dissipation, and maximum operating frequency. None of these ratings are printed directly on the transistor.

Signal transistors are available in either plastic or metal cases. The plastic kind is suitable for most uses, but some precision applications require the metal variety. Transistors that use metal cases (or "cans") are less susceptible to stray radio frequency interference. They also dissipate heat more readily. *Power* transistors come in metal cases, though a portion of the case (the back or sides) may be made of plastic. Fig. 5.5*a* shows the most common varieties of transistor cases. You'll often encounter the TO-220 and TO-3 style in your hobby electronics ventures.

Transistors have three or four wire leads. The leads in the typical three-lead transistor are base, emitter, and collector, as shown in Fig. 5.5*b*. A few transistors, most notably the field-effect transistor (or FET), have a fourth lead. This is for grounding the case to the chassis of the circuit.

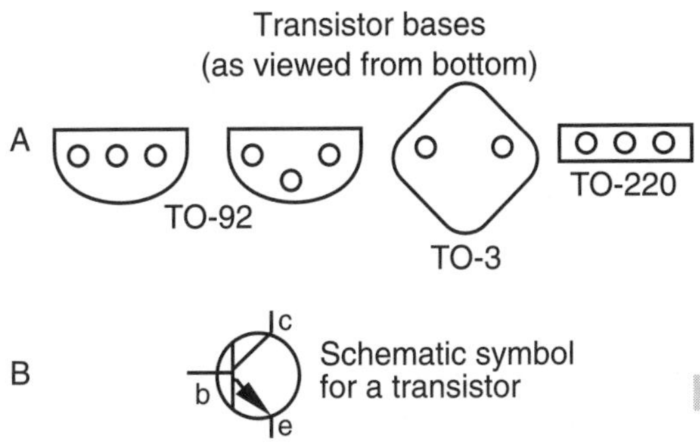

FIGURE 5.5 The most common transistor bases.

Transistors can be either NPN or PNP devices. This nomenclature refers to the sandwiching of semiconductor materials inside the device. You can't tell the difference between an NPN and PNP transistor just by looking at it. However, the difference is indicated in the catalog specifications sheet as well as schematically.

Some semiconductor devices look and act like transistors and are actually called transistors, but in reality they use a different technology. For example, the MOSFET (for metal-oxide semiconductor field-effect transistor) is often used in circuits that demand high current and high tolerance. MOSFET transistors don't use the standard base-emitter-collector connections. Instead, they call them "gate," "drain," and "source." Note, too, that the schematic diagram for the MOSFET is different than for the standard transistor.

Integrated Circuits

The integrated circuit forms the backbone of the electronics revolution. The typical integrated circuit comprises many transistors, diodes, resistors, and even capacitors. As its name implies, the integrated circuit, or IC, is a discrete and wholly functioning circuit in its own right. ICs are the building blocks of larger circuits. By merely stringing them together you can form just about any project you envision.

Integrated circuits are most often enclosed in dual in-line packages (DIPs), as shown in Fig. 5.6. The illustration shows several sizes of DIP ICs, from 8-pin to 40-pin. The most common are 8-, 14-, and 16-pin. The IC can either be soldered directly into the circuit board or mounted in a socket. As with transistors, ICs are identified by a unique code, such as 7400 or 4017. This code indicates the type of device. You can use this code to look up the specifications and parameters of the IC in a reference book. Many ICs also contain other written information, including manufacturer catalog number and date code. Do not confuse the date code or catalog number with the code used to identify the device.

Schematics and Electronic Symbols

Electronics use a specialized road map to tell you what components are being used in a device and how they are connected together. This pictorial road map is the schematic,

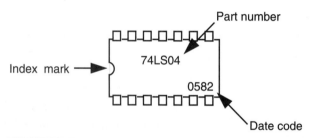

FIGURE 5.6 Integrated circuits (ICs) are common in most any electronic system, including robotics.

a kind of blueprint that tells you just about everything you need to know to build an electronic circuit. Schematics are composed of special symbols that are connected with intersecting lines. The symbols represent individual components and the lines the wires that connect these components together. The language of schematics, while far from universal, is intended to enable most anyone to duplicate the construction of a circuit with little more information than a picture.

The experienced electronics experimenter knows how to read a schematic. This entails recognizing and understanding the symbols used to represent electronic components and how these components are connected. All in all, learning to read a schematic is not difficult.

The following are the most common symbols:

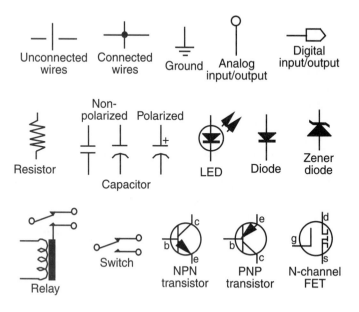

From Here

To learn about...   *Read*

Finding electronic components   Chapter 4, "Buying Parts"

Working with electronic components   Chapter 6, "Electronic Construction Techniques"

Using electronic components   Chapter 28, "An Overview of Robot 'Brains'"
with robot control computers

6

ELECTRONIC CONSTRUCTION
TECHNIQUES

To operate, all but the simplest robots require an electronic circuit of one type or another. The way you construct these circuits will largely determine how well your robot functions and how long it will last. Poor performance and limited life inevitably result when hobbyists use so-called rat's nest construction techniques such as soldering together the loose leads of components.

Using proper construction techniques will ensure that your robot circuits work well and last as long as you have a use for them. This chapter covers the basics of several types of construction techniques, including solderless breadboard, breadboard circuit board, point-to-point wiring, wire-wrapping, and printed circuit board. We will consider only the fundamentals. For more details, consult a book on electronic construction techniques. See Appendix A contains a list of suggested information sources.

Using a Solderless Breadboard

Solderless breadboards are not designed for permanent circuits. Rather, they are engineered to enable you to try out and experiment with a circuit, without the trouble of soldering. Then, when you are assured that the circuit works, you may use one of the other four construction techniques described in this chapter to make the design permanent. A

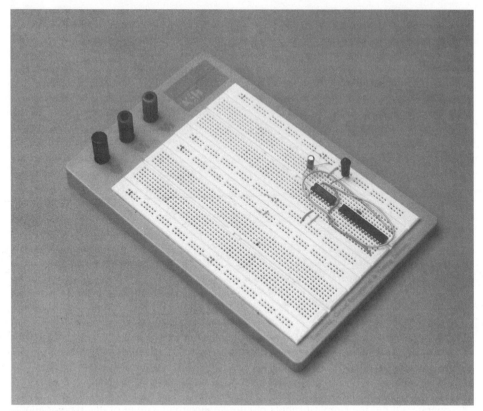

FIGURE 6.1 Use solderless breadboards to test out new circuit ideas. When they work, you can construct a permanent circuit.

typical solderless breadboard is shown in Fig. 6.1. Breadboards are available in many different sizes and styles, but most provide rows of common tie points that are suitable for testing ICs, resistors, capacitors, and most other components that have standard lead diameters.

I urge you to first test all the circuits you build on a solderless breadboard. You'll find that you can often improve the performance of the circuit just by changing a few component values. Such changes are easiest to make when you can simply remove one component and exchange it for another.

Permanent Circuits on Solder Breadboards

The breadboard circuit board—also called a *universal solder board* or *experimenters' PC board*—allows you to make permanent any design you create on a solderless breadboard. The universal solder board comes pre-etched with the same tie points as the solderless breadboard described in the last section. You simply solder the components into place, using jumper wires to connect components that can be directly tied together.

The main disadvantage of universal solder boards is that they don't provide for extremely efficient use of space. Unless you cram the components onto the board, you are limited to building circuits with only two or four ICs and a handful of discrete components. I therefore recommend that you reserve universal solder boards for small circuits—those that require only one or two ICs and a few parts. Simply cut the board to the desired size. Drill new mounting holes to secure the board in whatever enclosure you are using.

Point-to-Point Perforated Board Construction

Point-to-point perf board construction refers to the process of mounting the components on a predrilled board and connecting the leads together directly with solder. This technique was used extensively in the pre-IC days and was even found on commercial products. With the proliferation of ICs, transistors, and other high-speed electronics, however, the point-to-point wiring method has been all but abandoned. Circuits that depend on close tolerances for timing and amplification cannot tolerate point-to-point wiring. Unless you are careful and use insulated wire, point-to-point construction invites short circuits and burnouts.

Wire-Wrapping

When you are working only with low-voltage DC, which is typical of any digital circuit, you can mount the components on a perf board and connect them using special wire-wrapping posts and wire. No soldering is involved; you just wrap the wire around the posts with a tool. The advantage of wire-wrapping is that it's relatively easy to make changes. Simply unwrap the wire and reroute to another post.

Wire-wrapping is used most in IC-intensive circuits. You mount each IC in a wire-wrap socket, which you then cement or solder in place on the board. The sockets use extra-long posts that can accommodate up to about five wrapped wires. As shown in Fig. 6.2, each wire is wrapped around the post like the figures on a totem pole. Most designers try to limit the number of wires on each post to two or three in case they have to make changes to the circuit. Once the wire is wrapped six to eight times around the post, the connection is solid and secure as a soldered joint.

Wire-wrap posts are square shaped so they firmly grip the wire. You can wrap the wire onto rounded posts, but to make the connection solid and permanent add a dab of solder to the wire. You can use this method to directly connect wires to discrete components, such as resistors or capacitors.

A better approach is to cement wire-wrap IC sockets to the board and insert the components into the sockets. Bend and cut the leads so they fit into the socket. If the component is large or wide, use a 24-, 28-, or 40-pin socket. Special component sockets that have

FIGURE 6.2 Wire-wrapping creates circuits by literally wrapping wire around metal posts.

solder terminals are also available. Unless the board is designed for very rugged use (where the components may jiggle loose), you don't need solder terminal sockets.

Successful wire-wrapping takes practice. Before you build your first circuit using wire-wrapping techniques, first try your hand on a scrap socket and board. Visually inspect the wrapped connections and look for loose coils, broken wires, and excessive uninsulated wire at the base of the post. Most wire-wrap tools are designed so one end is used for wrapping wire and the other end for unwrapping. Undo a connection by inverting the tool and try again.

Wire-wrap wire comes in several lengths and gauges. For most applications, you want 30-gauge wire in either long spools or precut or prestripped packages (I prefer the latter). When using the spools, you cut the wire to length then strip off the insulation using the stripper attached to the wrapping tool (a regular wire stripper does a poor job). When you use the precut or prestripped packages the work is already done for you. Buy a selection of different lengths, and always try to use the shortest length possible. Precut and pre-stripped can be expensive ($5 or more for a canister of 200 pieces), but it will save you a great deal of time and effort.

When you get the hang of manual wire-wrapping, you can try one of the motorized tools that are available. Some even allow you to use a continuous spool of wire without the hassle of cutting and stripping. These tools are expensive (over $50), so I do not recommend them for beginners. One tool I like to use is the Vector P184 Slit-N-Wrap. It is a manual tool that permits easy daisy-chaining—going from one post to the next with the same length of wire. Like all special wire-wrapping tools, this one takes a while to get used to but saves time in the long run.

There are a variety of other prototyping systems besides wire-wrap. You may wish to visit a well-stocked electronics store in your area to see what they have available. Because the tools and supplies for prototyping systems tend to be expensive, see if you can get a hands-on demonstration first. That way, you'll know the system is for you before you invest in it.

Making Your Own Printed Circuit Boards

The electronic construction technique of choice is the printed circuit board (PCB). PCBs are made by printing or applying a special resist ink to a piece of copper clad (thin copper sheet over a plastic, epoxy, or phenolic base). The board is then immersed in etchant fluid, which removes all the copper except those areas covered with resist. The resist is washed off, leaving copper traces for the actual circuit. Holes are drilled through the board for mounting the components.

You've probably built a kit or two using a PCB supplied by the manufacturer. You can also make your own printed circuit boards using your own designs as well as the board layouts found in this book and a number of electronics magazines.

Understanding Wire Gauge

The thickness, or gauge, of the wire determines its current-carrying capabilities. Generally, the larger the wire, the more current it can pass without overheating and burning up. See Appendix E, "Reference," for common wire gauges and the maximum accepted current capacity, assuming reasonable wire lengths of 50 feet or less. When you are constructing circuits that carry high currents, be sure to use the proper gauge wire.

Using Headers and Connectors

Robots are often constructed from subsystems that may not be located on the same circuit board. You must therefore know how to connect together subsystems on different circuit boards. Avoid the temptation to directly solder wires between boards. This makes it much harder to work with your robot, including testing variations of your designs with different subsystems.

Instead, use connectors whenever possible, as shown in Fig. 6.3. In this approach you connect the various subsystems of your robot together using short lengths of wire. You terminate each wire with a connector of some type or another. The connectors attach to mating pins on each circuit board.

You don't need fancy cables and cable connectors for your robots. In fact, these can add significant weight to your 'bot. Instead, use ordinary 20- to 26-gauge wire, terminated with single- or double-row plastic connectors. You can use ribbon cable for the wire or

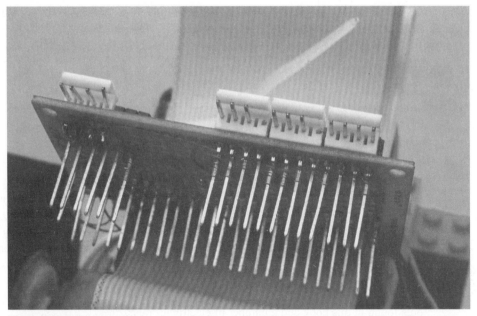

FIGURE 6.3 Using connectors makes for more manageable robots. Use connectors on all subsystems of your robot.

individual insulated strips of wire. Use plastic ties to bundle the wires together. The plastic connectors are made to mate with single- and double-row headers soldered directly on the circuit board. You can buy connectors and headers that have different numbers of pins or you can salvage them from old parts (the typical VCR is chock full of them!).

When making interconnecting cables, cut the wires to length so there is a modest amount of slack between subsystems, but not too much. You don't want, or need, gobs and gobs of excess wire. Nor do you want the wire lengths so short that the components are put under stress when you connect them together.

Eliminating Static Electricity

The ancient Egyptians discovered static electricity when they rubbed animal fur against the smooth surface of amber. Once the materials were rubbed together, they tended to cling to one another. Similarly, two pieces of fur that were rubbed against the amber tended to separate when they were drawn together. While the Egyptians didn't understand this mysterious unseen force—better known now as *static electricity*—they knew it existed.

Today, you can encounter static electricity by doing nothing more than walking across a carpeted floor. As you walk, your feet rub against the carpet, and your body takes on a static charge. Touch a metal object, like a doorknob or a metal sink, and that static is quickly discharged from your body. You feel the discharge as a shock.

Carpet shock has never been known to kill anyone. The amount of voltage and current is far too low to cause great bodily harm. But the same isn't true of electronic circuits. Considering how your body can develop a 10,000- to 50,000-volt charge when you walk across a carpet, you can imagine what that might do to electrical components rated at just 5 or 15 volts. The sudden crash of static can burn holes right through a sensitive transistor or integrated circuit, rendering it completely useless.

Many semiconductor devices are not so forgiving. Transistors and integrated circuits designed around a metal-oxide substrate can be particularly sensitive to high voltages, regardless of the current level. These components include MOSFET transistors, CMOS integrated circuits, and most computer microprocessors.

STORING STATIC-SENSITIVE COMPONENTS

Plastic is one of the greatest sources of static electricity. Storage and shipping containers are often made of plastic, and it's a great temptation to dump your static-sensitive devices into these containers. Don't do it. Invariably, static electricity will develop, and the component could be damaged. Unfortunately, there's no way to tell if a static-sensitive part has become damaged by electrostatic discharge just by looking at it, so you won't know things are amiss until you actually try to use the component. At first, you'll think the circuit has gone haywire or that your wiring is at fault. If you're like most, you won't blame the transistors and ICs until well after you've torn the rest of the circuit apart.

It's best to store static-sensitive components using one of the following methods. All work by grounding the leads of the IC or transistor together, which diminishes the effect of a strong jolt of static electricity. Note that none of these storage methods is 100 percent foolproof.

■ *Antistatic mat*. This mat looks like a black sponge, but it's really conductive foam. You can (and should) test this by placing the leads of a volt-ohm meter on either side of a length of the foam. Dial the meter to ohms. You should get a reading instead of an open circuit. The foam can easily be reused, and large sheets make convenient storage pads for many components.

■ *Antistatic pouch or bag*. Antistatic pouches are made of a special plastic (which generates little static) and are coated on the inside with a conductive layer. The bags are available in a variety of forms. Many are a smoky black or gray color; others are pink or jet black. As with mats, you should never assume a storage pouch is antistatic just from its color. Check the coating on the inside with a volt-ohm meter.

■ *Antistatic tube*. The vast majority of chips are shipped and stored in convenient plastic tubes. These tubes help protect the leads of the IC and are well suited to automatic manufacturing techniques. The construction of the tube is similar to the antistatic pouch: plastic on the outside, a thin layer of conductive material on the inside.

Remove the chip or transistor from its antistatic storage protection only when you are installing it in your project. The less time the component is unprotected the better.

TIPS TO REDUCE STATIC

Consider using any and all of the following simple techniques to reduce and eliminate the risk of electrostatic discharge:

■ *Wear low-static clothing and shoes*. Your choice of clothing can affect the amount of static buildup in your body. Whenever possible, wear natural fabrics such as cotton or wool. Avoid wearing polyester and acetate clothing, as these tend to develop copious amounts of static.

■ *Use an antistatic wrist strap*. The wrist strap grounds you at all times and prevents static buildup. The strap is one of the most effective means for eliminating electrostatic discharge, and it's one of the least expensive.

■ *Ground your soldering iron*. If your soldering pencil operates from AC current, it should be grounded. A grounded iron not only helps prevent damage from electrostatic discharge; it also lessens the chance of your receiving a bad shock should you accidentally touch a live wire.

■ *Use component sockets*. When you build projects that use ICs install sockets first. When the entire circuit has been completely wired, you can check your work, then add the chips. Note that some sockets are polarized so the component will fit into them one way only. Be sure to observe this polarity when wiring the socket.

Good Design Principles

While building circuits for your robots, observe the good design principles described in the following sections, even if the schematic diagrams you are working from don't include them.

PULL-UP/PULL-DOWN RESISTORS

When a device is unplugged, the state might waver back and forth, which can influence the proper functioning of your program. Use pull-up or pull-down resistors on all such inputs (6.8K to 10K should do it). In this way, the input always has a "default" state, even if nothing is connected to it. With a pull-up resistor, the resistor is connected between the input and the +V supply of the circuit; with a pull-down resistor, the resistor is connected between the input and ground, as shown in Fig. 6.4.

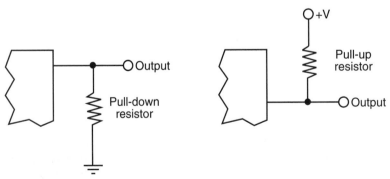

FIGURE 6.4 Use a pull-up or pull-down resistor to ensure that an input never "floats."

TIE UNUSED INPUTS LOW

Unless the instructions for the component you are using specify otherwise, you should tie unused inputs to ground to keep them from "floating." This is especially important for CMOS integrated circuits. A floating input can cause the circuit to go into oscillation, rendering it practically unusable.

USE BYPASS CAPACITORS

Some electronic components, especially fast-acting logic chips, generate a lot of noise in the power supply lines. You can reduce or eliminate this noise by using *bypass* capacitors. These are capacitors of between 0.1 and 10 µF that are positioned between the +V and ground terminals of the noisy chip. Some designers like to use a bypass capacitor on every integrated circuit, while others place them beside every third or fourth chip on the board.

It's also a good idea to put bypass (so-called *decoupling*) capacitors between the +V and ground rails of any circuit at the point of entry of the power supply wires, as shown in Fig. 6.5. Many engineering texts suggest the use of 1 to 10 µF tantalum capacitors for this job. Be sure to orient the capacitor with proper polarization.

KEEP LEAD LENGTHS SHORT

Long leads on components can introduce noise in other parts of a circuit. The long leads also act as a virtual antenna, picking up stray signals from the circuit, from overhead lighting, and even from your own body. When designing and building circuits, strive for the shortest lead lengths on all components. This means soldering the components close to the board and clipping off any excess lead length.

AVOID GROUND LOOPS

A ground loop is when the ground wire of a circuit comes back and meets itself. The +V and ground of your circuits should always have "dead ends" to them. Ground loops can cause erratic behavior and excessive noise in the circuit.

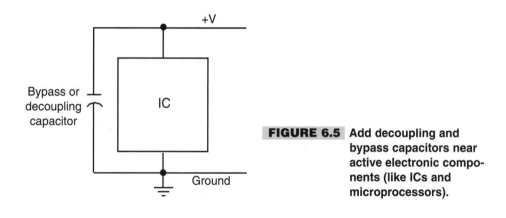

FIGURE 6.5 Add decoupling and bypass capacitors near active electronic components (like ICs and microprocessors).

Soldering Tips and Techniques

Few electronic projects can be assembled without soldering wires together. Soldering sounds and looks simple enough, but there really is a science to it. If you are unfamiliar with soldering or want a quick refresher course, read the primer on soldering fundamentals provided in the next several sections.

TOOLS AND EQUIPMENT

Good soldering means having the proper tools. If you don't have them already, you can purchase them at Radio Shack and most any electronics store. Let's take a quick look at the tools you will need.

Soldering iron and tip   You'll need a *soldering iron*, of course, but not just any old soldering iron. Get a soldering "pencil" with a low-wattage heating element, such as the model in Fig. 6.6. For electronics work, the heating element should not be higher than about 30 watts. Most soldering pencils are designed so you can change the heating elements as easily as changing a light bulb. Avoid using a soldering iron that is not grounded, or you will risk damaging sensitive electronic components by subjecting them to electrostatic discharge. *Do not* use the instant-on type soldering guns favored in the old tube days. They create far too much unregulated heat.

   If your soldering iron has a temperature control and readout, dial it to between 665 and 680° Fahrenheit. This provides maximum heat while posing the minimum danger of

FIGURE 6.6 Use a low-wattage (25 to 30 watts) soldering pencil for all electronics work.

damage to the electronic components. If your iron has just the control and lacks a heat readout, set it to low initially. Wait a few minutes for the iron to heat up, then try one or two test connections. Adjust the heat control so that solder flows onto the connection in under five seconds. When you are not using your soldering iron, keep it in an insulated stand. Don't rest the iron in an ashtray or precariously on the carpet. You or some precious belonging is sure to be burned.

Which *soldering tip* you choose is important. For best results, use a fine tip designed specifically for printed circuit board use (unless you are soldering larger wires; in that case, use a larger tip). Tips are made to fit certain types and brands of heating elements, so make sure you get the kind appropriate for your iron.

Sponge   Keep a damp sponge (just about any type of kitchen sponge will do) by the soldering station and use it to wipe off extra solder. Do not allow globs of solder to remain on the tip. The glob may come off while you're soldering and ruin the connection. The excess solder can also draw away heat from the tip of the pencil, causing a poor soldering job. You'll have to rewet the sponge now and then if you are doing lots of soldering. Never wipe solder off onto a dry sponge, as the sponge could melt or catch on fire.

Solder   You should use only rosin core solder. It comes in different thicknesses; for best results, use the thin type (0.050 inch) for most electronics work, especially if you're working with printed circuit boards. *Never* use acid core or silver solder on electronic equipment. (Note: certain "silver-bearing" solders are available for specialty electronics work, and they are acceptable to use.)

Solder should be kept clean and dry. Avoid tossing your spool of solder into your electronics junk bin. It can collect dust, grime, oil, grease, and other contaminants. Dirty solder requires more heat to melt. In addition, the grime fuses with the solder and melds into the connection. If your solder becomes dirty, wipe it off with a damp paper towel soaked in alcohol, and let it dry.

Soldering tools   Basic soldering tools include a good pair of small *needle-nose pliers*, *wire strippers*, and *wire cutters* (sometimes called side or diagonal cutters). The stripper should have a dial that lets you select the gauge of wire you are using. A pair of "nippy" cutters, which cut wire leads flush to the surface of the board, are also handy.

A *heat sink* is used to draw heat away from components during soldering. To use the heat sink, clip it onto the lead of the component you are soldering. For best results, place the heat sink as close to the component as possible.

Cleaning supplies   It is often necessary to clean up before and after you solder. *Isopropyl alcohol* makes a good, all-around soldering cleaner. You can also use *contract cleaner* (without lubrication). After soldering, and when the components and board are cool, you should spray or brush on some *rosin flux remover* to clean the components and circuit board.

Solder vacuum   A *solder vacuum* is a suction device used to pick up excess solder. It is often used when desoldering—that is, removing a wire or component on the board. Solder can also be removed using a length of copper braid. Most electronics stores sell a spool of solder vacuum specifically for removing solder.

BOARD AND COMPONENT LEAD CLEANING

Before soldering, make sure all parts of the connection are clean. If you're soldering a component onto a printed circuit board, clean the board first by following these steps:

1. Fill a bowl with lukewarm water.
2. Add just a drop or two or liquid dish detergent. Mix the detergent to produce a small amount of suds.
3. Dip the board into the bowl and scrub the copper (or tin-plated) traces with a non-metallic scrubbing pad (such as 3M ScotchBrite). Do not rub too vigorously.
4. Rinse the board under cold water and blot it dry with a paper towel. If desired, use compressed air to drive the trapped water out of the component holes. Let the board dry completely for 10 to 15 minutes.

When you are ready to solder, wet a cotton ball with isopropyl alcohol (or other cleaner) and wipe off all the connection points. Wait a few seconds for the alcohol to evaporate, then commence soldering.

HOW TO SOLDER

The basis of successful soldering is to use the soldering iron to heat up the work, whether it is a component lead, a wire, or whatever. You then apply the solder to the work. Do not apply solder directly to the soldering iron. If you take the shortcut of melting the solder on the iron, you might end up with a "cold" solder joint. A cold joint doesn't adhere well to the metal surfaces of the part or board, so electrical connection is impaired.

Once the solder flows around the joint (and some will flow to the tip), remove the iron and let the joint cool. Avoid disturbing the solder as it cools; a cold joint might be the result. Do not apply heat any longer than necessary. Prolonged heat can permanently ruin electronic components. A good rule of thumb is that if the iron is on any one spot for more than five seconds, it's too long. If at all possible, you should keep the iron at a 30° to 40° angle for best results. Most tips have a beveled tip for this purpose.

Apply only as much solder to the joint as is required to coat the lead and circuit board pad. A heavy-handed soldering job may lead to soldering bridges, which is when one joint melds with joints around it. At best, solder bridges cause the circuit to cease working; at worst, they cause short circuits that can burn out the entire board.

When the joint is complete and has cooled, test it to make sure it is secure. Wiggle the component to see if the joint is solid. If the lead moves within the joint, you'll have to resolder it. The cause may be that the joint is dirty, so you may have to remove the old solder first, clean the connection, and try again. When soldering on printed circuit boards, you'll need to clip off the excess leads that protrude beyond the solder joint. Use a pair of diagonal or nippy cutters for this task. Be sure to protect your eyes when cutting the lead; a bit of metal could fly off and lodge in one of your eyes.

SOLDER TIP MAINTENANCE AND CLEANUP

After soldering, let the iron cool. Loosen the tip from the heating element (use a pair of pliers and grip the tip by the shaft). Then store the iron until you need it again. After several soldering sessions, you should clean the tip using a soft brush. Don't file it or sand it down with emery paper.

After many hours of use, the soldering tip will become old, pitted, and deformed. This is a good time to replace the tip. Old or damaged tips impair the transfer of heat from the iron to the connection, and that can lead to poor soldering joints. Be sure to replace the tip with one made for specifically your soldering iron. Tips are generally not interchangeable between brands.

SOLDER SAFETY

Keep the following points in mind when soldering:

- A hot soldering iron can seriously burn you. Keep your fingers away from the tip of the soldering pencil.
- Never touch a solder joint until after it has cooled.
- While using the soldering iron, always place it in a properly designed iron stand.
- While the fumes produced during soldering are not particularly offensive, they are mildly toxic. Avoid inhaling the fumes for any length of time. Work only in a well-ventilated area.
- Protect your eyes when clipping leads off components. If possible, wear eye protection such as safety glasses or optically clear goggles.
- Keep a fire extinguisher handy, just in case.

From Here

To learn about...	Read
Tools for circuit construction	Chapter 3, "Tools and Supplies"
Where to find electronic components and other parts	Chapter 4, "Buying Parts"
Understanding components used in electronic circuitry	Chapter 5, "Common Electronic Components"

7

PROGRAMMING CONCEPTS:
THE FUNDAMENTALS

Back in the "olden days," all you needed to build a robot were a couple of motors, a switch or relay, a big-o' battery, and some wire. Today, many robots, including the hobby variety, are equipped with a computational brain of one type or another that is told what to do through programming. The brain and programming are typically easier and less expensive to implement than are discrete circuitry, which is one reason why it's so popular to use a computer to power a robot.

The nature of the programming depends on what the robot does. If the robot is meant to play tennis, then its programming is designed to help it recognize tennis balls, move in all lateral directions, perform a classic backhand, and maybe a special function for jumping over the net when it wins.

But no matter what the robot is supposed to do, all its actions boil down to a relatively small set of instructions in whatever *programming language* you are using. If you're new to programming or haven't practiced it in several years, read through this chapter to learn the basics of programming for controlling your robots. This chapter discusses rudimentary stuff so you can better understand the more technical material in the chapters that follow. Of course, if you're already familiar with programming, feel free to skip this chapter.

Important Programming Concepts

There are 11 important concepts to understanding programming, whether for robots or otherwise. In this chapter, we'll talk about each of the following in greater detail:

- Flow control
- Subroutines
- Variables
- Expressions
- Strings
- Numerical values
- Conditional statements
- Branching
- Looping
- Inputting data
- Outputting data

GOING WITH THE FLOW

You can create simple one-job programs for robot control without a predefined blueprint or flowchart. For more complex programs, you may find it helpful to draw a programming *flowchart*, which includes the basic steps of the program. Each box of the chart contains a complete step; arrows connect the boxes to indicate the progress or sequence of steps throughout the program. Flowcharts are particularly handy when you are creating programs that consist of many self-contained "routines" (see the next section) because their graphical format helps you visualize the function and flow of your entire program.

THE BENEFIT OF ROUTINE

One of the first things a programmer does when he or she starts on a project is to map out the individual segments, or *subroutines* (sometimes simply called "routines"), that make up the software. Even the longest, most complex program consists of little more than bite-sized subroutines. The program progresses from one subroutine to the next in an orderly and logical fashion.

A subroutine is any self-contained segment of code that performs a basic action. In the context of robot control programs, a subroutine can be a single command or a much larger action. Suppose your program controls a robot that you want to wander around a room, reversing its direction whenever it bumps into something. Such a program could be divided into three distinct subroutines:

- *Subroutine 1: Drive forward.* This is the default action or behavior of the 'bot.
- *Subroutine 2: Sense bumper collision.* This subroutine checks to see if one of the bumper switches on the robot has been activated.
- *Subroutine 3: Reverse direction.* This occurs after the robot has smashed into an object, in response to a bumper collision.

Figure 7.1 shows how this program might be mapped out in a flowchart. While there's no absolute need to physically separate these subroutines within a program, it often helps to think of the program as being composed of these more basic parts. If the program

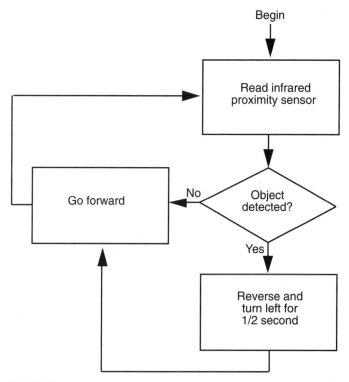

FIGURE 7.1 Use a flowchart to map out the functionality of your robot control programs.

doesn't work properly, you can more readily analyze the problem if you can narrow it down to a specific segment of the code.

Note that there are many approaches to writing the actual code that performs these three actions or subroutines. The two most common approaches are *procedural* and *multitasking*. Which of these you use will depend on the capabilities of the programming environment you're using. Traditional programming simply starts at the beginning and advances one step at a time to the end, taking each instruction in turn and acting upon it. This is procedural programming, and it is common in many robot systems that use either computers or microcontrollers (a microcontroller is a miniature computer, often contained in just one integrated circuit).

Since subroutine 2 in our example depends on the robot sensing some external event, we need a procedural program that constantly "branches off" to subroutine 2 to test the activation of the robot's touch sensor. This is most often done by using a polling loop (we talk about loops later in this chapter). If the bumper switch is triggered, subroutine 3 is activated. Fig. 7.2a shows a typical implementation of a procedural program.

In multitasking programming, you set up the routines and their relationships to one another, then set them all running. An external stimulus of some type triggers a subroutine. In the case of the robot-wanderer, you would likely combine routines 2 and 3, and colliding into an object would trigger the code in routine 3. After the robot has had a chance to back up and get out of the way, routine 1 would take over again. Multitasking programming of the kind shown in Fig. 7.2b is used in the LEGO Mindstorms Robotic Invention System (as just one example), which enables the robot's "operating system" to support up to 10 simultaneous routines (called *tasks*) at one time.

VARIABLES

A *variable* is a special holding area for information, a kind of storage box for data. In most programming languages you can make up the name of the variables as you write the program. The contents of the variable is either specified by you or filled in from some source when the program is run. Since the information is in a variable, it can be used someplace else in the program as many times as you need.

Variables can hold different kinds of data, depending on the number of bits or bytes that are needed to store that data. A numeric value from 0 to 255, for example, requires eight bits, that is, one *byte*, of storage space. Therefore, the variable for such a value needs to be at least one byte large, or else the number won't be stored properly. If you define a variable type that stores fewer than eight bits, its contents will be invalid. If you define a variable type that stores more than eight bits, you'll waste memory space that might otherwise be used for additional programming code.

EXPRESSIONS

An *expression* is a mathematical or logical "problem" that a program must solve before it can continue. Expressions are often simple math statements, such as 2 + 2. When you ask a program to perform some type of calculation or thinking process, you're asking it to *evaluate an expression*.

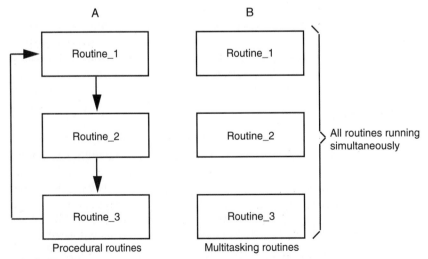

FIGURE 7.2 Procedural programming and multitasking programming use subroutines in different ways.

One of the most common expressions is evaluating if a statement is True or False (we've capitalized the words *true* or *false* to show that you're dealing with logical functions). Here's a good example of a True/False expression that must be evaluated by a program:

```
If Number=10 Then End
```

This expression reads: "If the contents of the *Number* variable is equal to 10, then end the program." Before proceeding, your robot must pause, take a peek inside the *Number* variable, and apply it to the logical expression. If the result is True, then the program ends. If it's False (*Number* has a value other than 10), then something else happens.

STRINGS

A common term encountered in programming circles is the *string*. A string is simply a sequence of alphabetic or numeric characters. These characters are stored within the computer's memory one right after the other, like beads on a string. In the context of programming languages, strings are most often used in variables. Once the string is stored in a variable, it can be acted upon by the program. For example, if the string is text, you can compare it with another string to see if the two are the same. Strings are also often used to allow the robot to communicate with you via a liquid crystal display (LCD) panel.

NUMERICAL VALUES

Computers, and the programs that run on them, are designed from the ground up to work with *numerical values* (numbers). As with strings, numerical values are often used in variables. Numbers differ from strings in one important way, however: the program can perform calculations on two numbers and provide you with the result. Math calculations are not possible with strings because you can't multiply or divide strings since they usually consist of text. (Actually, you *can* perform calculations with the numeric equivalent of a string character, but this is not typically done.)

Note that many programming languages can store digits as either numbers or as strings. The way the value is stored will determine what kind of operations you can perform on it. If the digits *12345* are stored as a number, you can use them in a math calculation. But if the digits *12345* are stored as a string, you can't use them in a calculation.

CONDITIONAL STATEMENTS

Programs can be constructed so that they perform certain routines in one instance and other routines in another. Which routine the program performs depends on specific conditions, set either by the programmer, by some sensor, or by a variable. A *conditional statement* is a fork in the road that offers the program a choice of two directions to take, depending on how it responds to a simple True/False question. The types and styles of conditional statements differ in robot programming languages, but they all have one thing in common: they activate a certain routine (or group of routines) depending on external data.

The most common conditional statement is built using the *If* command, which we've already seen in action in our wall-bumping robot example. Here's an example: "If it's cold outside, I'll wear my jacket. Otherwise, I'll leave the jacket at home." The statement can be broken down into three segments:

■ The condition to be met (if it's cold)
■ The result if the condition is True (wear the jacket)
■ The result if the condition is False (leave the jacket at home)

To be useful, a condition is based on input that may differ each time the robot's program is run. In the preceding example, the robot uses some sort of sensor to determine if it's cold. Based on the results provided by this sensor, the robot will then either put its jacket on, or leave it and enjoy a brisk roll on the grass.

BRANCHING

Akin to the conditional statement is the *branch*, where the program can take two or more paths, depending on external criteria. A good example of a branch is when a robot senses a collision while moving as in our earlier example (see the section "The Benefit of Routine"). The robot normally just drives forward, but many times each second it's program branches off to another subroutine that quickly checks to see if a collision sensor switch has been activated. If the switch has not been activated, nothing special happens, and the robot continues its forward movement. But if the switch has been activated, the program branches to a different subroutine and performs a special action to get away from the obstacle the robot has just struck.

LOOPING

A *loop* is programming code that repeats two or more times. A typical loop is an "entry validator" where the program checks to make sure some condition is met, and if it is the contents of the loop are processed. When the program gets to the bottom of the loop, it goes back to the top and starts all over again. This process continues until the test condition is no longer met. At that point, the program skips to the end of the loop and performs any commands that follow it, as shown in Fig. 7.3.

INPUTTING DATA

When you sit at a computer, you use a keyboard and a mouse to enter data into the machine. While some robots also have keyboards or keypads (and a few have mice), *data input* for automatons tends to be a bit more specialized, involving, for example, touch switches or a sonar ranging system. In all cases, the program uses the information fed to it to complete its task.

The reverse-on-collision robot described earlier (see "The Benefit of Routine") is once again a good example. The data to be input is simple: it is the state of a bumper switch on the front of the robot. When the switch is activated, it provides data—"Hey, I hit something!!" With that data your 'bot can (for example) back up, get out of the way, and head toward some other wall.

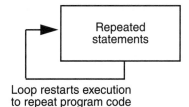

Loop restarts execution
to repeat program code

FIGURE 7.3 **A program loop repeats one or more commands.**

OUTPUTTING DATA

In the realm of robot control programs, data output is most often used to turn motors on and off, to activate a sonar chirp for sensing distance, and to blink a light-emitting diode on and off to communicate with you in a crude form of Morse code. Data output provides a means for the robot to either interact with its environment or interact with you, the Robot Master (*RM* for short).

On robots equipped with liquid crystal display panels, data output can provide a way for the machine to display its current condition. Simpler data output is possible with a single light-emitting diode (LED). If the LED is on, for example, it might mean your robot has run into trouble and needs your help.

Understanding Data Types

At their most basic level programs manipulate data. Data comes in many forms, including a funny-looking guy with platinum-colored skin on *Star Trek: The Next Generation*. Actually, the kind of data we're interested in is strictly numbers of one type or another. Those numbers might represent a value like 1 or 127, the numeric equivalent of a text character (65 is "A" using the ASCII standard), or binary 00010001, which could mean "turn on both motors" on your robot.

In a program, data types can take the following common forms:

■ *Literals*. A literal is, literally, a value "hard-coded" into the program. For example, in the statement *MyVariable = 10*, the *10* is literal data.
■ *Variables*. We've already seen what a variable is: it's a place where data can be stored and referenced elsewhere in the program. It's the *MyVariable* in the statement *MyVariable = 10*.
■ *Constants*. Depending on the design of the programming language, a constant can be just another name for a literal or it can be a special kind of variable that—once set—is never meant to be changed.
■ *Expressions*. The result of a math or logical expression can "return" a data type. For example, the expression 2 + 2 returns the value 4.

No matter what form the data type is in, the programming language expects to see its data follow predefined types. This is necessary so the data can be properly stored in memory. The most common data type is the 8-bit integer, so-called because the value stores an integer (a whole number) using eight bits. With eight bits, the program can work with a number from 0 to 255 (or -128 to $+127$, depending on how it uses the eighth bit). The basic data types are as follows:

- 8-bit integer, or *byte* (can hold a number, a True/False value, or a string value)
- 16-bit integer, or *word*
- 32-bit integer, or *long* or *double word* (dword)
- 32-bit floating point, or *single* ("floating point" means a number with a decimal point)

A few programming languages can manipulate smaller pieces of data, including single bits at a time or a nibble (four bits). In many cases, the language provides for either or both signed and unsigned values. The first bit (called the *most significant bit*, or MSB) is either 0 or 1, which means a positive or negative value. With a 16-bit unsigned integer, for example, the program can store values from 0 to 65535. With a 16-bit signed integer, the program can store values from -32768 to $+32767$.

In each case, the number of bits required to store the data is defined in the data type. Programming languages like Microsoft Visual Basic insulate you from needing to know the number of bits required to hold any particular kind of data. But as most robot programming languages are designed to be more streamlined (read: simpler), you often need to know how many bits are required to store a piece of information.

A Look at Common Programming Statements

In this section we discuss seven of the most common statements you'll encounter in most any programming language. The examples shown are those for the Basic language, but the fundamentals are the same for other languages.

COMMENTS

Comments are used by a programmer as remarks on or reminders of what a particular line of code or subroutine is for. The comments are especially helpful if the program is shared among many people. When the program is *compiled* (made ready) for the computer or microcontroller, the comments are ignored. They only appear in the human-ready *source code* of your programs.

To make a comment using Basic, use the ' (apostrophe) character. Any text to the right is treated as a comment and is ignored by the compiler. For example:

```
' this is a comment
```

Note that the symbol used for comments differs between languages. In the C programming language, for instance, the characters // are used (also /* and */ are used to mark a *comment block*).

IF

The *If* statement is used to build an "if" or "conditional" expression. It is called a conditional expression because it tests for a specific condition:

- If the expression is True, the program performs the commands following the If statement.
- If the expression is False, the program performs the commands after the Else statement, if any.

Here's a sample conditional statement in Basic:

```
If ExampleVar = 10 Then
      Call (Start)
Else
      End
End If
```

The *If* statement evaluates a condition to determine if it's True or False. The most common evaluation is for *equality*. The = (equals) sign is used to test if one value, such as the contents of a variable, is equal to another value. However, there are other forms of evaluation the *If* statement can use, such as "not equal to," "greater than," "less than," and several others. See the "Variables, Expressions, and Operators" section later in this chapter for more information.

SELECT CASE

The *Select Case* statement is used when you want to test a value (usually in a variable) against several possibilities. For example, you may want to test if the contents of the *MyVar* variable contain the values 1, 2, 3, or 4. The *Select Case* statement lets you test each number individually and tell the program specifically what you want to happen should there be a match.

The basic syntax for the *Select Case* statement is as follows:

```
Select Case (TestVar)
      Case x
            ' do if x
      Case y
            ' do if y
      Case z
            ' do if z
End Select
```

TestVar is the test expression and is almost always a variable. *TestVar* is evaluated against each of the *Case arguments* that follow. If the value in *TestVar* is equal to *x*, then the program performs the action that follows *Case x*. If the value in *TestVar* is equal to *y*, then the program performs the action following *Case y*, and so forth.

CALL

The *Call* statement tells the program to temporarily branch elsewhere in the program. This is a subroutine that is identified by name, using a *label*. The program expects to find a *Return* statement at the end of the subroutine. When it encounters the *Return* statement, the program jumps back to the *Call* statement and continues executing the rest of the program. A typical *Call* statement and label looks like this:

```
Call (Loop)
...
Loop:
' stuff goes here
Return        ' Program goes back to the Call, and continues
```

The example calls the subroutine that is named *Loop*. After *Loop* is run the program goes back up to the *Call* statement and executes the other statements that follow it.

GO

The *Go* statement is used to jump to the specified label. In programming parlance, using *Go* to go to a label is called *unconditional branching*. The *Go* statement uses one "argument," namely, the name of the destination label. For example:

```
Go (Loop)
...
Loop:
      ' stuff goes here
      ' Program doesn't go back to the Go
```

FOR/NEXT

The *For/Next* statement is actually a pair of commands. They repeat other programming instructions a specified number of times. The *For/Next* structure is perhaps the most commonly used loop. The *For* portion of the *For/Next* loop uses an expression that tells the program to count from one value to another. For each count, any programming code contained within the *For/Next* structure is repeated:

```
For x = 1 to 10
      ' ... stuff repeated 10 times
Next
```

x is a variable that the program uses to keep track of the current loop iteration. The first time the loop is run, x contains 1. The next time through x contains 2, and so on. When x contains 10, the program knows it has run the loop 10 times and skips to the *Next* statement. From there it executes any additional code that may be in the remainder of the program.

WHILE/WEND

The *While/Wend* statements also form a loop structure. But unlike *For/Next*, which repeats a set number of times, the loop of a *While/Wend* structure repeats while a condition is met.

When the *While* condition is no longer met, the loop is broken and the program continues. Code that falls between the *While* and the *Wend* statements is considered part of the loop that is repeated.

```
While x
     ' ... stuff to repeat
Wend
```

x is an expression that is evaluated each time the loop is repeated. For instance, the expression might be *Switch = 0*, which tests to see if the *Switch* variable contains a 0, signifying perhaps that some switch has not been activated. When *Switch* is no longer 0, the loop breaks.

Note that in many robot control programs the *While/Wend* loop is set up to run indefinitely. This so-called *infinite loop* repeats a process over and over again until the power to the robot is turned off.

Variables, Expressions, and Operators

Earlier in this chapter you read that variables are temporary holders of information. A variable can hold numeric values ("numbers") or characters. As we learned, another term for characters is a "string," so named because each letter follows another, like beads on a string. Placing numbers and strings in a variable is referred to as *assigning* or *assigning a value to a variable*. When you see a phrase such as "Assign the value of 10 to variable *Num*," you know that it means to store the number 10 in a variable referred to as *Num*.

ASSIGNING A VALUE TO A VARIABLE

The most common way to assign a value to a variable is to use the *assignment operator*. In most programming languages used for robot control, this is done with the = (equals) symbol (in some languages, such as Pascal, you'd use :=). It is most often necessary (or at least advisable) to define the data type that will be stored in the variable before assigning a value to it. Here's an example for the Basic language:

```
Dim X As Integer
X = 10
```

Dim stands for "Dimension," which tells the program that you are defining the type of the variable you wish to use. *X* is the name of the variable you wish to assign. Note that many languages restrict the names you can use. Specifically, the variable name must start with a letter character and cannot contain spaces or other punctuation. *Reserved words*—special identifiers such as *If*, *While*, and *Select*—are also unavailable for use as variable names. The = (equals) sign is the variable assignment symbol, and *10* is the value you are placing inside the *X* variable.

Once you define a variable as containing a certain kind of data it's important that that you do not then assign a different type to the variable. For example, all of the following would be incorrect:

```
Dim X As Byte
X = 290        ' data overflow; byte range is 0 to 255
```

```
Dim X As String
X = 15         ' 15 is not a string
```

```
Dim X As Integer
X = "hello"    ' "hello" is not an integer
```

Most robot control languages let you assign a variety of values to variables, as long as the values match the variable type you are using. These values are reviewed in the following five sections.

Literal values   As mentioned earlier, a literal value is a value you specify when you write the program. For example:

```
X = 15         ' X is the variable; 15 is the literal value
```

Contents of another variable   In this value, you can copy the contents of one variable to another. For example:

```
X = Y          ' X is the newly assigned variable;
               ' Y contains some value you're copying
```

(Careful with this one! In most languages, what gets copied is the contents of the variable, not the variable itself. If you later change the contents of Y, X stays the same. Some programming languages let you specify a "pointer" to a variable, so that if you change Y, X changes too. Not all languages for robot programming provide for this, however.)

Memory location   Many programming languages let you reference specific portions of physical memory. For example:

```
X = Peek (1024)            ' read value of data starting at memory
                           ' location 1024
```

Port reference   A port reference is a value maintained by your robot's hardware. A good example is the parallel printer port on your PC, which can be used to control robot parts like motors. For example:

```
X = Inp (889)              ' read value at port 889
```

Evaluated expression   The evaluated expression variable stores a value that is the result of an expression. For example:

```
X = 2 + 2                  ' X holds 4
```

COMBINING DATA TYPES IN A VARIABLE

Modern high-level languages such as Visual Basic or JavaScript are designed to take the drudgery out of mundane programming chores. Visual Basic and JavaScript both support (at least as an option) *weak data typing*, which basically means that the language lets you write programs without requiring that you manage the data types stored in variables. In Visual Basic, for example, you can just throw all data into a "variant" data type, which accepts numbers of different sizes, text strings, you name it. Visual Basic takes care of managing the underlying memory requirements for the data you provide.

The typical robot control language is designed to be small and fast because it's meant to be run on a small single-board computer, or an even smaller microcontroller. The Visual Basic programming environment takes up megabytes of hard disk space; the average robot control program is under 1 kilobyte in size.

Because of their simplicity, programming languages for robot control require *strong data typing*, where you—the programmer—do all the data-typing work yourself. It's not as hard as it seems. But if you're used to languages such as Visual Basic and JavaScript, learning the requirements of strong data typing may require a period of adjustment. One of the most difficult aspects of strong data typing is that you cannot directly mix two data types together in a single variable. The reason?—the result of the mixing probably won't fit in the memory space allocated for the variable. If you add a byte and a string together in an integer variable, the memory will only hold the one byte.

When you try to mix data types the programming environment you're using will either display an error or the robot will not function correctly. In fact, it could function erratically, possibly damaging itself, so exercise care! Here's an example of two data types that cannot be mixed when used in a programming environment that requires strong data typing:

```
Dim X as Byte
Dim Y as Word
Dim Z as Byte

X = 12
Y = 1025
Z = X + Y
```

Both *X* and *Z* are byte-length variables, each of which holds eight-bit values (e.g., 0 to 255). *Y* is a word-length variable, which can hold a 16-bit value (e.g., 0 to 65535). The statement $Z = X + Y$ will fail because *Z* cannot hold the contents of *Y*. If the programming environment doesn't catch this error, it'll create a bug in your robot. (At best, *Z* will hold the value 13 and not 1037, as you'd otherwise expect. The value 13 is what's left of 1037 when that number is stored in a space of only eight bits.)

Just because strong data typing restricts you from mixing and matching different data types doesn't mean you can't do it. The trick is to use the data conversion statements provided by the programming language. The most common data conversion is between integer numbers and text. Assuming the programming language you're using supports them, use *Val* to convert a string to a numeric type and *Str* to convert a numeric into a string:

```
Dim X as String
Dim Y As Integer

X = "1"
Y = Val (X)
```

You must also exercise care when using numeric types of different sizes. Data conversion statements are typically provided in strong data-typing programming languages to convert 8-, 16-, 32-, and (sometimes) 64-bit numbers from one form to another. If the programming language supports "floating-point" numbers (numbers that have digits to the right of the decimal point), then there will likely be data conversion statements for those as well.

CREATING EXPRESSIONS

An expression tells the program what you want it to do with information given to it. An expression consists of two parts:

- One or more values
- An operator that specifies what you want to do with these values

In most programming languages, expressions can be used when you are defining the contents of variables, as in the following:

```
Test1 = 1 + 1
Test2 = (15 * 2) + 1
Test3 = "This is" & "a test"
```

The program processes the expression and places the result in the variable.

Expressions can also be part of a more elaborate scheme that uses other program commands. Used in this way, expressions provide a way for your programs to think on their own (although they may seem to be more independent than you'd like them to be!).

The following sections present the commonly used operators and how they are used to construct expressions. Some operators work with numbers only, and some can also be used with strings. The list is divided into two parts: math operators (which apply to number values) and relational operators (which work with both numbers and in some programming languages strings as well). Depending on the language, you can use literal values or variables for *v1* and *v2*.

Math operators

OPERATOR	FUNCTION
− value	Treats the value as a negative number.
v1 + *v2*	Adds values *v1* and *v2* together.
v1 - *v2*	Subtracts value *v2* from *v1*.
v1 * *v2*	Multiplies values *v1* and *v2*.
v1 / *v2*	Divides value *v1* by *v2*. Sometimes also expressed as *v1* DIV *v2*.
v1 % *v2*	Divides value *v1* by *v2*. The result is the floating-point remainder of the division. Sometimes also given as *v1* MOD *v2*.

Relational operators

OPERATOR	FUNCTION
Not *value*	Evaluates the logical *Not* of *value*. The logical *Not* is the inverse of an expression: True becomes False, and vice versa.
v1 **And** *v2*	Evaluates the logical And of *v1* and *v2*.
v1 **Or** *v2*	Evaluates the logical Or of *v1* and *v2*.
v1 = *v2*	Tests that *v1* and *v2* are equal.
v1 <> *v2*	Tests that *v1* and *v2* are not equal.
v1 > *v2*	Tests that *v1* is greater than *v2*.
v1 >= *v2*	Tests that *v1* is greater than or equal to *v2*.
v1 < *v2*	Tests that *v1* is less than *v2*.
v1 <= *v2*	Tests that *v1* is less than or equal to *v2*.

Relational operators are also known as Boolean or True/False operators. Whatever they test, the answer is either yes (True) or no (False). For example, the expression *2 = 2* would be True, but the expression *2 = 3* would be False.

Using *And* and *Or* Relational Operators

The *And* and *Or* operators work with numbers (depending on the language) and expressions that result in a True/False condition. The *And* operator is used to determine if two values in an expression are *both* True (equal to something other than zero). If both A and B are True, then the result is True. But if A *or* B is False (equal to 0), then the result of the *And* is False.

The *Or* operator works in a similar way, except that it tests if *either* of the values in an expression is True. If at least one of them is True, then the result is True. Only when both values are False is the result of the *Or* expression False. It's often helpful to view the action of the *And* and *Or* operators by using a "truth table" like the two that follow. The tables show all the possible outcomes given to values in an expression:

AND TRUTH TABLE		
0 means False, 1 means True		
Val1	Val2	Result
0	0	0
0	1	0
1	0	0
1	1	1

OR TRUTH TABLE

0 means False, 1 means True

Val1	Val2	Result
0	0	0
0	1	1
1	0	1
1	1	1

USING BITWISE OPERATORS

Many programming languages for robot control support unique forms of operators that work with numbers only and let you manipulate the individual bits that make up those numbers. These are called *bitwise* operators. The following list shows the common characters or names used for bitwise operators. Note that some languages (namely Visual Basic) do not distinguish between bitwise and relational (logical) operators.

OPERATOR SYMBOL	NAME	FUNCTION
Not or ~	Bitwise Not	Reverses the bits in a number; 1s become 0s and vice versa
And or &	Bitwise And	Performs *And* on each bit in a number
Or or |	Bitwise Or	Performs *Or* on each bit in a number
Xor or ^	Bitwise Xor	Performs *Xor* on each bit in a number
<<	Shift left	Shifts the values of the bits 1 or more bits to the left
>>	Shift right	Shifts the values of the bits 1 or more bits to the right

Let's use some bitwise operators in an example. Suppose you assign 9 to variable *This* and 14 to variable *That*. If you *bitwise And* (& symbol) the two together you get 8 as a result. Here are the binary equivalents of the values 9 and 15 (only four bits are shown since 9 and 14 can be expressed in just four bits):

DECIMAL NUMBER	BINARY EQUIVALENT
9	1001
14	1110

Using the *And* truth table given earlier, manually compute what happens when you *bitwise And* these two numbers together:

$$1001 = 9$$
$$1110 = 14$$

& ____

$$1000 = 8$$

Using the same numbers for a *bitwise Or* computation, we get:

$$1001 = 9$$
$$1110 = 14$$

| ____

$$1111 = 15$$

Here's just one example of how this bitwise stuff is handy in robotics: *bitwise And* and *bitwise Or* are commonly used to control the output state of a robot's computer or microcontroller data port. The port typically has eight output pins, each of which can be connected to something, like a motor. Suppose the following four bits control the activation of four motors. When "0" the motor is off; when "1" the motor is on.

0000   all motors off

0010   second motor on

1100   fourth and third motor on

…and so forth. Note that the rightmost bit is motor 1, and the leftmost is motor 4.

Suppose your program has already turned on motors 1 and 4 at some earlier time, and you want them to stay on. You also want to turn motor 3 on (motor 2 stays off). All you have to do is get the current value of the motor port:

1001   motors 1 and 4 on (decimal value equivalent is 9)
You then *bitwise Or* this result with

0100   motor 3 on (decimal value equivalent is 4)
Here's how the *bitwise Or* expression turns out:

$$1001 = 9$$
$$0100 = 4$$

| ____

$$1101 = 13$$

In this example, *Or*'ing 9 and 4 happens to be the same as adding the decimal values of the numbers together. This is not always the case. For example, *Or*'ing 1 and 1 results in 1, not 2.

For your reference, the following table shows the binary equivalents of the first 16 binary digits (counting zero as the first digit and ending with 15). You can count higher by adding an extra 1 or 0 digit on the left. The extra digit increases the count by a power of two—31, 63, 127, 255, 511, 1023, and so on.

DECIMAL NUMBER	BINARY EQUIVALENT
0	0000
1	0001
2	0010
3	0011
4	0100
5	0101
6	0110
7	0111
8	1000
9	1001
10	1010
11	1011
12	1100
13	1101
14	1110
15	1111

USING OPERATORS WITH STRINGS

Recall that a string is an assortment of text characters. You can't perform math calculations with text, but you can compare one string of text against another. The most common use of operations involving strings is to test if a string is equal, or not equal, to some other string, as in:

```
If "MyString" = "StringMy" Then
```

This results in False because the two strings are not the same. In a working program, you'd no doubt construct the string comparison to work with variables, as in:

```
If StringVar1 = StringVar2 Then
```

Now the program compares the *contents* of the two variables and reports True or False accordingly. Note that string comparisons are almost always case-sensitive:

STRING 1	STRING 2	RESULT
hello	hello	Match
Hello	hello	No Match
HELLO	hello	No Match

While the = (equals) operator is used extensively when comparing strings, in many programming languages you can use <>, <, >, <=, and >= as well.

MULTIPLE OPERATORS AND ORDER OF PRECEDENCE

All but the oldest or very simple programming languages can handle more than one operator in an expression. This allows you to combine three or more numbers, strings, or variables together to make complex expressions, such as 5 + 10 / 2 * 7.

This feature of multiple operators comes with a penalty, however. You must be careful of the *order of precedence*, that is, the order in which the program evaluates an expression. Some languages evaluate expressions using a strict left-to-right process, while others follow a specified pattern where certain operators are dealt with first, then others. When the latter approach is used a common order of precedence is as follows:

ORDER	OPERATOR
1	- (unary minus), + (unary plus), ~ (bitwise *Not*), Not (logical *Not*)
2	* (multiply), / (divide), % or MOD (mod), DIV (integer divide)
3	+ (add), - (subtract)
4	<<(shift left)>>(shift right)
5	< (less than), <= (less than or equal to), > (greater than), >= (greater than or equal to), <> (not equal), = (equal)
6	& (bitwise *And*), I (bitwise *Or*), ^ (bitwise *Xor*)
7	And (logical *And*), Xor (logical *Xor*)
8	Or (logical *Or*)

The programming language usually does not distinguish between operators that are on the same level or precedence. If it encounters a + for addition and a - for subtraction, it will evaluate the expression by using the first operator it encounters, going from left to right. You can often specify another calculation order by using parentheses. Values and operators inside the parentheses are evaluated first.

From Here

To learn more about...	*Read*
Programming a robot computer or microcontroller	Chapter 28, "An Overview of Robot 'Brains'"
Programming a PC parallel port for robot interaction	Chapter 30, "Computer Control Via PC Printer Port"
Programming popular high-level language microcontroller chips	Chapters 31-33
Ideas for working with bitwise values for input/output ports	Chapter 29, "Interfacing with Computers and Microcontrollers"
Programming using infrared remote control	Chapter 34, "Remote Control Systems"

PART 2

ROBOT CONSTRUCTION

8

BUILDING A
PLASTIC ROBOT PLATFORM

It all started with billiard balls. In the old days, billiard balls were made from elephant tusks. By the 1850s, the supply of tusk ivory was drying up and its cost had skyrocketed. So, in 1863 Phelan & Collender, a major manufacturer of billiard balls, offered a $10,000 prize for anyone who could come up with a suitable substitute for ivory. A New York printer named John Wesley Hyatt was among several folks who took up the challenge.

Hyatt didn't get the $10,000. His innovation, celluloid, was too brittle to be used for billiard balls. But while Hyatt's name won't go down in the billiard parlor hall of fame, he will be remembered as the man who started the plastics revolution. Hyatt's celluloid was perfect for such things as gentlemen's collars, ladies' combs, containers, and eventually even motion picture film.

In the more than 100 years since the introduction of celluloid, plastics have taken over our lives. Plastic is sometimes the object of ridicule—from plastic money to plastic furniture—yet even its critics are quick to point out its many advantages:

- Plastic is cheaper per square inch than wood, metal, and most other construction materials.
- Certain plastics are extremely strong, approaching the tensile strength of such light metals as copper and aluminum.
- Some plastic is "unbreakable."

Plastic is an ideal material for use in hobby robotics. Its properties are well suited for numerous robot designs, from simple frame structures to complete assemblies. Read this chapter to learn more about plastic and how to work with it. At the end of the chapter, we'll show you how to construct an easy-to-build "turtle robot"—the Minibot—from inexpensive and readily available plastic parts.

Types of Plastics

Plastics represent a large family of products. Plastics often carry a fancy trade name, like Plexiglas, Lexan, Acrylite, Sintra, or any of a dozen other identifiers. Some plastics are better suited for certain jobs, so it will benefit you to have a basic understanding of the various types of plastics. Here's a short rundown of the plastics you may encounter:

- *ABS*. Short for "acrylonitrile butadiene styrene," ABS is most often used in sewer and wastewater plumbing systems. The large black pipes and fittings you see in the hardware store are made of ABS. ABS is a glossy, translucent plastic that can take on just about any color and texture. It is tough and hard and yet relatively easy to cut and drill. Besides plumbing fittings, ABS also comes in rods, sheets, and pipes—and as LEGO plastic pieces!
- *Acrylic*. Acrylic is clear and strong, the mainstay of the decorative plastics industry. It can be easily scratched, but if the scratches aren't too deep they can be rubbed out. Acrylic is somewhat tough to cut because it tends to crack, and it must be drilled carefully. The material comes mostly in sheets, but it is also available in extruded tubing, in rods, and in the coating in pour-on plastic laminate.
- *Cellulosics*. Lightweight and flimsy but surprisingly resilient, cellulosic plastics are often used as a sheet covering. Their uses in robotics are minor. One useful application, however, stems from the fact that *cellulosics* soften at low heat, and thus they can be slowly formed around an object. These plastics come in sheet or film form.
- *Epoxies*. Very durable clear plastic, epoxies are often used as the binder in fiberglass. Epoxies most often come in liquid form, so they can be poured over something or onto a fiberglass base. The dried material can be cut, drilled, and sanded.
- *Nylon*. Nylon is tough, slippery, self-lubricating stuff that is most often used as a substitute for twine. Plastics distributors also supply nylon in rods and sheets. Nylon is flexible, which makes it moderately hard to cut.
- *Phenolics*. An original plastic, phenolics are usually black or brown in color, easy to cut and drill, and smell terrible when heated. The material is usually reinforced with wood or cotton bits or laminated with paper or cloth. Even with these additives, phenolic plastics are not unbreakable. They come in rods and sheets and as pour-on coatings. The only application of phenolics in robotics is as circuit board material.
- *Polycarbonate*. Polycarbonate plastic is a close cousin of acrylic but more durable and resistant to breakage. Polycarbonate plastics are slightly cloudy in appearance and are easy to mar and scratch. They come in rods, sheets, and tube form. A common, inexpensive window-glazing material, polycarbonates are hard to cut and drill without breakage.
- *Polyethylene*. Polyethylene is lightweight and translucent and is often used to make flexible tubing. It also comes in rod, film, sheet, and pipe form. You can reform the

material by applying low heat, and when the material is in tube you can cut it with a knife.

- *Polypropylene*. Like polyethylene, polypropylene is harder and more resistant to heat.
- *Polystyrene*. Polystyrene is a mainstay of the toy industry. This plastic is hard, clear (though it can be colored with dyes), and cheap. Although often labeled "high-impact" plastic, polystyrene is brittle and can be damaged by low heat and sunlight. Available in rods, sheets, and foamboard, polystyrene is moderately hard to cut and drill without cracking and breaking.
- *Polyurethane*. These days, polyurethane is most often used as insulation material, but it's also available in rod and sheet form. The plastic is durable, flexible, and relatively easy to cut and drill.
- *PVC*. Short for polyvinyl chloride, PVC is an extremely versatile plastic best known as the material used in freshwater plumbing and in outdoor plastic patio furniture. Usually processed with white pigment, PVC is actually clear and softens in relatively low heat. PVC is extremely easy to cut and drill and almost impervious to breakage. PVC is supplied in film, sheet, rod, tubing, even nut-and-bolt form in addition to being shaped into plumbing fixtures and pipes.
- *Silicone*. Silicone is a large family of plastics all in its own right. Because of their elasticity, silicone plastics are most often used in molding compounds. Silicone is slippery to the touch and comes in resin form for pouring.

How to Cut Plastic

Soft plastics may be cut with a sharp utility knife. When cutting, place a sheet of cardboard or artboard on the table. This helps keep the knife from cutting into the table, which could ruin the tabletop and dull the knife. Use a carpenter's square or metal rule when you need to cut a straight line. Prolong the blade's life by using the rule against the knife holder, not the blade.

Harder plastics can be cut in a variety of ways. When cutting sheet plastic less than 1/8-inch thick, use a utility knife and metal carpenter's square to score a cutting line. If necessary, use clamps to hold down the square. Most sheet plastic comes with a protective peel-off plastic on both sides. Keep it on when cutting.

Carefully repeat the scoring two or three times to deepen the cut. Place a 1/2-inch or 1-inch dowel under the plastic so the score line is on the top of the dowel. With your fingers or the palms of your hands, carefully push down on both sides of the score line. If the sheet is wide, use a piece of 1-by-2 or 2-by-4 lumber to exert even pressure. Breakage and cracking is most likely to occur on the edges, so press on the edges first, then work your way toward the center. Don't force the break. If you can't get the plastic to break off cleanly, deepen the score line with the utility knife.

Thicker sheet plastic, as well as extruded tubes, pipes, and bars, must be cut with a saw. If you have a table saw, outfit it with a plywood-paneling blade. Among other applications, this blade can be used to cut plastics. You cut through plastic just as you do with wood, but the feed rate—the speed at which the material is sawed in two—must be slower. Forcing the plastic or using a dull blade, heats the plastic, causing it to deform and melt. A band saw is ideal for cutting plastics less than 1/2-inch thick, especially if you need to cut corners. Keep the protective covering on the plastic while cutting. When working with a power saw, use fences or pieces of wood held in place by C-clamps to ensure a straight cut.

You can use a handsaw to cut smaller pieces of plastic. A hacksaw with a medium- or fine-tooth blade (24 or 32 teeth per inch) is a good choice. You can also use a coping saw (with a fine-tooth blade) or a razor saw. These are good choices when cutting angles and corners as well as when doing detail work.

You can use a motorized scroll (or saber) saw to cut plastic, but you must take care to ensure a straight cut. If possible, use a piece of plywood held in place by C-clamps as a guide fence. Routers can be used to cut and score plastic, but unless you are an experienced router user you should not attempt this method.

How to Drill Plastic

Wood drill bits can be used to cut plastics, but I've found that bits designed for glass drilling yield better, safer results. If you use wood bits, you should modify them by blunting the tip slightly (otherwise the tip may crack the plastic when it exits the other side). Continue the flute from the cutting lip all the way to the end of the bit (see Fig. 8.1). Blunting the tip of the bit isn't hard to do, but grinding the flute is a difficult proposition. The best idea is to invest in a few glass or plastic bits, which are already engineered for drilling plastic.

Drilling with a power drill provides the best results. The drill should have a variable speed control. Reduce the speed of the drill to about 500 to 1000 rpm when using twist bits, and to about 1000 to 2000 rpm when using spade bits. When drilling, always back the plastic with a wooden block. Without the block, the plastic is almost guaranteed to crack. When using spade bits or brad-point bits, drill partially through from one side, then

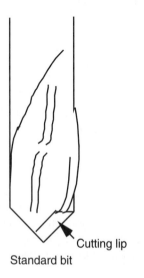

Cutting lip

Standard bit   Modified bit

FIGURE 8.1 Suggested modifications for drill bits used with plastic. The end is blunted and the flutes are extended to the end of the cutting tip.

complete the hole by drilling from the other side. As with cutting, don't force the hole and always use sharp bits. Too much friction causes the plastic to melt.

To make holes larger than 1/4 inch you should first drill a smaller, pilot hole. If the hole is large, over 1/4 inch in diameter, start with a small drill and work your way up several steps. Practice drilling on a piece of scrap until you get the technique right.

How to Bend and Form Plastic

Most rigid and semirigid plastics can be formed by applying low localized heat. A sure way to bend sheet plastic is to use a strip heater. These are available ready-made at some hardware and plastics supply houses, or you can build your own. A narrow element in the heater applies a regulated amount of heat to the plastic. When the plastic is soft enough, you can bend it into just about any angle you want.

There are two important points to remember when using a strip heater. First, be sure that the plastic is pliable before you try to bend it. Otherwise, you may break it or cause excessive stress at the joint (a stressed joint will looked cracked or crazed). Second, bend the plastic *past* the angle that you want. The plastic will "relax" a bit when it cools off, so you must anticipate this. Knowing how much to overbend will come with experience, and it will vary depending on the type of plastic and the size of the piece you're working with.

You can mold thinner sheet plastic around shapes by first heating it up with a hair dryer or heat gun, then using your fingers to form the plastic. Be careful that you don't heat up the plastic too much. You don't want it to melt, just conform to the underlying shape. You can soften an entire sheet or piece by placing it into an oven for 10 or so minutes (remove the protective plastic before baking). Set the thermostat to 300°F and be sure to leave the door slightly ajar so any fumes released during the heating can escape. Ventilate the kitchen and avoid breathing the fumes, as they can be noxious. All plastics release gases when they heat up, but the fumes can be downright toxic when the plastic actually ignites. Therefore, avoid heating the plastic so much that it burns. Dripping, molten plastic can also *seriously* burn you if it drops on your skin.

How to Polish the Edges of Plastic

Plastic that has been cut or scored usually has rough edges. You can file the edges of cut PVC and ABS using a wood or metal file. You should polish the edges of higher-density plastics like acrylics and polycarbonates by sanding, buffing, or burnishing. Try a fine-grit (200 to 300), wet-dry sandpaper and use it wet. Buy an assortment of sandpapers and try several until you find the coarseness that works best with the plastic you're using. You can apply jeweler's rouge, available at many hardware stores in large blocks, by using a polishing wheel. The wheel can be attached to a grinder or drill motor.

Burnishing involves using a very low-temperature flame (a match or lighter will do) to melt the plastic slightly. You can also use a propane torch kept some distance from the plastic. Be extremely careful when using a flame to burnish plastic. Don't let the plastic ignite, or you'll end up with an ugly blob that will ruin your project, not to mention filling the room with poisonous gas.

How to Glue Plastic

Most plastics aren't really glued together; they are cemented. The cement contains one or more solvents that actually melt the plastic at the joint. The pieces are then fused together—that is, made one. Household adhesives can be used for this, of course, but you get better results when you use specially formulated cements.

Herein lies a problem. The various plastics we have described rarely use the same cement formulations, so to make your project, er, stick you've got to use the right mixture. That means you have to know the type of plastic used in the material you are working with. See our earlier discussion in this chapter about the various types of plastics. Also refer to Table 8.1, which lists the major types of plastics and how they are used in common household and industrial products. Table 8.2 indicates the adhesives I recommend for cementing several popular plastics.

When using solvent for PVC or ABS plumbing fixtures, apply the goop in the recommended manner by spreading a thin coat on the pieces to be joined. A cotton applicator is included in the can of cement. Plastic sheet, bars, and other items require more careful cementing, especially if you want the end result to look nice.

With the exception of PVC solvent, the cement for plastics is watery thin and can be applied in a variety of ways. One method is to use a small painter's brush, with a #0 or #1 tip. Joint the pieces to be fused together and "paint" the cement onto the joint with the brush. Capillary action will draw the cement into the joint, where it will spread out. Another method is to fill a special syringe applicator with cement. With the pieces butted together, squirt a small amount of cement into the joint line.

In all cases, you must be sure that the surfaces at the joint of the two pieces are perfectly flat and that there are no voids where the cement may not make ample contact. If you are joining pieces whose edges you cannot make flush, apply a thicker type of glue, such as contact cement or white household glue. You may find that you can achieve a better bond by first roughing up the joints to be mated. You can use coarse sandpaper or a file for this purpose.

After applying the cement, wait at least 15 minutes for the plastic to re-fuse and the joint to harden. Disturbing the joint before it has a time to set will permanently weaken it. Remember that you cannot apply cement to plastics that have been painted. If you paint the pieces before cementing them, scrape off the paint and refinish the edges so they're smooth again.

Using Hot Glue with Plastics

Perhaps the fastest way to glue plastic pieces together is to use hot glue. You heat the glue to a viscous state in a glue gun, then spread it out over the area to be bonded. Hot-melt

TABLE 8.1 Plastics in Everyday Household Articles

HOUSEHOLD ARTICLE	TYPE OF PLASTIC
Bottles, containers	
Clear	Polyester, PVC
Translucent or opaque	Polyethylene, polypropylene
Buckets, washtubs	Polyethylene, polypropylene
Foam cushions	Polyurethane foam, PVC foam
Electrical circuit boards	Laminated epoxies, phenolics
Fillers	
Caulking compounds	Polyurethane, silicone, PVAC
Grouts	Silicone, PVAC
Patching compounds	Polyester, fiberglass
Putties	Epoxies, polyester, PVAC
Films	
Art film	Cellulosics
Audio tape	Polyester
Food wrap	Polyethylene, polypropylene
Photographic	Cellulosics
Glasses (drinking)	
Clear, hard	Polystyrene
Flexible	Polyethylene
Insulated cups	Styrofoam (polystyrene foam)
Hoses, garden	PVC
Insulation foam	Polystyrene, polyurethane
Lubricants	Silicones
Plumbing pipes	
Fresh water	PVC, polyethylene, ABS
Gray water	ABS
Siding and paneling	PVC
Toys	
Flexible	Polyethylene, polypropylene
Rigid	Polystyrene, ABS
Tubing (clear or translucent)	Polyethylene, PVC

Table 8.2   Plastic Bonding Guide

PLASTIC	CEMENTED TO... ITSELF, USE	...OTHER PLASTIC, USE	...METAL, USE
ABS	ABS-ABS solvent	Rubber adhesive	Epoxy cement
Acrylic	Acrylic solvent	Epoxy cement	Contact cement
Cellulosics	White glue	Rubber adhesive	Contact cement
Polystyrene	Model glue	Epoxy	CA glue*
Polystyrene foam	White glue	Contact cement	Contact cement
Polyurethane	Rubber adhesive	Epoxy, contact cement	Contact cement
PVC	PVC-PVC solvent	PVC-ABS (to ABS)	Contact cement

(*CA stands for cyanoacrylate ester, sometimes known as "Super Glue," after a popular brand name.)

glue and glue guns are available at most hardware, craft, and hobby stores in several different sizes. The glue is available in a "normal" and a low-temperature form. Low-temperature glue is generally better with most plastics because it avoids the "sagging" or softening of the plastic sometimes caused by the heat of the glue.

One caveat when working with hot-melt glue and guns (other than the obvious safety warnings) is that you should always rough up the plastic surfaces before trying to bond them. Plastics with a smooth surface will not adhere well when you are using hot-melt glue, and the joint will be brittle and perhaps break off with only minor pressure. By roughing up the plastic, the glue has more surface area to bond to, resulting in a strong joint. Roughing up plastic before you join pieces is an important step when you are using most any glue or cement, but it is particularly important when using hot-melt glue.

How to Paint Plastics

Sheet plastic is available in transparent or opaque colors, and this is the best way to add color to your robot projects. The colors are impregnated in the plastic and can't be scraped or sanded off. However, you can also add a coat of paint to the plastic to add color or to make it opaque. Most all plastics accept brush or spray painting.

Spray painting is the preferred method for all jobs that don't require extra-fine detail. Carefully select the paint before you use it, and always apply a small amount to a scrap piece of plastic before painting the entire project. Some paints contain solvents that may soften the plastic.

One of the best all-around paints for plastics are the model and hobby spray cans made by Testor. These are specially formulated for styrene model plastic, but I've found that the paint adheres well without softening on most all plastics. I've used it successfully on ABS, PVC, acrylic, polycarbonate, and many other plastics. You can purchase this paint in a variety of colors, in either gloss or flat finish. The same colors are available in bottles with self-contained brush applicators. If the plastic is clear, you have the option of painting on the front or back side (or both for that matter). Painting on the front side will produce the paint's standard finish: gloss colors come out gloss; flat colors come out flat. Flat-finish paints tend to scrape off easier, however, so exercise care.

Painting on the back side with gloss or flat paint will only produce a glossy appearance because you look through the clear plastic to the paint on the back side. Moreover, painting imperfections will more or less be hidden, and external scratches won't mar the paint job.

Buying Plastic

Some hardware stores carry plastic, but you'll be sorely frustrated at their selection. The best place to look for plastic—in all its styles, shapes, and chemical compositions—is a plastics specialty store or plastics sign-making shop. Most larger cities have at least one plastic supply store or sign-making shop that's open to public. Look in the yellow pages under *Plastics - Retail*.

Another useful source is the plastics fabricator. There are actually more of these than retail plastic stores. They are in business to build merchandise, display racks, and other plastic items. Although they don't usually advertise to the general public, most will sell to you. If the fabricator doesn't sell new material, ask to buy the leftover scrap.

Plastics around the House

You need not purchase plastic for all your robot needs at a hardware or specialty store. You may find all the plastic you really need right in your own home. Here are a few good places to look:

- *Used compact discs (CDs)*. These are a typical denizen of the modern-day landfill. CDs, made from polycarbonate plastics, are usually just thrown away and not recycled. What a waste of resources! With careful drilling and cutting, you can adapt them to serve as body parts and even wheels for your robots. Exercise caution when working with CDs: they can *shatter* when you drill and cut them, and the pieces are very sharp and dangerous.
- *Used LaserVision discs*. These are "grownup" versions of CDs. With the advent of Digital Versatile Disc (DVD), 12-inch-diameter laser discs are falling out of favor, and they are often available at garage sales and second-hand stores for just a dollar or two each. Make sure the movie isn't a classic, then drill and cut the plastic as needed. As with CDs, use care to avoid shattering the plastic.
- *Old phonograph records*. Found in the local thrift store, records—particularly the thicker 78-rpm variety—can be used in much the same way as CDs and laser discs. The older records made from the 1930s through 1950s used a thicker plastic that is very heavy and durable. Thrift stores are your best bet for old records no one wants anymore (who is that Montovani guy, anyway?). Note that some old records, like the V-Discs made

during World War II, are collector's items, so don't wantonly destroy a record unless you're sure it has no value.

■ *Salad bowls, serving bowls, and plastic knickknacks.* They can all be revived as robot parts. I regularly prowl garage sales and thrift stores looking for such plastic material.

■ *PVC irrigation pipe.* This can be used to construct the frame of a robot. Use the short lengths of pipe left over from a weekend project. You can secure the pieces with glue or hardware or use PVC connector pieces (Ts, "ells," etc.)

Build the Minibot

Now let's take a hands-on look at how plastic can be used to construct a robot, the Minibot. You can use a small piece of scrap sheet acrylic to build the foundation and frame of the Minibot. The robot is about six inches square and scoots around the floor or table on two small rubber tires. The basic version is meant to be wire-controlled, although in upcoming chapters you'll see how to adapt the Minibot to automatic electronic control, even remote control. The power source is a set of flashlight batteries. I used four AA batteries because they are small, lightweight, and provide more driving power than 9-volt transistor batteries. The parts list for the Minibot can be found in Table 8.3.

FOUNDATION OR BASE

The foundation is clear or colored Plexiglas or some similar acrylic sheet plastic. The thickness should be at least 1/8 inch, but avoid very thick plastic because of its heavy

Table 8.3 Minibot Parts List	
Minibot:	
1	6-inch-by-6-inch acrylic plastic (1/16-inch or 1/8-inch thick)
2	Small hobby motors with gear reduction
2	Model airplane wheels
1	3 1/2-inch (approx.) 10/24 all-thread rod
1	6-inch-diameter (approx.) clear plastic dome
1	Four-cell AA battery holder
Assorted	1/2-inch-by-8/32 bolts, 8/32 nuts, lock washers, 1/2-inch-by-10/24 bolts, 10/24 nuts, lock washers, capnuts
Motor Control Switch:	
1	Small electronic project enclosure
2	Double-pole, double-throw (DPDT) momentary switches, with center-off return
Misc.	Hookup wire

weight. The prototype Minibot used 1/8-inch thick acrylic, so there was minimum stressing caused by bending or flexing.

Cut the plastic as shown in Fig. 8.2. Remember to keep the protective paper cover on the plastic while you cut. File or sand the edges to smooth the cutting and scoring marks. The corners are sharp and can cause injury if the robot is handled by small children. You can easily fix this by rounding off the corners with a file. Find the center and drill a hole with a #10 bit. Fig. 8.2 also shows the holes for mounting the drive motors. These holes are spaced for a simple clamp mechanism that secures hobby motors that are commonly available on the market.

MOTOR MOUNT

The small DC motors used in the prototype Minibot were surplus gear motors with an output speed of about 30 rpm. The motors for your Minibot should have a similar speed because even with fairly large wheels, 30 rpm makes the robot scoot around the floor or a table at about four to six inches a second. Choose motors small enough so they don't crowd the base of the robot and add unnecessary weight. Remember that you have other items to add, such as batteries and control electronics.

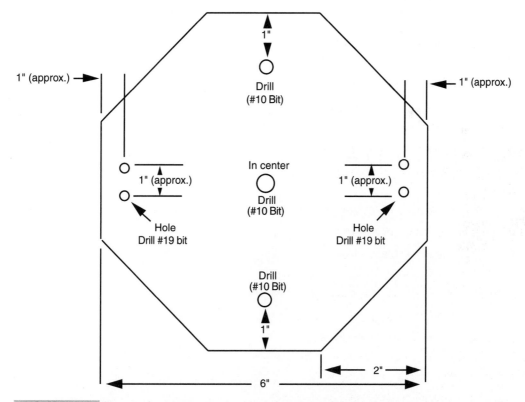

FIGURE 8.2 The cutting guide for the base of the plastic Minibot. The sets of two holes on either side are for the motor mount, and they should be spaced according to the specific mount you are using. Motors of different sizes and types will require different mounting holes.

Use 3/8-inch-wide metal mending braces to secure the motor (the prototype used plastic pieces from an old Fastech toy construction kit; you can use these or something similar). You may need to add spacers or extra nuts to balance the motor in the brace. Drill holes for 8/32 bolts (#19 bit), spaced to match the holes in the mending plate. Another method is to use small U-bolts, available at any hardware store. Drill the holes for the U-bolts and secure them with a double set of nuts.

Attach the tires to the motor shafts. Tires designed for a radio-controlled airplane or race car are good choices. The tires are well made, and the hubs are threaded in standard screw sizes (the threads may be metric, so watch out!). I threaded the motor shaft and attached a 4-40 nut on each side of the wheel. Fig. 8.3 shows a mounted motor with a tire attached.

Installing the counterbalances completes the foundation-base plate. These keep the robot from tipping backward and forward along its drive axis. You can use small ball bearings, tiny casters, or—as I did in the prototype—the head of a 10/24 locknut. The

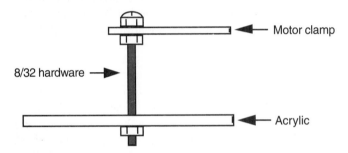

FIGURE 8.3 How the drive motors of the Minibot look when mounted. In the prototype Minibot the wheels were threaded directly onto the motor shaft. Note the gear-reduction system built onto the hobby motor.

locknut is smooth enough to act as a kind of ball bearing and is about the right size for the job. Attach the locknut with a 10/24-by-1/2-inch bolt (if the bolt you have is too long to fit in the locknut, add washers or a 10/24 nut as a spacer).

TOP SHELL

The top shell is optional, and you can leave it off if you choose. The prototype used a round display bowl six inches in diameter that I purchased from a plastics specialty store. Alternatively, you can use any suitable half sphere for your robot, such as an inverted salad bowl. Feel free to use colored plastic.

Attach the top by measuring the distance from the foundation to the top of the shell, taking into consideration the gap that must be present for the motors and other bulky internal components. Cut a length of 10/24 all-thread rod to size. The length of the prototype shaft was 3 1/2 inches. Secure the center shaft to the base using a pair of 10/24 nuts and a tooth lock washer. Secure the center shaft to the top shell with a 10/24 nut and a 10/24 locknut. Use a tooth lock washer on the inside or outside of the shell to keep the shell from spinning loose.

BATTERY HOLDER

You can buy battery holders that hold from one to six dry cells in any of the popular battery sizes. The Minibot motors, like almost all small hobby motors, run off 1.5 to 6 volts. A four-cell, AA battery holder does the job nicely. The wiring in the holder connects the batteries in series, so the output is 6 volts. Secure the battery holder to the base with 8/32 nuts and bolts. Drill holes to accommodate the hardware. Be sure the nuts and bolts don't extend too far below the base or they may drag when the robot moves. Likewise, be sure the hardware doesn't interfere with the batteries.

WIRING DIAGRAM

The wiring diagram in Fig. 8.4 allows you to control the movement of the Minibot in all directions. This simple two-switch system, which will be used in many other projects in this book, uses double-pole, double-throw (DPDT) switches. The switches called for in the circuit are spring-loaded so they return to a center-off position when you let go of them.

From Here

To learn more about...	*Read*
Wooden robots	Chapter 9, "Building a Basic Wooden Platform"
Metal robots	Chapter 10, "Building a Metal Platform"
Using batteries	Chapter 15, "All About Batteries and Robot Power Supplies"
Selecting the right motor	Chapter 17, "Choosing the Right Motor for the Job"
Using a computer or microcontroller	Chapter 28, "An Overview of Robot 'Brains'"

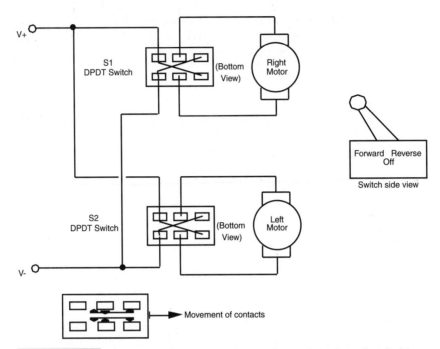

FIGURE 8.4 Use this schematic for wiring the motor control switches for the Minibot. Note that the switches are double-pole, double-throw (DTDP), with a spring return to center-off.

9

BUILDING A BASIC
WOODEN PLATFORM

Wood may not be high-tech, but it's an ideal building material for hobby robots. Wood is available just about everywhere. It's relatively inexpensive, it's easy to work with, and mistakes can be readily covered up or painted over. In this chapter, we'll take a look at using wood in robots and how you can apply simple woodworking skills to construct a basic wooden robot platform. This platform can then serve as the foundation for a number of robot designs you may want to explore.

Choosing the Right Wood

There is good wood and there is bad wood. Obviously, you want the good stuff, but you have to be willing to pay for it. For reasons you'll soon discover, you should buy only the best stock you can get your hands on. The better woods are available at specialty wood stores, particularly the ones that sell mostly hardwoods and exotic woods. Your local lumber and hardware store may have great buys on rough-hewn redwood planking, but it's hardly the stuff of robots.

PLYWOOD

The best overall wood for robotics use, especially for foundation platforms, is *plywood*. In case you are unfamiliar with plywood, this common building material comes in many grades and is made by laminating thin sheets of wood together. The cheapest plywood is called "shop grade," and it is the kind often used for flooring and projects where looks aren't too important. The board is full of knots and knotholes, and there may be considerable *voids* inside the board.

The remaining grades specify the quality of both sides of the plywood. Grade N is the best and signifies "natural finish" veneer. The surface quality of grade N really isn't important to us, so we can settle for grade A. Since we want both sides of the board to be in good shape, a plywood with a grade of A-A (grade A on both sides) is desired. Grades B and C are acceptable, but only if better plywoods aren't around. Depending on the availability of these higher grades, you may have to settle for A-C grade plywood (grade A on one side, grade C on the other).

Most plywoods you purchase at the lumber stores are made of softwoods—usually fir and pine. You can get hardwood plywood as well through a specialty wood supplier or from hobby stores. Hardwoods are more desirable because they are denser and less likely to chip. Don't confuse *hardwood* plywood with *hardboard*. The latter is made of sawdust epoxied together under high pressure. Hardboard has a smooth finish; its close cousin, particleboard, does not. Both types are unsuitable for robotics because they are too heavy and brittle.

Plywood comes in various thicknesses starting at about 5/16 inch and going up to over 1 inch. Thinner sheets are acceptable for use in a robotics platform if the plywood is made from hardwoods. When using construction-grade plywoods (the stuff you get at the home improvement store), a thickness in the middle of the range—1/2 inch or 3/8 inch—is ideal.

Construction plywood generally comes in 4-by-8-foot panels. Hardwood plywoods, particularly material for model building, come in smaller sizes, such as 2 feet by 2 feet. You don't need a large piece of plywood; the smaller the board, the easier it will be to cut it to the exact size you need.

PLANKING

An alternative to working with plywood is *planking*. Use ash, birch, or some other solid hardwood. Stay away from the less meaty softwoods such as fir, pine, and hemlock. Most hardwood planks are available in widths of no more than 12 or 15 inches, so you must take this into consideration when designing the platform. You can butt two smaller widths together if absolutely necessary. Use a router to fashion a secure joint, or attach metal mending plates to mate the two pieces together. (This latter option is not recommended; it adds a lot of unnecessary weight to the robot.)

When choosing planked wood, be especially wary of warpage and moisture content. Take along a carpenter's square, and check the squareness and levelness of the lumber in every possible direction. Reject any piece that isn't perfectly square; you'll regret it otherwise. Defects in milled wood go by a variety of colorful names, such as crook, bow, cup, twist, wane, split, shake, and check, but they all mean "headache" to you.

Wood with excessive moisture may bow and bend as it dries, causing cracks and warpage. These can be devastating in a robot you've just completed and perfected. Buy only seasoned lumber stored inside the lumberyard, not outside. Watch for green specks or grains—these indicate trapped moisture. If the wood is marked, look for an "MC" specification. An "MC-15" rating means that the moisture content doesn't exceed 15 percent, which is acceptable.

Good plywoods and hardwood planks meet or exceed this requirement. Don't get anything marked MC-20 or higher or marked S-GREEN—bad stuff for robots.

DOWELS

Wood dowels come in every conceivable diameter, from about 1/16 inch to over 1½ inches. Wood dowels are three or four feet in length. Most dowels are made of high-quality hardwood, such as birch or ash. The dowel is always cut lengthwise with the grain to increase strength. Other than choosing the proper dimension, there are few considerations to keep in mind when buying dowels.

You should, however, inspect the dowel to make sure that it is straight. At the store, roll the dowel on the floor. It should lie flat and roll easily. Warpage is easy to spot. Dowels can be used either to make the frame of the robot or as supports and uprights.

The Woodcutter's Art

You've cut a piece of wood in two before, haven't you? Sure you have; everyone has. You don't need any special tools or techniques to cut wood for a robot platform. The basic shop cutting tools will suffice: a handsaw, a backsaw, a circular saw, a jigsaw (if the wood is thin enough for the blade), a table saw, a radial arm saw, or—you name it.

Whatever cutting tool you use, make sure the blade is the right one for the wood. The combination blade that probably came with your power saw isn't the right choice for plywood and hardwood. Outfit the saw with a cutoff blade or a plywood-paneling blade. Both have many more teeth per inch. Handsaws generally come in two versions: crosscut and ripsaw. You need the crosscut kind.

Deep Drilling

You can use a hand or motor drill to drill through wood. Electric drills are great and do the job fast, but I prefer the older-fashioned hand drills for applications that require precision. Either way, it's important that you use only sharp drill bits. If your bits are dull, replace them or have them sharpened.

It's important that you drill straight holes, or your robot may not go together properly. If your drill press is large enough, you can use it to drill perfectly straight holes in plywood and other large stock. Otherwise, use a portable drill stand. These attach to the drill or work in a number of other ways to guarantee you a straight hole.

Finishing

You can easily shape wood using rasps and files. If the shaping you need to do is extensive—like creating a circle in the middle of a large plank—you may want to consider getting a rasp for your drill motor. They come in various sizes and shapes. You'll want the shaped wood to be as smooth as possible, both for appearance's sake and so it's easier to attach other wood pieces or components of the robot to it. A medium-grit sandpaper is fine for the job.

Add a coat of paint to the wood when you're done. Silver, gloss black, or some other high-tech opaque color are good choices. The paint also serves to protect the wood against the effects of moisture and aging.

Building a Wooden Motorized Platform

Figures 9.1 and 9.2 show approaches for constructing a basic square and round motorized wooden platform, respectively. See Table 9.1 for a list of the parts you'll need.

To make the square plywood platform shown in Fig. 9.1, cut a piece of 3/8- or 1/2-inch plywood to 10 by 10 inches (the thinner 3/8-inch material is acceptable if the plywood is the heavy-duty hardwood variety, such as that used for model ships). Make sure the cut is square. Notch the wood as shown to make room for the robot's wheels. The notch should be large enough to accommodate the width and diameter of the wheels, with a little "breathing room" to spare. For example, if the wheels are 6 inches in diameter and 1.5 inches wide, the notch should be about 6.5 by 1.75 inches.

To make a motor control switch for this platform, see the parts list given in Table 8.3 of in Chap. 8.

Figure 9.2 shows the same 10-inch-square piece of 3/8- or 1/2-inch plywood cut into a circle. Use a scroll saw and circle attachment for cutting. As you did for the square platform, make a notch in the center beam of the circle to allow room for the wheels—the larger the wheel, the larger the notch.

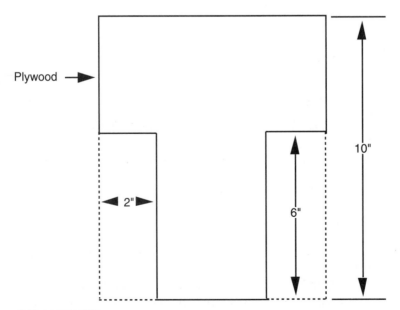

FIGURE 9.1   Cutting plan for a square plywood base.

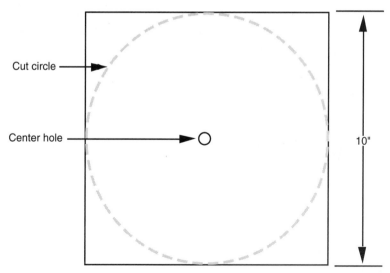

FIGURE 9.2 Cutting plan for a round plywood base.

TABLE 9.1 PARTS LIST FOR WOOD BASE	
Robot Base:	
1	10-inch-by-10-inch plywood (3/8- or 1/2-inch thickness)
Misc	1-inch-by-10/24 stove bolts, 10/24 nuts, flat washers
Motor/drive:	
2	DC gear motors
2	5- to 7-inch rubber wheels
1	1 1/4-inch caster
2	2-by-4-inch lumber (cut to length to fit motor)
1	Four-cell "D" battery holder
Misc	3-inch-by-10/24 stove bolts, 10/24 nuts, flat washers (for motor mount)
	1 1/4-inch-by-8/32 stove bolts, 8/32 nuts, flat washers, lock washers (for caster mount)

ATTACHING THE MOTORS

The wooden platform you have constructed so far is perfect for a fairly sturdy robot, so the motor you choose should be too. Use heavy-duty motors, geared down to a top speed of no more than about 75 rpm; 30 to 40 rpm is even better. Anything faster than 75 rpm will cause the robot to dash about at speeds exceeding a few miles per hour, which is unacceptable unless you plan on entering your creation in the Robot Olympics.

Note: You can use electronic controls to reduce the speed of the gear motor by 15 or 20 percent without losing much torque, but you should not slow the motor too much, or you'll lose power. The closer the motor operates at its rated speed, the better results you'll have.

If the motors have mounting flanges and holes on them attach them using corner brackets. Some motors do not have mounting holes or hardware, so you must fashion a hold-down plate for them. You can make an effective hold-down plate, as shown in Fig. 9.3, out of wood. Round out the plate to match the cylindrical body of the motor casing. Then secure the plate to the platform. Last, attach the wheels to the motor shafts. You may need to thread the shafts with a die so you can secure the wheels. Use the proper size nuts and washers on either side of the hub to keep the wheel in place. You'll make your life much easier if you install wheels that have a setscrew. Once they are attached to the shaft, tighten the setscrew to screw the wheels in place.

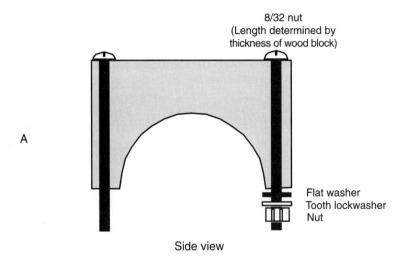

Side view

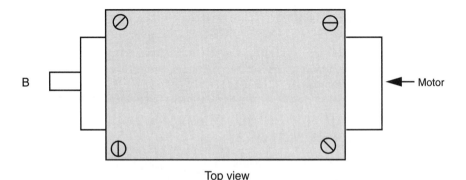

Top view

FIGURE 9.3   One way to secure the motors to the base is to use a wood block hollowed out to match the shape of the motor casing. *a.* **Side view;** *b.* **Top view.**

STABILIZING CASTER

Using two motors and a centered caster, as depicted in Fig. 9.4, allows you to have full control over the direction your robot travels. You can make the robot turn by stopping or reversing one motor while the other continues turning. Attach the caster using four 8/32-by-1-inch bolts. Secure the caster with tooth lock washers and 8/32 nuts.

Note that the caster must be level with (or a little higher than) the drive motors. You may, if necessary, use spacers to increase the distance from the baseplate of the caster to the bottom of the platform. If the caster is already lower than the wheels, you'll have trouble because the motors won't adequately touch the ground. You can rectify this problem by simply using larger-diameter wheels.

BATTERY HOLDER

You can purchase battery holders that contain from one to six dry cells in any of the popular battery sizes. When using 6-volt motors, you can use a four-cell "D" battery holder.

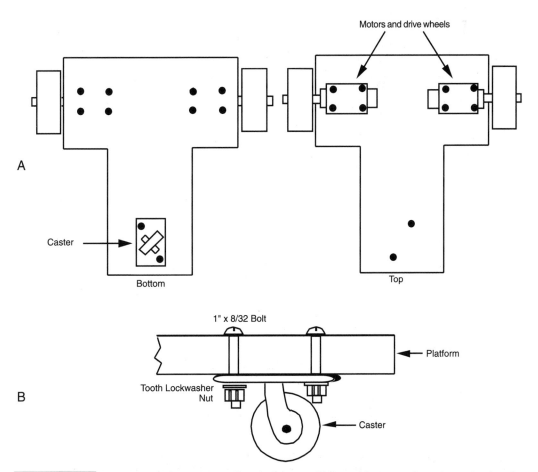

FIGURE 9.4 Attaching the caster to the platform. *a.* Top and bottom view; *b.* caster hardware assembly detail.

You can also use a single 6-volt lantern or rechargeable battery. Motors that require 12 volts will need two battery holders, two 6-volt batteries, or one 12-volt battery (the latter is somewhat hard to find unless you're looking for a car battery, which you are not). For the prototype, I used 6-volt motors and a four-cell "D" battery holder.

Secure the battery holder(s) to the base with 8/32 nuts and bolts. Drill holes to accommodate the hardware. Be sure the nuts and bolts don't extend too far below the base or they may drag when the robot moves. Likewise, be sure the hardware doesn't interfere with the batteries.

Wire the batteries and wheels to the DPDT through control switches, as shown in the Minibot project described at the end of Chapter 8, "Building a Plastic Robot Platform." One switch controls the left motor; the other switch controls the right motor.

From Here

To learn more about...	*Read*
Plastic robots	Chapter 8, "Building a Plastic Robot Platform"
Metal robots	Chapter 10, "Building a Metal Platform"
Using batteries	Chapter 15, "All about Batteries and Robot Power Supplies"
Selecting the right motor	Chapter 17, "Choosing the Right Motor for the Job"
Using a computer or microcontroller	Chapter 28, "An Overview of Robot 'Brains'"

10

BUILDING A METAL PLATFORM

Metal is perhaps the best all-around material for building robots because if offers extra strength that other materials, such as plastic and wood, cannot. If you've never worked with metal before, you shouldn't worry; there is really nothing to it. The designs outlined in this chapter and the chapters that follow will show you how to construct robots both large and small out of readily available metal stock, without resorting to welding or custom machining.

Metal Stock

Metal stock is available from a variety of sources. Your local hardware store is the best place to start. However, some stock may only be available at the neighborhood sheet metal shop. Look around and you're sure to find what you need.

EXTRUDED ALUMINUM

Extruded stock is made by pushing molten metal out of a shaped orifice. As the metal exits it cools, retaining the exact shape of the orifice. Extruded aluminum stock is readily available at most hardware and home improvement stores. It generally comes in 12-foot sections, but many hardware stores will let you buy cut pieces if you don't need all 12 feet.

Extruded aluminum is available in more than two dozen common styles, from thin bars to pipes to square posts. Although you can use any of it as you see fit, the following standard sizes may prove to be particularly beneficial in your robot-building endeavors:

- 1-by-1-by-1/16-inch angle stock
- 57/64-by-9/16-by-1/16-inch channel stock
- 41/64-by-1/2-by-1/16-inch channel stock
- Bar stock, in widths from 1 to 3 inches and thicknesses of 1/16 to 1/4 inch

SHELVING STANDARDS

You've no doubt seen those shelving products where you nail two metal rails on the wall and then attach brackets and shelves to them. The rails are referred to as "standards," and in a pinch they are well suited to be girders in robot frames. The standards come in either aluminum or steel and measure 41/64 by 1/2 by 1/16 inch. The steel stock is cheaper but considerably heavier, a disadvantage you will want to carefully consider. Limit its use to structural points in your robot that need extra strength.

Another disadvantage of using shelving standards instead of extruded aluminum are all the holes and "slots" you'll find on the standards. The holes are for mounting the standards to a wall; the slots are for attaching shelving brackets. Both can be troublesome when you are drilling the metal for bolt holes. The drill can slip into a hole or slot, and as a result the hole may not end up where you want it. For this reason, use extruded aluminum pieces when possible. It will yield more professional results.

MENDING PLATES

Galvanized mending plates are designed to strengthen the joint of two or more pieces of lumber. Most of these plates come preformed in all sorts of weird shapes and so are pretty much unusable for building robots. But flat plates are available in several widths and lengths. You can use the plates as is or cut them to size. The plates are made of galvanized iron and have numerous pre-drilled holes in them to help you hammer in nails. The material is soft enough so you can drill new holes, but if you do so only use sharp drill bits.

Mending plates are available in lengths of about 4, 6, and 12 inches. Widths are not as standardized, but 2, 4, 6, and 12 inches seem common. You can usually find mending plates near the rain gutter and roofing section in the hardware store. Note that mending plates are heavy, so don't use them for small, lightweight robot designs. Reserve them for medium- to large-sized robots where the plate can provide added structural support and strength.

RODS AND SQUARES

Most hardware stores carry a limited quantity of short extruded steel or zinc rods and squares. These are solid and somewhat heavy items and are perfect for use in some advanced projects, such as robotic arms. Lengths are typically limited to 12 or 24 inches, and thicknesses range from 1/16 to about 1/2 inch.

IRON ANGLE BRACKETS

You will need a way to connect all your metal pieces together. The easiest method is to use galvanized iron brackets. These come in a variety of sizes and shapes and have predrilled holes to facilitate construction. The 3/8-inch-wide brackets fit easily into the two sizes of channel stock mentioned at the beginning of the chapter: 57/64 by 9/16 by 1/16 inch and 41/64 by 1/2 by 1/16 inch. You need only drill a corresponding hole in the channel stock and attach the pieces together with nuts and bolts. The result is a very sturdy and clean-looking frame. You'll find the flat corner angle iron, corner angle ("L"), and flat mending iron to be particularly useful.

Working with Metal

If you have the right tools, working with metal is only slightly harder than working with wood or plastic. You'll have better-that-average results if you always use sharpened, well-made tools. Dull, bargain-basement tools can't effectively cut through aluminum or steel stock. Instead of the tool doing most of the work, you do.

CUTTING

To cut metal, use a hacksaw outfitted with a fine-tooth blade, one with 24 or 32 teeth per inch. Coping saws, keyhole saws, and other handsaws are generally engineered for cutting wood, and their blades aren't fine enough for metal work. You can use a power saw, like a table saw or reciprocating saw, but, again, make sure that you use the right blade.

You'll probably do most of your cutting by hand. You can help guarantee straight cuts by using an inexpensive miter box. You don't need anything fancy, but try to stay away from the wooden boxes. They wear out too fast. The hardened plastic and metal boxes are the best buys. Be sure to get a miter box that lets you cut at 45° both vertically and horizontally. Firmly attach the miter box to your workbench using hardware or a large clamp.

DRILLING

Metal requires a slower drilling speed than wood, and you need a power drill that either runs at a low speed or lets you adjust the speed to match the work. Variable speed power drills are available for under $30 these days, and they're a good investment. Be sure to use only sharp drill bits. If your bits are dull, replace them or have them sharpened. Quite often, buying a new set is cheaper than professional resharpening. It's up to you.

You'll find that when you cut metal the bit will skate all over the surface until the hole is started. You can eliminate or reduce this skating by using a punch prior to drilling. Use a hammer to gently tap a small indentation into the metal with the punch. Hold the smaller pieces in a vise while you drill.

When it comes to working with metal, particularly channel and pipe stock, you'll find a drill press is a godsend. It improves accuracy, and you'll find the work goes much faster. Always use a proper vise when working with a drill press. *Never* hold the work with your hands. Especially with metal, the bit can snag as it's cutting and yank the piece out of your hands. If you can't place the work in the vise, use a pair of Vise-Grips or other suitable locking pliers.

FINISHING

Cutting and drilling often leaves rough edges, called *flashing*, in the metal. These edges must be filed down using a medium- or fine-pitch metal file, or else the pieces won't fit together properly. Aluminum flash comes off quickly and easily; you need to work a little harder when removing the flash in steel or zinc stock.

Build the Buggybot

The Buggybot is a small robot built from a single 6-by-12-inch sheet of 1/16-inch thick aluminum, nuts and bolts, and a few other odds and ends. You can use the Buggybot as the foundation and running gear for a very sophisticated petlike robot. As with the robots built with plastic and wood we discussed in the previous two chapters, the basic design of the all-metal Buggybot can be enhanced just about any way you see fit. This chapter details the construction of the framework, locomotion, and power systems for a wired remote control robot. Future chapters will focus on adding more sophisticated features, such as wireless remote control, automatic navigation, and collision avoidance and detection. Refer to Table 10.1 for a list of the parts needed to build the Buggybot.

TABLE 10.1 PARTS LIST FOR BUGGYBOT	
Frame:	
1	6-by-12-inch sheet of 1/16-inch thick aluminum
Motor and Mount:	
2	Tamiya high-power gearbox motors (from kit); see text
2	3-inch-diameter "Lite Flight" foam wheels
2	5/56 nuts (should be included with the motors)
2	3/16-inch collars with setscrews
1	Two-cell "D" battery holder
Misc	1-inch-by-6/32 stove bolts, nuts, flat washers
Support Caster:	
1	1 1/2-inch swivel caster
Misc	1/2-inch-by-6/32 stove bolts, nuts, tooth lock washers, flat washers (as spacers)
	See parts list in Table 8.3 of Chapter 8 for motor control switch

FRAMEWORK

Build the frame of the Buggybot from a single sheet of 1/16-inch thick aluminum sheet. This sheet, measuring 6 by 12 inches, is commonly found at hobby stores. As this is a standard size, there's no need to cut it. Follow the drill-cutting template shown in Fig. 10.1.

After drilling, use a large shop vise or woodblock to bend the aluminum sheet as shown in Fig. 10.2. Accuracy is not all that important. The angled bends are provided to give the Buggybot its unique appearance.

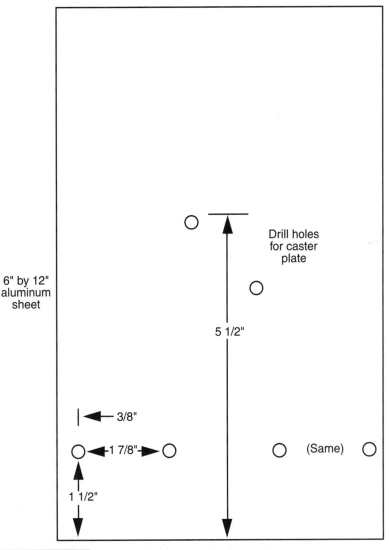

FIGURE 10.1 Drilling diagram for the Buggybot frame.

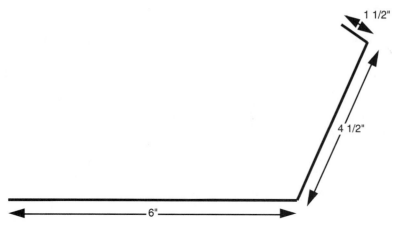

FIGURE 10.2 Bend the aluminum sheet at the approximate angles shown here.

MOTORS AND MOTOR MOUNT

The prototype Buggybot uses two high-power gearbox motor kits from Tamiya, which come in kit form and are available at many hobby stores (as well as Internet sites, such as TowerHobbies.com). These motors come with their own gearbox; choose the 1:64.8 gear ratio. An assembled motor is shown in Fig. 10.3. Note that the output shaft of the motor can be made to protrude a variable distance from the body of the motor. Secure the shaft (using the Allen setscrew that is included) so that only a small portion of the opposite end of the shaft sticks out of the gear box on the other side, as shown in Fig 10.3.

You should secure the gearboxes and motors to the aluminum frame of the Buggybot as depicted in Fig. 10.4. Use 6/32 bolts, flat washers, and nuts. Be sure that the motors are aligned as shown in the figure. Note that the shaft of each motor protrudes from the side of the Buggybot.

Figure 10.5 illustrates how to attach the wheels to the shafts of the motors. The wheels used in the prototype were 3-inch-diameter foam "Lite Flight" tires, commonly available at hobby stores. Secure the wheels in place by first threading a 3/16-inch collar (available at hobby stores) over the shaft of the motor. Tighten the collar in place using its Allen setscrew. Then cinch the wheel onto the shaft by tightening a 5/56 threaded nut to the end of the motor shaft (the nut should be included with the gearbox motor kit). Be sure to tighten down on the nut so the wheel won't slip.

SUPPORT CASTER

The Buggybot uses the two-wheel drive tripod arrangement. You need a caster on the other end of the frame to balance the robot and provide a steering swivel. The 1 1/2-inch swivel caster is not driven and doesn't do the actual steering. Driving and steering are taken care of by the drive motors. Refer to Fig. 10.6 on p. 131. Attach the caster using two 6/32 by 1/2-inch bolts and nuts.

Note that the mechanical style of the caster, and indeed the diameter of the caster wheel, is dependent on the diameter of the drive wheels. Larger drive wheels will require either a

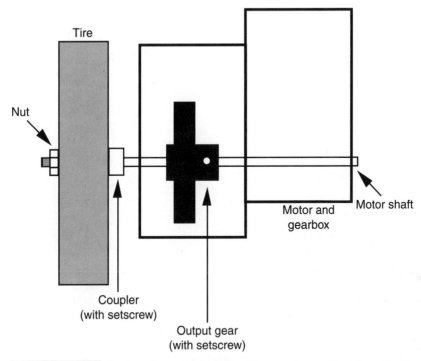

FIGURE 10.3 Secure the output shaft of the motor so that almost all of the shaft sticks out on one side of the motor.

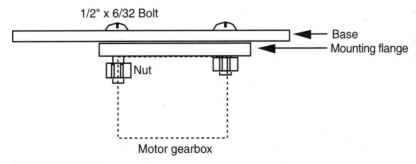

FIGURE 10.4 The gearboxes and motors are attached to the frame of the Buggybot using ordinary hardware.

different mounting or a larger caster. Small drive wheels will likewise require you to adjust the caster mounting and possibly use a smaller-diameter caster wheel.

BATTERY HOLDER

The motors require an appreciable amount of current, so the Buggybot really should be powered by heavy-duty "C"- or "D"-size cells. The prototype Buggybot used a two-cell "D" battery holder. The holder fits nicely toward the front end of the robot and acts as a

FIGURE 10.5 Attach the foam wheels (with plastic hubs) for the Buggybot onto the shafts of the motors.

good counterweight. You can secure the battery holder to the robot using double-sided tape or hook-and-loop (Velcro) fabric.

WIRING DIAGRAM

The basic Buggybot uses a manual wired switch control. The control is the same one used in the plastic Minibot detailed in Chapter 8, "Building a Plastic Robot Platform." Refer to the wiring diagram in Fig. 8.4 of that chapter for information on powering the Buggybot.

To prevent the control wire from interfering with the robot's operation, attach a piece of heavy wire (the bottom rail of a coat hanger will do) to the caster plate and lead the wire up it. Use nylon wire ties to secure the wire. The completed Buggybot is shown in Fig. 10.7.

Test Run

You'll find that the Buggybot is an amazingly agile robot. The distance it needs to turn is only a little longer than its length, and it has plenty of power to spare. There is room on the robot's front and back to mount additional control circuitry. You can also add control circuits and other enhancements over the battery holder. Just be sure that you can remove the circuit(s) when it comes time to change or recharge the batteries.

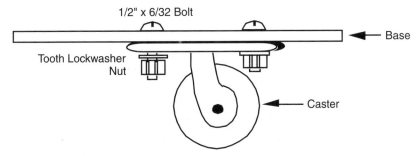

FIGURE 10.6   Mounting the caster to the Buggybot.

FIGURE 10.7   The completed Buggybot.

From Here

11

CONSTRUCTING HIGH-TECH
ROBOTS FROM TOYS

Ready-made toys can be used as the basis for more complex homebrew hobby robots. The toy industry is robot crazy, and you can buy a basic motorized or unmotorized robot for parts, building on it and adding sophistication and features. Snap or screw-together kits, such as the venerable Erector Set, let you use premachined parts for your own creations. And some kits, like LEGO and Robotix, are even designed to create futuristic motorized robots and vehicles. You can use the parts in the kits as is or cannibalize them, modifying them in any way you see fit. Because the parts already come in the exact or approximate shape you need, the construction of your own robots is greatly simplified.

About the only disadvantage to using toys as the basis for more advanced robots is that the plastic and lightweight metal used in the kits and finished products are not suitable for a homemade robot of any significant size or strength. You are pretty much confined to building small Minibot or Scooterbot-type robots from toy parts. Even so, you can sometimes apply toy parts to robot subsystems, such as a light-duty arm-gripper mechanism installed on a larger automaton.

Let's take a closer look at using toys in your robot designs in this chapter, and examine several simple, cost-effective designs using readily available toy construction kits.

Erector Set

Erector Set, now sold by Meccano, has been around since the Dawn of Time—or so it seems. The kits, once made entirely of metal but now commonly including many plastic pieces, come in various sizes and are generally designed to build a number of different projects. Many kits are engineered for a specific design with perhaps provisions for moderate variations. I've found the general-purpose sets to be the best bets. Among the useful components of the kits are prepunched metal girders, plastic and metal plates, tires, wheels, shafts, and plastic mounting panels. You can use any as you see fit, assembling your robots with the hardware supplied with the kit or with 6/32 or 8/32 nuts and bolts.

Several Erector Sets, such as those in the Action Troopers collection, come with wheels, construction beams, and other assorted parts that you can use to construct a robot base. Motors are typically not included in these kits, but you can readily supply your own. Because Erector Set packages regularly come and go, what follows is a general guide to building a robot base. You'll need to adapt and reconfigure based on the Erector Set parts you have on hand.

The prepunched metal girders included in the typical Erector Set make excellent motor mounts. They are lightweight enough that they can be bent, using a vise, into a U-shaped motor holder. Bend the girder at the ends to create tabs for the bolts, or use the angle stock provided in an Erector Set kit. The basic platform is designed for four or more wheels, but the wheel arrangement makes it difficult to steer the robot. The design presented in Fig. 11.1 uses only two wheels. The platform is stabilized using a miniature swivel caster at one end. You'll need to purchase the caster at the hardware store.

Note that the shafts of the motors are not directly linked to the wheels. The shaft of the wheels connect to the baseplate as originally designed in the kit. The drive motors are equipped with rollers, which engage against the top of the wheels for traction. You can use a metal or rubber roller, but rubber is better. The pinch roller from a discarded cassette tape player is a good choice, as is a 3/8-inch beveled bibb washer, which can be found in the plumbing section of the hardware store. You can easily mount a battery holder on the top of the platform. Position the battery holder in the center of the platform, toward the caster end. This will help distribute the weight of the robot.

The basic platform is now complete. You can attach a dual-switch remote control, as described in Chapter 8, "Building a Plastic Robot Platform," or connect automatic control circuitry as detailed in Part 5 of this book, "Computers and Electronic Control."

Do note that over the years the Erector Set brand has gone through many owners. Parts from old Erector Sets are unlikely to fit well with new parts. This includes but is not limited to differences in the threads used for the nuts and bolts. If you have a very old Erector Set (such as those made and sold by Gilbert), you're probably better off keeping them as collector's items rather than raiding them for robotic parts. The very old Erector Sets of the 1930s through 1950s fetch top dollar on the collector's market (when the sets are in good, complete condition, of course).

Similarly, today's Meccano sets are only passably compatible with the English-made Meccano sets sold decades ago. Hole spacing and sizes have varied over the years, and "mixing and matching" is neither practical nor desirable.

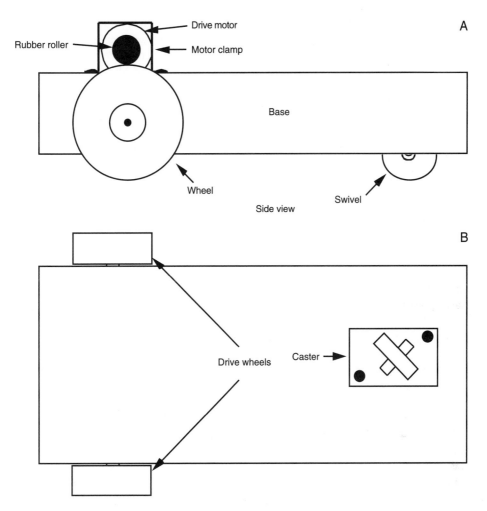

FIGURE 11.1 Constructing the motorized base for a robot using Erector Set (Meccano) parts. *a.* Attaching the motor and drive roller over the wheel; *b.* Drive wheel-caster arrangement.

Robotix

The Robotix kits, originally manufactured by Milton-Bradley and now sold by Learning Curve, are specially designed to make snap-together walking and rolling robots. Various kits are available, and many of them include at least one motor (additional motors are available separately). You control the motors using a central switch pad. Pushing the switch forward turns the motor in one direction; pushing the switch back turns the motor in the other direction. The output speed of the motors is about six rpm, which makes them a bit slow for moving a robot across the room but perfect for arm-gripper designs.

The structural components in the Robotix kits are molded from high-impact plastic. You can connect pieces together to form just about anything. One useful project is to build a robotic arm using several of the motors and structural components. The arm can be used by itself as a robotic trainer or attached to a larger robot. It can lift a reasonable eight ounces or so, and its pincher claw is strong enough to firmly grasp most small objects.

While the Robotix kit allows you to snap the pieces apart when you're experimenting, the design presented in this chapter is meant to be permanent. Glue the pieces together using plastic model cement or contact cement. Cementing is optional, of course, and you're free to try other, less permanent methods to secure the parts together, such as small nuts and bolts, screws, or Allen setscrews.

When cemented, the pieces hold together much better, and the arm is considerably stronger. Remember that, once cemented, the parts cannot be easily disassembled, so make sure that your design works properly before you commit to it. When used as a stand-alone arm, you can plug the shoulder motor into the battery holder or base. You don't need to cement this joint.

Refer to Fig. 11.2 as you build the arm. Temporarily attach a motor (we'll call it "motor 1") to the Robotix battery holder-baseplate. Position the motor so that the drive spindle points straight up. Attach a double plug to the drive spindle and the end connector of another motor, "motor 2." Position this motor so that the drive spindle is on one side. Next, attach another double plug and an elbow to the drive spindle of motor 2. Attach the other end of the elbow connector to a beam arm.

Connect a third motor, "motor 3," to the large connector on the opposite end of the beam arm. Position this motor so the drive spindle is on the other end of the beam arm. Attach a double plug and an elbow between the drive spindle of motor 3 and the connector opposite the drive spindle of the fourth motor, "motor 4." The two claw levers directly attach to the drive spindle of motor 4.

Motorize the joints by plugging in the yellow power cables between the power switch box and the motor connectors. Try each joint, and note the various degrees of freedom. Experiment with picking up various objects with the claw. Make changes now before disassembling the arm and cementing the pieces together.

After the arm is assembled, route the wires around the components, making sure there is sufficient slack to permit free movement. Attach the wires to the arm using nylon wire ties.

A Variety of Construction Sets

Toy stores are full of plastic put-together kits and ready-made robot toys that seem to beg you to use them in your own robot designs. Here are some toys you may want to consider for your next project.

LEGO

LEGO has become the premier construction toy, for both children and adults. The LEGO Company, parent company of the LEGO brand, has expanded the line as educational resources, making the ubiquitous LEGO "bricks" common in schools across the country

FIGURE 11.2 The robot arm constructed with parts from a Robotix construction kit.

and around the world. LEGO also makes the Mindstorms, a series of sophisticated computerized robots. Chapters 12 through 14 detail several LEGO-based robot creations, including the Mindstorms, as well as describing how you can use LEGO parts to build custom robots.

CAPSULA

Capsula is a popular snap-together motorized parts kit that uses unusual tube and sphere shapes. Capsula kits comes in different sizes and have one or more gear motors that can be attached to various components. The kits contain unique parts that other put-together toys don't, such as a plastic chain and chain sprockets or gears. Advanced kits come with remote control and computer circuits. All the parts from these kits are interchangeable. The links of the chain snap apart, so you can make any length of chain that you want. Combine the links from many kits and you can make an impressive drive system for an experimental lightweight robot.

FISCHERTECHNIK

The Fischertechnik kits, made in Germany and imported into North America by a few educational companies, are the Rolls-Royces of construction toys. Actually, "toy" isn't the proper term because the Fischertechnik kits are not just designed for use by small children. In fact, many of the kits are meant for high school and college industrial engineering students, and they offer a snap-together approach to making working electromagnetic, hydraulic, pneumatic, static, and robotic mechanisms.

All the Fischertechnik parts are interchangeable and attach to a common plastic base-plate. You can extend the lengths of the baseplate to just about any size you want, and the baseplate can serve as the foundation for your robot, as shown in Fig. 11.3. You can use the motors supplied with the kits or use your own motors with the parts provided. Because of the cost of the Fischertechnik kits, you may not want to cannibalize them for robot components. But if you are interested in learning more about mechanical theory and design, the Fischertechnik kits, used as is, provide a thorough and programmed method for jumping in with both feet.

INVENTA

U.K.-based Valiant Technologies offers the Inventa system, a reasonably priced construction system aimed at the educational market. Inventa is a good source for gears, tracks, wheels, axles, and many other mechanical parts. The beams used for construction are semiflexible and can be cut to size. Angles and brackets allow the beams to be connected in a variety of ways. It is not uncommon—and in fact Valiant encourages it—to find Inventa creations intermixed with other building materials, including balsa wood, LEGO pieces, you name it. Inventa isn't the kind of thing you'll find at the neighborhood Toys-R-Us. It's available mail order and through the Internet; see the Inventa Web site at www.valiant-technology.com.

FIGURE 11.3 A sampling of Fischertechnik parts.

K'NEX

K'Nex uses unusual half-round plastic spokes and connector rods (see Fig. 11.4) to build everything from bridges to Ferris wheels to robots. You can build a robot with just K'Nex parts or use the parts in a larger, mixed-component robot. For example, the base of a walking robot may be made from a thin sheet of aluminum, but the legs might be constructed from various K'Nex pieces.

A number of K'Nex kits are available, from simple starter sets to rather massive special-purpose collections (many of which are designed to build robots, dinosaurs, or robot-dinosaurs). Several of the kits come with small gear motors so you can motorize your creation. The motors are also available separately.

ZOOB

Zoob (made by Primordial) is a truly unique form of construction toy. A Zoob piece consists of stem with a ball or socket on either end. You can create a wide variety of construction projects by linking the balls and sockets together. The balls are dimpled so they connect securely within their sockets. One practical application of Zoob is to create "armatures" for human- or animal-like robots. The Zoob pieces work in a way similar to bone joints.

FIGURE 11.4 K'Nex sets let you create physically large robots that weigh very little. The plastic pieces form very sturdy structures when properly connected.

CHAOS

Chaos sets are designed for structural construction projects: bridges, buildings, working elevator lifts, and the like. The basic Chaos set provides beams and connectors, along with chutes, pulleys, winches, and other construction pieces. Add-on sets are available that contain parts to build elevators, vortex tubes, and additional beams and connectors.

FASTECH

No longer in production, Fastech construction kits used to offer among the best assortments of parts you could buy. All the parts were plastic, and the kits came with a plastic temporary riveting system that you probably wouldn't use in your designs. The plastic parts were predrilled and came in a variety of shapes and styles. The components can be used on their own to make a small, light-duty robot frame and body, or they can be used as parts in a larger robot. Fastech kits may not be manufactured (but who knows, they could come back), but they can sometimes be found at garage sales, thrift stores, and online auctions.

OTHER CONSTRUCTION TOYS

There are many other construction toys that you may find handy. Check the nearest well-stocked toy store or a toy retailer on the Internet for the following:

- Expandagon Construction System (Hoberman)
- Fiddlestix Gearworks (Toys-N-Things)
- Gears! Gears! Gears! (Learning Resources)
- PowerRings (Fun Source)
- Zome System (Zome System)
- Construx (no longer made, but sets may still be available for sale)

Perhaps the most frequently imitated construction set has been the Meccano/Erector Set line. Try finding these "imitators," either in new, used, or thrift stores:

- Exacto
- Mek-Struct
- Steel Tec

Specialty Toys for Robot Hacking

Some toys and kits are just made for "hacking" (retrofitting, remodeling) into robots. Some are already robots, but you may design them to be controlled manually instead of interfacing to control electronics. The following sections describe some specialty toys you may wish to experiment with.

TAMIYA

Tamiya is a manufacturer of a wide range of radio-controlled models. They also sell a small selection of gearboxes in kit form that you can use for your robot creations. One of the most useful is a dual-gear motor, which consists of two small motors and independent drive trains. You can connect the long output shafts to wheels, legs, or tracks.

They also sell a tracked tractor-shovel kit (see Fig. 11.5) that you control via a switch panel. You can readily substitute the switch panel with computerized control circuitry (relays make it easy) that will provide for full forward and backward movement of the tank treads as well as the up and down movement of the shovel.

OWIKITS and MOVITS

The OWIKITS and MOVITS robots are precision-made miniature robots in kit form. A variety of models are available, including manual (switch) and microprocessor-based versions. The robots can be used as is, or they can be modified for use with your own electronic systems. For example, the OWIKIT Robot Arm Trainer (model OWI007) is normally operated by pressing switches on a wired control pad. With just a bit of work, you can connect the wires from the arm to a computer interface (with relays, for example) and operate the arm via software control.

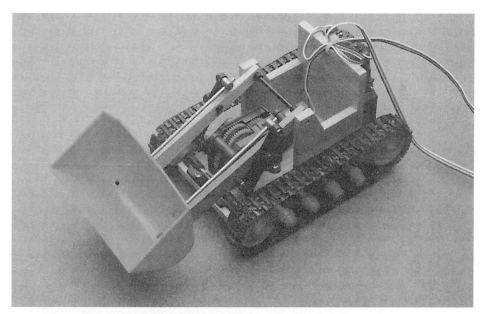

FIGURE 11.5 The Tamiya Bulldozer kit can be used as a lightweight robot platform. The kit comes with an interesting dual motor that operates the left and right treads.

Most of the OWIKITS and MOVITS robots come with preassembled circuit boards; you are expected to assemble the mechanical parts. Some of the robots use extremely small parts and require a keen eye and steady hand. The kits are available in three skill levels: beginner, intermediate, and advanced. If you're just starting out, try one or two kits in the beginner level.

Once (properly) constructed, the OWIKITS and MOVITS robots last a long, long time. I have several models I built in the mid-1980s, and with just the occasional nut tightening here and a dab of grease there they have continued to operate flawlessly.

FURBY

The Furby, from Tiger Electronics, is an animatronic creature that appears to have learning capabilities and even artificial intelligence. Actually, the Furby has neither; it's a clever collection of motors, gears, switches, sensors, and an on-board computer, as shown in the "deconstructed Furby" shown in Fig. 11.6. The Furby interacts with its master, as well as the environment, and contains thousands of behavioral permutations.

FIGURE 11.6 The Furby, from Tiger Electronics, is a marvel of engineering finesse.

Furby's on-board computer is a "blob," or chip-on-board (COB), and inherently not reprogrammable or hackable. However, you can connect the main Furby motor and its various sensors to a new control circuit, perhaps one made with a Basic Stamp II microcontroller.

ROKENBOK

Rokenbok toys are radio-controlled vehicles that are available in wheeled and tracked versions. Despite the European-sounding name, Rokenbok toys are made in the United States yet can be difficult to find because of their high cost (the basic set costs well over $180). Each vehicle is controlled by a game controller; all the game controllers are connected to a centralized radio transmitter station.

The Sony PlayStation-style controllers can be hacked and connected directly to the input/output (I/O) of a computer or microcontroller. The transmitter station also has a connector for computer control, but as of this writing Rokenbok has not released the specifications or an interface for this port.

ARMATRON

The Armatron, once a common find at Radio Shack, occasionally reappears at that store, ready to capture the imagination of anyone who plays with it. The Armatron (see Fig. 11.7) is a remotely controlled arm that you can use to manipulate small lightweight objects. You control the Armatron by moving two joysticks.

A rewarding but demanding project is to convert the Armatron to computer control. Because of the way the Armatron works, using a single motor and clutched drive shafts, you need to rebuild it with separate motors so a computer can individually control its various joints. One way to do this might be to connect small 12-volt DC solenoids to the joysticks. Each joystick requires six solenoids: four for the up, down, left, and right movement of the joystick and two to turn the post of the joystick clockwise or counterclockwise.

Making Robots with Converted Vehicles

Toy cars, trucks, and tractors can make ideal robot platforms—if they are properly designed. As long as the vehicle has separate drive motors for the left and right drive wheels, you have a fighting chance of being able to adapt it for use as a robot base.

MOTORIZED VEHICLES

The least expensive radio-controlled cars have a single drive motor and a separate steering servo or mechanism, which doesn't lend itself to robot conversion. In many cases, the steering mechanism is not separately controlled; you "steer" the car by making it go in reverse. The car drives forward in a straight line but turns in long arcs when reversed. These are impractical for use as a robot base and should not be considered.

On the other hand, most radio- and wire-controlled tractor vehicles are perfectly suited for conversion into a robot. Strip off the extra tractor stuff to leave the basic chassis, drive

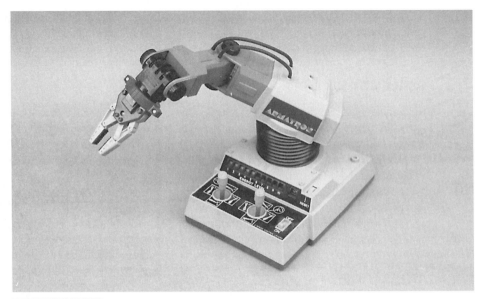

FIGURE 11.7 The indefatigable Armatron motorized robotic arm, still occasionally available at Radio Shack.

motors, and tracks. You can keep the remote control system as is or remove the remote receiver (or wires, if it's a wired remote) and replace it with new control circuitry. In the case of a wired remote, you can substitute relays for the switches in the remote. Of course, each toy is a little different, so you'll need to adapt this wiring diagram to suit the construction of the vehicle you are using.

Another option is to use two small motorized vehicles (mini "4-wheel-drive" trucks are perfect), remove the wheels on opposite sides, and mount them on a robot platform. Your robot uses the remaining wheels for traction. Each of the vehicles is driven by a single motor, but since you have two vehicles (see Fig. 11.8), you still gain independent control of both wheel sides. The trick is to make sure that, whatever vehicles you use, they are the same exact type. Variations in design (motor, wheel, etc.) will cause your robot to "crab" to one side as it attempts to travel a straight line. The reason: the motor in one vehicle will undoubtedly run a little slower or faster than the other, and the speed differential will cause your robot to veer off course.

MOTORIZING VEHICLES

Not all toy vehicles already have motors. If you find a motorless vehicle that's nevertheless perfect as a robot base, consider adding motors to it. You can cannibalize the motors (and gear train) from another toy and implant them in your new robot base. Or you can purchase one of the motor kits made by Tamiya (see "Specialty Toys for Robot Hacking," earlier in this chapter) and retrofit it onto your robot. The dual-motor kit described in Chapter 10 is small enough to be used on most toy vehicles, and it can be readily mounted using screws, double-sided tape, hook-and-loop (Velcro), or other techniques.

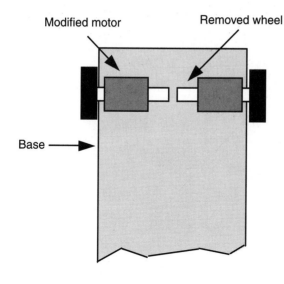

Modified motor

Removed wheel

Base

FIGURE 11.8 You can build a motorized robot platform by cannibalizing two small motorized toys and using each "half" of them.

From Here

To learn more about...	Read
Using LEGO to create robots	Chapter 12, "Build Custom LEGO-based Robots"
Building robots with the LEGO Mindstorms system	Chapter 13, "Creating Functionoids with LEGO Mindstorms Robotics Invention System"
Brains you can add to robots made from toys	Chapter 28, "An Overview of Robot 'Brains'"
Controlling a small robot using the IBM PC parallel port	Chapter 30, "Computer Control via PC Printer Port"

12

BUILD CUSTOM
LEGO-BASED ROBOTS

Once just a playtime toy for youngsters, LEGO parts are now used in schools to teach construction and mechanics, even robotics. In fact, LEGO 'bots are becoming so common that many schools and libraries around the world host LEGO robot-building races and competitions. The main benefit of a LEGO-based robot: The snap-together design of LEGO pieces allows you to build a robot in just an hour or two, with no cutting or drilling necessary. Changes and improvements are easy to make too, and you can do them "on the fly" as you experiment with new ideas.

There are plenty of books, magazine articles, and Internet sites that discuss the design and construction of all-LEGO robots. So, in this chapter we'll review another approach: using various LEGO parts as a *basis* for a robot and then augmenting those parts with other construction materials. For example, the robot described in this chapter uses modified hobby servo motors and not the more costly LEGO gear motors. Some gluing and other so-called hard construction is also called for.

Working with LEGO Parts

There are several dozen varieties of LEGO parts, which come in various sizes (not to mention colors). In your robotics work you'll likely encounter the primary construction pieces detailed in the following subsections.

BUILDING BLOCKS

The three primary building blocks are shown in Figure 12.1.

- *Bricks.* The basic LEGO building block is the brick. Bricks have one or more raised "nubs" (or "bumps") on the top and corresponding sockets on the bottom. The nubs and sockets mate, allowing you to stack bricks one on top of another.
- *Plates.* Plates are like bricks but are half their height. Like bricks, plates have one or more nubs on the top and corresponding sockets on the bottom, and they are made to be stacked together.
- *Beams.* Beams are specific to the Technic brand of LEGO parts, such as the LEGO Mindstorms Robotics Invention System, which is detailed in the next two chapters. Beams are bricks that have one row of nubs and then holes down their sides. Beam variations include the L and hooked shape, which can be used for such things as building robotic grippers.

The arrangement of the nubs on bricks and plates define their size. For example, a "2 by 8" brick or plate has two rows and eight columns of nubs. In some LEGO instruction books you'll see the nubs referred to as "units," such as "2u-by-8u" (the *u* stands for units). I'll stick to this style as a quasi-standard throughout the rest of the chapter.

Building plates are very large versions of plates, which are available in the 8u-by-16u size. Thin baseplates are also available; these lack sockets underneath.

CONNECTION PIECES

Figure 12.2 shows several common LEGO parts that are used for connecting things, such as one beam to another or a wheel to a beam. The four most common are as follows:

- *Connector peg.* Connector pegs are typically used with the LEGO Technic line to attach beams together. Connector pegs are labeled either 1/2, 3/4, or full—the dif-

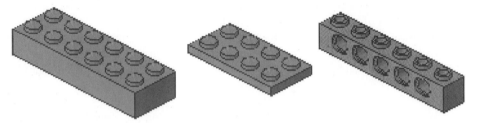

FIGURE 12.1 Bricks, plates, and beams are the primary LEGO pieces used for construction. They are available in various sizes.

FIGURE 12.2 Connection pieces allow you to snap various LEGO parts together.

ference lies in the length of one side of the connector's middle collar. Some connector pegs are half connector and half axle (we'll discuss axles at the end of this list).

■ *Friction connector.* Available in a wide variety of sizes and shapes, friction connectors are used with both LEGO Technic and standard parts to connect such things as beams and wheels.

■ *Bushing.* Bushings hold things, such as axles. So-called half bushings are about half the width of a standard bushing. Some can also be used as miniature pulleys.

PULLEYS, TIRE HUBS, AND GEARS

LEGO parts are replete with a variety of wheels, pulleys, and gears. Several common varieties are shown in Fig. 12.3. In almost all cases, you attach pulleys, tire hubs, and gears using stud axles and/or friction connectors.

■ *Wheel hubs.* Wheel hubs are used with any of a number of different-sized rubber or plastic wheels. There are several sizes of wheel hubs.

■ *Pulleys.* Pulleys are used with string, rubber bands, or O-rings to create a kind of belt. Several sizes of pulleys are available.

■ *Gears.* There are more than a dozen kinds of LEGO gears. All are designated by the number of teeth they have on their outside circumference. For example, the 8-tooth spur gear has eight teeth; the 14-tooth bevel (slanted) gear has 14 teeth. All gears of the same type (e.g., spur or bevel or crown) are made to mesh with another, and you can often mix one type with another. For example, a spur gear can be effectively used with a crown gear.

■ *Stud axle.* Stud (or "cross") axles come in various lengths, which are designated by number (e.g., 4, 6, 8, etc.). Axles connect to gears, wheels, and other parts. They have a spine of four studs along their length.

There are also toggle joints, sloped beams, catches, cross blocks, axle connectors, cams, and a raft of others. Check out any good book on LEGO, or see the LEGO Web site for more information on LEGO parts.

FIGURE 12.3 Use LEGO wheels, pulleys, and gears to enable your robot creations to scoot around the floor.

Securing Parts

LEGO parts are made to snap together. When properly constructed, the snap-together system provides for relatively strong joints. However, the active nature of most robots—machines that are often in motion and perhaps bumping into things—makes snap-on joints less than ideal. If you're not just experimenting with a new, perhaps temporary, design, you may want to think about using alternative means to secure LEGO parts together. You'll also need a different method of securing if you're combining LEGO and non-LEGO parts.

In the next five subsections we discuss methods for securing LEGO parts together. Take note: except for cementing, gluing, and hardware (nuts, bolts, and screws) the remaining methods discussed here—taping, tie wraps, and Velcro—are best used for nonpermanent constructions. These materials do not provide a long-lasting bond, especially for a robot that's always on the go.

CEMENTING AND GLUING

You can cement together nonmovable pieces using ordinary household cement (such as Duco), epoxy, contact, or CA (cyanoacrylate) glues. If you want a very permanent bond, use clear ABS plastic solvent (LEGO parts are made of ABS plastic). ABS solvent cement is available at most home improvement stores, as well as many hobby stores. Another approach is to use hot-melt glue. You can use either the standard or low-temperature variety. (See Chapter 8 for more information on bonding plastic.)

When using glue or epoxy, be sure to rough up the surfaces to be joined. This is especially important when using hot-melt glues. There is less need to rough up the edges when you are using solvent-type cements since the adhesive actually melts the plastic to form a strong joint.

TAPING

Temporary constructions can be taped together using black electrical tape or even strapping tape normally used for box packaging. Do note that most tapes leave a sticky residue. Use acetone to remove the residue.

TIE WRAPS

Tie wraps are used to bundle wires and other loose items. They consist of a single piece of plastic with a catch on one end. You thread the loose end of the tie through the catch. A ratchet in the catch keeps the tie in place. Most ties have a one-way ratchet that cannot be undone; others have a catch "release" so you can reuse the tie. Tie wraps are available in a variety of lengths, thicknesses, and colors. Get an assortment, and use the size that best suits the job.

NUTS, SCREWS, AND BOLTS

You can use small hardware to attach LEGO pieces together. In most cases, you'll need to drill through the plastic to secure the screw or bolt. You can use self-tapping metal screws if the plastic material is thick. When you use self-tapping screws, be sure to use a drill size slightly smaller than the screw.

VELCRO (HOOK-AND-LOOP)

You can attach non-LEGO things to LEGO pieces with Velcro (otherwise known by the generic term "hook-and-loop"). Get the kind with self-sticking adhesive backing. Put the hook material on one piece and the loop on the other. Press them together for a very strong—but not permanent—bond.

Build the LEGO Pepbot

The Pepbot is a small, two-wheeled robot constructed from an assortment of LEGO parts. Motive power for the Pepbot comes from two radio-controlled (R/C) servos and two foam tires, both of which are available at most any hobby store that sells radio-controlled parts and accessories. The "brain" of the Pepbot is a microcontroller—just about any microcontroller will do. In this section, you'll learn how to construct the Pepbot body using common LEGO parts and how to attach the servo motors and tires. Apart from some special modifications you need to make to the servos, constructing the Pepbot consists of snapping together LEGO parts and cementing things in place with epoxy or hot glue.

We'll also briefly introduce the idea of using the OOPic object-oriented microcontroller to control the robot. This microcontroller, and others, are covered in more detail in Chapters 31 through 33.

CONSTRUCTING THE BODY

Begin by building the basic frame for the Pepbot. Use the following steps.

Frame

1. Begin by assembling the frame as shown in Fig. 12.4a. Attach two 16u beams to two 2u-by-12u plates, forming a rectangle.
2. On the underside of the frame attach the two 2u-by-16u plates lengthwise (see Fig. 12.4b).
3. Connect two additional 16u beams down the centerline of the bottom of the frame (Fig. 12.4c).
4. Attach two 2u-by-12u plates to the top of the frame, as shown in Fig. 12.4d.
5. Attach a 2u-by-6u plate down the centerline to the top of the chassis, as shown in Fig. 12.4e.
6. Attach a round pad with a 2u-by-2u round brick, then press this assembly into the ends of the centerline beams, on the bottom of the frame (Fig. 12.4f).

Side blocks

1. Create four sets of three 2u-by-4u bricks, stacked on top of one another (Figure 12.5a).
2. Attach the assembled side blocks to the frame, as shown in Figure 12.5b.

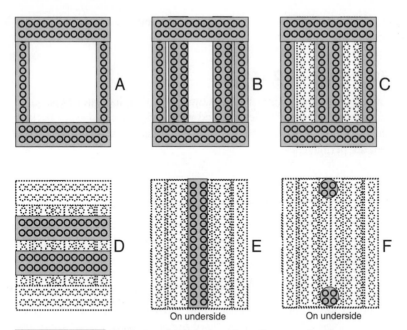

FIGURE 12.4 Pepbot main construction. See text for the construction sequence.

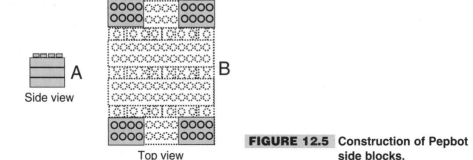

FIGURE 12.5 Construction of Pepbot side blocks.

ATTACHING THE SERVO MOTORS

The Pepbot uses two modified R/C servos for drive motors. In an ordinary servo, rotation is limited by an internal stop inside the motor. By modifying the servo, you can make the motor turn continuously. Modifying servos is not a particularly hard task, but the exact steps will vary depending on the model of servo you are using. See Chapter 20 for details on how to modify several popular R/C servos. For the rest of this section, we'll assume you've already modified two servos and are ready to use them on the Pepbot.

Use epoxy or hot-melt glue to affix two 2u-by-6u plates to the side of each servo casing, as shown in Fig. 12.6. Use only a moderate amount of epoxy or hot-melt glue, as you may need to remove the plate from the servo casing. For one servo, glue the plate to the right side (looking at the servo top down, with the output shaft on the top). For the other servo, glue the plate to the left side. Be careful to align the plates on the servos so they are

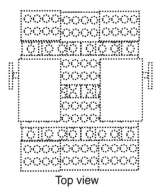

FIGURE 12.6 Attaching servo motors to LEGO plates.

FIGURE 12.7 Securing the mounted motors to the rest of the robot.

straight, and make sure the plate doesn't block the seam in the case of the servo (this will make it easier to disassemble the servo, if you need to for whatever reason).

Hot-melt glue is the preferred construction technique for this task, as the glue sets quickly—usually in under a minute in normal room temperature. Hold the servo in place until the glue has set. Again, be absolutely sure that the plate is squarely affixed to the servo. After the glue has set and dried, you may attach the servos (now on their motor mounts) to the frame of the robot, as shown in Fig. 12.7, using two 2u-by-12u plates.

ATTACHING THE WHEELS

The wheels of the Pepbot are lightweight foam tires, which are used in model R/C airplanes. I selected wheels with a 3-inch diameter, which makes the Pepbot travel fairly fast across the floor (hence the name Pepbot). You can use smaller wheels if you wish, but consider the following:

- The smaller the wheel, the slower the robot. R/C servo motors turn at about 1–2 revolutions per second, depending on the model (some are slower; some are faster). With 3-inch wheels and a 1.5 rps servo, the Pepbot will travel about 14 inches per second. This speed is calculated by multiplying the diameter of the wheel (3) by pi (3.14), then multiplying that number by the speed of the servo (1.5 rps). That is, 3 * 3.14 * 1.5, or approximately 14. Just as smaller wheels will make Pepbot a little less peppy, larger wheels will make Pepbot travel faster.
- The smaller the wheel, the less clearance there is between the bottom of the robot and the floor. Conversely, the larger the wheel, the more clearance there will be. This can be helpful if you run Pepbot over thick carpet or want it to travel over small bumps, like the threshold between a carpeted and a tile room.
- The design of the Pepbot will not allow tires smaller than about 2 3/4 inches because of the position of the servo motor output shafts. You'll have to redesign Pepbot if you want to use smaller wheels.

To attach the wheels, first insert the standard 1-inch servo plate over the output shaft of the servo. This plate connects firmly to the output shaft using a screw, which is provided with the servo. Then, use epoxy or hot-melt glue (hot-melt glue is preferred) to secure the hub of the wheel to the plate that is connected to the servo. *Be absolutely sure* that the wheel hub is exactly centered over the servo plate; if it is not, the robot will not travel in a straight line.

ATTACHING THE DECK

The "deck" is where you place the Pepbot's batteries and control circuitry. The deck is simply an 8u-by-16u LEGO building plate. A four-AA battery pack is affixed to the deck using double-sided foam tape. The microcontroller board (in this case an OOPic) is attached using small wire tires.

CONSTRUCTING THE INTERFACE BOARD

An interface board allows you to easily connect the microcontroller, batteries, and the two servos. It also provides adequate room if you want to add to the Pepbot, such as light sensors to detect light and dark, bumper switches to determine if the robot has hit an obstacle, or a small audio amplifier and speaker to allow the Pepbot to sound off when it wants to.

Construct the interface board by using a small 2-by-3 inch project board, available at Radio Shack. You can use most any construction technique that you like. I used wire-wrapping for the Pepbot prototype. See Fig. 12.8 for a construction diagram; Fig. 12.9 shows the schematic you should follow.

Note the 40-pin cable and connector. This cable attaches to the I/O (input/output) port of the OOPic microcontroller. While the OOPic provides for 31 inputs and outputs, the Pepbot uses only three of them. As you can see, there is plenty of room to expand the Pepbot with sensors, should you wish to do so. See Chapter 29, "Interfacing with

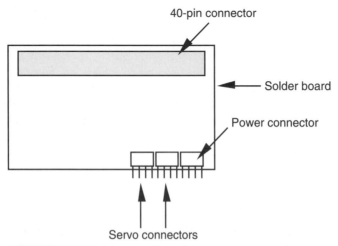

FIGURE 12.8 Construction plans for the Pepbot interface board. The board leaves plenty of
room for expansion.

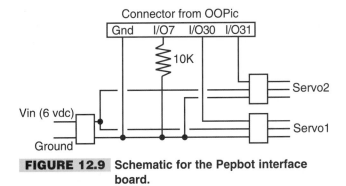

FIGURE 12.9 **Schematic for the Pepbot interface board.**

Computers and Microcontrollers," for some ideas on creating and interfacing sensors to robots.

BATTERY POWER

The Pepbot is powered by two separate battery sources, as follows:

■ A pack of four AA cells provides power to the two R/C servo motors. You may use 1.5-volt alkaline cells or 1.2-volt nickel-cadmium or nickel-metal hydride batteries. However, if you use 1.2-volt cells the servos will run a little slower than under 1.5-volt cells (4.8 volts for the four batteries, as opposed to 6 volts). Under normal use, this battery pack will last for about 30–60 minutes of "play" time, so you may wish to use rechargeable alkalines to save money.
■ A 9-volt "transistor" battery provides power to the microcontroller. The microcontroller consumes little current, so this battery should last for a long time under normal use.

The separate battery supplies serve an important purpose: R/C servo motors consume a lot of current when they are first turned on, so much, in fact, that the power for the four-battery pack can dip to under 4.5 volts. The OOPic, like most microcontrollers, employs a "brownout" circuit that resets the controller when the voltage falls below a certain level. This prevents the controller from operating in an unstable state. Using the same power pack for both servos and microcontroller can cause erratic behavior in your programs.

For ease of connection, attach a female connector to the end of the battery leads for both the four-AA pack and the 9-volt battery. Both connectors should be the common 0.100-inch pin type. Note that the OOPic microcontroller board already contains a +5 vdc (volts dc) regulator, so no outboard voltage regulator is required. The R/C servos do not require regulation. The ground connections of both battery packs should be connected together.

CONNECTING THE PIECES

Thanks to the interface board, connecting the servos, batteries, and microcontroller is a snap. The steps are as follows:

1. Connect the four-AA battery pack (observe proper polarity!) to the interface board.

2. Connect both the right and left servos to their respective connections on the interface board. Again, observe correct polarity. One or both servos may "jump" slightly when plugged in, but they should not run.

3. Connect the I/O cable between the OOPic and the interface board.

4. Connect the 9-volt battery to the OOPic microcontroller board.

Figure 12.10 shows the completed Pepbot.

PROGRAMMING THE MICROCONTROLLER

As mentioned earlier, the Pepbot uses an OOPic object-oriented microcontroller—a kind of computer on a single chip. The OOPic is just one of many microcontrollers you can use. Chapter 33 describes the OOPic in more detail, so in this section we'll just briefly outline the steps for connecting it to the Pepbot. By the way, feel free to use another microcontroller, if that's your wish. The popular Parallax Basic Stamp can also be used with the Pepbot. Chapter 31 is devoted entirely to using and programming the Basic Stamp.

Listing 12.1 shows the testing program for calibrating the modified servos. Calibration is required to find the "center position" of the servos. In an unmodified servo, the OOPic controls the position of the servo output shaft by using a number from 0 to 63. Zero turns the output shaft of the servo completely one direction, and 63 turns it completely the other direction.

Since the servos in the Pepbot have been modified for continuous rotation (see Chapter 20 for more detail on this technique), the values of 0 and 63 are used to control the direction of the motor. It stands to reason, then, that a value of 31 (midway between 0 and 63)

FIGURE 12.10 The Pepbot, ready for programming.

should stop the motor. This stop point is calibrated to the value of 31. As detailed in Chapter 20, the modified servo is equipped with an external 5K trimmer potentiometer.

To run the program you must first download it to the OOPic microcontroller. This process is outlined in Chapter 33, and so won't be repeated here. Once downloaded, the program will automatically run, so be sure you have the Pepbot's wheels raised off the floor or table for testing. (*Important! If the OOPic doesn't automatically run the program when downloaded, unplug the programming cable from it. The program should now immediately run.*)

With the servos calibrated, load the program as shown as Listing 1. This is a demonstration program than runs the Pepbot "through its paces," making it go forward and backward and in turns. Notice that subroutines such as *GoForward* and *HardLeft* are used for motion control. The Pepbot is able to move in all directions by controlling the right and left motors independently—either by turning the motors off or by turning the motors forward or backward. For example, to go forward both motors are told to turn on in the forward direction. To make a "hard" left (spin in place toward the left), the left motor goes in reverse, and the right motor goes forward.

Listing 12.1

```
Dim S1 As New oServo
Dim S2 As New oServo
Dim CenterPos as New oByte
Dim Button As New oDio1
Dim x as New oByte
Dim y as New oWord


'_____ --
Sub Main()
CenterPos = 31                  ' Set centering of servos
Call Setup
Do
        If Button = cvPressed Then
                ' Special program to calibrate servos
                S1 = CenterPos
                S2 = CenterPos
        Else
                ' Main program (IO line is held low)
                Call GoForward
                Call LongDelay

                Call HardRight
                Call LongDelay

                Call HardLeft
                Call LongDelay

                Call GoReverse
                Call LongDelay
        End If
Loop
End Sub


'_____ --
Sub Setup()
Button.Ioline = 7               ' Set IO Line 7 for function input
Button.Direction = cvInput      ' Make IO Line 7 input
S1.Ioline = 30                  ' Servo 1 on IO line 30
```

```
S1.Center = CenterPos           ' Set center of Servo 1
S1.Operate = cvTrue             ' Turn on Servo 1
S2.Ioline = 31                  ' Servo 2 on IO line 31
S2.Center = CenterPos           ' Set center of Servo 2
S2.Operate = cvTrue             ' Turn on Servo 2
S2.InvertOut = cvTrue           ' Reverse direction of Servo 2
End Sub


'_____--

Sub LongDelay()
    For x = 1 To 200:Next x
End Sub

Sub GoForward()
S1 = 0
S2 = 0
End Sub

Sub GoReverse()
S1 = 63
S2 = 63
End Sub

Sub HardRight()
S1 = 0
S2 = 63
End Sub

Sub HardLeft()
S1 = 63
S2 = 0
End Sub
```

Save, compile, and download the program in Listing 12.1. You may need to depress the reset button on the OOPic microcontroller board to prevent the Pepbot from activating prematurely. Place the robot on the floor, then release the reset button. The Pepbot robot should come to life, first going forward, then spinning both to the left and right. It should then turn right, then left, and then finally back up. The program will repeat itself until you disconnect the power.

If you have constructed your Pepbot properly, the robot should trace more or less the same area of the ground for each iteration through the *Do* loop of the program in Listing 12.1. If the Pepbot veers in one direction when it's not supposed to, it could indicate that either one or both wheels were not properly attached to the servo plates. Or it could indicate that the servos are not properly aligned on their mounting plate. Assuming you didn't apply too much epoxy or hot-melt glue, you should be able to use moderate force to remove the wheel and/or mounting plate. Clean off the old epoxy or hot-melt glue, and try again.

Once the Pepbot test program is a success, you can play around with coding your own actions for the robot. With sensors that you can add to the interface board, you can have the Pepbot respond to external stimulus, such as light or touch.

From Here

To learn more about...	*Read*
Building and programming robots with the Mindstorms Robotics System	Chapter 13, "Creating Functionoids with LEGO Mindstorms Robotics Invention System"

Using and modifying R/C servos	Chapter 20, "Working with Servo Motors"
Connecting robot motors to a computer, micro-controller, or other circuitry	Chapter 29, "Interfacing with Computers and Microcontrollers"
Controlling a LEGO robot using the Basic Stamp	Chapter 31, "Using the Basic Stamp"
Controlling a LEGO robot using the OOPic microcontroller	Chapter 33, "Using the OOPic Microcontroller"

13

CREATING FUNCTIONOIDS
WITH LEGO MINDSTORMS
ROBOTICS INVENTION SYSTEM

Say the word *LEGO* and most adults think of a million tiny plastic pieces strewn across a floor, waiting to be stepped on with bare feet. Anyone who has ever had one of those little angle pieces jab into a tender arch or break the skin on a heel knows how painful owning a LEGO set can be!

Still, apart from that occasional torture of walking over hard plastic, LEGO sets are wonderful amusements for both young and old. You can build most anything with LEGO parts. And with the help of your PC and the LEGO Mindstorms *Robotics Invention System* you can even create your own *functionoids*—functional (and sometimes useful!) computer programmed robots. In this chapter, we'll look at the popular LEGO Mindstorms Robotics Invention System, discuss what it has to offer, and give you detailed information on how the Mindstorms system works.

The LEGO Mindstorms Robotics Invention System comes with a detailed "Constructapedia" of practical robotic experiments. Moreover, other projects available on the LEGO Mindstorms Web page—and indeed hundreds of other independent Web pages—ensure that you'll have plenty to experiment with. Because of the wealth of printed project designs available, we won't get into the general use of the Mindstorms set in this chapter. Instead, we'll talk about the internals of the LEGO Mindstorms robot and how to "hack" it to extend its usefulness.

Be sure to also check out Chapter 14, "Programming the LEGO Mindstorms RCX: Advanced Methods," for additional details on programming the Mindstorms robot using third-party tools and utilities.

What Makes Up the Mindstorms System

The LEGO Mindstorms Robotics Invention System (RIS) consists of three major parts:

- The *RCX* (Robotic Command Explorer) controller. The RCX is otherwise known as a "brick" or "programmable brick." The term comes from robotics researchers at the Massachusetts Institute of Technology, who were the original developers of the concept behind the RCX. You can attach a collection of motors and sensors to the RCX and create a mobile automaton. Because the RCX uses standard LEGO "bumps," you can attach regular LEGO parts to it and build your own robot.
- The Mindstorms *programming environment,* otherwise known as *RCX Code,* allows you to create, store, and download programs from your personal computer and into the RCX. While the Mindstorms programming environment is the standard method for writing programs for the RCX, it is not the only one. Chapter 14 addresses two popular alternative programming environments for the LEGO RCX.
- A two-way *communications tower* for transmitting signals between your computer and the RCX. The tower uses modulated infrared (IR) light rather than radio signals, so two or more RCX units can be programmed in the same room (you can adjust the power output of the IR tower to avoid interference).

A Look Inside the RCX

The LEGO Mindstorms RCX contains an Hitachi H8/3292 *microcontroller,* running at 16 MHz. Hitachi calls its product a "single-chip microcomputer," but many others in the chip industry refer to such devices as "microcontrollers." Microcontrollers are like miniature computers but are designed for "embedded applications" for controlling hardware. The main advantage microcontrollers have over full computers is that they are fairly inexpensive—just a few dollars, as opposed to several hundred dollars.

The H8 supports several memory types, including both ROM (read-only memory) and RAM (random access memory). It also comes with its own built-in timers—three to be exact, though the RCX splits one of them to create a total of four. It also has eight 10-bit analog-to-digital converters. In all, it is a highly capable chip, which is one reason why the RCX can pack so many features in such a small package. (By the way, Chapters 31 through 33 of this book deal with several off-the-shelf microcontrollers, including the venerable Basic Stamp, that you can use with your robot creations. Be sure to check these chapters out if you are interested in creating your own version of the LEGO RCX.)

Figure 13.1 shows several different "layers" of the program instruction used in the RCX. At the bottom is a form of hardware *BIOS* (basic input/output system). This hardware level is a permanent part of the H8 processor and provides for very low-level functionality, including the downloading of programs. The hardware BIOS is stored in 16K bytes of ROM. It cannot be changed or erased.

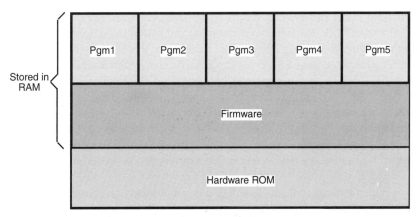

FIGURE 13.1 The Mindstorms RCX uses a hardware BIOS in ROM, along with firmware and data in RAM.

The *firmware* layer contains what could be considered the RCX's operating system. This operating system can be periodically updated. In fact, when you first install the Mindstorms system on your computer, part of the setup process involves downloading the firmware from a file on your computer's hard disk and into the RCX. Whenever LEGO releases updates for the RCX you need merely to return to the setup portion of the Mindstorms program and download the new version of the firmware.

Finally, the *data* layer contains the actual programs that you run on the RCX. Data programs are stored in random access memory. The batteries in the RCX continuously apply a small amount of power to the memory so your programs remain, even when the RCX is turned off.

Both the firmware and the data are stored in 32K bytes of RAM. Being RAM, the data cannot only be replaced; it can be erased (including accidentally). The RCX can store up to five separate programs. There is enough RAM capacity for the firmware and no more than 1.2K for each of the five programs. Program data is stored in a 6K-byte region; the rest of the RAM is allocated to the firmware. Compared to your desktop computer, that's not very much storage space. However, the RCX needs relatively little RAM to run its programs. Since there are only five "slots" for programs, once you've downloaded a program into each slot you have to overwrite one of the old programs in order to download a new one.

Because data and firmware are held in RAM, all your programs will be wiped out if the batteries are removed or are allowed to become depleted. When the memory is swept clean you must also redownload the firmware. This involves running the basic setup section of the Mindstorms installation program.

Brick Variations

The RCX that comes with the LEGO Mindstorms Robotic Invention System isn't the only programmable brick that LEGO makes. The CyberMaster, for example, is a programmable robot designed primarily for use in schools.

Included in the LEGO Mindstorms Robot Discovery Set (RDS) is the Scout, another programmable brick that supports two motors and two sensors. You can also program the Scout via a computer (though the original RDS lacked this feature). The official Scout programming language from LEGO is PBrick Assembler, which is said to be a common language for LEGO's future products. You can find information on this programming language at the LEGO Mindstorms Web site at *www.legomindstorms.com.* Note that in this chapter we concentrate on the RCX, but that doesn't mean you should turn a blind eye toward the Scout. Feel free to explore the RCX, Scout, or preferably both!

LEGO sells a version of the LEGO Mindstorms Robotics Invention System that has different software than the retail version. The school version uses a programming platform known as *Robolab,* while the commercial or home version uses a fully graphical programming environment called RCX Code. The hardware is the same, but the software—the way to program the robot—is different.

The Origins of the Mindstorms RCX

Thanks to its unusual design and almost limitless potential, there has been much interest in the genesis of the Mindstorms RCX. The idea of integrating a small computer into a generic, programmable device goes back some years, and it has been the subject of very active research at the Massachusetts Institute of Technology. MIT first demonstrated so-called programmable bricks during the early 1990s, and these prototypes clearly influenced the design of the LEGO Mindstorms RCX. Researchers at MIT are quick to point out that the internals of the RCX were developed entirely by LEGO designers. Still, a quick look at *http://fredm.www.media.mit.edu/people/fredm/* and other Web pages hosted by MIT demonstrates the impact of this pioneering work.

An important aspect of the *"brick"-based microcontroller* is that it extends the programmability and flexibility of the microcontroller to nontechnical users. As you'll discover in later chapters, wiring and programming microcontrollers is not a simple task for the newcomer. But the RCX, and its MIT ancestors, make working with microcontrollers much easier.

It is clear that the RCX is but the first of a new kind of universal, consumer-oriented microcontroller. Many others will follow. Because they are fully programmable, these microcontroller bricks will be useful in scores of projects, including home automation, home security, automotive applications, personal care and exercise equipment, toys, tools—you name it!

Basic Robots

As with most any LEGO assortment you buy, the LEGO Mindstorms RIS comes with a booklet of suggested project plans, but you're free to design most anything you want. And because you can use standard LEGO parts, you can cannibalize other kits to extend your Mindstorms creations. Figure 13.2 shows the PathFinder 1, the basic Mindstorms robot built with the RIS. Using two motors and two wheels, the robot vehicle is able to move

FIGURE 13.2 The basic rover robot is equipped with two drive motors and a light sensor.

forward and back, and it can turn in place—that is, it has no turning radius like a car; it can just spin to turn.

While building the PathFinder is fun in itself, the real enjoyment—and challenge—comes in programming the thing. Pop the Mindstorms CD-ROM into your computer, and you can design your own programs to control the RCX. The Mindstorms CD-ROM comes with a programming tutorial, but the whole technique is so simple and straightforward that even nonprogrammers will easily master the basics. To program the RCX you merely click and drag predefined program blocks, connecting them on the screen like links of a chain. You can move the blocks around and add additional blocks in between.

And, of course, you are not limited to building just the standard two-wheel roving robot. With just the parts included in the Mindstorms RIS kit you can construct a simple robot arm, or even a walking robot. The RCX has three motor output ports; you can add a third motor (available separately) to create more complex robot creations.

Robotic Sensors

One light and two touch sensors are included in the Mindstorms RIS box. These allow the RCX to interact with its environment (without the sensors, all you really have is an expensive R/C toy). You get two touch sensors—they're really miniature spring-loaded momentary switches enclosed in a LEGO block and a light emitter/detector pair that can control the action of a motor when a change of light occurs. Additional sensor types are available, and we will discuss them in the following section.

SENSOR PROGRAMMING

You control the sensors of the RCX using the graphical Mindstorms programming environment. The programming environment treats the input sensors as "events": when a sensor event occurs, the RCX can be programmed to take some action. For example, suppose you've created the basic two-wheeled Pathfinder 1 roverbot. Your program starts by activating both motors so the robot travels in a forward direction. A touch sensor is attached to the front of the RCX. If the sensor is activated—when the RCX strikes an object—your program can reverse the motors so the robot travels in the opposite direction.

Similarly, a touch sensor mounted on the back of the RCX can be programmed to make the robot travel forward again. In a crowded room, the RCX would likely ping-pong back and forth between objects—fun for a while, but Mindstorms can do more. You can program your robot with time delays to create sophisticated movements. For instance, instead of just reversing both motors when a touch sensor is activated, you might activate just one motor for a brief moment. You can then command both the motors to turn on again in the forward direction. This would have the effect of turning the robot by an arbitrary amount, so that its travel around a room is less predictable.

USING THE LIGHT AND TOUCH SENSORS

The light sensor can be used to enable your robot to detect the presence or absence of light. It is a fun gadget to use when constructing a flashlight-controlled robot. With such a robot, the RCX can be commanded to stop, turn, or reverse direction when a flashlight is directed at the sensor. The sensor includes its own light source, so you can also use it to construct a "line tracing" robot. The Mindstorms RIS kit comes with a large white pad with a black line that you can use as a "course" or track for the RCX to follow. You can draw your own line-following track on any light-colored surface.

OPTIONAL SENSORS

Additional sensors are available from LEGO that you can connect directly to the RCX. These include the following:

- *Temperature.* These sensors sense differences in temperature, such as operating indoors or outdoors or the direct touch of a human hand.
- *Rotation.* Used with the drive motors, these sensors sense the actual number of rotations of the motor shaft, allowing you to position of the RCX robot more accurately.

In addition to LEGO-made sensors for the RCX, you can also construct your own. See the section "Making Your Own RCX Sensors" later in this chapter for more information.

Downloading Programs

One noteworthy feature of the Mindstorms is that the RCX is a nontethered controller. By not being connected to a PC, the RCX robot appears much more like an autonomous machine, even though you use your PC as a programming station. There is no control wire for the RCX to get tangled with. This is actually typical of most microcontrollers used in robotics; read more about microcontrollers in Part 5.

You download the programs you create on your PC to the RCX via a two-way infrared (IR) transceiver. The IR communications tower sends program code to the RCX, and the RCX responds to indicate a proper download. For optimum performance, you should place the IR tower no more than about a foot from the RCX, though I've successfully used the tower to download programs to an RCX that was four to five feet across the room.

When you think you have a working program, you place the RCX near the infrared transmitter and click the "Download" button in the Mindstorms programming screen. Most programs download in less than 10 seconds. When downloading is complete, you merely depress the "Run" button on the RCX unit and watch your robot creation come to life.

If your robot doesn't behave quite like you expected, you can reexamine your program, make changes, and download the revised code. Once you've built a program you like, you can save it for future reference. The RCX can store five programs internally at a time, but you can keep hundreds or even thousands of programs on your computer's hard disk drive. Just download them again into any of the RCX's five program slots when you want to run them.

Remote Control

An optional accessory for the RCX is a handheld infrared remote. This remote lets you operate the RCX from up to 20 feet away. You can individually control the forward or reverse direction of any of the three motor outputs (A, B, or C). You can also start and stop any of the five programs stored in the RCX's internal memory as well as send sequences of RCX code to override the internal programs.

Hacking the Mindstorms

Not long after LEGO introduced the first Mindstorms kit, folks found ways to hack into the RCX and programming software. Among the first hacks on the scene were various Microsoft ActiveX and programming components for coding the RCX using Microsoft Visual Basic. LEGO itself now offers (but does not actively support) an RCX software developer's kit (SDK) using Visual Basic. You can download the free documentation and software for it at the LEGO Mindstorms Web page (*www.legomindstorms.com*).

The LEGO Visual Basic SDK works with both the RCX brick included with the RIS and the CyberMaster brick that accompanies the LEGO Technic CyberMaster, a product designed for classroom use. (Note: the Mindstorms Scout, used in the Robotics Discovery Set, has a separate SDK of its own.)

The SDK requires a special ActiveX (also called *OCX*) component, *spirit.ocx,* that serves as an interface between the Visual Basic programming platform and your PC's hardware. From there, you need only a copy of Visual Basic 5.0 or higher. In actuality, you can use most any programming platform that can interface to ActiveX modules with the SDK. However, the programming examples in the SDK are provided in Visual Basic, so if you use another language you'll need to do the language conversion yourself. Chapter 14, "Programming the LEGO Mindstorms RCX: Advanced Methods," discusses how to use Visual Basic with the Mindstorms robot.

Other RIS programming hacks are available as well. At *http://www.enteract.com/~dbaum/,* for instance, you can download NQC (Not Quite C), a development language that uses a C-like syntax for programming the LEGO RCX brick. Versions are available for use under Linux, Windows, and the Macintosh. The NQC language is provided using the "Mozilla Public License," a kind of open source license. NQC is discussed in detail in Chapter 14.

Making Your Own RCX Sensors

As we've mentioned, LEGO provides a number of sensors you can use with the Mindstorms RCX, including sensors for light, touch (simple switch), temperature, and wheel rotation. Several of these sensors—namely, the light sensor and the wheel rotation sensor—are powered; they require operating juice from the RCX to operate. At first glance, this may seem an impossibility: each input on the RCX has just two connections (there are four contact points on the connector brick, but each pair is wired together, so you can attach the connector with any orientation).

However, the RCX uses an interesting circuit connection to its sensor inputs so that a single pair of wires can serve both as outgoing power to run the circuit and as an input. The RCX's approach is to toggle the power to its sensors on and off very rapidly. During the on power times, the sensor receives current to operate. During the off times the sensor value is read. A capacitor in the sensor serves as a kind of voltage reservoir during the off times.

The RCX directs power and input to their particular portions of the sensor circuit by using a diode bridge, shown in Fig. 13.3. Connect your circuit as shown, being sure to add a 33 to 47 μF capacitor across the +V and ground rails; the capacitor is required to keep voltage applied to the circuit during the periods when the RCX is reading the value at the input.

You can, of course, also create your own unpowered contact-type sensors. These are easily connected to the inputs, as shown in Fig. 13.4. You simply wire a 470-ohm resistor in series with the switch. The resistor is used because the terminals of each RCX output are powered at 5 volts continuous. The resistor prevents a dead short across the terminals.

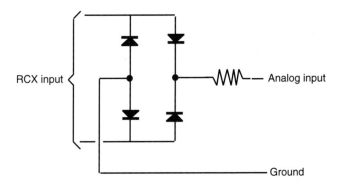

FIGURE 13.3 The basic bridge diode network for RCX active sensors supplies power to the sensor electronics while providing the output signal back to the RCX.

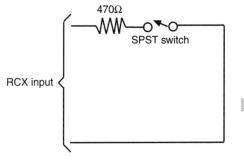

FIGURE 13.4 Nonactive sensors such as switches can be connected to the RCX via a 470-ohm resistor.

There are countless examples of homebrew RCX sensors on the Internet. Rather than repeat these excellent designs, I've provided a few quick samples here, and I refer you to several worthwhile pages on the Internet in Appendix C, "Robot Information on the Internet."

REPLACEMENT TOUCH SENSOR

Figure 13.5 shows a replacement whisker-type touch sensor for the RCX that is made from a surplus leaf switch (often called a "Microswitch," after the brand name that made this kind of switch popular). Use the schematic in Fig. 13.4 to connect the switch to the RCX input.

SENSOR INPUT TECHNICAL DETAILS

Here are some useful technical details about the RCX inputs:

■ You must set the sensor type in the software before a sensor can be used. This is done with the RCX Code software that comes with the RCX unit or through a substitute programmer, such as Not Quite C.

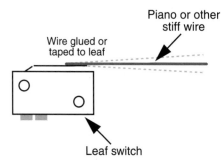

Figure 13.5 A leaf switch and small diameter wire ("piano wire" or "music wire") makes for a good whisker-type bump sensor for the LEGO RCX.

- When used with a nonpowered sensor (e.g., a touch switch), the sensor outputs 5 volts and can drive up to 10 milliamps (mA).
- When used with a powered sensor (e.g., a light sensor), the sensor nominally outputs 7.5 to 9 volts (depending on battery capacity), with an "on/off" square wave. The square wave has a period of 2.8 milliseconds; the off period is 0.1 milliseconds in duration (meaning the RCX applies power for a much longer time than it detects the sensor reading). Note that the "off" voltage is 5 volts, not 0 volts.

Using Alternative Motors and Output Devices

Caution! What follows should be considered for *experimental use only*. Connecting non-LEGO devices to your RCX can damage the RCX, and possibly the device you've attached to it. *Proceed at your own risk!*

Like many microcontroller-based electronics, the LEGO Mindstorms RCX uses motor driver circuitry to boost the current-handling capabilities required to drive motors. There are a number of ways to do this, including using bipolar transistors, power MOSFET transistors, and specially made H-bridge motor drivers (all of these technologies are fully explored in Part 3).

As of this writing (this kind of thing can change now and then), the RCX uses a trio of MLX10402 motor driver circuits, made by Melexis Microelectronic Integrated Systems, a company that specializes in automotive sensors and control. The MLX10402 includes overload protection against temperature and current extremes and has a maximum rating of about 500 mA, at 9 volts. The chip can handle motor voltages of 5 to 12 volts (with an absolute maximum of 16 volts), though it is designed to be controlled by +5 vdc, which is typical of computers and microcontrollers. Because the device is designed for use in automotive applications, it has excellent thermal ratings: a storage temperature of −55°C to 125°C and a maximum die operating temperature of +150°C. The chip goes into protective fail-safe mode at temperatures exceeding this.

You control the motor attached to the MLX10402 by altering just two input lines (set Mode to HIGH), according to the following truth table, which gives you an idea of the capabilities of the RCX:

IN1	IN2	FUNCTION	DESCRIPTION
1	0	Forward	Turning
0	1	Backward	Turning opposite
1	1	Brake	Motor is shorted (fast stop)
0	0	Off	Motor is disabled (coast)

HARDWARE HACKING WITH THE MLX10402 CHIP

While the RCX, or more specifically the MLX10402 chip, is primarily designed for operating a motor, it can also control a number of other devices, including relays and solenoids. And, of course, you can control other kinds of DC motors, not just the ones that come with the Mindstorms set. The factors to keep in mind are as follows:

- The RCX provides 9-vdc power to the three motor outputs, so the motors you use should be rated for 9-volt operation, "more or less." Many 6- or 12-volt DC motors will run at 9 volts; 6 vdc motors will run fast, and 12 vdc motors will run slow. Damage *could* result to a 6-vdc motor that is operated at 9 vdc for long periods of time. If you are using relays or solenoids, look for 5–6 vdc units that will work acceptably. You can use diodes or resistors to drop the 9 volts from the RCX to the 5 or 6 volts expected by the relay coil.
- The MLX10402 can provide up to 500 mA to each motor (this is the specification rating of the MLX10402; a more conservative rating you should go by is 400 mA). The standard 9 vdc motors that come with the LEGO Mindstorms consume no more than about 320 mA each, at 9 vdc. (This rating was determined by stalling the motor—clamping its output shaft so it will not move—and measuring the current draw when powered by 9 vdc.) Assuming the RX was designed to adequately handle up to three motors at a time (960 mA total), then the motors, relays, solenoids, and other devices you attach to the RCX should consume no more than about 960 mA total, worst case. If you are using motors, the "worst case" is the current rating of the motors when they are stalled (i.e., their shafts locked tight so they won't turn).
- When you are using a reduced power mode, the voltage to the outputs is "chopped" at a frequency of 8 milliseconds between pulses, as shown in Fig. 13.6. The ratio of on-time versus off-time determines the power delivered through the output. Note that the output has two modes: float and on/off, as shown in Fig 13.6. When in float mode, the output ramps from full voltage to 0 volts. In on/off mode, the output toggles from HIGH to LOW with no ramping.

More and More LEGO

The LEGO Mindstorms community is a rather large one, and it's growing. LEGO actively sponsors Mindstorms experimenters and hackers. The LEGO Mindstorms Web page, at

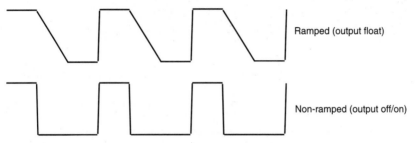

Ramped (output float)

Non-ramped (output off/on)

FIGURE 13.6 The RCX uses pulse-width modulation to vary the power level of its three outputs. The pulses can be steep square waves, or they can be ramped.

*/www.legomindstorms.com,*provides links for message boards, contests, tips and tricks, and a LEGO community called "First LEGO League," for 9- through 14-year-olds interested in exploring robotics.

There are several Web forums devoted to LEGO that contain special sections for using and programming the Mindstorms RCX. Check out Lugnet at *www.lugnet.com*This non-commercial Web forum is divided into several dozen discussion groups. Find the group that's of interest to you and read through the postings by other visitors. Feel free to post your own message if you have a question or comment.

From Here

To learn more about...	*Read*
Constructing custom robots using standard LEGO parts	Chapter 12, "Build Custom LEGO-based Robots"
Additional ways to program the LEGO RCX brick	Chapter 14, "Programming the LEGO Mindstorms RCX: Advanced Methods"
Using a microcontroller for the brains of a robot	Chapters 31–33
Using stepper motors	Chapter 19, "Working with Stepper Motors"
Constructing sound sensors	Chapter 40, "Sound Output and Input"

14

PROGRAMMING THE LEGO
MINDSTORMS RCX

Advanced Methods

The LEGO Mindstorms Robotic Invention System has become a watershed for hobby robotics. The Mindstorms set allows both child and adult alike to experiment with robotics, without the usual requirements of constructing a frame and body or soldering electronic circuits. As such, Mindstorms provides a quick and simple introduction to robotics, and especially the programming behind it.

Yet the robots you can build with the Mindstorms go far beyond simple automated toys. There is a surprising amount of power under the yellow plastic of the Mindstorms robot module (the RCX). Much of this power is either not easily recognizable in the standard programming environment that comes on the Mindstorms CD-ROM or is not available, for whatever technical reasons.

Fortunately, you can tap the full potential of the Mindstorms robotics system by using alternative programming environments. This chapter discusses two popular alternative environments, including one that works with Microsoft Visual Basic.

Using Visual Basic to Program the RCX

Microsoft Visual Basic provides a convenient method for programming the LEGO Mindstorms RCX. You don't even need the full Visual Basic package (some $250 or more retail). You can also use any program that supports Visual Basic for Applications—such as Microsoft Word 97 or Corel WordPerfect 9 or later versions.

The text that follows will work equally well when using Visual Basic 5.0 or later or Visual Basic for Applications. However, some menu commands may be different between the two products as well as between different versions. For the sake of brevity, from here on we'll refer to Visual Basic as "VB" and Visual Basic for Applications as "VBA."

Note: by necessity, this chapter does not discuss programming with VB or VBA. It is assumed that you are already familiar with at least the basics of VB or VBA and that you know how to create and work with user forms and code modules. If VB and/or VBA are new to you, pick up a good introductory book on the subject at your library or neighborhood bookstore.

Before attempting to program in VB/VBA visit the main LEGO Mindstorms Web page (*www.mindstorms.com*) and look for the Software Developer's Kit page. Download the informational document on the "PBrick" programming syntax for the *spirit.ocx* ActiveX control. As of this writing, the document is available only in Adobe Acrobat format, so you will need a copy of the Adobe Acrobat reader to view the file. The Acrobat reader is available free at *www.adobe.com* and many other places; see the link on the LEGO Mindstorms page to download this software. You may also wish to download the sample VB program for later review.

When you use RCX with Visual Basic, you program the RCX by using a "middleware" component. This component is *spirit.ocx,* a Windows file that serves as an interface between the programming environment (VB or VBA) and the RCX itself. Spirit.ocx comes with the LEGO Mindstorms installation CD and is placed on your computer when you install the software. Keep this in mind: you will not be able to perform any programming until the Mindstorms software has been loaded. If you haven't done so already, use the standard Mindstorms programmer software to run the RCX through a couple of its basic paces. Once you know the RCX works with the standard programmer software you're ready to forge ahead with VB!

RUNNING THE TEST PROGRAM

To begin, start Visual Basic in the usual way. If you are using VBA, start the Visual Basic Editor (for example, in Word 97 and later you would choose *Tools, Macro, Visual Basic Editor*). Once you are in the Editor, follow these steps:

1. Create a new form by choosing *Insert, UserForm.*
2. Add the spirit.ocx component by choosing *Tools, Additional Controls.* Locate the "Spirit Control" and click on it to add a checkmark beside it. Choose OK to close the Additional Controls dialog box. (Note: this step need only be done once; thereafter the spirit.ocx control will be registered with VB/VBA for use in other projects.)
3. A new LEGO icon should appear in the Toolbox (choose *View, Toolbox,* if the Toolbox is not visible).
4. Click on the LEGO icon (this is the Spirit Control) and drag anywhere over the user form you created in step 1.
5. For ease of use, make the LEGO icon about a quarter-inch square and place it in the lower-right corner, as shown in Fig. 14.1.
6. Verify the proper settings of the Spirit Control by clicking on it and examining its properties in the Properties box. Specifically, check that the serial communications port is set properly (usually either *COM1* or *COM2*), that the LinkType is *Infrared* (assuming you're using the standard infrared tower that comes with the Mindstorms set), and that the PBrick type is RCX.

7. Change the name of the Spirit Control you've added to *PBrickCtrl*. (This step is optional; however, it conveniently conforms to the examples provided in the PBrick documentation provided by LEGO.)

8. Click on any blank area of the form, and change the name of the form (in the Properties box) to *RCXFrm*. While you can choose any name for the form, the sample programs that follow later in this chapter use the name RCXFrm to reference the PBrickCtrl control.

Adding the Spirit Control (spirit.ocx) component to the form allows you to write VB code so as to interface with the RCX. You are now ready to begin programming:

1. Create a new code module by choosing *Insert, Module.*

2. Type the BasicTest code shown below. Be on the lookout for typographical errors.

```
Sub BasicTest()
RCXFrm.PBrickCtrl.InitComm
RCXFrm.PBrickCtrl.PlaySystemSound (2)
End Sub
```

3. Verify that your RCX is on and that it is positioned no more than about a foot from the infrared tower.

4. In VB/VBA, run the BasicText program (choose *Run, Run Sub,* or press F5). (Note: you do not need to depress the Run button on the RCX in order to execute the BasicTest code.)

If all is working properly, the RCX should emit a short tone. If you get an error or the tone doesn't sound, recheck that the RCX is operating properly. Verify that the IR tower is functioning by verifying that the dim green light is on when the BasicTest program is being downloaded. This light will extinguish a few seconds after downloading is complete.

PROGRAMMING MOTOR ACTIONS

Sounding tones is hardly the life's work of the LEGO RCX unit, so let's try some more advanced programming techniques, including running two motors. For the following test, we'll assume that your RCX robot has two motors, attached to outputs A and C. Type the following

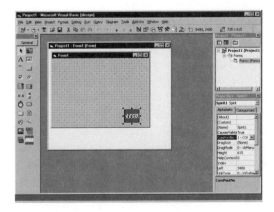

FIGURE 14.1 The test form in Visual Basic, with the spirit.ocx control added.

code, either in the same module in which you created BasicTest earlier or in a new module. Again, be watchful for typographical errors. We'll discuss what the code does in a bit.

Listing 14.1   Testmotors

```
Option Explicit

Public Const SWEEP_DOWN_SOUND = 2
Public Const SWEEP_UP_SOUND = 3
Public Const SWEEP_FAST_SOUND = 5

Sub TestMotors()
With RCXFrm.PBrickCtrl
 .InitComm
 .SelectPrgm 0
 .BeginOfTask 0
  .Wait 2, 30
  .SetPower "02", 2, 7
  .SetFwd "02"
  .On "02"
  .Wait 2, 50
  .SetRwd "02"
  .Wait 2, 50
  .Off "02"
  .PlaySystemSound SWEEP_FAST_SOUND
 .EndOfTask
End With
MsgBox "Download complete"
End Sub
```

Running the TestMotors program   When you are done typing, run the TestMotors program in VB/VBA. A message box appears when downloading is complete. For this test, you must select program 1 on the RCX, using the *Pgm* button. Press *Run* when you're ready to run the program. The RCX should spin its motors forward and reverse for a short burst each way. When done, the RCX will emit its "up-sweep" tone to tell you it's finished.

Examining the TestMotors program   The TestMotors program is actually straight-forward. You may want to increase your understanding of what the program does by reviewing the PBrick documentation (described earlier in this chapter) from LEGO. Here is the first line of the program:

```
With RCXFrm.PBrickCtrl
```

The *With* statement is a standard VB/VBA command. It allows you to reference an object—in this case, RCXFrm.PBrickCtrl—using a shorthand syntax. Refer to the VB/VBA documentation for additional information on using *With*. All of the statements that follow, except for MsgBox, are commands built into the spirit.ocx component:

```
.InitComm
```

.InitComm (note the period prefix) sets up communications between the IR tower and the RCX. You must always include this statement before sending other commands to the RCX.

```
.SelectPrgm 0
.BeginOfTask 0
```

The *.SelectPrgm 0* statement selects program 1 in the RCX (e.g., press the *Pgm* button until program 1 appears in the RCX's LCD display). Much of the programming with spirit.ocx involves zero-based values, so *SelectPrgm 0* is program 1, *SelectPrgm 1* is program 2, and so forth. Recall that you can store up to five programs in the RCX at any one time.

As a point of reference, the 0 after the *.SelectPrgm* statement is known as a *parameter.* Many of the statements used to program the RCX with the spirit.ocx component require that you use of one or more parameters.

The *.BeginOfTask 0* statement tells the RCX that the code that follows is its main task. This functionality will occur when you press the *Run* button on the RCX. Each program can have up to 10 tasks. The RCX is designed to run each task simultaneously, which allows your programs to be multithreaded. For example, you might have your RCX play a tune while driving a zigzag course. Each of these actions is contained in its own task in one program.

```
.Wait 2, 30
```

The *.Wait* statement tells the RCX to wait a brief period of time. Wait uses two parameters: the *2* tells the RCX that the parameter that follows is a literal "constant". The *30* indicates 30/100s of a second, or about a third of a second. Other *Wait* statements in the remainder of the program perform a similar function.

```
.SetPower "02", 2, 7
.SetFwd "02"
.On "02"
```

These three statements set up the drive motors and turn them on. *.SetPower* sets the power to motors 0 and 2 (labeled A and C on the RCX) to full. The *2* indicates a literal constant, and the *7* indicates the value (1 is slow, 7 is fast, and there are several speeds in between). Similarly, *.SetFwd* sets the direction of motors 0 and 2, and *.On* turns them on.

```
.SetRwd "02"
```

Similar to *.SetFwd, .SetRwd* sets the direction of motors 0 and 2 in reverse.

```
.Off "02"
.PlaySystemSound SWEEP_FAST_SOUND
```

The *.Off* statement turns motors 0 and 2 off. The *.PlaySystemSound* statement, previously used in the BasicTest program earlier in this chapter, sounds a tone. Note the use of the *SWEEP_FAST_SOUND* constant variable, which is defined at the top of the program (a constant is a variable whose value does not change throughout the execution of the program). You can—and should—get into the habit of using constants instead of literal numeric values. It's a lot easier to see what *PlaySystemSound(SWEEP_FAST_SOUND)* means than *PlaySystemSound(5),* though both do exactly the same thing.

Consult the documentation for VB/VBA if you're new to the concept of using constants.

```
.EndOfTask
```

The *.EndOfTask* statement tells the RCX that the task begun earlier is now complete.

CLOSING THE COMMUNICATIONS PORT

In the program examples given in the previous section, the communications port, such as COM1 or COM2, is opened so the spirit.ocx component can send signals out of the Mindstorm's IR tower. This is done with the *.InitComm* statement. In each of the program examples we just examined, the communications port is left open. This is to simplify download timing, but in general it's not an advisable practice because it leaves the communications port opened, and therefore locked against use by other programs on your computer.

Use the *.CloseComm* statement to close the communications port. You can integrate this statement with your programs—for example, as the last command sent to the RCX. However, you must be careful not to close the communications port before downloading is complete or else your program will not function properly. One way to get around this is to use a waiting loop in VB; another is to use the *DownloadDone* event, which is a signal sent by the RCX when it has received all of its programming. The PBrick documentation from LEGO explains how to use the *DownloadDone* event.

Yet another approach is to specifically run a small program that closes the communications port. Here's all the code you really need for the job:

```
Sub CloseComm()
RCXFrm.PBrickCtrl.CloseComm
End Sub
```

GOING FURTHER

There are many more things you can do with the spirit.ocx component and VB/VBA, including reading sensor values (this is done with the Poll statement), adjusting the power output of the IR tower, even turning the RCX off remotely. Sadly, we don't have the room to delve into these subjects in more detail.

Fortunately, you can turn to the PBrick documentation provided by LEGO on the Mindstorms Web site (*www.mindstorms.com*) for additional information. Be sure to also check out the additional examples and resources on RCX programming at the support site for this book, *www.robotoid.com*.

Using Not Quite C (NQC) to Program the RCX

At last count, there were over a *dozen* programming alternatives for the LEGO Mindstorms RCX. One of the first, and still one of the most popular, is NQC—the letters stand for "Not Quite C." NQC is a stand-alone, text-based programming environment for the RCX. It is capable of performing the same basic programming functions as the Visual Basic approach, described earlier, but the programming language is radically different.

NQC is a freely distributed program available at *http://www.enteract.com/~dbaum/nqc/*. Versions of NQC are available for Windows-based systems, as well as the Apple Macintosh and Linux. Fetch the version you want, and place the NQC files in their own directory on your computer's hard disk drive. Assuming the IR tower is properly connected to your computer and the Mindstorms software has been previously installed, you're ready to go!

Note that the following steps assume you're using a Windows-based PC. Consult the documentation that comes with NQC if you are using a different computer.

CREATING A NQC PROGRAM

As its name suggests, Not Quite C uses a C-language syntax for programming. For those unfamiliar with C, the syntax can look daunting. However, with just a little bit of study, you'll find NQC is not difficult to use. If anything, it's often easier to interact with the RCX using NQC than with Visual Basic.

You may use any text editor to prepare an NQC file. In Windows, for example, you can use the Notepad program. You should store your NQC program files in the same directory as the *nqc.exe* program itself. Listing 14.2 shows a simple NQC program that does an amazing amount of computational work. Run this program on an RCX with two motors attached to the A and C outputs and with a light sensor connected to Input 1 and pointing forward. When you do, the RCX will seek out any bright light in the room. Aim a flashlight at the light sensor, for example, and the robot will come toward the light.

Figure 14.2 shows the prototypical RCX "rover" robot, set up for the sample programs in this section.

Listing 14.2 photophile.nqc.

```
#define LIGHT    SENSOR_1
#define MOTOR   OUT_A+OUT_C

task main()
{
     SetPower(MOTOR, 7);
     SetDirection (MOTOR, OUT_REV);
     SetSensorType(LIGHT, SENSOR_TYPE_LIGHT);
     while(true) {
           if(LIGHT > 60)
                 On(MOTOR);
           else
                 Off(MOTOR);
     }
}
```

If you key in this program in order to try it out, name it *photophile.nqc* (*photophile* means "lover of light"). Be on guard for typographical errors, and do not omit any of the brace characters! As with most C-based languages, capitalization is important. For example, *while* is correct, but not *While.*

EXAMINING THE NCQ PROGRAM

Let's take a closer look at this program. The first two lines:

```
#define LIGHT        SENSOR_1
#define MOTOR        OUT_A+OUT_C
```

FIGURE 14.2 The basic "rover" robot to be used with the programs in this chapter.

These two statements define constants (unchanging variables) used elsewhere in the program. Constants are defined by using the *#define* keyword followed by the name of the constant and finally by the value of that constant. Note that the value of the constants *LIGHT* and *MOTOR* are themselves constants! In this case, the constants *SENSOR_1, OUT_A,* and *OUT_B* are built-in constants, with values already defined by NQC. We use our own constants to make working with the RCX even easier. See Table 14.1 for a list of the most commonly used built-in constants.

You will note that the *MOTOR* constant refers to two outputs, both A and C (*OUT_A+OUT_C*). This allows us to operate both motors together, which will make it a little easier to command the robot to go forward or backward. The next two lines of the program are:

```
task main()
{
```

Each NQC program can have up to 10 tasks. Each task can be run simultaneously. The task that is run when you press the Run button on the RCX is called *main*. You can create your own names for other tasks you add in your program, but main always refers to the, well, "main" task that the RCX automatically runs.

Note the open brace (the "{" character) that follows the task *main()* statement. In NQC, as in C, multiple program statements are defined in *blocks* or *compound statements*. For

TABLE 14.1 NQC STANDARD CONSTANTS

CONSTANT NAME	FUNCTION	EQUIVALENT VALUE
SetSensor		
SENSOR_1	Input 1	0
SENSOR_2	Input 2	1
SENSOR_3	Input 3	2
SetSensorMode		
SENSOR_MODE_RAW	Raw value from sensor (0 to 1023)	hex 0x00
SENSOR_MODE_BOOL	Return Boolean (0 or 1) value	hex 0x20
SENSOR_MODE_EDGE	Count number of rising/ falling edges	hex 0x40
SENSOR_MODE_PULSE	Count number of pulses	hex 0x60
SENSOR_MODE_PERCENT	Show value as percentage	hex 0x80
SENSOR_MODE_CELSIUS	Temperature sensor Celsius reading	hex 0xa0
SENSOR_MODE_FAHRENHEIT	Temperature sensor Fahrenheit reading	hex 0xc0
SENSOR_MODE_ROTATION	Rotation encoder	hex 0xe0
SENSOR_TYPE_TOUCH	Pushbutton switch	1
SENSOR_TYPE_TEMPERATURE	Temperature sensor	2
SENSOR_TYPE_LIGHT	Powered light detector	3
SENSOR_TYPE_ROTATION	Rotation encoder	4
Outputs		
OUT_A	Select motor A	1 << 0
OUT_B	Select motor B	1 << 1
OUT_C	Select motor C	1 << 2
Output modes		
OUT_FLOAT	Let motors coast	0
OUT_OFF	Stop motors	hex 0x40
OUT_ON	Run motors	hex 0x80
Output directions		
OUT_REV	Motors in reverse	0
OUT_TOGGLE	Motors change direction	0x40
OUT_FWD	Motors forward	0x80

TABLE 14.1 NQC STANDARD CONSTANTS (*Continued*)

CONSTANT NAME	FUNCTION	EQUIVALENT VALUE
Output power levels		
OUT_LOW	Motors at low speed	0
OUT_HALF	Motors at medium speed	3
OUT_FULL	Motors at full speed	7
Sounds for PlaySound		
SOUND_CLICK	Short beep	0
SOUND_DOUBLE_BEEP	Two beeps	1
SOUND_DOWN	Tone scale down	2
SOUND_UP	Tone scale up	3
SOUND_LOW_BEEP	"Error" beep	4

each { character there must always be a } character, indicating the end of the block. You will use blocks in *if, while,* and other statements. The one thing you need to remember about blocks and the brace characters that define them, is this: always make sure you have a close brace for every open brace. The next two lines of the program are as follows:

```
SetPower(MOTOR, 7);
SetDirection (MOTOR, OUT_REV);
```

The *SetPower* statement sets the power to the motors. The *7* means full power; use *1* for low power or other values in between. *SetDirection* sets the direction of the outputs, in this case reverse. This will make the robot move toward the light.

```
SetSensorType(LIGHT, SENSOR_TYPE_LIGHT);
```

A single light sensor, connected to input 1, is used for the robot. The *SetSensorType* statement sets the input—specified here as *LIGHT*—to accept a powered light.

```
while(true) {
    if(LIGHT > 60)
            On(MOTOR);
    else
            Off(MOTOR);
}
```

The main body of the program is a *while* loop, which thanks to the *true* expression repeats the program until you depress the *Run* button on the RCX a second time or turn the RCX off. The important part of the program is the *if* statement, which reads (in "human" terms):

"if the output of the light sensor is greater than 60, turn the motors on;

   otherwise

turn the motors off"

Light sensors on the RCX return a value of 0 to 100, with 0 being absolute darkness and 100 being fairly bright light. The value of 60 was selected as a kind of threshold. If you operate the RCX in dim room light, pointing a flashlight at the sensor will cause the motors to run. Turning the flashlight off will cause the motors to stop.

DOWNLOADING THE NQC PROGRAM

Now that the program has been written (and saved), you can download it to the RCX by using the main *nqc.exe* program. This program does two things: it compiles your text programs into a form that is suitable for the RCX, and it transfers the code to the RCX via the Mindstorms' IR tower.

NQC is a command-line program. To use it, open a new MS-DOS window by choosing *Start, Programs, MS-DOS Prompt.* (Note: if you don't have this option, choose *Start,* select *Run,* type *command.com,* and then press OK.) If necessary, switch to the NQC program directory using the *CD* (change directory) command, such as:

```
cd \nqc
```

This assumes that *nqc.exe* and your programs are in a directory named *NQC*. Then compile and download your program with the following command:

```
nqc -d program.nqc
```

where *program.nqc* is the actual program name you want to use. If you saved the sample program as *photophile.nqc,* for example, type the following:

```
nqc -d photophile.nqc
```

and press the Enter key. NQC will then compile the program and download it to the RCX. If there are syntax errors in your program, NQC will alert you of them and display the approximate line where the error occurs (usually the actual error is a line or two above). Assuming the program compiles correctly, NQC displays "Downloading program…" and then finally "Complete" when downloading is finished. Run the downloaded program by pressing the *Run* button on the RCX.

Note: Unless you tell it otherwise, NQC assumes that the IR tower is connected to COM1 of your computer. Use the set command in DOS to set a different communications port, such as:

```
set RCX_PORT=COM2
```

This tells NQC to use COM2 instead. Valid values are COM1, COM2, COM3, and COM4. If you type one of these in the DOS window you have opened for NQC, the value will remain only until you close the window. If you want to make the setting permanent, edit the *autoexec.bat* file (it's in the root of the C drive) and add the *set* command there. It will take effect the next time you start your computer.

ALTERING THE BEHAVIOR OF THE ROBOT

It's easy to alter the behavior of your NQC-controlled robot creations. One small change you can make is to have the motors turn the other way when the light shines on the sensor. This has the effect of creating a "photophobic" robot—a robot that appears to run away from the light. The complete code example is shown in Listing 14.3. If you retype this program, name it *photophobe.nqc.*

Listing 14.3   photophobe.nqc.

```
#define LIGHT        SENSOR_1
#define MOTOR        OUT_A+OUT_C

task main()
{
      SetPower(MOTOR, 7);
      SetDirection (MOTOR,OUT_FWD);
      SetSensorType(BUTTON, SENSOR_TYPE_LIGHT);
      while(true) {
            if(LIGHT > 60)
                  On(MOTOR);
            else
                  Off(MOTOR);
      }
}
```

CREATING A MULTITASKING CONTROL PROGRAM

One of the most important capabilities of the RCX is that it is a multitasking device. You can run up to 10 tasks "simultaneously" in one program. The microcontroller in the RCX divvies up a little bit of its processing time to each task, so in human terms things appear to happen simultaneously. Of course, in microcontroller terms it's handling one instruction at a time, but at very fast speeds.

The following program is a rudimentary but fully functional example of an RCX program with multiple concurrent tasks. The program is based on the *photophobe.ncq* example in the previous section. We have added separate tasks, one to play a little song (the first notes of something that sounds like "Mary Had a Little Lamb") and another to reverse the motors and spin if a touch sensor is activated.

When the program is run, the robot exhibits three events (sometimes called "behaviors" in modern robot artificial intelligence efforts):

Event 1.   A song is played every few seconds. This event is free running, and no other event in the program affects it.

Event 2.   When a strong enough light strikes the light sensor, the robot backs away from the light source (of course, "backs away" depends on how the motors and light sensor are mounted on the RCX, but you get the idea). The motors will continue to run as long as enough light strikes the sensor.

Event 3.   When the touch sensor—mounted on the side of the RCX opposite the light sensor—is activated, Event 2 is suspended ("subsumed"). The robot reverses direction for a brief moment, then spins on its axis. Finally, it stops moving, and it is more than likely no longer facing in the same direction. At this time, Event 2 is reactivated so that the robot will "run away" from any light that shines into the light sensor.

See Listing 14.4 (let's call it *multitask.ncq*), which contains short comments that are indicated by the double slash ("//") characters. These comments serve to describe the main functionality of the program.

Listing 14.4   multitask.ncq.

```
// Constants definitions
#define LIGHT       SENSOR_1
#define SWITCH      SENSOR_2
#define MOTOR       OUT_A+OUT_C

// Main task; run when Run button is pressed on RCX
// starts all tasks
task main()
{
    start play_song;
    start run_from_light;
    start timed_backup;
}

// Task for running away from the light (same as photophobe.ncq,
// except that motors run a little slower)
task run_from_light()
{
    while (true) {
            SetPower(MOTOR, 3);
            SetDirection (MOTOR, OUT_FWD);
            SetSensorType(LIGHT, SENSOR_TYPE_LIGHT);
            if(LIGHT > 60)
                    On(MOTOR);
            else
                    Off(MOTOR);
    }
}

// Task for backing up and spinning in response to switch touch
task timed_backup()
{
    while (true) {
            SetPower(MOTOR, 3);
            SetSensor(SWITCH, SENSOR_TOUCH);
            if (SWITCH == 1) {
                stop run_from_light;       // disallow run_from_light task
                SetDirection (MOTOR, OUT_REV);
                On(MOTOR);
                Wait (50);
                SetDirection (OUT_A, OUT_FWD);
                Wait (150);
                SetDirection (MOTOR, OUT_FWD);
                Off(MOTOR);
                start run_from_light;      // allow run_from_light task
```

```
        }
          }
}

// Task for playing a little tune
task play_song()
{
     while (true) {
             PlayTone(392,25);
             PlayTone(349,25);
             PlayTone(330,25);
             PlayTone(349,25);
             PlayTone(392,25);
             PlayTone(0,2);
             PlayTone(392,25);
             PlayTone(0,2);
             PlayTone(392,25);
             PlayTone(0,2);
             Wait (500);
     }
}
```

Feel free to experiment with the code for *multitask.ncq.* The only real caveat is that if you want a task to continue it should have its own loop. The *While* statement is one method for doing this, but NQC provides other looping statements you may wish to try. Also, remember that the RCX supports up to 10 tasks.

GOING FURTHER

Of course, there's far more to Not Quite C than we have discussed here. The NQC download includes complete documentation on its capabilities. For example, NQC supports a wide variety of programming statements, loops, variable assignments, conditional expressions, and more. With NQC you can develop highly sophisticated programs for the RCX robot, and with a surprisingly small amount of code. Look for additional NQC samples and resources at the support site for this book, *www.robotoid.com.*

From Here

To learn more about...	Read
Introduction to programming concepts	Chapter 7, "Programming Concepts—The Fundamentals"
Using LEGO parts to create custom robots	Chapter 12, "Build Custom LEGO-based Robots"
Using the LEGO Mindstorms Robotics Invention System	Chapter 13, "Creating Functionoids with the LEGO Mindstorms Robotics Invention System"
Computer control of robots	Part 5, Chapters. 28–34

PART 3

POWER, MOTORS, AND
LOCOMOTION

15

ALL ABOUT BATTERIES AND
ROBOT POWER SUPPLIES

The robots in science fiction films are seldom like the robots in real life. Take the robot power supply. In the movies, robots almost always have some type of advanced nuclear drive or perhaps a space-age solar cell that can soak up the sun's energy, then slowly release it over two or three days. Nuclear power supplies are out of the question, except in some top-secret robotic experiment conducted by the Army. And solar cells don't provide enough power for the typical motorized robot, and as yet they have no power storage capabilities.

Most self-contained real-life robots are powered by batteries, the same kind of batteries used to provide juice to a flashlight, cassette radio, portable television, or other electrical device. Batteries are an integral part of robot design, as important as the frame, motor, and electronic brain—those components we most often think of when the discussion turns to robots. To robots, batteries are the elixir of life, and without them, robots cease to function.

While great strides have been made in electronics during the past 20 years—including entire computers that fit on a chip—battery technology is behind the times. On the whole, today's batteries don't pack much wallop for their size and weight, and the rechargeable ones take hours to come back to life. The high-tech batteries you may have heard about exist, but they are largely confined to the laboratories and a few high-priced applications, such as space or medical science. That leaves us with the old, run-of-the-mill batteries used in everyday applications.

With judicious planning and use, however, combined with some instruction in how to make do with the limitations, these common everyday batteries can provide more than adequate power to all of your robot creations.

Types of Batteries

There are seven main types of batteries, which come in a variety of shapes, sizes, and configurations.

ZINC

Zinc batteries are the staple of the battery industry and are often referred to simply as "flashlight" cells. The chemical makeup of zinc batteries takes two forms: carbon zinc and zinc chloride. Carbon zinc, or "regular-duty," batteries die out the quickest and are unsuited to robotic applications. Zinc chloride, or "heavy-duty," batteries provide a little more power than regular carbon zinc cells and last 25 to 50 percent longer. Despite the added energy, zinc chloride batteries are also unsuitable for most robotics applications.

Both carbon zinc and zinc chloride batteries can be "rejuvenated" a few times after being drained. See the section "Battery Recharging" later in the chapter for more information on recharging batteries. Zinc batteries are available in all the standard flashlight (D, C, A, AA, and AAA) and lantern battery sizes.

ALKALINE

Alkaline cells use a special alkaline manganese dioxide formula that lasts up to 800 percent longer than carbon zinc batteries. The actual increase in life expectancy ranges from about 300 percent to 800 percent, depending on the application. In robotics, where the batteries are driving motors, solenoids, and electronics, the average increase is a reasonable 450 to 550 percent.

Alkaline cells, which come in all the standard sizes (as well as 6- and 12-volt lantern cells), cost about twice as much as zinc batteries. But the increase in power and service life is worth the cost. Unlike zinc batteries, however, ordinary alkaline batteries cannot be rejuvenated by recharging without risking fire (though some people try it just the same). During recharging, alkaline batteries generate considerable internal heat, which can cause them to explode or catch fire. So when these batteries are dead, just throw them away.

Recently, however, a new rechargeable alkaline cell has reached the market. These provide many of the benefits of ordinary alkaline cells but with the added advantage of being rechargeable. A special low-current recharger is required (don't use the recharger on another battery type or you may damage the recharger or the batteries). While rechargeable alkalines cost more than ordinary alkaline cells, over time your savings from reusing the batteries can be considerable.

NICKEL-CADMIUM

When you think "rechargeable battery," you undoubtedly think nickel-cadmium—or "Ni-Cad" for short. Ni-Cads aren't the only battery specifically engineered to be recharged, but they are among the least expensive and easiest to get. Ni-Cads are ideal for most all robotics applications.

The Ni-Cad cells are available in all standard sizes, plus special-purpose "sub" sizes for use in sealed battery packs (as in rechargeable handheld vacuum cleaners, photoflash equipment, and so forth). Most sub-size batteries have solder tabs, so you can attach wires directly to the battery instead of placing the cells in a battery holder. Ni-Cads don't last nearly as long as zinc or alkaline batteries, but you can easily recharge them when they wear out. Most battery manufacturers claim their Ni-Cad cells last for 500 or more recharges.

A new, higher capacity Ni-Cad battery is available that offers two to three times the service life of regular Ni-Cads. More importantly, these high-capacity cells provide considerably more power and are ideally suited for robotics work. Of course, they cost more.

Note that Ni-Cads can suffer from "memory effect" whereby the useful capacity of the battery is reduced if the cell is not fully discharged before it is recharged. To be fair, newer Ni-Cad batteries don't exhibit this memory effect nearly as much as the older kind (and some Ni-Cad makers insist memory effect is no longer an issue). Read more about memory effect and other problems with Ni-Cad batteries in "Ni-Cad Disadvantages," later in this chapter.

NICKEL METAL HYDRIDE

Nickel metal hydride (NiMH) batteries represent one of the best of the affordable rechargeable battery technologies. NiMH batteries can be recharged 400 or more times and have a low internal resistance, so they can deliver high amounts of current (read more about internal resistance and current in "Battery Ratings," later in the chapter). Nickel metal hydride batteries are about the same size and weight as Ni-Cads, but they deliver about 50 percent more operating juice than Ni-Cads. In fact, NiMH batteries work best when they are used in very high current situations. Unlike Ni-Cads, NiMH batteries do not exhibit any memory effect, nor do they contain cadmium, a highly toxic material.

While NiMH batteries are discharging, especially at high currents, they can get quite hot. You should consider this when you place the batteries in your robot. If the NiMH pack will be pressed into high-current service, be sure it is located away from any components that may be affected by the heat. This includes any control circuitry or the microcontroller.

NiMH batteries should be recharged using a recharger specially built for them. According to NiMH battery makers, NiMH batteries should be charged at an aggressive rate. A by-product of this kind of high-current recharging is that NiMH can be recycled back into service more quickly than Ni-Cads. You can deplete your NiMH battery pack, put it under charge for an hour or two, and be back in business.

One disadvantage of NiMH relative to other rechargeable battery technologies is that the battery does not hold the charge well. That is, over time (weeks, even days) the charge in the battery is depleted, even while the battery is in storage. For this reason, it's always a good idea to put NiMH batteries on the recharger at regular intervals, even when they haven't been used. Because NiMHs don't exhibit a memory effect, however, this disadvantage will not cause a change in the discharge curve of the batteries.

LITHIUM AND LITHIUM-ION

Lithium and rechargeable lithium-ion batteries are popular in laptop computers. They are best used at a steady discharge rate and tend to be expensive. Lithium batteries of various types provide the highest "energy density" of most any other commercially available battery, and they retain their charge for months, even years. Like other rechargeable battery types, rechargeable lithium-ion batteries require their own special recharging circuitry, or overheating and even fire could result.

LEAD-ACID

The battery in your car is a lead-acid battery. It is made up of not much more than lead plates crammed in a container that's filled with an acid-based electrolyte. These brutes pack a wallop and have an admirable between-charge life. When the battery goes dead, recharge it, just like a Ni-Cad.

Not all lead-acid batteries are as big as the one in your car. You can also get—new or surplus—6-volt lead-acid batteries that are about the size of a small radio. The battery is sealed, so the acid doesn't spill out (most automotive batteries are now sealed as well). The sealing isn't complete though: during charging gases develop inside the battery and are vented out through very small pores. Without proper venting, the battery would be ruined after discharging and recharging. These batteries are often referred to as sealed lead-acid, or SLA.

Lead-acid batteries typically come in self-contained packs. Six-volt packs are the most common, but you can also get 12- and 24-volt packs. The packs are actually made by combining several smaller cells. The cells are wired together to provide the rated voltage of the entire pack. Each cell typically provides 2.0 volts, so three cells are required to make a 6-volt pack. You can, if you wish, take the pack apart, unsolder the cells, and use them separately.

Although lead-acid batteries are powerful, they are heavy. A single 6-volt pack can weigh four or five pounds. Lead-acid batteries are often used as a backup or emergency power supply for computers, lights, and telephone equipment. The cells are commonly available on the surplus market, and although used they still have many more years of productive life. The retail price of new lead-acid cells is about $25 for a 6-volt pack. Surplus prices are 50 to 80 percent lower.

Motorcycle batteries make good power cells for robots. They are easy to get, compact, and relatively lightweight. The batteries come in various amp-hour capacities (discussed later in this chapter), so you can choose the best one for your application. Motorcycle batteries are somewhat pricey, however. You can also use car batteries, as long as your robot is large and sturdy enough to support it. It's not unusual for a car battery to weigh 20 pounds.

Gelled electrolyte batteries (commonly called "gel-cell," after a popular trade name) use a special gelled electrolyte and are the most common form of SLA batteries. They are rechargeable and provide high current for a reasonable time, which makes them perfect for robots. Fig. 15.1 shows a typical sealed lead-acid battery.

FIGURE 15.1 A sealed lead-acid battery.

Battery Ratings

Batteries carry all sorts of ratings and specifications. The two most important specifications are per-cell voltage and amp-hour current. We will discuss these and others in the following sections.

VOLTAGE

The voltage rating of a battery is fairly straightforward. If the cell is rated for 1.5 volts, it puts out 1.5 volts, give or take. That "give or take" is more important than you may think because few batteries actually deliver their rated voltage throughout their life span. Most rechargeable batteries are recharged 20 to 30 percent higher than their specified rating. For example, the 12-volt battery in your car, a type of lead-acid battery, is charged to about 13.8 volts.

Standard zinc and alkaline flashlight batteries are rated at 1.5 volts per cell. Assuming you have a well-made battery in the first place, the voltage may actually be 1.65 volts when the cell is fresh, and dropping to 1.3 volts or less, at which point the battery is considered "dead." The circuit or motor you are powering with the battery must be able to operate sufficiently throughout this range.

Most batteries are considered dead when their power level reaches 80 percent of their rated voltage. That is, if the cell is rated at 6 volts, it's considered dead when it puts out only 4.8 volts. Some equipment may still function at levels below 80 percent, but the efficiency of the battery is greatly diminished. Below the 80 percent mark, the battery no longer provides the rated current (discussed later), and if it is the rechargeable type, the cell is likely to be damaged and unable to take a new charge. When experimenting with your robot systems, keep a volt-ohm meter handy and periodically test the output of the batteries. Perform the test while the battery is in use. The test results may be erroneous if you do not test the battery under load.

It is often helpful to know the battery's condition when the robot is in use. Using a volt-ohm meter to periodically test the robot's power plant is inconvenient. But you can build a number of "battery monitors" into your robot that will sense voltage level. The output of the monitor can be a light-emitting diode (LED), which will allow you to see the relative voltage level, or you can connect the output to a circuit that instructs the robot to seek a recharge or turn off. Several monitor circuits are discussed later in this chapter.

If your robot has an on-board computer, you want to avoid running out of juice midway through some task. Not only will you lose the operating program and have to rekey or reload it, but the robot may damage itself or its surroundings if the power to the computer is suddenly turned off.

CAPACITY

The capacity of a battery is rated as amp-hour current. This is the amount of power, in amps or milliamps, the battery can deliver over a specified period of time. The amp-hour current rating is a little like the current rating of an AC power line, but with a twist. AC power is considered to be never ending, available night and day, always in the same quantity. But a battery can only store so much energy before it poops out, so the useful service life must be taken into account. The current rating of a battery is at least as important as the voltage rating because a battery that can't provide enough juice won't be able to turn a motor or sufficiently power all the electronic junk you've stuck onto your robot.

What exactly does the term *amp-hour* mean? Basically, the battery will be able to provide the rated current for one hour before failing. If a battery has a rating of 5 amp-hours (expressed as "AH"), it can provide up to five amps continuously for one hour, one amp for five hours, and so forth, as shown in Fig. 15.2. So far, so good, but the amp-hour rating is not that simple. The 5 AH rating is actually taken at a 10- or 20-hour discharge interval. That is, the battery is used for 10 or 20 hours, at a low or medium discharge rate. After the specified time, the battery is tested to see how much juice it has left. The rating of the

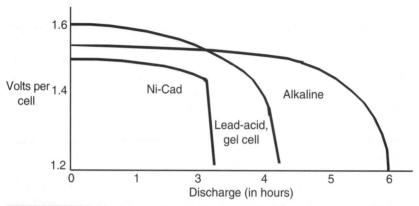

FIGURE 15.2 Representative discharge curves for several common battery types.

battery is then calculated taking the difference between the discharge rate and the reserve power and multiplying it by the number of hours under test.

What all this means is that it's an unusual battery that provides the stated amps in the one-hour period. The battery is much more likely to fail after 30 or 45 minutes of heavy-duty use and won't be able to supply the specified current for more than about 15 to 20 minutes. Discharging at or above the amp-hour rating may actually cause damage to the battery. This is especially true of Ni-Cad cells.

The lesson to be learned is that you should always choose a battery that has an amp-hour rating 20 to 40 percent more than what you need to power your robot. Figuring the desired capacity is nearly impossible until the entire robot is designed and built (or unless you are very good at computing current consumption). The best advice is to design the robot with the largest battery you think practical. If you find that the battery is way too large for the application, you can always swap it out for a smaller one. It's not so easy to do the reverse.

Note that some components in your robot may draw excessive current when they are first switched on, then settle down to a more reasonable level. Motors are a good example of this. A motor that draws one amp under load may actually require several amps at start up. The period is very brief, on the order of 100 to 200 milliseconds. No matter; the battery should be able to accommodate the surge. This means that the 20 to 40 percent overhead in using the larger battery is a necessity, not just a design suggestion. A rough comparison of the discharge curve at various discharge times is shown in Fig. 15.3.

RECHARGE RATE

Most (but not all) batteries are recharged slowly, over a 12- to 24-hour period. The battery can't take on too much current without breaking down and destroying itself, so the current from the battery charger must be kept at a safe level.

A good rule of thumb to follow when recharging any battery is to limit the recharging level to one tenth the amp-hour rating of the cell. For a 5 AH battery, then, a safe recharge level is 500 milliamps. Limiting current is extremely important when recharging Ni-Cads, which can be permanently damaged if charged too quickly. Lead-acid and gel-cell batteries can take an occasional "fast-charge"—a quickie at 25 to 50 percent of the rated amp-hour capacity of the battery. However, repeated quick-charging will warp the plates and disturb the electrolyte action in the battery, and is not recommended.

The recharge period, the number of hours the battery is recharged, varies depending on the type of cell. A recharge interval of 2 to 10 times the discharge rate is recommended. Fig. 15-4 shows a typical discharge/recharge curve for a lead-acid or gel-cell battery.

Most manufacturers specify the recharge time for their batteries. If no recharge time is specified, assume a three- or four-to-one discharge/recharge ratio and place the battery under charge for that period of time. Continue experimenting until you find an optimum recharge interval.

NOMINAL CELL VOLTAGE

Each battery type generates different nominal (normal, average) output voltages. The traditional rating for zinc, alkaline, and similar nonrechargeable batteries is 1.5 volts per cell. The actual voltage delivered by the cell can vary from a high of around 1.7 volts (fresh and fully charged) to around 1.2 or 1.3 volts ("dead"). Other battery types, most notably

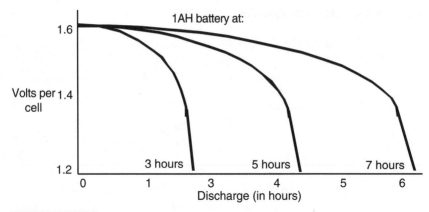

FIGURE 15.3 Discharge curves of a 1-AH battery at three-, five-, and seven-hour rates.

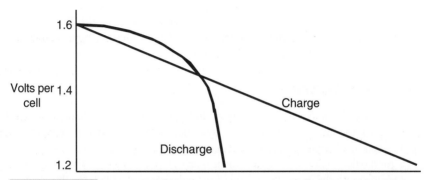

FIGURE 15.4 The charge/discharge curves of a typical rechargeable battery. Note that the charge time is longer than the discharge time.

Ni-Cads, provide different nominal cell voltages. Specifically, Ni-Cad and nickel metal hydride batteries provide 1.2 volts per cell, and lead-acid batteries provide 2.0 volts per cell.

To achieve higher voltages, you can link cells internally or externally (see the next section, "Increasing Battery Ratings," for more information). By internally linking together six 1.5-volt cells, for example, the battery will output 9 volts.

Nominal cell voltage is important when you are designing the battery power supplies for your robots. If you are using 1.5-volt cells, a four-cell battery pack will nominally deliver 6 volts, an eight-cell pack will nominally deliver 12 volts, and so forth. Conversely, if you are using 1.2-volt cells, a four-cell battery pack will nominally deliver 4.8 volts and an eight-cell pack will nominally deliver 9.6 volts. The lower voltage will have an effect on various robotic subsystems. For example, many microcontrollers used with robots (see Part 5 of this book) are made to operate at 5 volts and will reset—restart their programming—at 4.5 volts. A battery pack that delivers only 4.8 volts will likely cause problems with the microcontroller. You either need to add more cells or change the battery type to a kind that provides a higher per-cell voltage.

INTERNAL RESISTANCE

The internal resistance of a battery determines the maximum rate at which power can be drawn from the cells. A lower internal resistance means more power can be drawn out in less time. Lead-acid and nickel metal hydride cells are good examples of batteries that have a very low internal resistance.

When comparing batteries, you don't really need to know the actual internal resistance of the cells you use. Rather, you'll be more concerned with the discharge curve and the maximum amp-hour ratings of the battery. Still, knowing that the battery's internal resistance dictates the discharge curve and capacity of the battery will help you to design power packs for your robots.

Increasing the Battery Ratings

You can obtain higher voltages and current by connecting several cells together, as shown in Fig. 15.5. There are two basic approaches:

- To increase voltage, connect the batteries in series. The resultant voltage is the sum of the voltage outputs of all the cells combined.
- To increase current, connect the batteries in parallel. The resultant current is the sum of the current capacities of all the cells combined.

Take note that when you connect cells together not all cells may be discharged or recharged at the same rate. This is particularly true if you combine two half-used batteries with two new ones. The new ones will do the lion's share of the work and won't last as long. Therefore, you should always replace or recharge all the cells at once. Similarly, if one or more of the cells in a battery pack is permanently damaged and can't deliver or take on a charge like the others, you should replace it.

Battery Recharging

Most lead-acid and gel-cell batteries can be recharged using a 200- to 800-mA battery charger. The charger can even be a DC adapter for a video game or other electronics. Standard Ni-Cad batteries can't withstand recharge rates exceeding 50 to 100 mA, and if you use a charger that supplies too much current you will destroy the cell. Use only a battery charger designed for Ni-Cads. High-capacity Ni-Cad batteries can be charged at higher rates, and there are rechargers designed specially for them.

Nickel metal hydride, rechargeable alkalines, and rechargeable lithium-ion batteries all require special rechargers. Avoid substituting the wrong charger for the battery type you are using, or you run the risk of damaging the charger and/or the battery (and perhaps causing a fire).

You can rejuvenate zinc batteries by placing them in a recharger for a few hours. The process is not true recharging since the battery is not restored to its original power or

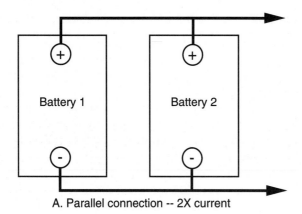

A. Parallel connection -- 2X current

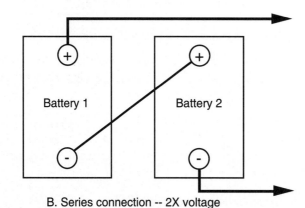

B. Series connection -- 2X voltage

FIGURE 15.5 **Wiring batteries to increase ratings.** *a.* **Parallel connection increases current;** *b.* **Series connection increases voltage.**

voltage level. The rejuvenated battery lasts about 20 to 30 percent as long as it did during its initial use. Most well-built zinc batteries can be rejuvenated two or three times before they are completely depleted.

Rechargeable batteries should be periodically recharged whether they need it or not. Batteries not in regular use should be recharged every two to four months, more frequently for NiMH batteries. Always observe polarity when recharging batteries. Inserting the cells backward in the recharger will destroy the batteries and possibly damage the recharger.

You can purchase ready-made battery chargers for the kind of battery you are using or build your own. The task of building your own is fairly easy because several manufacturers make specialized integrated circuits just for recharging batteries. These ICs provide all the necessary voltage and current protective mechanisms to ensure that the battery is properly charged. For example, you can use the Unitrode UC/2906 and UC/3906 from Texas Instruments to build an affordable charger for sealed lead-acid and gelled electrolyte batteries. Similarly, the MAX712 from Maxim lets you construct a flexible fast recharger for NiMH batteries. These and other specialty ICs are not always widely available, so you may need to check several sources before you find them. However, the search is well worth the time because of the cost and construction advantages these chips can provide.

Ni-Cad Disadvantages

Despite their numerous advantages, Ni-Cad batteries have a few peculiarities you'll want to consider when designing your robot power system. The most annoying problem is the "memory effect" we discussed earlier in this chapter. Not all battery experts agree that Ni-Cads still suffer from this problem, but most anyone who has tried to use Ni-Cads has experienced it in one form or another. For various reasons we won't get into, the discharge curve of Ni-Cad batteries is sometimes altered. The net effect is that the battery won't last as long on a full charge as it should. This so-called memory effect can be altered in two ways:

- *The dangerous way.* Short the battery until it's dead. Recharge it as usual. Some batteries may be permanently damaged by this technique.
- *The safe way.* Use the battery in a low-current circuit, like a flashlight, until it is dead. Recharge the battery as usual. You must repeat this process a few times until the memory effect is gone.

The best way to combat memory effect is to avoid it in the first place. Always fully discharge Ni-Cad batteries before charging them. If you don't have a flashlight handy, build yourself a discharge circuit using a battery holder and a flashlight bulb. The bulb acts as a "discharge" indicator. When it goes out, the batteries are fully discharged.

The other disadvantage is that the polarity of Ni-Cads can change—positive becomes negative and vice versa—under certain circumstances. Polarity reversal is common if the battery is left discharged for too long or if it is discharged below 75 or 80 percent capacity. Excessive discharging can occur if one or more cells in a battery pack wears out. The adjacent cells must work overtime to compensate, and discharge themselves too fast and too far.

You can test for polarity reversal by hooking the battery to a volt-ohm meter (remove it from the pack if necessary). If you get a negative reading when the leads are connected properly, the polarity of the cell is reversed. You can sometimes correct polarity reversal by fully charging the battery (connecting it in the recharger in reverse), then shorting it out. Repeat the process a couple of times if necessary. There is about a fifty-fifty chance that the battery will survive this. The alternative is to throw the battery out, so you actually stand to lose very little.

Recharging the Robot

You'll probably want to recharge the batteries while they are inside the robot. This is no problem as long as you install a connector for the charger terminals on the outside of the robot. When the robot is ready for a charge, connect it to the charger.

Ideally, the robot should be turned off during the charge period, or the batteries may never recharge. However, turning off the robot during recharging may not be desirable, as this will end any program currently running in the robot. There are several schemes you can employ that will continue to supply current to the electronics of the robot yet allow the batteries to charge. One way is to use a relay switchout. In this system, the external power plug on your robot consists of four terminals: two for the battery and two for the electronics. When the recharger is plugged in, the batteries are disconnected from the robot. You can use relays to control the changeover or heavy-duty open-circuit jacks and plugs (the ones for audio applications may work). While the batteries are switched out and being recharged, a separate power supply provides operating juice to the robot.

Battery Care

Batteries are rather sturdy little creatures, but you should follow some simple guidelines when using them. You'll find that your batteries will last much longer, and you'll save yourself some money.

■ Store new batteries in the fresh food compartment of your refrigerator (not the freezer). Put them in a plastic bag so if they leak they won't contaminate the food. Remove them from the refrigerator for several hours before using them.

■ Avoid using or storing batteries in temperatures above 75°F or 80°F. The life of the battery will be severely shortened otherwise. Using a battery above 100°F to 125°F causes rapid deterioration.

■ Unless you're repairing a misbehaving Ni-Cad, avoid shorting out the terminals of the battery. Besides possibly igniting fumes exhausted by the battery, the sudden and intense current output shortens the life of the cell.

■ Keep rechargeable batteries charged. Make a note when the battery was last charged.

■ Fully discharge Ni-Cads before charging them again. This prevents memory effect. Other rechargeable battery types (nickel metal hydride, rechargeable alkaline, lead-acid, etc.) don't exhibit a memory effect and can be recharged at your convenience.

■ Given the right circumstances all batteries will leak, even the "sealed" variety. When they are not in use, keep batteries in a safe place where leaked electrolyte will not cause damage. Remove batteries from their holder when they are not being used.

Power Distribution

Now that you know about batteries, you can start using them in your robot designs. The most simple and straightforward arrangement is to use a commercial-made battery holder. Holders are available that contain from two to eight AA, C, or D batteries. The wiring in these holders connects the batteries in series, so a four-cell holder puts out 6 volts (1.5 times 4). You attach the leads of the holder (red for positive and black for ground or negative) to the main power supply rail in your robot. If you are using a gel-cell or lead-acid battery you would follow a similar procedure.

FUSE PROTECTION

Flashlight batteries don't deliver extraordinary current, so fuse protection is not required on the most basic robot designs. Gel-cell, lead-acid, and high-capacity Ni-Cad batteries

can deliver a most shocking amount of current. In fact, if the leads of the battery accidentally touch each other or there is a short in the circuit the wires may melt and a fire could
erupt.

Fuse protection helps eliminate the calamity of a short circuit or power overload in your
robot. As illustrated in Fig. 15.6, connect the fuse in line with the positive rail of the
battery, as near to the battery as possible. You can purchase fuse holders that connect
directly to the wire or that mount on a panel or printed circuit board.

Choosing the right value of fuse can be a little tricky, but it is not impossible. It does
require that you know how much current your robot draws from the battery during normal
and stalled motor operation. You can determine the value of the fuse by adding up the
current draw of each separate subsystem, then tack on 20 to 25 percent overhead.

Let's say that the two drive motors in the robot draw 2 amps each, the main circuit
board draws 1 amp, and the other small motors draw 0.5 amp each (for a total of, perhaps,
2 amps). Add all these up and you get 7 amps. Installing a fuse with a rating of at least 7
amps at 125 volts will help assure that the fuse won't burn out prematurely during normal
operation. Adding that 20 to 25 percent margin calls for an 8- to 10-amp fuse.

Recall from earlier in this chapter that motors draw excessive current when they are
first started. You can still use that 8- to 10-amp fuse, but make sure it is the slow-blow type.
Otherwise, the fuse will burn out every time one of the heavy-duty motors kick in.

Fuses don't come in every conceivable size. For the sake of standardization, choose the
regular 1 1/4-inch-long-by-1/4-inch-diameter bus fuses. You'll have an easier job finding
fuse holders for them and a greater selection of values. Even with a standard fuse size,
there is not much to choose from past 8 amps, other than 10, 15, and 20 amps. For values
over 8 amps, you may have to go with ceramic fuses, which are used mainly for microwave
ovens and kitchen appliances.

MULTIPLE VOLTAGE REQUIREMENTS

Some advanced robot designs require several voltages if they are to operate properly. The
drive motors may require 12 volts, at perhaps two to four amps, whereas the electronics
require +5, and perhaps even −5 volts. Multiple voltages can be handled in several ways.
The easiest and most straightforward is to use a different set of batteries for each main

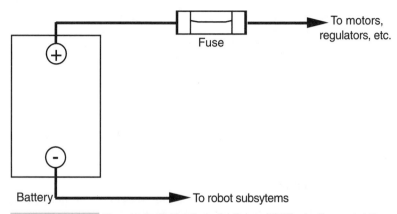

FIGURE 15.6 How to install a fuse in line with the battery and the
robot electronics or motor.

subsection. The motors operate off one set of large lead-acid or gel-cell batteries; the electronics are driven by smaller capacity Ni-Cads.

This approach is actually desirable when the motors used in the robot draw a lot of current. Motors naturally distribute a lot of electrical noise throughout the power lines, noise that electronic circuitry is extremely sensitive to. The electrical isolation that is provided when you use different batteries nearly eliminates problems caused by noise (the remainder of the noise problems occur when the motor commutators arc, causing RF interference). In addition, when the motors are first started the excessive current draw from the motors may zap all the juice from the electronics. This "sag" can cause failed or erratic behavior, and it could cause your robot to lose control.

The other approach to handling multiple voltages is to use one main battery source and "step" it down (sometimes up) so it can be used with the various components in the system. This is called DC-DC conversion, and you can accomplish it by using circuits of your own design or by purchasing specialty integrated circuit chips that make the job easier. One 12-volt battery can be regulated (see "Voltage Regulation" later in this chapter) to just about any voltage under 12 volts. The battery can directly drive the 12-volt motors and, with proper regulation, supply the +5-volt power to the circuit boards.

Connecting the batteries judiciously can also yield multiple voltage outputs. By connecting two 6-volt batteries in series, as shown in Fig. 15.7, you get +12 volts, +6 volts, and −6 volts. This system isn't nearly as foolproof as it seems, however. More than likely, the two batteries will not be discharged at the same rate. This causes extra current to be drawn from one to the other, and the batteries may not last as long as they might otherwise.

If all of the subsystems in your robot use the same batteries, be sure to add sufficient filtering capacitors across the positive and negative power rails. The capacitors help soak up excessive current spikes and noise, which are most often contributed by motors. Place the capacitors as near to the batteries and the noise source as possible. Exact values are not critical, but they should be over 100 μF—even better is 1000 to 3000 μF. Be certain the capacitors you use are rated at the proper voltage (25 to 35 volts is fine). Using an underrated capacitor will burn it out and possibly cause a short circuit.

You should place smaller value capacitors, such as 0.1 μF, across the positive and negative power rails wherever power enters or exits a circuit board. As a general rule, you

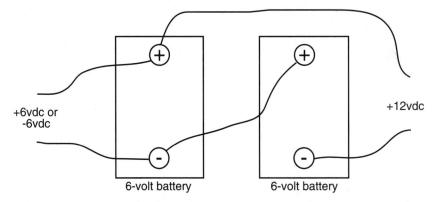

FIGURE 15.7 Various voltage tap-offs from two 6-volt batteries. This is not an ideal approach (the batteries will discharge at different rates), but it works in a pinch.

should add these "decoupling" capacitors beside clocked logic ICs, particularly flip-flops and counters. A few linear ICs, such as the 555 timer, need decoupling capacitors, or the noise they generate through the power lines can ripple through to other circuits. If many ICs are on the board, you can usually get by with adding one 0.1 µF decoupling capacitor for every three or four chips.

SEPARATE BATTERY SUPPLIES

Most hobby robots now contain computer-based control electronics of some type. The computer requires a specific voltage (called regulation, discussed in the next section), and it expects the voltage to be "clean" and free of noise and other glitches. A common problem in robotic systems is that the motors cause so-called sags and noise in the power supply system, which can affect the operation of the control electronics. You can largely remedy this by using separate battery supplies for the motors and the electronics. Simply join the ground connection for the supplies together.

With this setup, the motors have one unregulated power supply, and the control electronics have their own regulated power supply. Even if the motors turn on and off very rapidly this approach will minimize sags and noise on the electronics side. It's not always possible to have separate battery supplies, of course. In these cases, use the capacitor filtering techniques described in the earlier "Multiple Voltage Requirements" section. The large capacitors that are needed to achieve good filtering between the electronics and motor sections will increase the size of your robot. A 2200 µF capacitor, for example, may measure 3/4 inch in diameter by over an inch in height. You should plan for this in your design.

Voltage Regulation

Many types of electronic circuits require a precise voltage or they may be damaged or act erratically. Generally, you provide voltage regulation only to those components and circuit boards in your robot that require it. It is impractical to regulate the voltage for the entire robot as it exits the battery. You can easily add solid-state voltage regulators to all your electronic circuits. They are easy to obtain, and you can choose from among several styles and output capacities. Two of the most popular voltage regulators, the 7805 and 7812, provide +5 volts and +12 volts, respectively. You connect them to the "+" and "−" (ground) rails of your robot, as shown in Fig. 15.8 (refer to the parts list in Table 15.1).

Other 7800 series power regulators are designed for +15, +18, +20, and +24 volts. The 7900 series provide negative power supply voltages in similar increments. The current capacity of the 7800 and 7900 series that come in the TO-220 style transistor packages (these can often be identified as they have no suffix or use a "T" suffix in their part number), is limited to less than one amp. As a result, you must use them in circuits that do not draw in excess of this amount.

Other regulators are available in a more traditional TO-3-style transistor package ("K" suffix) that offers current output to several amps. The "L" series regulators come in the small TO-92 transistor packages and are designed for applications that require less than about 500 mA. Other regulators of interest:

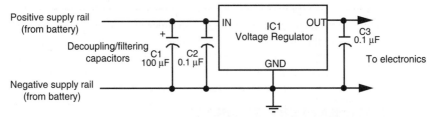

FIGURE 15.8 Three-terminal linear voltage regulators, like the 7805, can be used to provide stable voltages for battery-powered robots. The capacitors help filter (smooth out) the voltage.

TABLE 15.1 PARTS LIST FOR +5-VOLT BATTERY REGULATOR.

IC1	7805 linear voltage regulator
C1	100 μF electrolytic capacitor
C2, C3	0.1 μF tantalum capacitor

- The 328K provides an adjustable output to 5 volts, with a maximum current of 5A (amperes).
- The 78H05K offers a 5-volt output at 5A.
- The 78H12K offers a 12-volt output at 5A.
- The 78P05K delivers 5 volts at 10 amps.

SWITCHING VOLTAGE REGULATION

All of the regulators described in the last section are the *linear* variety. They basically take an incoming voltage and clamp it to some specific value. Linear regulation isn't very efficient; a lot of energy is wasted in heat from the regulator. This inefficiency is particularly notable in battery-powered systems, where the current capacity and the battery life are limited.

An alternative to linear regulators is to use a *switching* (or switching-mode) voltage regulator, which exhibits better efficiencies. Most high-tech electronics equipment now use switching power supplies, especially since single-IC switching voltage regulators are now so common and inexpensive. Maxim, Texas Instruments, Dallas Semiconductor, and many other companies are actively involved in the design and sale of switching voltage regulators. See Appendix B, "Sources," and Appendix C, "Robot Information on the Internet," for more information on these and other companies offering power supply ICs and circuits.

A good example of a switching voltage regulator is the MAX638, from Maxim. With just a few added parts (a typical circuit, taken from the MAX638's data sheet, is shown in Fig. 15.9; refer to the parts list in Table 15.2), you can build a simple, compact, inexpensive, and efficient voltage regulator. The chip can also be used as a low-battery detector. See "Battery Monitors" later in this chapter for more information on low-battery detection.

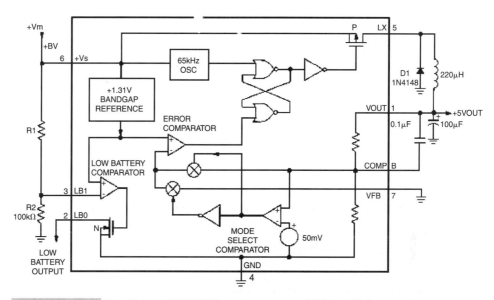

FIGURE 15.9 The Maxim MAX638 is among several high-efficiency voltage regulators available. The MAX638 is most commonly used to provide regulated +5 volts, but it can also be adjusted using external components to provide other voltages.

TABLE 15.2 PARTS LIST FOR MAXIM MAX638 SWITCHING POWER SUPPLY.	
IC1	MAX 638 (Maxim)
R1	120K resistor
R2	47K–100K resistor
C1	0.1 μF ceramic capacitor
C2	100 μF electrolytic capacitor
D1	1N4148 diode
L1	220 μH inductor

All resistors have 5 or 10 percent tolerance, 1/4-watt.

ZENER VOLTAGE REGULATION

A quick and inexpensive method for providing a semiregulated voltage is to use zener diodes, as shown in Fig. 15.10. With a zener diode, current does not begin to flow through the device until the voltage exceeds a certain level (called the breakdown voltage). Voltage over this level is then "shunted" through the zener diode, effectively limiting the voltage to the rest of the circuit. Zener diodes are available in a variety of voltages, such as 3.3 volts, 5.1 volts, 6.2 volts, and others.

Zener diodes are also rated by their tolerance (1 percent and 5 percent are common) and their power rating, in watts. For low-current applications, a 0.3- or 0.5-watt zener should be sufficient; higher currents require larger 1-, 5-, and even 10-watt zeners. Note the

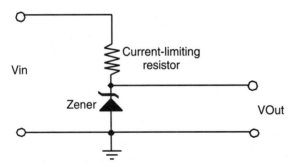

FIGURE 15.10 A zener diode and resistor can make a simple and inexpensive voltage regulator. Be sure to select the proper wattage for the zener and the proper wattage and resistance for the resistor.

resistor R1 in the schematic shown in Fig. 15.10. This resistor limits the current through the zener, and its value (and wattage) is determined by the current draw from the load, as well as the input and output voltages.

POWER DISTRIBUTION

You may choose to place all or most of your robot's electronic components on a single board. You can mount the regulator(s) directly on the board. You can also have several smaller boards share one regulator as long as the boards together don't pull power in excess of what the regulator can supply. Fig. 15.11 shows how to distribute the power from a single battery source to many separate circuit boards. The individual regulators provide power for one or two large boards or to a half dozen or so smaller ones.

Voltage regulators are great devices, but they are somewhat "wasteful." To work properly, the regulator must be provided with several volts more than the desired output voltage. For example, the 7812 +12-volt regulator needs 13 to 15 volts to deliver the full voltage and current specified for the device. Well-regulated 12-volt robotic systems may require you to use an 18-volt supply.

Voltage Double and Inverters

If your robot is equipped with 12-volt motors and uses circuitry that requires only +5 and/or +12 volts, then your work is made easy for you. But if you require negative supply voltages for some of the circuits, you're faced with a design dilemma. Do you add more batteries to provide the negative supply? That's one solution, and it may be the only one available to you if the current demand of the circuits is moderate to high.

Another approach is to use a polarity-reversal circuit, such as the one in Fig. 15.12 (refer to the parts list in Table 15.3). The current at the output is limited to less than 200 mA, but this is often enough for devices like op amps, CMOS analog switches, and other small devices that require a −5 or −12 vdc voltage. The negative output voltage is proportional to the positive output voltage. So, +12 volts in means roughly −12 volts out.

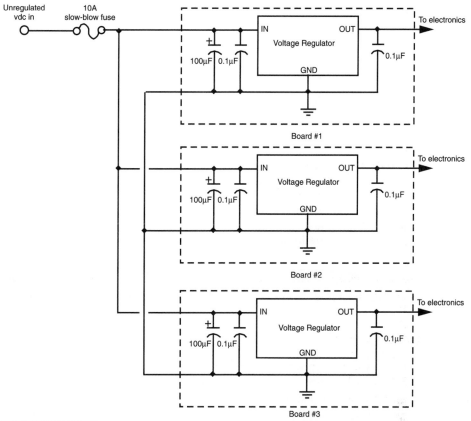

FIGURE 15.11 Parallel connection of circuit boards from a single power source. Each board has its own voltage regulator.

Some computer components require voltages above the normal supply. An example are the "high-side" power MOSFET transistors in an H-bridge driver circuit. You can use the circuit shown in Fig. 15.13 (refer to the parts list in Table 15.4) to double the supply voltage to provide the desired voltage. The current level at the output is very low, but it should be enough to be used as a high-voltage pulse.

The circuits described in the previous sections of this chapter are quickly and inexpensively built, but because of their simplicity they can leave a lot to be desired. If you need more precise power supplies, the power management ICs from Maxim, Texas Instruments, and others are probably the better bet. See Appendix B, "Sources," for more information.

Battery Monitors

Quick! What's the condition of the battery in your robot? With a battery monitor, you'd know in a flash. A battery monitor continually samples the output voltage of the battery during operation of the robot (the best time to test the battery) and provides a visual or logic output. In this section we'll profile the most common types.

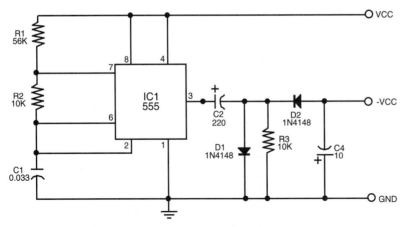

FIGURE 15.12 A polarity inverter circuit. Current output is low but enough for a few op amps.

TABLE 15.3	PARTS LIST FOR POLARITY INVERTER.
IC1	555 timer IC
R1	56K resistor
R2	10K resistor
R3	10K resistor
C1	0.033 µF ceramic capacitor
C2	220 µF electrolytic capacitor
C3	10 µF electrolytic capacitor
D1,D2	1N4148 diode

All resistors have 5 or 10 percent tolerance, 1/4-watt; all capacitors have 10 percent tolerance.

4.3-VOLT ZENER BATTERY MONITOR

Figure 15.14 (refer to the parts list in Table 15.5) shows a simple battery monitor using a 4.3-volt quarter-watt zener diode. R1 sets the trip point. When in operation, the LED winks off when the voltage drops below the setpoint. To use the monitor, set R1 (which should be a precision potentiometer, 1 or 3 turn) when the batteries to your robot are low. Adjust the pot carefully until the LED just winks off. Recharge the batteries. The LED should now light. Another, more "scientific" way to adjust R1 is to power the circuit using an adjustable power supply. While watching the voltage output on a meter, set the voltage at the trip point (e.g., for a 12-volt robot, set it to about 10 volts).

ZENER/COMPARATOR BATTERY MONITOR

A microprocessor-compatible battery monitor is shown in Fig. 15.15 (refer to the parts list in Table 15.6). This monitor uses a 5.1-volt quarter-watt zener as a voltage reference for a

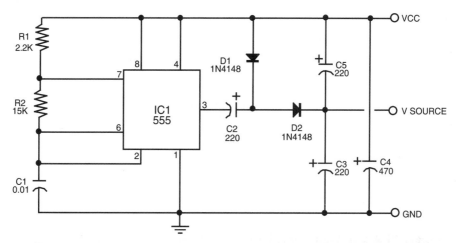

FIGURE 15.13 A voltage doubler. The voltage at the V Source terminal is roughly double the voltage applied to the Vcc terminal.

TABLE 15.4	PARTS LIST FOR VOLTAGE DOUBLER
IC1	555 timer IC
R1	2.2K resistor
R2	15K resistor
C1	0.01 µF ceramic capacitor
C2,C3	220 µF electrolytic capacitor
C4	470 µF electrolytic capacitor
C5	220 µF electrolytic capacitor
D1,D2	1N4148 diode

All resistors have 5 or 10 percent tolerance, 1/4-watt; all capacitors have 10 percent tolerance.

339 quad comparator IC. Only one of the comparator circuits in the IC is used; you are free to use any of the remaining three for other applications. The circuit is set to trip when the voltage sags below the (approximate) 5-volt threshold of the zener (in my test circuit the comparator tripped when the supply voltage dipped to under 4.5 volts). When this happens, the output of the comparator immediately drops to 0 volts. One advantage of this circuit is that the voltage drop at the output of the comparator is fairly steep (see Fig. 15.16).

USING A BATTERY MONITOR WITH A MICROPROCESSOR

You can usually connect battery monitors to a microprocessor or microcontroller input. When in operation, the microprocessor is signaled by the interrupt when the LED is triggered. Software running on the computer interprets the interrupt as "low battery; quick get a recharge." The robot can then place itself into nest mode, where it seeks out its own battery charger. If the charger terminals are constructed properly, it's possible for the robot to plug itself in. Fig. 15.17 shows a simplified flow chart illustrating how this kind of behavior might work.

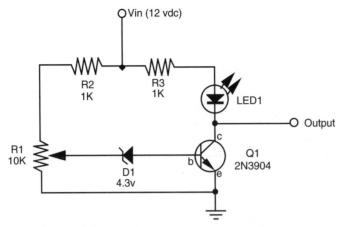

FIGURE 15.14 Battery monitor using 4.3-volt zener diode. This circuit is designed to be used with a 12-volt battery.

TABLE 15.5 PARTS LIST FOR 4.3-VOLT ZENER BATTERY MONITOR.	
R1	10K potentiometer
R2,R3	1K resistor
D1	4.3-volt zener diode (1/4-watt)
Q1	2N3904 NPN transistor
LED1	Light-emitting diode

All resistors have 5 or 10 percent tolerance, 1/4-watt.

Build a Robot Testing Power Supply

Using a battery while testing or experimenting with new robot designs is both inconvenient and counter productive. Just when you get a circuit perfected, the battery goes dead and must be recharged. A stand-alone power supply, which operates off of 117 VAC house current, can supply your robot designs with regulated DC power, without requiring you to install, replace, or recharge batteries. You can buy a ready-made power supply (they are common in the surplus market) or make your own.

One easy-to-use and inexpensive power supply is the DC wall transformer, or "wall-wart." Wall-warts convert AC power into low-voltage DC, usually 6–18 volts (note: some wall-warts only reduce the voltage, but do not convert it from AC to DC—don't use these!). Wall-warts have no voltage regulation, and most cannot provide more than a few hundred milliamps of current. Use them only when you don't need regulation (or when it is provided elsewhere in your circuit) and for nondemanding current applications.

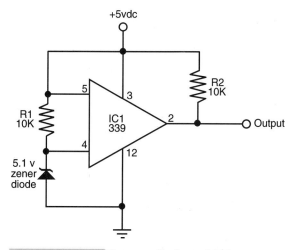

FIGURE 15.15 A zener diode and 339 comparator can be used to construct a fairly accurate 5-volt battery monitor.

TABLE 15.6 PARTS LIST FOR 339 COMPARATOR BATTERY MONITOR.

IC1	339 comparator IC
R1,R2	10K resistor
D1	5.1-volt zener diode (1/4- or 1/2-watt)

All resistors have 5 or 10 percent tolerance, 1/4-watt.

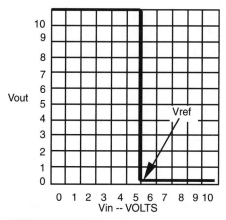

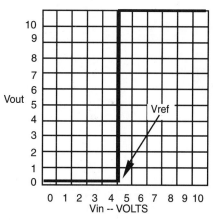

FIGURE 15.16 The output of a 339 comparator has a sharp cutoff as the voltage goes above or below the setpoint. The voltages shown here are representative only.

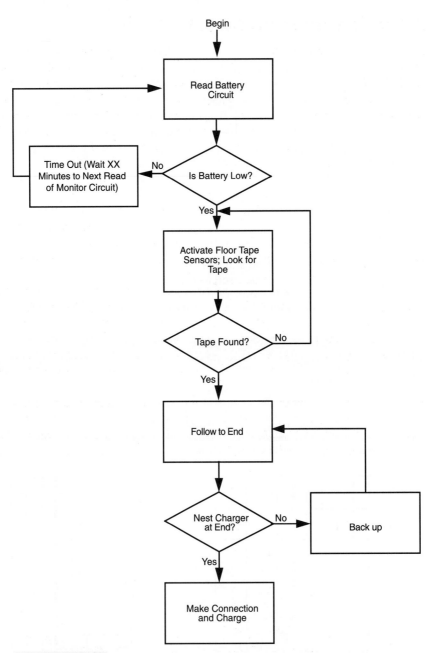

FIGURE 15.17 Software can be used to command the robot to return to its battery recharger nest should the battery exceed a certain low point.

BUILDING THE POWER SUPPLY

Refer to Fig. 15.18 for the schematic of the 5-volt regulated power supply (refer to the parts list in Table 15.7). For safety reasons, you must enclose the power supply in a plastic or metal chassis (plastic is better because there is less chance of a short circuit). Use a perforated board to secure the components, and solder them together using 18- or 16-gauge insulated wire. Alternatively, you can make your own circuit board using a home etching kit. Before constructing the board, collect all the parts and design the board to fit the specific parts you have. There is little size standardization when it comes to power supply components and large value electrolytic capacitors, so presizing is a must.

HOW THE CIRCUIT WORKS

Here's how the circuit works. The incoming AC is routed to the AC terminals of the transformer. The "hot" side of the AC is connected through a 2-amp slow-blow fuse and a single-pole, single-throw (SPST) toggle switch. With the switch in the off (open) position, the transformer receives no power so the supply is off.

The 117 VAC is stepped down to the secondary voltage of the transformer (12 to 18 volts, depending on the exact voltage of the transformer you use). The transformer specified here is rated at 2 amps, which is sufficient for the task at hand. Remember that the power supply is limited to delivering the capacity of the transformer (and later the voltage regulator), no more. A bridge rectifier, BR1 (shown schematically in the box in Fig. 15.18), converts the AC to DC. You can also construct the rectifier using discrete diodes, and connect them as shown within the dotted box.

When using the bridge rectifier, be sure to connect the leads to the proper terminals. The two terminals marked with a "~" connect to the transformer. The "+" and "−" terminals are the output and must connect as shown in the schematic in Fig. 15.18. Use a 5-volt, 1-amp regulator—a 7805—to maintain the voltage output at a steady 5 volts.

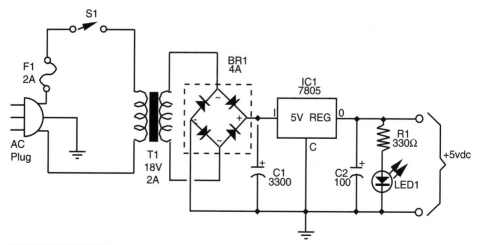

FIGURE 15.18 Five-volt regulated power supply using 117 VAC input.

TABLE 15.7 PARTS LIST FOR 5-VOLT POWER SUPPLY	
IC1	LM7805 +5 vdc voltage regulator
BR1	4 amp bridge rectifier
T1	12 to 18 volt, 2 amp AC transformer
R1	330 ohm resistor
C1	3000 µF electrolytic capacitor, 35 volt min.
C2	100 µF electrolytic capacitor, 35 volt min.
LED1	Light-emitting diode
F1	2 amp slow—blow fuse
S1	SPST toggle switch
Misc	Fuse holder, heat sink for U1, binding posts, AC cord with plug, chassis

All resistors have 5 or 10 percent tolerance, 1/4-watt; all capacitors have 10 percent tolerance.

Note that the transformer supplies a great deal more voltage than is needed. There are two reasons for this. First, lower-voltage 6.3- or 9-volt transformers are available, but most do not deliver more than 0.5 amps. It is far easier to find 12- or 18-volt transformers that deliver sufficient power. Second, to operate properly the regulator requires a few extra volts as "overhead."

The 7805 regulator comes in different styles. The 7805T, which comes in a TO-220-style transistor case, delivers up to 1 amp of current. You can increase the current rating, as long as the transformer will comply, with a 7805K regulator. This regulator comes in a TO-3-style transistor case, and mounted on a heat sink it can deliver in excess of 3 amps. Unless you plan to use the power supply to test only small circuits, why not opt for the larger-capacity regulator. You never know when the extra current will come in handy.

Capacitors C1 and C2 filter the ripple that is inherent in the rectified DC found at the outputs of the bridge rectifier. With the capacitors installed as shown in the schematic (note the polarity), the ripple at the output of the power supply is negligible. LED1 and R1 form a simple indicator. The LED will glow when the power supply is on. Remember the 220-ohm resistor. The LED will burn up without it.

The output terminals are insulated binding posts. Don't leave the output wires bare. The wires may accidentally touch one another and short the supply. Solder the output wires to the lug on the binding posts, and attach the posts to the front of the power supply chassis. The posts accept bare wires, alligator clips, even banana plugs.

REMEMBER: SAFETY FIRST!

This circuit is the only one in the book that requires you to work with 117 VAC current. Be extra careful when wiring the power supply circuits and triple-check your work. Never operate the supply when the top of the cabinet is off, unless you are testing it. Even then, stay away from the incoming AC. Touching it can cause a serious shock, and depending on the circumstances it could kill you. Be extra certain that no wires touch the chassis or front panel. Use an all-plastic enclosure whenever possible.

Do not operate the power supply if it has gotten wet, has been dropped, or shows signs of visible damage. Fix any problems *before* plugging it in. Do not assemble the supply without a fuse, and don't use a fuse rated much higher than 2 amps. Defeating the fuse protection diminishes or eliminates the only true safety net in the circuit.

From Here

To learn more about...	Read
Understanding motor current ratings	Chapter 17, "Choosing the Right Motor for the Job"
Other power systems (e.g., hydraulic)	Chapter 24, "An Overview of Arm Systems"
Power requirements for computers and microcontrollers	Chapter 28, "An Overview of Robot 'Brains'"

16

ROBOT LOCOMOTION PRINCIPLES

As you graduate to building larger mobile robots, you should consider the physical properties of your creations, including their size, weight, and mode of transport. A robot that is too heavy for its frame, or a locomotion mechanism that doesn't provide sufficient stability, will greatly hinder the usefulness of your mechanical invention.

In this chapter you'll find a collection of assorted tips, suggestions, and caveats for designing the locomotion systems for your robots. Because the locomotion system is intimately related to the frame of the robot, we'll cover frames a little bit as well, including their weight and weight distribution. Of course, there's more to the art and science of robot locomotion than we can possibly cover here, but what follows will serve as a good introduction.

First Things First: Weight

"He ain't heavy; he's my robot...." Okay, so the old song lyrics don't *quite* apply to robots, but the sentiment has some merit. Most personal robots weight under 20 pounds, and a high percentage of those weigh under 10 pounds. Weight is one of the most important factors affecting the mobility of a robot. A heavy robot requires larger motors and higher-capacity batteries—both of which add even more pounds to the machine. At some point, the robot becomes too heavy to even move.

On the other hand, robots designed for heavy-duty work often need some girth and weight. Your own design may call for a robot that needs to weigh a particular amount in order for it to do the work you have envisioned. The parts of a robot that contribute the most to its weight are the following, in (typical) descending order:

- Batteries
- Drive motors
- Frame

A 12-volt battery pack can weigh one pound; larger-capacity sealed lead-acid batteries can weigh five to eight pounds. Heavier-duty motors will be needed to move that battery ballast. But bigger and stronger motors weigh more because they must be made of metal and use heavier-duty bushings. And they cost more. Suddenly, your "little robot" is not so little anymore; it has become overweight and expensive.

TIPS FOR REDUCING WEIGHT

If you find that your robot is becoming too heavy, consider putting it on a diet, starting with the batteries. Nickel-cadmium and nickel metal hydride batteries weigh less, volt for volt, than their lead-acid counterparts. While nickel-cadmium and nickel metal hydride batteries may not deliver the amp-hour capacity that a large sealed lead-acid battery will, your robot will weigh less and therefore may not require the same stringent battery ratings as you had originally thought.

If your robot must use a lead-acid battery, consider carefully whether you truly need the capacity of the battery or batteries you have chosen. You may be able to install a smaller battery with a lower amp-hour rating. The battery will weigh less, but, understandably, it will need to be recharged more often. An in-use time of 60 to 120 minutes is reasonable (that is, the robot's batteries must be recharged after an hour or two of continual use).

If you require longer operational times but still need to keep the weight down, consider a replaceable battery system. Mount the battery where it can be easily removed. When the charge on the battery goes down, take it out and replace it with a fully charged one. Place the previously used battery in the recharger. The good news is that smaller, lower-capacity batteries tend to be significantly less expensive than their larger cousins, so you can probably buy two or three smaller batteries for the price of a single big one.

Drive motors are most often selected because of their availability and cost, not because of their weight or construction. In fact, many robots are designed *around* the requirements of the drive motor. The motors are selected (often they're purchased surplus), and from these the frame of the robot is designed and appropriate batteries are added. Still, it's important to give more thought to the selection of the motors for the robot that you have in mind. Avoid motors that are obviously overpowered in relation to the robot in which they are being used. Motors that are grossly oversized will add unnecessary weight, and they will require larger (and therefore heavier and more expensive) batteries to operate.

BEWARE OF THE HEAVY FRAME

The frame of the robot can add a surprising amount of weight. An 18-inch square, 2-foot high robot constructed from extruded aluminum and plastic panels might weigh in excess of 20 or 30 pounds, *without* motors and batteries. The same robot in wood (of sufficient strength and quality) could weigh even more.

Consider ways to lighten your heavy robots, but without sacrificing strength. This can be done by selecting a different construction material and/or by using different construction techniques. For example, instead of building the base of your robot using solid 1/8-inch (or thicker) aluminum sheet, consider an aluminum frame with crossbar members for added stability. If you need a surface on which to mount components (the batteries and motors will be mounted to the aluminum frame pieces), add a 1/16-inch acrylic plastic sheet as a "skin" over the frame. The plastic is strong enough to mount circuit boards, sensors, and other lightweight components on it.

Aluminum and acrylic plastic aren't your only choices for frame materials. Other metals are available as well, but they have a higher weight-to-size ratio. Both steel and brass weigh several times more per square inch than aluminum. Brass sheets, rods, and tubes (both round and square) are commonly available at hobby stores. Unless your robot requires the added strength that brass provides, you may wish to avoid it because of its heavier weight.

Ordinary acrylic plastic is rather dense and therefore fairly heavy, considering its size. Lighter-weight plastics are available but not always easy to find. For example, ABS and PVC plastic—popular for plumbing pipes—can be purchased from larger plastics distributors in rod, tube, and sheet form. There are many special-purpose plastics available that boast both structural strength and light weight. Look for Sintra plastic, for example, which has an expanded core and smooth sides and is therefore lighter than most other plastics. Check the availability of glues and cements before you purchase or order any material. (See Chapter 8 for more on these and other plastics.)

CONSTRUCTING ROBOTS WITH MULTIPLE "DECKS"

For robots that have additional "decks," like the 'bot shown in Fig. 16.1, select construction materials that will provide rigidity but the lowest possible weight. One technique, shown in the figure, is to use 1/2-inch thin-wall (Schedule 125) PVC pipe for uprights and attach the "decks" using 8/32 or 10/24 all-thread rod. The PVC pipe encloses the all-thread; both act as a strong support column. You need three such columns for a circular robot, four columns for a square robot.

Unless your robot is heavy, be sure to use the thinner-walled Schedule 125 PVC pipe. Schedule 80 pipe, commonly used for irrigation systems, has a heavier wall and may not be needed. Note that PVC pipe is always the same diameter outside, no matter how thick its plastic walls. The thicker the wall, the smaller the *inside* diameter of the pipe. You can readily cut PVC pipe to length using a PVC pipe cutter or a hacksaw, and you can paint it if you don't like the white color. Use Testor model paints for best results, and be sure to spray *lightly*. For a bright white look, you can remove the blue marking ink on the outside of the PVC pipe with acetone, which is available in the paint department of your local home improvement store.

FRAME SAGGING CAUSED BY WEIGHT

A critical issue in robot frame design is excessive weight that causes the frame to "sag" in the middle. In a typical robot, a special problem arises when the frame sags: the wheels on either side pivot on the frame and are no longer perpendicular to the ground. Instead, they bow out at the bottom and in at the top (this is called "negative camber"). Depending on which robot tires you use, this can cause traction errors because the contact area of the wheel is no longer consistent. As even more weight is added, the robot may have a tendency to veer off to one side or the other.

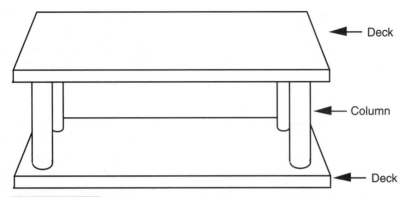

FIGURE 16.1 "Decked" robots provide extra space for batteries and electronics, but they can also add considerably to the weight. Use lightweight construction materials to avoid unduly increasing the weight of the robot.

There are three general fixes for this problem: reduce the weight, strengthen the frame, or add cross braces to prevent the wheels from cambering. Strengthening the frame usually involves adding even more weight. So if you can, strive for the first solution instead—reduce the weight.

If you can't reduce weight, look for ways to add support beams or braces to prevent the sagging. An extra cross brace along the wheelbase (perhaps stretched between the two motors) may be all that's required to prevent the problem. The cross-brace can be made of lightweight aluminum tubing or even from a wooden dowel. The tubing or dowel does not need to support any weight; it simply needs to act as a brace to prevent compression when the frame sags and the wheels camber.

Yet another method is to apply *extreme* camber to the wheels, as shown in Fig. 16.2. This minimizes the negative effects of any sagging, and if the tires have a high frictional surface traction is not diminished. However, don't do this with smooth, hard plastic wheels as they don't provide sufficient traction. You can camber the wheels outward or inward. Inward (negative) camber was used in the old Topo and Bob robots made by Nolan Bushnell's failed Androbot company of the mid-1980s. The heavy-duty robot in Fig. 16.2 uses outward (positive) camber. The robot can easily support over 20 pounds in addition to its own weight, which is about 10 pounds, with battery, which is slung under the frame using industrial-strength hook-and-loop (Velcro or similar) fasteners.

Horizontal Center of Balance

Your robot's horizontal center of balance (think of it as a balance scale) indicates how well the weight of the robot is distributed on its base. If all the weight of a robot is to one side, for example, then the base will have a lopsided horizontal center of balance. The result is an unstable robot: the robot may not travel in a straight line and it might even tip over.

Ideally, the horizontal center of balance of a robot should be the center of its base (see Fig. 16.3a). Some variation of this theme is allowable, depending on the construction of

FIGURE 16.2 This "Tee-Bot" (so named because it employs the T-braces used for home construction) uses extreme camber to avoid the frame sagging that results from too much weight.

the robot. For a robot with a single balancing caster, as shown in Fig. 16.3b, it is usually acceptable to place more weight over the drive wheels and less on the caster. This increases traction, and as long as the horizontal center of balance isn't extreme there is no risk that the robot will tip over.

Unequal weight distribution is the most troublesome result if the horizontal center of balance favors one wheel or track over the other—the right side versus the left side, for example. This can cause the robot to continually "crab" toward the heavier side. Since the heavier side has more weight, traction is improved, but motor speed may be impaired because of the extra load.

Vertical Center of Gravity

City skyscrapers must be rooted firmly in the ground or else there is a risk they will topple over in the slightest wind. The taller an object is, the higher its center of gravity. Of

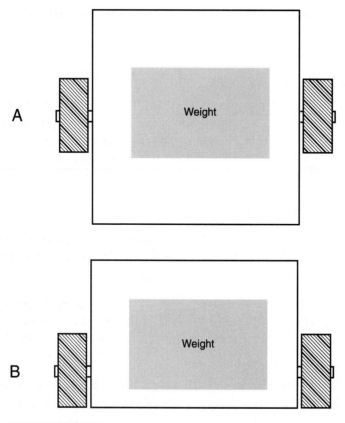

FIGURE 16.3 The distribution of weight on a robot affects its stability and traction. *a*. Centering the weight down the middle in a robot with two balancing casters; *b*. Sliding the center of balance toward the drive wheel in a single-caster 'bot.

critical importance to vertical center of gravity is the "footprint" or base area of the object—that is, the amount of area contacting the ground. The ratio between the vertical center of gravity and the area of the base determines how likely it is that the object will fall over. A robot with a small base but high vertical center of gravity risks toppling over. You can correct such a design in either of two ways:

- Reduce the height of the robot to better match the area of the base, or
- Increase the area of the base to compensate for the height of the robot.

(There is also a third method called *dynamic balance*. Here, mechanical weight is dynamically repositioned to keep the robot on even kilter. These systems are difficult to engineer and, in any event, are beyond the scope of this book.)

Which method you choose will largely depend on what you plan to use your robot for. For example, a robot that must interact with people should be at least toddler height. For a "pet-size" robot you'll probably not want to reduce the height, but rather increase the base area to prevent the robot from tipping over.

Locomotion Issues

The way your robot gets from point A to point B is called *locomotion.* Robot locomotion takes many forms, but wheels and tracks are the most common. Legged robots are also popular, especially among hobbyists, as designing them represents a challenge both in construction and weight-balance dynamics.

WHEELS AND TRACKS

Wheels, and to a lesser extent tracks, are the most common means chosen to scoot robots around. However, some wheels are better for mobile robots than others. Some of the design considerations you may want keep in mind include the following:

- The wider the wheels, the more the robot will tend to stay on course. With very narrow wheels, the robot may have a tendency to favor one side or the other and will trace a slow curve instead of a straight line. Conversely, if the wheels are too wide, the friction created by the excess wheel area contacting the ground may hinder the robot's ability to make smooth turns.
- Two driven wheels positioned on either side of the robot (and balanced by one or two casters on either end) can provide full mobility. This is the most common drive wheel arrangement.
- Tracks turn by skidding or slipping, and they are best used on surfaces such as dirt that readily allow low-friction steering.
- Four or more driven wheels, mounted in sets on each side, will function much like tracks. In tight turns, the wheels will experience significant skidding, and they will therefore create friction over any running surface. If you choose this design, position the wheel sets close together.
- You should select wheel and track material to reflect the surface the robot will be used on. Rubber and foam are common choices; both provide adequate grip for most kinds of surfaces. Foam tires are lighter in weight, but they don't skid well on hard surfaces (such as hardwood or tile floors).

LEGS

Thanks to the ready availability of smart microcontrollers, along with the low cost of R/C (radio-controlled) servos, legged automatons are becoming a popular alternative for robot builders. Robots with legs require more precise construction than the average wheeled robot. They also tend to be more expensive. Even a "basic" six-legged walking robot requires a minimum of 2 or 3 servos, with some six- and eight-leg designs requiring 12 or more motors. At about $12 per servo (more for higher-quality ones), the cost can add up quickly!

Obviously, the first design decision is the number of legs. Robots with one leg ("hoppers") or two legs are the most difficult to build because of balance issues, and we will not address them here. Robots with four and six legs are more common. Six legs offer a *static balance* that ensures that the robot won't easily fall over. At any one time, a minimum of three legs touch the ground, forming a stable tripod.

In a four-legged robot, either the robot must move one leg at a time—keeping the other three on the ground for stability—or else employ some kind of dynamic balance when only two of its legs are on the ground at any given time. Dynamic balance is often accomplished by repositioning the robot's center of gravity, typically by moving a weight (such as the robot's head or tail, if it has one). This momentarily redistributes the center of balance to prevent the robot from falling over. The algorithms and mechanisms for achieving dynamic balance are not trivial. Four-legged robots are difficult to steer, unless you add additional degrees of freedom for each leg or articulate the body of the beast like those weird segmented city buses you occasionally see.

The movement of the legs with respect to the robot's body is often neglected in the design of legged robots. The typical six-legged (hexapod) robot uses six identical legs. Yet the crawling insect a hexapod robot attempts to mimic is designed with legs of different lengths and proportions—the legs are made to do different things. The back legs of an insect, for example, are often longer and are positioned near the back for pushing (this is particularly true of insects that burrow through dirt). The front legs may be similarly constructed for digging, carrying food, fighting, and walking. You may wish to replicate this design, or something similar, for your own robots. Watch some documentaries on insects and study how they walk and how their legs are articulated. Remember that the cockroach has been around for over a million years and represents a very advanced form of biological engineering!

Motor Drives

Next to the batteries, the drive motors are probably the heaviest component in your robot. You'll want to carefully consider where the drive motor(s) are located and how the weight is distributed throughout the base.

One of the most popular mobile robot designs uses two identical motors to spin two wheels on opposite sides of the base. These wheels provide forward and backward locomotion, as shown in Fig. 16.4, as well as left and right steering. If you stop the left motor, the robot turns to the left. By reversing the motors relative to one another, the robot turns by spinning on its wheel axis ("turns in place"). You use this forward-reverse movement to make "hard" or sharp right and left turns.

CENTER-LINE DRIVE MOTOR MOUNT

You can place the wheels—and hence the motors—just about anywhere along the length of the platform. If they are placed in the middle, as shown in Fig. 16.5, you should add two casters to either end of the platform to provide stability. Since the motors are in the center of the platform, the weight is more evenly distributed across it. You can place the battery or batteries above the center line of the wheel axis, which will maintain the even distribution.

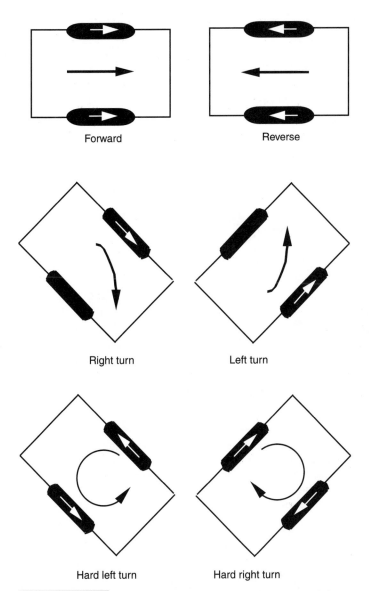

FIGURE 16.4 Two motors mounted on either side of the robot can power two wheels. Casters provide balance. The robot steers by changing the speed and direction of each motor.

A benefit of center-line mounting is that the robot has no "front" or "back," at least as far as the drive system is concerned. Therefore, you can create a kind of *multidirectional* robot that can move forward and backward with the same ease. Of course, this approach also complicates the sensor arrangement of your robot. Instead of having bump switches only in the front of your robot, you'll need to add additional ones in the back in case the robot is reversing direction when it strikes an object.

FRONT-DRIVE MOTOR MOUNT

You can also position the wheels on one end of the platform. In this case, you add one caster on the other end to provide stability and a pivot for turning, as shown in Fig. 16.6. Obviously, the weight is now concentrated more on the motor side of the platform. You should place more weight over the drive wheels, but avoid putting all the weight there since maneuverability and stability may be diminished.

One advantage of front-drive mounting is that it simplifies the construction of the robot. Its "steering circle," the diameter of the circle in which the robot can be steered, is still the same diameter as the center-line drive robot. However, it extends beyond the front/back

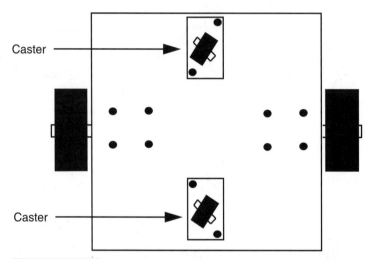

Caster

Caster

FIGURE 16.5 A robot with a centerline motor mount uses two casters (very occasionally one) for balance. When using one caster, you may need to shift the balance of weight toward the caster end to avoid having the robot tip over.

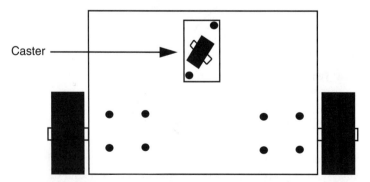

Caster

FIGURE 16.6 A robot with a front-drive motor mount uses a single opposing caster for balance. Steering is accomplished using the same technique as a centerline motor mount.

dimension of the robot (see Fig. 16.7). This may or may not be a problem, depending on the overall size of your robot and how you plan to use it. Any given front-drive robot may be smaller than its centerline drive cousin. Because of the difference in their physical size, the diameter of the steering circle for both may be about the same.

CASTER CHOICES

As we mentioned earlier, most robots employing the two motor drive system use at least one unpowered caster, which provides support and balance. Two casters are common in robots that use center-line drive-wheel mounting. Each caster is positioned at opposite ends of the robot. When selecting casters it is important to consider the following factors:

- The size of the caster wheel should be in proportion to the drive wheels (see Fig. 16.8). When the robot is on the ground the drive motors must firmly touch terra firma. If the caster wheels are too large, the drive motors may not make adequate contact, and poor traction will result. You might also consider using a suspension system of your own design on the casters to compensate for uneven terrain.

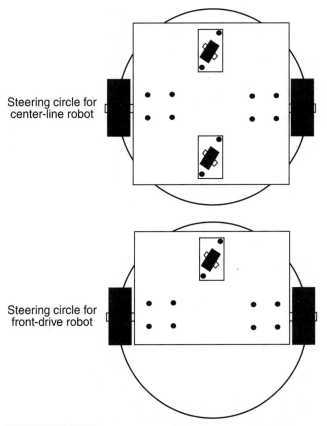

FIGURE 16.7 The steering circle of a robot with centerline and front-drive mounted motors.

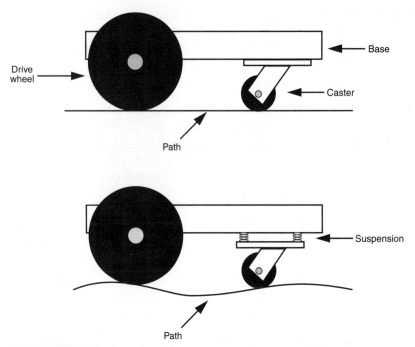

FIGURE 16.8 The height of the caster with respect to the drive wheels will greatly influence the robot's traction and maneuverability. A spring-loaded caster (a kind of suspension) can improve functionality of the robot on semirough terrain.

■ The casters should spin and swivel freely. A caster that doesn't spin freely will impede the robot's movement.

■ In most cases, since the caster is provided only for support and not traction you should construct the caster from a hard material to reduce friction. A caster made of soft rubber will introduce more friction, and it may affect a robot's movements.

■ Consider using "ball casters" (also called "ball transfers"), which are primarily designed to be used in materials processing (conveyor chutes and the like). Ball casters (see Fig. 16.9) are made of a single ball—either metal or rubber—held captive in a housing, and they function as omnidirectional casters for your robot. The size of the ball varies from about 11/16 inch to over 3 inches in diameter. Look for ball casters at mechanical surplus stores and also at industrial supply outlets, such as Grainger and McMaster-Carr.

Steering Methods

A variety of methods are available to steer your robot. The following sections describe several of the more common approaches.

DIFFERENTIAL

For wheeled and tracked robots, *differential steering* is the most common method for getting the machine to go in a different direction. The technique is exactly the same as steering a military tank: one side of wheels or treads stops or reverses direction while the other side keeps going. The result is that the robot turns in the direction of the stopped or reversed wheel or tread. Because of friction effects, differential steering is most practical with two-wheel-drive systems. Additional sets of wheels, as well as rubber treads, can increase friction during steering.

■ If you are using multiple wheels ("dually"), position the wheels close together, as shown in Fig. 16.10. The robot will pivot at a virtual point midway between the two wheels on each side.

FIGURE 16.9 Ball casters (or ball transfers) are omnidirectional. For medium- to large-sized robots consider using them instead of wheeled casters.

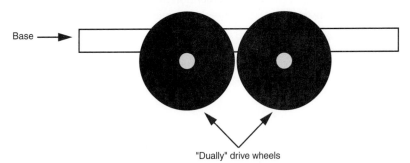

FIGURE 16.10 "Dually" wheels should be placed close to one another. If they are spaced farther apart the robot cannot steer as easily.

■ If you are using treads, select a relatively low-friction material such as cloth or hard plastic. Very soft rubber treads will not steer well on smooth surfaces. If this cannot be helped, one approach is to always steer by reversing the tread directions. This will reduce the friction.

CAR-TYPE

Pivoting the wheels in the front is yet another method for steering a robot (see Fig. 16.11). Robots with *car-type* steering are not as maneuverable as differentially steered robots, but they are better suited for outdoor uses, especially over rough terrain. You can obtain somewhat better traction and steering accuracy if the wheel on the inside of the turn pivots more than the wheel on the outside. This technique is called *Ackerman steering* and is found on most cars but not on as many robots.

TRICYCLE

One of the greatest drawbacks of the differentially steered robot is that the robot will veer off course if one motor is even a wee bit slow. You can compensate for this by monitoring the speed of both motors and ensuring that they operate at the same rpm. This typically requires a control computer, as well as added electronics and mechanical parts for sensing the speed of the wheels.

Car-type steering, described in the last section, is one method for avoiding the problem of "crabbing" as a result of differences in motor speed simply because the robot is driven by just one motor. But car-type steering makes for fairly cumbersome indoor mobile robots. A better approach is to use a single drive motor powering two rear wheels and a single steering wheel in the front. This arrangement is just like a child's tricycle, as shown in Fig. 16.12. The robot can be steered in a circle just slightly larger than the width of the machine. Be careful about the wheelbase of the robot (distance from the back wheels to the front steering wheel). A short base will cause instability in turns, and the robot will tip over opposite the direction of the turn.

Tricycle-steered robots must have a very accurate steering motor in the front. The motor must be able to position the front wheel with subdegree accuracy. Otherwise, there is no guarantee the robot will be able to travel a straight line. Most often, the steering wheel is controlled by a servo motor. Servo motors use a "closed-loop feedback" system that provides a high degree of positional accuracy (depending on the quality of the

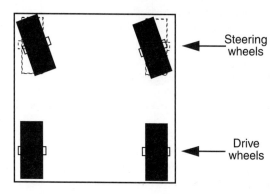

Steering wheels

Drive wheels

FIGURE 16.11 Car-type steering offers a workable alternative for an outdoors robot, but it is less useful indoors or in places where there are many obstructions that must be steered around.

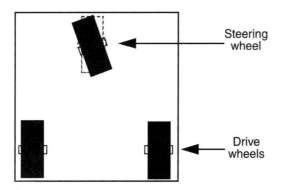

FIGURE 16.12 In tricycle steering, one drive motor powers the robot; a single wheel in front steers the robot. Be wary of short wheelbases as this can introduce tipping when the robot turns.

motor, of course). Read more about servo motors in Chapter 20, "Working with Servo Motors."

OMNIDIRECTIONAL

To have the highest tech of all robots, you may want omnidirectional drive. It uses steerable drive wheels, usually at least three, as shown in Fig. 16.13. The wheels are operated by two motors: one for locomotion and one for steering. In the usual arrangement, the drive/steering wheels are "ganged" together using gears, rollers, chains, or pulleys. Omnidirectional robots exhibit excellent maneuverability and steering accuracy, but they are technically more difficult to construct.

Calculating the Speed of Robot Travel

The speed of the drive motors is one of two elements that determines the travel speed of your robot. The other is the diameter of the wheels. For most applications, the speed of the drive motors should be under 130 rpm (under load). With wheels of average size, the resultant travel speed will be approximately four feet per second. That's actually pretty fast. A better travel speed is one to two feet per second (approximately 65 rpm), which requires smaller diameter wheels, a slower motor, or both.

How do you calculate the travel speed of your robot? Follow these steps:

1. Divide the rpm speed of the motor by 60. The result is the revolutions of the motor per second (rps). A 100-rpm motor runs at 1.66 rps.
2. Multiply the diameter of the drive wheel by *pi,* or approximately 3.14. This yields the circumference of the wheel. A 7-inch wheel has a circumference of about 21.98 inches.
3. Multiply the speed of the motor (in rps) by the circumference of the wheel. The result is the number of linear inches covered by the wheel in one second.

With a 100-rpm motor and 7-inch wheel, the robot will travel at a top speed of 35.168 inches per second, or just under three feet. That's about two miles per hour! You can readily see that you can slow down a robot by decreasing the size of the wheel. By reducing the wheel to 5 inches instead of 8, the same 100-rpm motor will propel the robot at about 25 inches per second. By reducing the motor speed to, say, 75 rpm, the travel speed falls even more, to 19.625 inches per second. Now that's more reasonable.

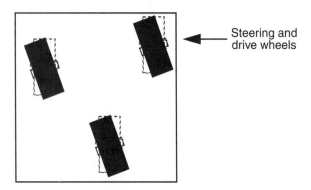

FIGURE 16.13 An omnidirectional robot uses the same wheels for drive and steering.

Bear in mind that the actual travel speed once the robot is all put together may be lower than this. The heavier the robot, the larger the load on the motors, so the slower they will turn.

Round Robots or Square?

Robots can't locomote where they can't fit. Obviously, a robot that's too large to fit through doorways and halls will have a hard time of it. In addition, the overall shape of a robot will also dictate how maneuverable it is, especially indoors. If you want to navigate your robot in tight areas, you should consider its basic shape: round or square.

■ A round robot is generally able to pass through smaller openings, no matter what its orientation when going through the opening (see Fig. 16.14). To make a round robot, you must either buy or make a rounded base or frame. Whether you're working with metal, steel, or wood, a round base or frame is not as easy to construct as a square one.
■ A square robot must orient itself so that it passes through openings straight ahead rather than at an angle. Square-shaped robot bases and frames are easier to construct than round ones.

While you're deciding whether to build a round- or square-shaped robot, consider that a circle of a given diameter has less surface area than a square of the same width. For example, a 10-inch circle has a surface area of about 78 square inches. Moreover, because the surface of the base is circular, less of it will be useful for your robot (unless your printed circuit boards are also circular). Conversely, a 10-inch-by-10-inch square robot has a surface area of 100 inches. Such a robot could be reduced to about 8.5 inches square, and it would have about the same surface area as a 10-inch round robot, and its surface area would be generally more usable.

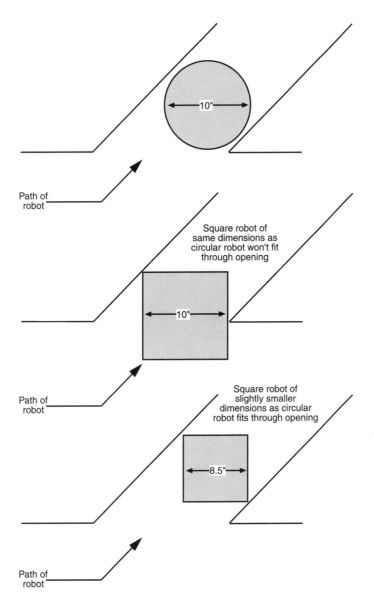

10"

Square robot of
same dimensions as
circular robot won't fit
through opening

10"

Path of
robot

Square robot of
slightly smaller
dimensions as circular
robot fits through opening

8.5"

Path of
robot

Path of
robot

FIGURE 16.14 A round robot versus a square robot. All
things being equal, a round robot is better
able to navigate through small openings.
However, rounded robots also have less
usable surface area, so a square-shaped
robot can be made smaller and still support
the same onboard "real estate."

From Here

To learn more about...	Read
Selecting wood, plastic, or metal to construct your robot	Chapter 8–10
Choosing a battery for your robot	Chapter 15, "All about Batteries and Robot Power Supplies"
Selecting motors	Chapter 17, "Choosing the Right Motor for the Job"
Building a walking robot	Chapter 22, "Build a Heavy-duty, Six-legged Walking Robot"
Constructing a treaded robot	Chapter 23, "Advanced Locomotion Systems"
Controlling the speed of a two-motor-driven robot	Chapter 38, "Navigating through Space"

17

CHOOSING THE RIGHT
MOTOR FOR THE JOB

Motors are the muscles of robots. Attach a motor to a set of wheels and your robot can scoot around the floor. Attach a motor to a lever, and the shoulder joint for your robot can move up and down. Attach a motor to a roller, and the head of your robot can turn back and forth, scanning its environment. There are many kinds of motors; however, only a select few are truly suitable for homebrew robotics. In this chapter, we'll examine the various types of motors and how they are used.

AC or DC?

Direct current—DC—dominates the field of robotics, either mobile or stationary. DC is used as the main power source for operating the onboard electronics, for opening and closing solenoids, and, yes, for running motors. Few robots use motors designed to operate from AC, even those automatons used in factories. Such robots convert the AC power to DC, then distribute the DC to various sub-systems of the machine.

DC motors may be the motors of choice, but that doesn't mean you should use just any DC motor in your robot designs. When looking for suitable motors, be sure the ones you buy are reversible. Few robotic applications call for just unidirectional (one-direction) motors. You must be able to operate the motor in one direction, stop it, and change its direction. DC motors are inherently bidirectional, but some design limitations may prevent reversibility.

The most important factor is the commutator brushes. If the brushes are slanted, the motor probably can't be reversed. In addition, the internal wiring of some DC motors prevents them from going in any but one direction. Spotting the unusual wiring scheme by just looking at the exterior or the motor is difficult, at best, even for a seasoned motor user.

The best and easiest test is to try the motor with a suitable battery or DC power supply. Apply the power leads from the motor to the terminals of the battery or supply. Note the direction of rotation of the motor shaft. Now, reverse the power leads from the motor. The motor shaft should rotate in reverse.

Continuous or Stepping

DC motors can be either continuous or stepping. Here is the difference: with a *continuous motor,* like the ones in Fig. 17.1, the application of power causes the shaft to rotate continually. The shaft stops only when the power is removed or if the motor is stalled because it can no longer drive the load attached to it.

With *stepping motors,* shown in Fig. 17.2, the application of power causes the shaft to rotate a few degrees, then stop. Continuous rotation of the shaft requires that the power be pulsed to the motor. As with continuous DC motors, there are subtypes of stepping motors. Permanent magnet steppers are the ones you're likely to encounter, and they are also the easiest to use.

The design differences between continuous and stepping DC motors need to be addressed in detail. Chapter 18, "Working with DC Motors," focuses entirely on continu-

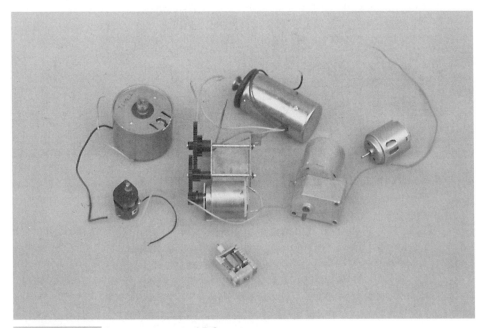

FIGURE 17.1 An assortment of DC motors.

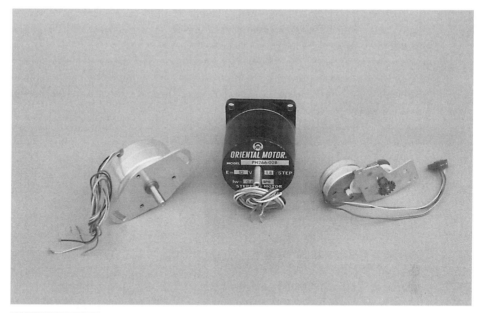

FIGURE 17.2 An assortment of stepper motors.

ous motors. Chapter 19, "Working with Stepper Motors," focuses entirely on the stepping variety. Although these two chapters focus on the main drive motors of your robot, you can apply the information to motors used for other purposes as well.

Servo Motors

A special "subset" of continuous motors is the servo motor, which in typical cases combines a continuous DC motor with a "feedback loop" to ensure the accurate positioning of the motor. A common form of servo motor is the kind used in model and hobby radio-controlled (R/C) cars and planes.

R/C servos are in plentiful supply, and their cost is reasonable (about $10–12 for basic units). Though R/C servos are continuous DC motors at heart, we will devote a separate chapter just to them. See Chapter 20, "Working with Servo Motors," for more information on using R/C servo motors not only to drive your robot creations across the floor but to operate robot legs, arms, hands, heads, and just about any other appendage.

Other Motor Types

There are many other types of motors, some of which may be useful in your hobby robot, some of which will not. DC, stepper, and servo motors are the most common, but you may also see references to some of the following:

- *Brushless DC.* This is a kind of DC motor that has no brushes. It is controlled electronically. Brushless DC motors are commonly used in fans inside computers and for motors in VCRs and videodisc players.
- *Switched reluctance.* This is a DC motor without permanent magnets.
- *Synchronous.* Also known as brushless AC, this motor operates synchronously with the phase of the power supply current. These motors function much like stepper motors, which will be discussed in Chapter 19.
- *Synchro.* These motors are considered distinct from the synchronous variety, described above. Synchro motors are commonly designed to be used in pairs, where a "master" motor electrically controls a "slave" motor. Rotation of the master causes an equal amount of rotation in the slave.
- *AC induction.* This is the ordinary AC motor used in fans, kitchen mixers, and many other applications.
- *Sel-Syn.* This is a brand name, often used to refer to synchronous AC motors.

Note that AC motors aren't always operated at 50/60 Hz, which is common for household current. Motors for 400-Hz operation, for example, are common in surplus stores and are used for both aircraft and industrial applications.

Motor Specifications

Motors come with extensive specifications. The meaning and purpose of some of the specifications are obvious; others aren't. Let's take a look at the primary specifications of motors—voltage, current draw, speed, and torque—and see how they relate to your robot designs.

OPERATING VOLTAGE

All motors are rated by their *operating voltage*. With small DC "hobby" motors, the rating is actually a range, usually 1.5 to 6 volts. Some high-quality DC motors are designed for a specific voltage, such as 12 or 24 volts. The kinds of motors of most interest to robot builders are the low-voltage variety—those that operate at 1.5 to 12 volts.

Most motors can be operated satisfactorily at voltages higher or lower than those specified. A 12-volt motor is likely to run at 8 volts, but it may not be as powerful as it could be, and it will run slower (an exception to this is stepper motors; see Chapter 19, "Working with Stepper Motors," for details). You'll find that most motors will refuse to run, or will not run well, at voltages under 50 percent of the specified rating.

Similarly, a 12-volt motor is likely to run at 16 volts. As you may expect, the speed of the shaft rotation increases, and the motor will exhibit greater power. I do not recommend that you run a motor continuously at more than 30 or 40 percent its rated voltage, however. The windings may overheat, which may cause permanent damage. Motors designed for high-speed operation may turn faster than their ball-bearing construction allows.

If you don't know the voltage rating of a motor, you can take a guess at it by trying various voltages and seeing which one provides the greatest power with the least amount of heat dissipated through the windings (and felt on the outside of the case). You can also listen to the motor. It should not seem as if it is straining under the stress of high speeds.

CURRENT DRAW

Current draw is the amount of current, in milliamps or amps, that the motor requires from the power supply. Current draw is more important when the specification describes motor loading, that is, when the motor is turning something or doing some work. The current draw of a free-running (no-load) motor can be quite low. But have that same motor spin a wheel, which in turn moves a robot across the floor, and the current draw jumps 300, 500, even 1000 percent.

With most permanent magnet motors (the most popular kind), current draw increases with load. You can see this visually in Fig. 17.3. The more the motor has to work to turn the shaft, the more current is required. The load used by the manufacturer when testing the motor isn't standardized, so in your application the current draw may be more or less than that specified.

A point is reached when the motor does all the work it can do, and no more current will flow through it. The shaft stops rotating; the motor has "stalled." Some motors, but not many, are rated (by the manufacturer) by the amount of current they draw when stalled. This is considered the worse-case condition. The motor will never draw more than this current unless it is shorted out, so if the system is designed to handle the stall current it can handle anything. Motors rated by their stall current will be labeled as such. Motors designed for the military, available through surplus stores, are typically rated by their stall current. When providing motors for your robots, you should always know the approximate current draw under load. Most volt-ohm meters can test current. Some special-purpose amp meters are made just for the job.

Be aware that some volt-ohm meters can't handle the kind of current pulled through a motor. Most digital meters (discussed more completely in Chapter 3, "Tools and Supplies") can't deal with more than 200 to 400 milliamps of current. Even small hobby motors can draw in excess of this. Be sure your meter can accommodate current up to 5 or 10 amps.

If your meter cannot register this high without popping fuses or burning up, insert a 1- to 10-ohm power resistor (10 to 20 watts) between one of the motor terminals and the positive supply rail, as shown in Fig. 17.4. With the meter set on DC voltage, measure the voltage developed across the resistor.

A bit of Ohm's law, I = E/R (*I* is current, *E* is voltage, *R* is resistance) reveals the current draw through the motor. For example, if the resistance is 10 ohms and the voltage is 2.86 volts, the current draw is 286 mA. You can watch the voltage go up (and therefore the current too) by loading the shaft of the motor.

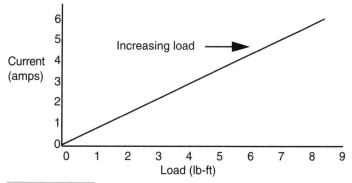

FIGURE 17.3 The current draw of a motor increases in proportion to the load on the motor shaft.

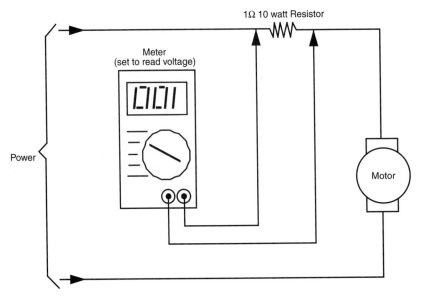

FIGURE 17.4 How to test the current draw of a motor by measuring the voltage developed across an in-line resistor. The actual value of the resistor can vary, but it should be under about 20 ohms. Be sure the resistor is a high-wattage type.

SPEED

The rotational speed of a motor is given in revolutions per minute (rpm). Most continuous DC motors have a normal operating speed of 4000 to 7000 rpm. However, some special-purpose motors, such as those used in tape recorders and computer disk drives, operate as slow as 2000 to 3000 rpm. For just about all robotic applications, these speeds are much too high. You must reduce the speed to no more than 150 rpm (even less for motors driving arms and grippers) by using a gear train. You can obtain some reduction by using electronic control, as described in Part 5 of this book, "Computers and Electronic Control." However, such control is designed to make fine-tuned speed adjustments, not reduce the rotation of the motor from 5000 rpm to 50 rpm. See the later sections of this chapter for more details on gear trains and how they are used.

Note that the speed of stepping motors is not rated in rpm but in steps (or pulses) per second. The speed of a stepper motor is a function of the number of steps that are required to make one full revolution plus the number of steps applied to the motor each second. As a comparison, the majority of light- and medium-duty stepper motors operate at the equivalent of 100 to 140 rpm. See Chapter 19, "Working with Stepper Motors," for more information.

TORQUE

Torque is the force the motor exerts upon its load. The higher the torque, the larger the load can be and the faster the motor will spin under that load. Reduce the torque, and the motor slows down, straining under the workload. Reduce the torque even more, and the load may prove too demanding for the motor. The motor will stall to a grinding halt, and in doing so eat up current (and put out a lot of heat).

Torque is perhaps the most confusing design aspect of motors. This is not because there is anything inherently difficult about it but because motor manufacturers have yet to settle on a standard means of measurement. Motors made for industry are rated one way, motors for the military another.

At its most basic level, torque is measured by attaching a lever to the end of the motor shaft and a weight or gauge on the end of that lever, as depicted in Fig. 17.5. The lever can be any number of lengths: one centimeter, one inch, or one foot. Remember this because it plays an important role in torque measurement. The weight can either be a hunk of lead or, more commonly, a spring-loaded scale (as shown in the figure). Turn the motor on and it turns the lever. The amount of weight it lifts is the torque of the motor. There is more to motor testing than this, of course, but it'll do for the moment.

Now for the ratings game. Remember the length of the lever? That length is used in the torque specification. If the lever is one inch long, and the weight successfully lifted is two ounces, then the motor is said to have a torque of two ounce-inches, or oz-in. (Some people reverse the "ounce" and "inches" and come up with "inch-ounces.")

The unit of length for the lever usually depends on the unit of measurement given for the weight. When the weight is in grams, the lever is in centimeters (gm-cm). When the weight is in ounces, as already seen, the lever used is in inches (oz-in). Finally, when the weight is in pounds, the lever used is commonly in feet (lb-ft). Like the ounce-inch measurement, gram-centimeter and pound-foot specifications can be reversed—"centimeter-gram" or "foot-pound." Note that these easy-to-follow conventions aren't always used. Some motors may be rated by a mixture of the standards—ounces and feet or pounds and inches.

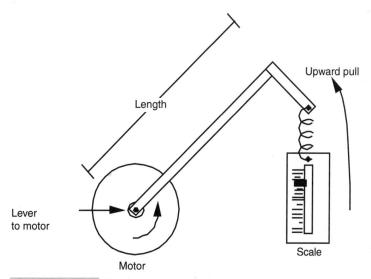

FIGURE 17.5 The torque of a motor is measured by attaching a weight or scale to the end of a lever and mounting the lever of the motor shaft.

STALL OR RUNNING TORQUE

Most motors are rated by their *running torque,* or the force they exert as long as the shaft continues to rotate. For robotic applications, it's the most important rating because it determines how large the load can be and still guarantee that the motor turns. How running torque tests are conducted varies from one motor manufacturer to another, so results can differ. The tests are impractical to duplicate in the home shop, unless you have an elaborate slip-clutch test stand, precision scale, and sundry other test jigs.

If the motor(s) you are looking at don't have running torque ratings, you must estimate their relative strength. This can be done by mounting them on a makeshift wood or metal platform, attaching wheels to them, and having them scoot around the floor. If the motor supports the platform, start piling on weights. If the motor continues to operate with, say, 40 or 50 pounds of junk on the platform, you've got an excellent motor for driving your robot.

Some motors you may test aren't designed for hauling heavy loads, but they may be suitable for operating arms, grippers, and other mechanical components. You can test the relative strength of these motors by securing them in a vise, then attaching a large pair of Vise-Grips or other lockable pliers to them. Use your own hand as a test jig, or rig one up with fishing weights. Determine the rotational power of the motor by applying juice to the motor and seeing how many weights it can successfully handle.

Such crude tests make more sense if you have a "standard" by which to judge others. If you've designed a robotic arm before, for example, and are making another one, test the motors that you successfully used in your prototype. If subsequent motors fail to match or exceed the test results of the standard, you know they are unsuitable for the test.

Another torque specification, *stall torque,* is sometimes provided by the manufacturer instead of or in addition to running torque (this is especially true of stepping motors). Stall torque is the force exerted by the motor when the shaft is clamped tight. There is an indirect relationship between stall torque and running torque, and although it varies from motor to motor you can use the stall torque rating when you select candidate motors for your robot designs.

Gears and Gear Reduction

We've already discussed the fact that the normal running speed of motors is far too fast for most robotics applications. Locomotion systems need motors with running speeds of 75 to 150 rpm. Any faster than this, and the robot will skim across the floor and bash into walls and people. Arms, gripper mechanisms, and most other mechanical subsystems need even slower motors. The motor for positioning the shoulder joint of an arm needs to have a speed of less than 20 rpm; 5 to 8 rpm is even better.

There are two general ways to decrease motor speed significantly: build a bigger motor (impractical) or add gear reduction. Gear reduction is used in your car, on your bicycle, in the washing machine and dryer, and in countless other motor-operated mechanisms.

GEARS 101

Gears perform two important duties. First, they can make the number of revolutions applied to one gear greater or lesser than the number of revolutions of another gear that is connected to it. They also increase or decrease power, depending on how the gears are oriented. Gears can also serve to simply transfer force from one place to another.

Gears are actually round levers, and it may help to explain how gears function by first examining the basic mechanical lever. Place a lever on a fulcrum so the majority of the lever is on one side. Push up on the long side, and the short side moves in proportion. Although you may move the lever several feet, the short side is moved only a few inches. Also note that the force available on the short end is proportionately larger than the force applied on the long end. You use this wonderful fact of physics when you dig a rock out of the ground with your shovel or jack up your car to replace a tire.

Now back to gears. Attach a small gear to a large gear, as shown in Fig. 17.6. The small gear is directly driven by a motor. For each revolution of the small gear, the large gear turns one half a revolution. Expressed another way, if the motor and small gear turn at 1000 rpm, the large gear turns at 500 rpm. The gear ratio is said to be 2:1.

Note that another important thing happens, just as it did with the lever and fulcrum. *Decreasing the speed of the motor also increases its torque.* The power output is approximately twice the input. Some power is lost in the reduction process due to the friction of the gears. If the drive and driven gears are the same size, the rotation speed is neither increased nor decreased, and the torque is not affected (apart from small frictional losses). You can use same-size gears in robotics design to transfer motive power from one shaft to another, such as driving a set of wheels at the same speed and in the same direction.

ESTABLISHING GEAR REDUCTION

Gears are an old invention, going back to ancient Greece. Today's gears are more refined, and they are available in all sorts of styles and materials. However, they are still based on

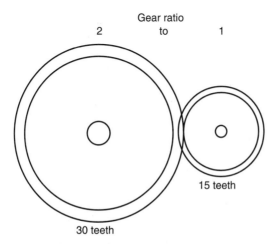

FIGURE 17.6 A representation of a 2:1 gear reduction ratio.

the old Greek design in which the teeth from the two mating gears mesh with each other. The teeth provide an active physical connection between the two gears, and the force is transferred from one gear to another.

Gears with the same size teeth are usually characterized not by their physical size but by the number of teeth around their circumference. In the example in Fig. 17.6, the small gear contains 15 teeth, the large gear 30 teeth. And, you can string together a number of gears one after the other, all with varying numbers of teeth (see Fig. 17.7). Attach a tachometer to the hub of each gear, and you can measure its speed. You'll discover the following two facts:

- The speed always decreases when going from a small to a large gear.
- The speed always increases when going from a large to a small gear.

There are plenty of times when you need to reduce the speed of a motor from 5000 rpm to 50 rpm. That kind of speed reduction requires a reduction ratio of 100:1. To accomplish that with just two gears you would need, as an example, a drive gear that has 10 teeth and a driven gear that has 1000 teeth. That 1000-tooth gear would be quite large, bigger than the drive motor itself.

You can reduce the speed of a motor in steps by using the arrangement shown in Fig. 17.8. Here, the driver gear turns a larger "hub" gear, which in turn has a smaller gear permanently attached to its shaft. The small hub gear turns the driven gear to produce the final output speed, in this case 50 rpm. You can repeat this process over and over again until the output speed is but a tiny fraction of the input speed. This is the arrangement most often used in motor gear reduction systems.

USING MOTORS WITH GEAR REDUCTION

It's always easiest to use DC motors that already have a gear reduction box built onto them, such as the motor in Fig. 17.9. R/C servo motors already incorporate gear reduction, and stepper motors may not require it. This fact saves you from having to find a gear reducer that fits the motor and application and attach it yourself. When selecting gear motors, you'll be most interested in the output speed of the gearbox, not the actual running speed

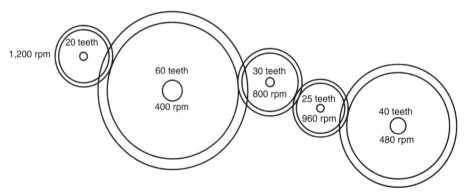

FIGURE 17.7 Gears driven by the 20-tooth gear on the left rotate at different speeds, depending on their diameter.

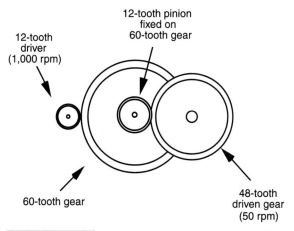

12-tooth pinion
fixed on
60-tooth gear

12-tooth
driver
(1,000 rpm)

60-tooth gear

48-tooth
driven gear
(50 rpm)

FIGURE 17.8 True gear reduction is achieved by
ganging gears on the same shaft.

of the motor. Note as well that the running and stall torque of the motor will be greatly
increased. Make sure that the torque specification on the motor is for the output of the
gearbox, not the motor itself.

With most gear reduction systems, the output shaft is opposite the input shaft (but usually off center). With other boxes, the output and input are on the same side of the box.
When the shafts are at 90 degrees from one another, the reduction box is said to be a "right-
angle drive." If you have the option of choosing, select the kind of gear reduction that best
suits the design of your robot. I have found that the "shafts on opposite sides" is the all-
around best choice. Right-angle drives also come in handy, but they usually carry high
price tags.

When using motors without built-in gear reduction, you'll need to add reduction boxes,
such as the model shown in Fig. 17.10, or make your own. Although it is possible to do
both of these yourself, there are many pitfalls:

- Shaft diameters of motors and ready-made gearboxes may differ, so you must be sure
  that the motor and gearbox mate.
- Separate gear reduction boxes are hard to find. Most must be cannibalized from salvage
  motors. Old AC motors are one source of surplus boxes.
- When designing your own gear reduction box, you must take care to ensure that all the
  gears have the same hub size and that meshing gears exactly match each other.
- Machining the gearbox requires precision, since even a small error can cause the gears
  to mesh improperly.

ANATOMY OF A GEAR

Gears consist of teeth, but these teeth can come in any number of styles, sizes, and orien-
tations. Spur gears are the most common type. The teeth surround the outside edge of the
gear, as shown in Fig. 17.11. Spur gears are used when the drive and driven shafts are parallel. Bevel gears have teeth on the surface of the circle rather than the edge. They are used

FIGURE 17.9 A motor with an enclosed gearbox. These are ideal for robotics use.

FIGURE 17.10 A gear reduction box, originally removed from an open-frame AC motor. On this unit, the input and output shafts are on the same side.

to transmit power to perpendicular shafts. Miter gears serve a similar function but are designed so that no reduction takes place. Spur, bevel, and miter gears are reversible. That is, unless the gear ratio is very large, you can drive the gears from either end of the gear system, thus increasing or decreasing the input speed.

Worm gears transmit power perpendicularly, like bevel and miter gears, but their design is unique. The worm (or lead screw) resembles a threaded rod. The rod provides the power. As it turns, the threads engage a modified spur gear (the modification takes into consideration the cylindrical shape of the worm).

Worm gear systems are specifically designed for large-scale reduction. The gearing is not usually reversible; you can't drive the worm by turning the spur gear. This is an important point because it gives worm gear systems a kind of automatic locking capability. Work gears are particularly well suited for arm mechanisms in which you want the joints to remain where they are. With a traditional gear system, the arm may droop or sink back due to gravity once the power from the drive motor is removed.

Rack gears are like spur gears unrolled into a flat rod. They are primarily intended to transmit rotational motion to linear motion. Racks have a kind of self-locking characteristic as well, but it's not as strong as that found in worm gears.

The size of gear teeth is expressed as *pitch,* which is roughly calculated by counting the number of teeth on the gear and dividing it by the diameter of the gear. For example, a gear that measures two inches and has 48 teeth has a tooth pitch of about 24. Common pitches are 12 (large), 24, 32, and 48. Some gears have extra-fine 64-pitch teeth, but these are

FIGURE 17.11 Spur gears. These particular gears are made of nylon and have aluminum hubs. It's better to use metal hubs in which the gear is secured to the shaft with a setscrew.

usually confined to miniature mechanical systems, such as radio-controlled models. Odd-sized pitches exist, of course, as do metric sizes, so you must be careful when matching gears that the pitches are exactly the same. Otherwise, the gears will not mesh properly and may cause excessive wear.

The degree of slope of the face of each tooth is cailed the *pressure angle.* The most common pressure angle is 20°, although some gears, particularly high-quality worms and racks, have a 14 1/2° pressure angle. Textbooks claim that you should not mix two gears with different pressure angles even if the pitch is the same, but it can be done. Some excessive wear may result because the teeth aren't meshing fully.

The orientation of the teeth on the gear can differ. The teeth on most spur gears are perpendicular to the edges of the gear. But the teeth can also be angled, as shown in Fig. 17.12, in which case it is called a helical gear.

A number of other unusual tooth geometries are in use. These include double-teeth, where two rows of teeth offset one another, and herringbone, where there are two sets of helical gears at opposite angles. These gears are designed to reduce the backlash phenomenon. The space (or "play") between the teeth when meshing can cause the gears to rock back and forth.

Pulleys, Belts, Sprockets, and Roller Chain

Akin to the gear are pulleys, belts, sprockets, and roller chains. Pulleys are used with belts, and sprockets are used with roller chain. The pulley and sprocket are functionally identical to the gear. The only difference is that pulley and sprockets use belts and roller chain, respectively, to transfer power. With gears, power is transferred directly.

A benefit of using pulleys-belts or sprockets-chain is that you don't need to be as concerned with the absolute alignment of the mechanical parts of your robot. When using gears you must mount them with high precision. Accuracies to the hundredths of an inch are desirable to avoid "slop" in the gears as well as the inverse—binding caused by gears that are meshing too tightly. Belts and roller chain are designed to allow for slack; in fact, if there's no slack you run the risk of breaking the pulley or chain!

Standard tooth spur gear

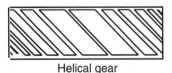

Helical gear

FIGURE 17.12 Standard spur gear versus diagonal helical spur gear. The latter is used to decrease backlash—the play inherent when two gears mesh. Some helical gears are also made for diverting the motion at right angles.

MORE ABOUT PULLEYS AND BELTS

Pulleys come in a variety of shapes and sizes. You're probably familiar with the pulleys and belts used in automotive applications. These are likely to be too bulky and heavy to be used with a robot. Instead, look for smaller and lighter pulleys and belts used for copiers, fax machines, VCRs, and other electronic equipment. These are available for salvage from whole units or in bits and pieces from surplus outlets.

Pulleys can be either the V type (the pulley wheel has a V-shaped groove in it) or the cog type. Cog pulleys require matching belts. You need to ensure that the belt is not only the proper width for the pulley you are using but also has the same cog pitch.

MORE ABOUT SPROCKETS AND ROLLER CHAIN

Sprockets and roller chain are preferred when you want to ensure synchronism. For large robots you can use 3/8-inch bicycle chain. Most smaller robots will do fine with 1/4-inch roller chain, which can frequently be found in surplus stores. Metal roller chain is commonly available in preset lengths, though you can sometimes shorten or lengthen the chain by adding or removing links. Plastic roller chain, while not as strong, can be adjusted more easily by using snap-on links.

Mounting the Motor

Every motor requires a different mounting arrangement. It's easier for you when the motor has its own mounting hardware or holes; you can use these to mount the motor in your robot. Remember that Japanese- and European-made motors often have metric threads, so be sure to use the proper-sized bolt.

Other motors may not be as cooperative. Either the mounting holes are in a position where they don't do you much good, or the motor is completely devoid of any means for securing it to your robot. You can still mount these motors successfully by using an assortment of clamps, brackets, woodblocks, and homemade angles.

For example, to secure the motor shown in Fig. 17.13, mounting brackets were fashioned using six-inch galvanized iron mending "T" plates. A large hole was drilled for the drive shaft and gear to poke through, and the two halves of the mounting bracket were joined together with nuts, bolts, and spacers. The bracket was then attached to the frame of the robot using angle irons and standard hardware. This motor arrangement was made a little more difficult by the addition of a drive gear and sprocket. Construction time for each motor bracket was about 90 minutes. You can read more about this particular design in Chapter 22, "Build a Heavy-duty, Six-legged Walking Robot."

Another example is shown in Fig. 17.14. Here, the motor has mounting holes on the end by the shaft, but these holes are in the wrong position for the design of the robot. Two commonly available flat corner irons were used to mount the motor. This is just one approach; a number of other mounting schemes might have worked satisfactorily as well. This design is more thoroughly discussed in Chapter 25, "Build a Revolute Coordinate Arm."

You can also fashion your own mounting brackets using metal or plastic. Cut the bracket to the size you need, and drill mounting holes. This technique works well when you are using servo motors for model radio-controlled cars and airplanes.

FIGURE 17.13 One approach to mounting a large motor in a robot. The motor is sandwiched inside two large hardware plates and is secured to the frame of the motor with angle irons.

FIGURE 17.14 Another approach to mounting a motor to a robot. Flat corner irons secure the motor flange to the frame.

If the motor lacks mounting holes, you can to use clamps to hold it in place. U-bolts, available at the hardware store, are excellent solutions. Choose a U-bolt that is large enough to fit around the motor. The rounded shape of the bolt is perfect for motors with round casings. If desired, you can make a holding block out of plastic or wood to keep the bottom of the motor from sliding. Cut the plastic or wood to size, and round it out with a router, rasp, or file so it matches the shape of the motor casing.

Connecting to the Motor Shaft

Connecting the shaft of the motor to a gear, wheel, lever, or other mechanical part is probably the most difficult task of all. There is one exception to this, however: R/C servomotors are easier to mount, which is one reason they are so popular in hobby robotics. Motor shafts come in many different sizes, and because most—if not all—of the motors you'll use will come from surplus outlets, the shaft may be peculiar to the specific application for which the motor was designed.

Common shaft sizes are 1/16- and 1/8-inch for small hobby motors and 1/4-, 3/8-, or 5/16-inch for larger motors and gearboxes. Gear hubs are generally 1/4-, 1/2-, or 5/8-inch, so you'll need to find reducing bushings at an industrial supply store. Surplus is also a good source. The same goes for wheels, sprockets (for roller chain and timing pulleys), and bearings.

To attach things like gears and sprockets the gear or sprocket must usually be physically secured to the shaft by way of a setscrew, as depicted in Fig. 17.15. Sometimes a "press fit" is all that's required. Most better-made gears and sprockets have the setscrews in them or have provisions for inserting them. If the gear or sprocket has no setscrew and there is no hole for one, you'll have to drill and tap the hole for the screw.

There are two common alternatives if you can't use a setscrew. The first method is to add a spline, or key, to secure the gear or sprocket to the shaft. This requires some careful machining, as you must make a slot for the spline in the shaft as well as for the hub of the gear or sprocket. Another method is to thread the gear shaft, and mount the gear or sprocket using nuts and split lock washers (the split in the washer provides compression that keeps the assembly from working loose). Shaft threading is also sometimes necessary when you are attaching wheels. I've successfully used both methods, but I have found that threading the shaft is easier. Threading requires you to lock the shaft so it won't turn, which can be a problem with some motors. Also, be careful that the shavings from the threading die do not fall into the motor.

Attaching two shafts to one another is a common, but not insurmountable, problem. The best approach is to use a coupler. You tighten the coupler to the shaft using setscrews. Couplers are available from industrial supply houses and can be expensive, so shop carefully. Some couplers are flexible; that is, they give if the two shafts aren't perfectly aligned. These are the best, considering the not-too-close tolerances inherent in home-built robots. Some couplers are available that accept two shafts of different sizes.

Finally, there are dozens of other methods for attaching wheels, gears, and other objects to motor shafts. Several of these will be detailed in context in the chapters to come.

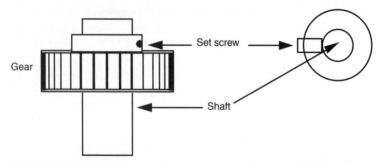

Gear

Set screw

Shaft

FIGURE 17.15 Use a setscrew to secure the gear to the shaft.

From Here

To learn more about...	*Read*
Learn more about robot locomotion systems	Chapter 16, "Robot Locomotion Principles"
All about DC motors	Chapter 18, "Working with DC Motors"
All about stepper motors	Chapter 19, "Working with Stepper Motors"
All about servo motors	Chapter 20, "Working with Servo Motors"
Operating motors by computer	Chapter 28, "An Overview of Robot 'Brains'"
Interfacing motors to electronic circuitry	Chapter 29, "Interfacing with Computers and Microcontrollers"

18

WORKING WITH DC MOTORS

DC motors are the mainstay of robotics. A surprisingly small motor, when connected to wheels through a gear reduction system, can power a 25-, 50-, even 100-pound robot seemingly with ease. A flick of a switch, a click of a relay, or a tick of a transistor, and the motor stops in its tracks and turns the other way. A simple electronic circuit enables you to gain quick and easy control over speed—from a slow crawl to a fast sprint.

This chapter shows you how to apply open-loop continuous DC motors (as opposed to stepping or servo DC motors) to power your robots. The emphasis is on using motors to propel a robot across your living room floor, but you can use the same control techniques for any motor application, including gripper closure, elbow flexion, and sensor positioning.

The Fundamentals of DC Motors

There are a many ways to build a DC motor. By their nature, all DC motors are powered by direct current—hence the name *DC*—rather than the alternating current (AC) used by most motorized household appliances. By and large, AC motors are less expensive to manufacture than DC motors, and because their construction is simpler they tend to last longer than DC motors.

Perhaps the most common form of DC motor is the permanent magnet type, so called because it uses two or more permanent magnet pole pieces (called the *stator*). The turning shaft of the motor, or the *rotor,* is composed of windings that are connected to a mechanical commutator. Internally, metal brushes (which can wear out!) supply the contact point for the current that turns the motor.

Other types of DC motors exist as well, including the series wound (or universal) and the shunt wound DC motors. These differ from the permanent magnet motor in that no magnets are used. Instead, the stator is composed of windings that, when supplied with current, become electromagnets.

One of the prime benefits of most, but not all, DC motors is that they are inherently reversible. Apply current in one direction (the + and − on the battery terminals, for example), and the motor may spin clockwise. Apply current in the other direction, and the motor spins counterclockwise. This capability makes DC motors well suited for robotics, where it is often desirable to have the motors reverse direction, such as to back a robot away from an obstacle or to raise or lower a mechanical arm.

Not all DC motors are reversible, and those that are typically exhibit better performance (though often just slightly better) in one direction over the other. For example, the motor may turn a few revolutions per minute faster in one direction. Normally, this is not observable in the typical motor application, but robotics isn't typical. In a robot with the common two-motor drive (see Chapter 16, "Robot Locomotion Principles"), the motors will be facing opposite directions, so one will turn clockwise while the other turns counterclockwise. If one motor is slightly faster than the other, it can cause the robot to steer off course. Fortunately, this effect isn't usually seen when the robot just travels short distances, and in any case, it can often be corrected by the control circuitry used in the robot.

Reviewing DC Motor Ratings

Motor ratings, such as voltage and current, were introduced in Chapter 17, "Choosing the Right Motor for the Job." Here are some things to keep in mind when considering a DC motor for your robot:

- DC motors can often be effectively operated at voltages above and below their specified rating. If the motor is rated for 12 volts, and you run it at 6 volts, the odds are the motor will still turn but at reduced speed and torque. Conversely, if the motor is run at 18 to 24 volts, the motor will turn faster and will have increased torque. This does not mean that you should intentionally under- or overdrive the motors you use. Significantly overdriving a motor may cause it to wear out much faster than normal. However, it's usually fairly safe to run a 10-volt motor at 12 volts or a 6-volt motor at 5 volts.
- DC motors draw the most current when they are "stalled." Stalling occurs if the motor is supplied current, but the shaft does not rotate. Any battery, control electronics, or drive circuitry you use with the motor must be able to deliver the current at stall, or major problems could result.
- DC motors vary greatly in efficiency. Many of the least expensive motors you may find are meant to be used in applications (such as automotive) where brute strength, rather

than the conservation of electricity, is the most important factor. Since the typical mobile robot is powered by a battery, strive for the most efficient motors you can get. It's best to stay away from automotive starter, windshield wiper, power window, and power seat motors since these are notoriously inefficient.

■ The rotational speed of a DC motor is usually too fast to be directly applied in a robot. Gear reduction of some type is necessary to slow down the speed of the motor shaft. Gearing down the output speed has the positive side effect of increasing torque.

Direction Control

As we noted earlier, it's fairly easy to change the rotational direction of a DC motor. Simply switch the power lead connections to the battery, and the motor turns in reverse. The small robots discussed in earlier chapters performed this feat by using a double-pole, double-throw (DPDT) switch. Two such switches were used, one for each of the drive motors. The wiring diagram for these robot motors is duplicated in Fig. 18.1 for your convenience. The DPDT switches used here have a center-off position. When they are in the center position, the motors receive no power so the robot does not move.

You can use the direction control switch for experimenting, but you'll soon want to graduate to more automatic control of your robot. There are a number of ways to accomplish the electronic or electrically assisted direction control of motors. All have their advantages and disadvantages. We'll see what they are in the sections to follow.

RELAY CONTROL

Perhaps the most straightforward approach to the automatic control of DC motors is to use relays. It may seem rather daft to install something as old-fashioned and cumbersome as relays in a high-tech robot, but it is still a useful technique. You'll find that while relays may wear out in time (after a few hundred thousand switchings), they are fairly inexpensive and easy to use.

You can accomplish basic on/off motor control with a single-pole relay. Rig up the relay so that current is broken when the relay is not activated. Turn on the relay, and the switch closes, thus completing the electrical circuit. The motor turns.

How you activate the relay is something you'll want to consider carefully. You *could* control it with a pushbutton switch, but that's no better than the manual switch method just described. Relays can easily be driven by digital signals. Fig. 18.2 shows the complete driver circuit for a relay-controller motor. Logical 0 (LOW) turns the relay off; logical 1 (HIGH) turns it on (refer to the parts list in Table 18.1). The relay can be operated from any digital gate, including a computer or microprocessor port. The chapters in Part 5 deal more extensively with using a computer for external control. Use the wiring diagram in Fig. 18.2 to prepare yourself for these later chapters.

Controlling the direction of the motor is only a little more difficult. This requires a double-pole, double-throw (DPDT) relay, wired in series after the on/off relay just described (see Fig. 18.3; refer to the parts list in Table 18.2). With the contacts in the relay in one

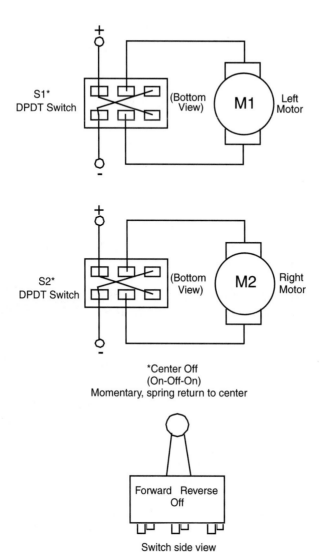

FIGURE 18.1 The basic wiring diagram for controlling twin robot drive motors. Note that the switches are DPDT and the spring return is set to center-off.

position, the motor turns clockwise. Activate the relay, and the contacts change positions, turning the motor counterclockwise. Again, you can easily control the direction relay with digital signals. Logical 0 makes the motor turn in one direction (let's say forward), and logical 1 makes the motor turn in the other direction. Both on/off and direction relay controls are shown combined in Fig. 18.4.

You can quickly see how to control the operation and direction of a motor using just two data bits from a computer. Since most robot designs incorporate two drive motors, you can control the movement and direction of your robot with just four data bits. When selecting

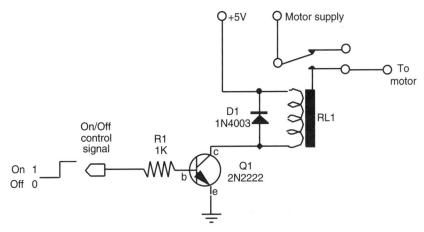

FIGURE 18.2 Using a relay to turn a motor on and off. The input signal is TTL/microprocessor compatible.

TABLE 18.1   PARTS LIST FOR ON-OFF RELAY CONTROL.

RL1	SPDT relay, 5 volt coil, contacts rated 2 amps or more
Q1	2N2222 NPN transistor
R1	1K resistor
D1	1N4003 diode

All resistors have 5 or 10 percent tolerance, 1/4-watt.

relays, make sure the contacts are rated for the motors you are using. All relays carry contact ratings, and they vary from a low of about 0.5 amp to over 10 amps, at 125 volts. Higher-capacity relays are larger and may require bigger transistors to trigger them (the very small reed relays can often be triggered by digital control without adding the transistor). For most applications, you don't need a relay rated higher than two or three amps.

BIPOLAR TRANSISTOR CONTROL

Bipolar transistors provide true solid-state control of motors. For the purpose of motor control, you use the bipolar transistor as a simple switch. By the way, note that when I refer to a "transistor" in this section I'm referring to a *bipolar* transistor. There are many kinds of transistors you can use, including the field effect transistor, or FET. In fact, we'll talk about FETs in the next section.

There are two common ways to implement the transistor control of motors. One way is shown in Fig. 18.5 (see the parts list in Table 18.3). Here, two transistors do all the work. The motor is connected so that when one transistor is switched on, the shaft turns clockwise. When the other transistor is turned on, the shaft turns counterclockwise. When both transistors are off, the motor stops turning. Notice that this setup requires a dual-polarity power supply. The schematic calls for a 6-volt motor and a +6-volt and −6-volt power source. This is known as a *split* power supply.

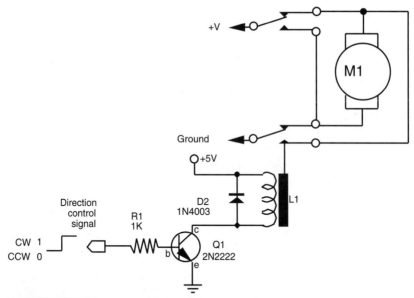

FIGURE 18.3 Using a relay to control the direction of a motor. The input signal is TTL/microprocessor compatible.

TABLE 18.2	PARTS LIST FOR DIRECTION RELAY CONTROL
RL1	DPDT relay, 5 volt coil, contacts rated 2 amps or more
Q1	2N2222 NPN transistor
R1	1K resistor
D1	1N4003 diode

All resistors have 5 or 10 percent tolerance, 1/4-watt.

Perhaps the most common way to control DC motors is to use the H-bridge network, as shown in Fig. 18.6 (see the parts list in Table 18.4). The figure shows a simplified H-bridge; some designs get quite complicated. However, this one will do for most hobby robot applications. The H-bridge is wired in such a way that only two transistors are on at a time. When transistor 1 and 4 are on, the motor turns in one direction. When transistor 2 and 3 are on, the motor spins the other way. When all transistors are off, the motor remains still.

Note that the resistor is used to bias the base of each transistor. These are necessary to prevent the transistor from pulling excessive current from the gate controlling it (computer port, logic gate, whatever). Without the resistor, the gate would overheat and be destroyed. The actual value of the bias resistor depends on the voltage and current draw of the motor, as well as the characteristics of the particular transistors used. For ballpark computations, the resistor is usually in the 1K- to 3K-ohm range. You can calculate the exact value of the resistor using Ohm's law, taking into account the gain and current output of the transistor, or you can experiment until you find a resistor value that works. Start high and work down, noting when the controlling electronics seem to get too hot. Don't go below 1K.

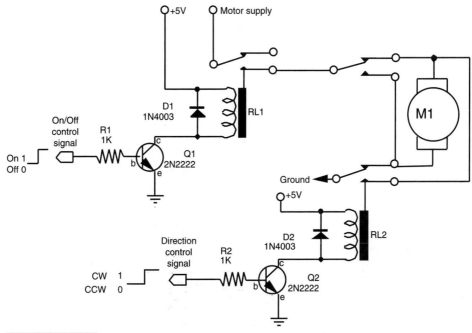

FIGURE 18.4 Both on/off and direction relay controls in one.

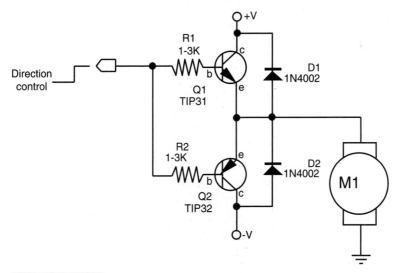

FIGURE 18.5 Using a complementary pair of transistors to control the direction of a motor. Note the double-ended (+ and −) power supply.

TABLE 18.3 PARTS LIST FOR TWO-TRANSISTOR MOTOR DIRECTION CONTROL.	
Q1	TIP31 NPN power transistor
Q2	TIP32 PNP power transistor
R1, R2	1–3K resistor
D1, D2	1N4002 diode
Misc.	Heat sinks for transistors

All resistors have 5 or 10 percent tolerance, 1/4-watt.

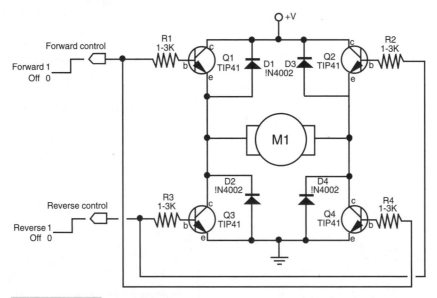

FIGURE 18.6 Four NPN transistors connected in an "H" pattern can be used to control the direction of a motor. The power supply is single ended.

TABLE 18.4 PARTS LIST FOR H-BRIDGE BIPOLAR TRANSISTOR MOTOR DIRECTION CONTROL.	
Q1–Q4	TIP41 NPN power transistor
R1–R4	1–3K resistor
D1–D4	1N4002 diode
Misc.	Heat sinks for transistors

All resistors have 5 or 10 percent tolerance, 1/4-watt.

The transistors you choose should comply with some general guidelines. First, they must be capable of handing the current draw demanded by the motors, but which specific transistor you finally choose will largely depend on your application and your design preference. Most large drive motors draw about one to three amps continuously, so the transistors you choose should be able to handle this. This immediately rules out the small-signal transistors, which are rated for no more than a few hundred milliamps.

A good NPN transistor for medium-duty applications is the TIP31, which comes in a TO-220 style case. Its PNP counterpart is the TIP32. Both of these transistors are universally available. Use them with suitable heat sinks. For high-power jobs, the NPN transistor that's almost universally used is the 2N3055 (get the version in the TO-3 case; it handles more power). Its close PNP counterpart is the MJ2955 (or 2N2955). Both transistors can handle up to 10 amps (115 watts), when used with a heat sink, such as the one in Fig. 18.7.

Another popular transistor to use in H-bridges is the TIP120, which is known as a Darlington transistor. Internally, it's actually two transistors: a smaller "booster" transistor and a larger power transistor. The TIP120 is preferred because it's often easier to interface it with control electronics. Some transistors, like the 2N3055, require a hefty amount of current in order to switch, and not all computer ports can supply this current. If you're not using a Darlington like the TIP120, it's sometimes necessary to use small-signal transistors (the 2N2222 is common) between the computer port and the power transistor.

The driving transistors should be located off the main circuit board—ideally directly on a large heat sink or at least on a heavy board with clip-on or bolt-on heat sinks attached to

FIGURE 18.7 Power transistors mounted on a heat sink.

the transistors. Use the proper mounting hardware when attaching transistors to heat sinks.

Remember that with most power transistors, the case is the collector terminal. This is particularly important when there is more than one transistor on a common heat sink and they aren't supposed to have their collectors connected together. It's also important when that heat sink is connected to the grounded metal frame of the robot. You can avoid any extra hassle by using the insulating washer provided in most transistor mounting kits.

The power leads from the battery and to the motor should be 12- to 16-gauge wire. Use solder lugs or crimp-on connectors to attach the wire to the terminals of T0-3-style transistors. Don't tap off power from the electronics for the driver transistors; get it directly from the battery or main power distribution rail. See Chapter 15, "All about Batteries and Robot Power Supplies," for more details about robot power distribution systems.

POWER MOSFET CONTROL

Wouldn't it be nice if you could use a transistor without bothering with bias resistors? Well, you can as long as you use a special brand of transistor, the power MOSFET. The MOSFET part stands for "metal oxide semiconductor field effect transistor." The "power" part means you can use them for motor control without worrying about them or the controlling circuitry going up in smoke.

Physically, MOSFETs look a lot like transistors, but there are a few important differences. First, like CMOS ICs, it is entirely possible to damage a MOSFET device by zapping it with static electricity. When handling it, always keep the protective foam around the terminals. Further, the names of the terminals are different than transistors. Instead of *base, emitter,* and *collector,* MOSFETs have a *gate, source,* and *drain.* You can easily damage a MOSFET by connecting it in the circuit improperly. Always refer to the pin-out diagram before wiring the circuit, and double-check your work.

A commonly available power MOSFET is the IRF-5xx series (such as the IRF-520, IRF-530, etc.), from International Rectifier, one of the world's leading manufacturers of power MOSFET components. These *N-channel* MOSFETs come in a T0-220-style transistor case and can control several amps of current (when on a suitable heat sink). A basic, semi-useful circuit that uses MOSFETs is shown in Fig. 18.8 (see the parts list in Table 18.5). Note the similarity between this design and the transistor design on Fig. 18.6.

An even better H-bridge with power MOSFETs uses two N-channel MOSFETs for the "low side" of the bridge and two complementary P-channel MOSFETs for the "high side." I won't get into the details about why this is better (the subject is adequately addressed in many books and Web sites). Suffice it to say that the use of complementary MOSFETs allows all four transistors in the H-bridge to turn completely on, thereby supplying the motor with full voltage. Fig. 18.9 shows a revised schematic (refer to the parts list in Table 18.6).

In both circuits, logic gates provide positive-action control. When the control signal is LOW, the motor turns clockwise. When the control signal is HIGH, the motor turns counterclockwise.

MOTOR BRIDGE CONTROL

The control of motors is big business, and it shouldn't come as a surprise that dozens of companies offer all-in-one solutions for controlling motors through fully electronic means. These products range from inexpensive $2 integrated circuits to sophisticated modules

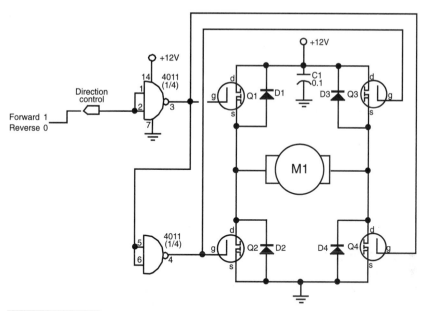

FIGURE 18.8 Four N-channel power MOSFET transistors in an "H" pattern can be used to control the direction of a motor. In a circuit application such as this, MOSFET devices do not strictly require biasing resistors, as do standard transistors.

TABLE 18.5 PARTS LIST FOR N-CHANNEL POWER MOSFET MOTOR CONTROL BRIDGE.

IC1	4011 CMOS Quad NAND Gate IC
Q1–Q4	IRF-5xx series (e.g., IRF-530 or equiv.) power MOSFET
D1–D4	1N4002 diode
Misc.	Heat sinks for transistors

costing tens of thousands of dollars. Of course, we'll confine our discussion to the low end of this scale!

The basic motor control is an H-bridge, as discussed earlier, an all-in-one integrated circuit package. Bridges for high-current motors tend to be physically large, and they may come with heat fins or have connections to a heat sink. A good example of a motor bridge is the Allegro Microsystems 3952, which provides in one single package a much improved version of the circuit shown in Fig. 18.9. A typical working circuit using the 3952 is shown in Fig. 18.10.

Motor control bridges have two or more pins on them for connection to control electronics. Typical functions for the pins are:

■ *Motor enable.* When enabled, the motor turns on. When disabled, the motor turns off. Some bridges let the motor "float" when disabled; that is, the motor coasts to a stop. On other bridges disabling the motor causes a full or partial short across the motor terminals, which acts as a brake to stop the motor very quickly.

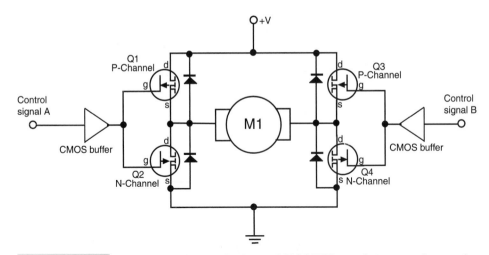

FIGURE 18.9 Combination N- and P-channel MOSFET transistors can be used to increase the voltage flowing to the motors. The MOSFETs should be "complementary pairs" (made to work with one another) that share the same voltage and current ratings. Most makers of MOSFET transistors provide complementary N- and P-channel products.

TABLE 18.6   PARTS LIST FOR COMPLEMENTARY N- AND P-CHANNEL POWER MOSFET MOTOR CONTROL BRIDGE.

IC1	4009 CMOS hex inverter IC
Q1, Q3	IRF-9530 (or equiv.) P-channel power MOSFET
Q2, Q4	IRF-530 (or equiv.) N-channel power MOSFET
D1–D4	1N4002 diode
Misc.	Heat sinks for transistors

■ *Direction.* Setting the direction pin changes the direction of the motor.
■ *Brake.* On bridges that allow the motor to float when the enable pin is disengaged, a separate brake input is used to specifically control the braking action of the motor.
■ *PWM.* Most H-bridge motor control ICs are used not only to control the direction and power of the motor, but its speed as well. The typical means used to vary the speed of a motor is with pulse width modulation, or PWM. This topic is described more fully in the next section.

The better motor control bridges incorporate overcurrent protection circuitry, which prevents them from being damaged if the motor pulls too much current and overheats the chip. Some even provide for *current sense,* an output that can be fed back to the control electronics in order to monitor the amount of current being drawn from the motor. This

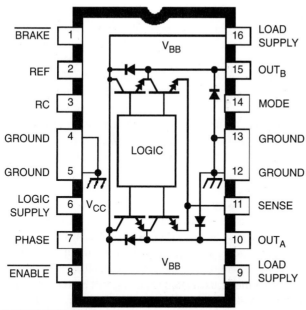

FIGURE 18.10 The Allegro 3952 is one of several all-in-one H-bridge motor control ICs.

can be useful when you need to determine if the robot has become stuck. Recall from earlier in this chapter that DC motors will draw the most current when they are stalled. If the robot gets caught on something and can't budge, the motors will stop, and the current draw will increase.

Because motor control bridges are so easy to use, and cost relatively little, we'll gravitate toward using them in the projects we describe rather than the "discrete" methods discussed in the previous sections. Of course, you're free to use whatever motor control methods you wish.

Some of the available motor control bridges include the L293D and L298N from SGS-Thomson; the 754410, an improved version of the L293 from Texas Instruments; and the LM18293 from National Semiconductor. Be sure to also check the listings in Appendix B, "Sources," and Appendix C, "Robot Information on the Internet," for additional sources of information on motor control bridges.

Motor Speed Control

There will be plenty of times when you'll want the motors in your robot to go a little slower, or perhaps track at a predefined speed. Speed control with continuous DC motors is a science in its own, and there are literally dozens of ways to do it. We'll cover some of the more popular methods in this and later chapters.

NOT THE WAY TO DO IT

Before exploring the right ways to control the speed of motors, let's examine how *not* to do it. Many robot experimenters first attempt to vary the speed of a motor by using a potentiometer. While this scheme certainly works, it wastes a lot of energy. Turning up the resistance of the pot decreases the speed of the motor, but it also causes excess current to flow through the pot. That current creates heat and draws off precious battery power.

Another similar approach is shown in Fig. 18.11. Here, a transistor is added to the basic circuit, but again, excess current flows through the transistor, and the energy is dissipated as lost heat. There are, fortunately, far better ways of doing it. Read on.

BASIC SPEED CONTROL

Figure 18.12 shows a schematic that is a variation of the MOSFET circuit shown in Fig. 18.8, above. This circuit provides rudimentary speed control. The 4011 NAND gate acts as an astable multivibrator, a pulse generator. By varying the value of R3, you increase or decrease the duration of the pulses emitted by the gates of the 4011. The longer the duration of the pulses, the faster the motor because it is getting full power for a longer period of time. The shorter the duration of the pulses, the slower the motor.

Notice that the power or voltage delivered to the motor does not change, as it does with the pot-only or pot-transistor scheme described earlier. The only thing that changes is the amount of time the motor is provided with full power. Incidentally, this technique is called *duty cycle* or *pulse width modulation (PWM),* and is the basis of most popular motor speed control circuits. There are a number of ways of providing PWM; this is just one of dozens. Fig. 18.13 shows a timing diagram of the PWM technique, from 100 percent duty cycle (100 percent on) to 0 percent duty cycle (0 percent on).

It is important to note that the frequency of the pulses does not change, just the relative on/off times. PWM frequencies of 2 kHz to over 25 kHz are commonly employed, depending on the motor. Unless you have a specification sheet from the manufacturer of the motor, you may have to do some experimentation to arrive at the "ideal" pulse frequency to use. You want to select the frequency that offers maximum power with minimum current draw.

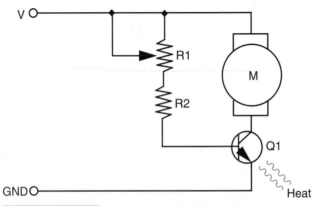

FIGURE 18.11   How not to vary the speed of a motor.
This approach is very inefficient.

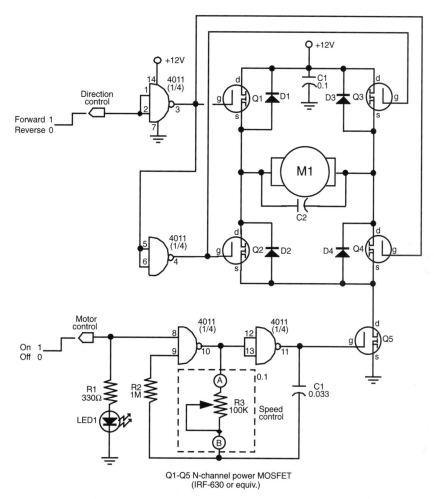

FIGURE 18.12 A rudimentary speed and direction control circuit using power MOSFETs. Resistor R1 and the LED serve to indicate that the motor is on.

Excessively high PWM frequencies may negate the speed control aspect, whereas excessively low frequencies may cause significant current draw and motor heating.

In the circuit shown in Fig. 18.12, R3 is shown surrounded by a dotted box. You can substitute R3 with a fixed resistor if you want to always use a certain speed, or you can use the circuit shown in Fig. 18.14. This circuit employs a 4066 CMOS analog switch IC. The 4066 allows you to select any of up to four speeds by computer or electronic control.

You connect resistors of various values to one side of the switches; the other side of the switches are collectively connected to the 4011. To modify the speed of the motor, activate one of the switches by bringing its control input to HIGH. The resistor connected to that switch is then brought into the circuit. You can omit the 3.3K pull-down resistors on the control inputs if your control circuitry is always activated and connected.

The 4066 is just one of several CMOS analog switches. There are other versions of this IC with different features and capabilities. We chose the 4066 here because it adds very

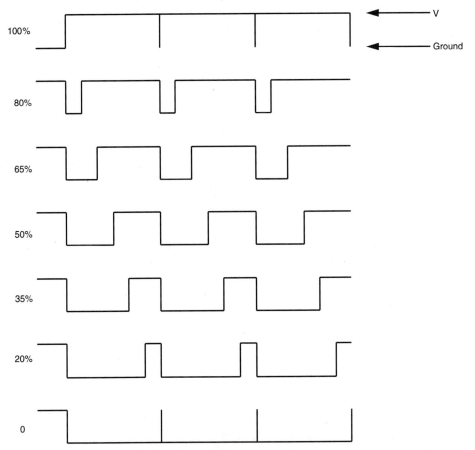

FIGURE 18.13 Pulse width modulation waveform. Note that the frequency of the pulses do not change, just the on and off times (duty cycle).

little resistance of its own when the switches are on. Note that the 4066 specifications sheet says that only one switch should be closed at a time.

PROCESSOR-BASED SPEED CONTROL

Using 4066 analog switches and individual resistors limits the number of speed choices you have. You may want to go from 90 percent to 88 percent duty cycle to control your motor, but the selection of resistors that you've used only provide for 90 percent and 80 percent, with no other values between. If you plan on controlling your robot via a computer or microcontroller (see Part 5 for more information on these topics), you can use software to provide any duty cycle you darn well please.

The computer or microcontroller cannot directly control a motor because the motor draws too much current. Instead, you connect the output of the computer or microcontroller to the control pin of an H-bridge or motor bridge IC, as shown in Fig. 18.15. Later chapters in Part 5 will detail the specific software you can use to vary the speed of a DC motor.

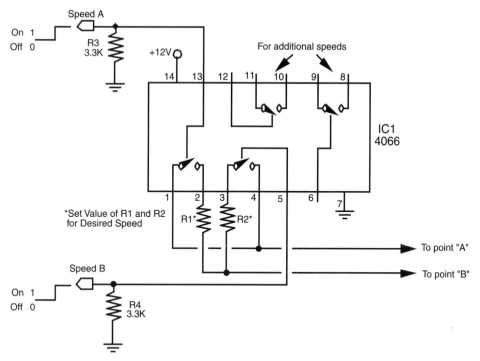

FIGURE 18.14 Using a 4066 CMOS analog switch to remotely control the speed of the motor. Use a device such as the 4051 for even more speed choices.

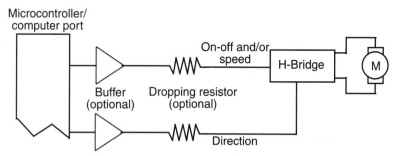

FIGURE 18.15 The basic connection between a computer or microcontroller and a DC motor. The computer or microcontroller operates the on/off control and the speed of the motor.

Odometry: Measuring Distance of Travel

Shaft encoders allow you to measure not only the distance of travel of the motors, but their velocity. By counting the number of transitions provided by the shaft encoder, the robot's control circuits can keep track of the revolutions of the drive wheels.

ANATOMY OF A SHAFT ENCODER

The typical shaft encoder is a disc that has numerous holes or slots along its outside edge. An infrared LED is placed on one side of the disc, so that its light shines through the holes. The number of holes or slots is not a consideration here, but for increased speed resolution, there should be as many holes around the outer edge of the disc as possible. An infrared-sensitive phototransistor is positioned directly opposite the LED (see Fig. 18.16) so that when the motor and disc turn, the holes pass the light intermittently. The result, as seen by the phototransistor, is a series of flashing light.

Instead of mounting the shaft encoders on the motor shafts, mount them on the wheel shafts (if they are different). The number of slots in the disk determines the maximum accuracy of the travel circuit. The more slots, the better the accuracy.

Let's say the encoder disc has 50 slots around its circumference. That represents a minimum sensing angle of 7.2°. As the wheel rotates, it provides a signal to the counting cir-

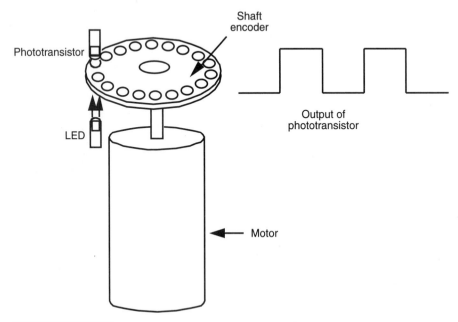

FIGURE 18.16 An optical shaft encoder attached to a motor. Alternatively, you can place a series of reflective strips on a black disc and bounce the LED light into the phototransistor.

cuit every 7.2°. Stated another way, if the robot is outfitted with a 7-inch wheel (circumference = 21.98 inches), the maximum travel resolution is approximately 0.44 linear inches. Not bad at all! This figure was calculated by taking the circumference of the wheel and dividing it by the number of slots in the shaft encoder.

The outputs of the phototransistor are conditioned by Schmitt triggers. This smooths out the wave shape of the light pulses so only voltage inputs above or below a specific threshold are accepted (this helps prevent spurious triggers). The output of the triggers is applied to the control circuitry of the robot.

THE DISTANCE COUNTER

The pulses from a shaft encoder do not in themselves carry distance measurement. The pulses must be counted and the count converted to distance. Counting and conversion are ideal tasks for a computer. Most single-chip computers and microprocessors, or their interface adapters, are equipped with counters. If your robot lacks a computer or microprocessor with a timer, you can add one using a 4040 12-stage binary ripple counter (see Fig. 18.17). This CMOS chip has 12 binary weighted outputs and can count to 4096. You'd probably use just the first eight outputs to count to 256.

Any counter with a binary or BCD output can be used with a 7485 magnitude comparator. A pinout of this versatile chip is shown in Fig. 18.18 and a basic hookup diagram in Fig. 18.19. In operation, the chip will compare the binary weighted number at its "A" and "B" inputs. One of the three LEDs will then light up, depending on the result of the difference between the two numbers. In a practical circuit, you'd replace the DIP switches (in the dotted box) with a computer port.

You can cascade comparators to count to just about any number. If counting in BCD, three packages can be used to count to 999, which should be enough for most distance

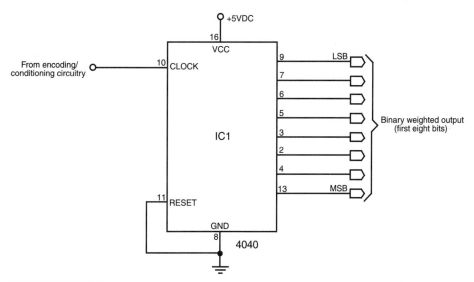

FIGURE 18.17 The basic wiring diagram of the 4040 CMOS 12-stage ripple counter IC.

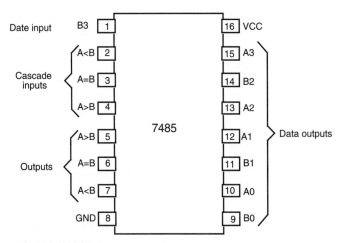

FIGURE 18.18 Pinout diagram of the 7485 magnitude comparator IC.

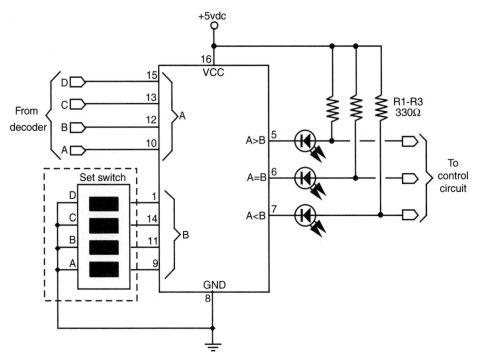

FIGURE 18.19 The basic wiring diagram of a single-state magnitude comparator circuit.

recording purposes. Using a disc with 25 slots in it and a 7-inch drive wheel, the travel resolution is 0.84 linear inches. Therefore, the counter system will stop the robot within 0.84 inches of the desired distance (allowing for coasting and slip between the wheels and ground) up to a maximum working range of 69.93 feet. You can increase the distance by building a counter with more BCD stages or decreasing the number of slots in the encoder disc.

MAKING THE SHAFT ENCODER

By far, the hardest part about odometry is making or adapting the shaft encoders. (You can also buy shaft encoders ready-made.) The shaft encoder you make may not have the fine resolution of a commercially made disc, which often have 256 or 360 slots in them, but the home-made versions will be more than adequate. You may even be able to find already machined parts that closely fit the bill, such as the encoder wheels in a discarded mouse (the computer kind, not the live rodent kind). Fig. 18.20 shows the encoder wheels from a surplus $5 mouse. The mouse contains two encoders, one for each wheel of the robot.

You can also make your own shaft encoder by taking a 1- to 2-inch disc of plastic or metal and drilling holes in it. Remember that the disc material must be opaque to infrared light. Some things that may look opaque to you may actually pass infrared light.

FIGURE 18.20 The typical PC mouse contains two shaft encoder discs. They are about perfect for the average small-or-medium-size robot.

When in doubt, add a coat or two of flat black or dark blue paint. That should block stray infrared light from reaching the phototransistor. Mark the disc for at least 20 holes, with a minimum size of about 1/16 inch. The more holes the better. Use a compass to scribe an exact circle for drilling. The infrared light will only pass through holes that are on this scribe line.

MOUNTING THE HARDWARE

Secure the shaft encoder to the shaft of the drive motor or wheel. Using brackets, attach the LED so that it fits snugly on the back side of the disc. You can bend the lead of the LED a bit to line it up with the holes. Do the same for the phototransistor. You must mask the phototransistor so it doesn't pick up stray light or reflected light from the LED, as shown in Fig. 18.21. You can increase the effectiveness of the phototransistor placing an infrared filter (a dark red filter will do in a pinch) between the lens of the phototransistor and the disc. You can also use the type of phototransistor that has its own built-in infrared filter.

If you find that the circuit isn't sensitive enough, check whether stray light is hitting the phototransistor. Baffle it with a piece of black construction paper if necessary. Or, if you prefer, you can use a "striped" disc of alternating white and black spokes as well as a reflectance IR emitter and detector. Reflectance discs are best used when you can control or limit the amount of ambient light that falls on the detector.

QUADRATURE ENCODING

So far we've investigated shaft encoders that have just one output. This output pulses as the shaft encoder turns. By using two LEDs and phototransistors, positioned 90° out of phase (see Fig. 18.22), you can construct a system that not only tells you the amount of travel, but the direction as well. This can be useful if the wheels of your robot may

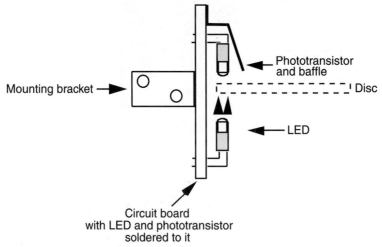

FIGURE 18.21 How to mount an infrared LED and phototransistor on a circuit board for use with an optical shaft encoder disc.

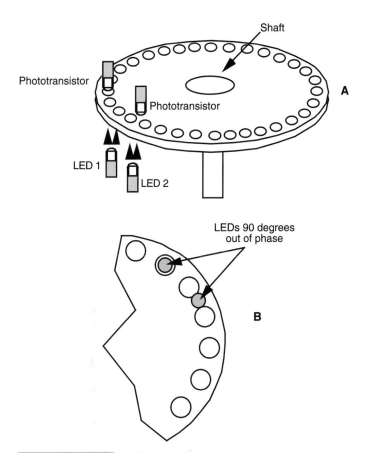

FIGURE 18.22 LEDs and phototransistors mounted on a two-channel optical disc. *a.* The LEDs and phototransistors can be placed anywhere about the circumference of the disc; *b.* The two LEDs and phototransistors must be 90° out of phase.

slip. You can determine if the wheels are moving when they aren't supposed to be, and you can determine the direction of travel. This so-called two-channel system uses *quadrature encoding*—the channels are out of phase by 90° (one quarter of a circle).

Use the flip-flop circuit in Fig. 18.23 to "separate" the distance pulses from the direction pulses. Note that this circuit will *only* work when you are using quadrature encoding, where the pulses are in the following format:

off/off

on/off

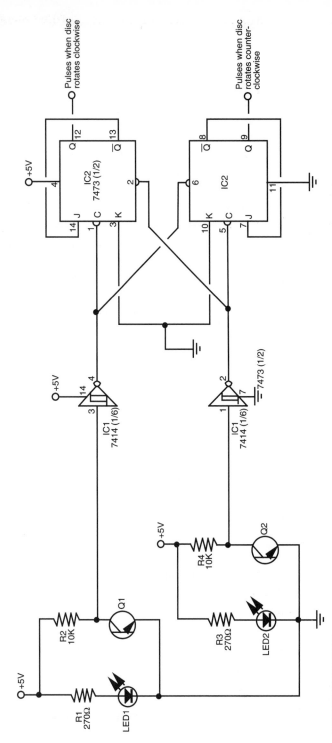

FIGURE 18.23 A two-channel shaft encoder circuit for use with a quadrature (also called two channel, 2-bit Gray code, or sine/cosine) encoder. One of the outputs of the flip-flop indicates distance (or relative speed) and the other the direction of rotation.

on/on

off/on

(... and repeat.)

From Here

To learn more about...	*Read*
Selecting the right motors for your robot	Chapter 17, "Choosing the Right Motor for the Job"
Using stepper motors	Chapter 19, "Working with Stepper Motors"
Interfacing motors to computers and microcontrollers	Chapter 29, "Interfacing with Computers and Microcontrollers"
More on odometry and measuring the distance of travel of a robot	Chapter 38, "Navigating through Space"

19

WORKING WITH STEPPER MOTORS

In past chapters we've looked at powering robots using everyday continuous DC motors. DC motors are cheap, deliver a lot of torque for their size, and are easily adaptable to a variety of robot designs. By their nature, however, the common DC motor is rather imprecise. Without a servo feedback mechanism or tachometer, there's no telling how fast a DC motor is turning. Furthermore, it's difficult to command the motor to turn a specific number of revolutions, let alone a fraction of a revolution. Yet this is exactly the kind of precision robotics work, particularly arm designs, often requires.

Enter the stepper motor. Stepper motors are, in effect, DC motors with a twist. Instead of being powered by a continuous flow of current, as with regular DC motors, they are driven by pulses of electricity. Each pulse drives the shaft of the motor a little bit. The more pulses that are fed to the motor, the more the shaft turns. As such, stepper motors are inherently "digital" devices, a fact that will come in handy when you want to control your robot by computer. By the way, there are AC stepper motors as well, but they aren't really suitable for robotics work and so won't be discussed here.

Stepper motors aren't as easy to use as standard DC motors, however, and they're both harder to get and more expensive. But for the applications that require them, stepper motors can solve a lot of problems with a minimum of fuss. Let's take a closer look at steppers and learn how to apply them to your robot designs.

Inside a Stepper Motor

There are several designs of stepper motors. For the time being, we'll concentrate on the most popular variety, the four-phase unipolar stepper, like the one in Fig. 19.1. A unipolar stepper motor is really two motors sandwiched together, as shown in Fig. 19.2. Each motor is composed of two windings. Wires connect to each of the four windings of the motor pair, so there are eight wires coming from the motor. The commons from the windings are often ganged together, which reduces the wire count to five or six instead of eight (see Fig. 19.3).

WAVE STEP SEQUENCE

In operation, the common wires of a unipolar stepper are attached to the positive (sometimes the negative) side of the power supply. Each winding is then energized in turn by grounding it to the power supply for a short time. The motor shaft turns a fraction of a revolution each time a winding is energized. For the shaft to turn properly, the windings must be energized in sequence. For example, energize wires 1, 2, 3, and 4 in sequence and the motor turns clockwise. Reverse the sequence, and the motor turns the other way.

FOUR-STEP SEQUENCE

The wave step sequence is the basic actuation technique of unipolar stepper motors. Another, and far better, approach actuates two windings at once in an on-on/off-off four-step sequence, as shown in Fig. 19.4. This enhanced actuation sequence increases the driving power of the motor and provides greater shaft rotation precision.

There are other varieties of stepper motors, and they are actuated in different ways. One you may encounter is bipolar. It has four wires and is pulsed by reversing the polarity of the power supply for each of the four steps. We will discuss the actuation technique for these motors later in this chapter.

Design Considerations of Stepper Motors

Stepping motors differ in their design characteristics over continuous DC motors. The following section discusses the most important design specifications for stepper motors.

STEPPER PHASING

A unipolar stepper requires that a sequence of four pulses be applied to its various windings for it to rotate properly. By their nature, all stepper motors are at least two-phase. Many are four-phase; some are six-phase. Usually, but not always, the more phases in a motor, the more accurate it is.

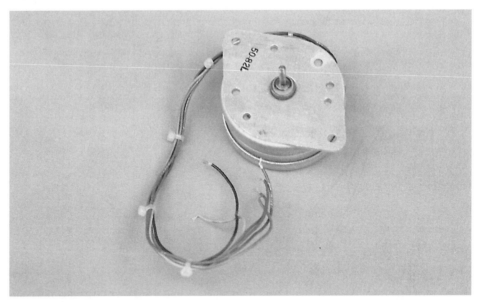

FIGURE 19.1 A typical unipolar stepper motor.

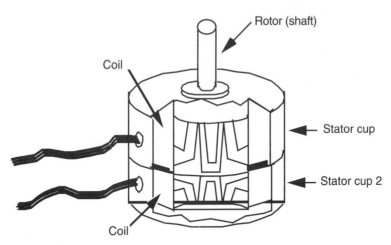

FIGURE 19.2 Inside a unipolar stepper motor. Note the two sets of coils and stators. The unipolar stepper is really two motors sandwiched together.

STEP ANGLE

Stepper motors vary in the amount of rotation of the shaft each time a winding is energized. The amount or rotation is called the *step angle* and can vary from as small as 0.9° (1.8° is more common) to 90°. The step angle determines the number of steps per revolution. A stepper with a 1.8° step angle, for example, must be pulsed 200 times for the shaft to turn one complete revolution. A stepper with a 7.5° step angle must be pulsed 48 times for one revolution, and so on.

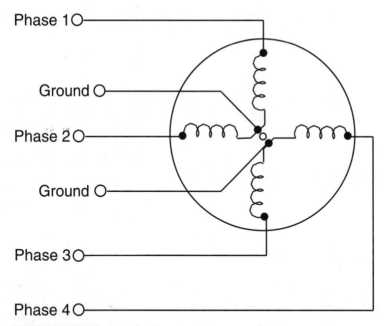

FIGURE 19.3 The wiring diagram of the unipolar stepper. The common connections can be separate or combined.

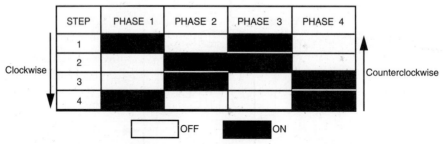

FIGURE 19.4 The enhanced on-on/off-off four-step sequence of a unipolar stepper motor.

PULSE RATE

Obviously, the smaller the step angle is, the more accurate the motor. But the number of pulses stepper motors can accept per second has an upper limit. Heavy-duty steppers usually have a maximum pulse rate (or step rate) of 200 or 300 steps per second, so they have an effective top speed of one to three revolutions per second (60 to 180 rpm). Some smaller steppers can accept a thousand or more pulses per second, but they don't usually provide very much torque and aren't suitable as driving or steering motors.

Note that stepper motors can't be motivated to run at their top speeds immediately from a dead stop. Applying too many pulses right off the bat simply causes the motor to freeze up. To achieve top speeds, you must gradually accelerate the motor. The acceleration can be quite swift in human terms. The speed can be one-third for the first few milliseconds, two thirds for the next few milliseconds, then full blast after that.

RUNNING TORQUE

Steppers can't deliver as much running torque as standard DC motors of the same size and weight. A typical 12-volt, medium-sized stepper motor may have a running torque of only 25 oz-in. The same 12-volt, medium-sized standard DC motor may have a running torque that is three or four times more.

However, steppers are at their best when they are turning slowly. With the typical stepper, the slower the motor revolves, the higher the torque. The reverse is usually true of continuous DC motors. Fig. 19.5 shows a graph of the running torque of a medium-duty, unipolar 12-volt stepper. This unit has a top running speed of 550 pulses per second. Since the motor has a step angle of 1.8°, that results in a top speed of 2.75 revolutions per second (165 rpm).

BRAKING EFFECT

Actuating one of the windings in a stepper motor advances the shaft. If you continue to apply current to the winding the motor won't turn any more. In fact, the shaft will be locked, as if you've applied brakes. As a result of this interesting locking effect, you never need to add a braking circuit to a stepper motor because it has its own brakes built in.

The amount of braking power a stepper motor has is expressed as *holding torque*. Small stepper motors have a holding torque of a few oz-in. Larger, heavier-duty models have holding torque exceeding 400 oz-in.

VOLTAGE, CURRENT RATINGS

Like DC motors, stepper motors vary in their voltage and current ratings. Steppers for 5-, 6-, and 12-volt operation are not uncommon. But unlike DC motors, if you use a higher voltage than specified for a stepper motor you don't gain faster operation, but more running

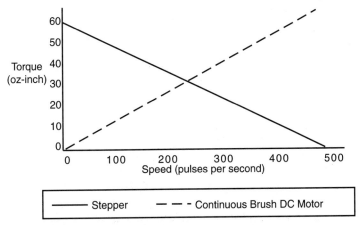

FIGURE 19.5 With a stepper motor, torque increases as the speed of the motor is reduced.

and holding torque. Overpowering a stepper by more than 80 to 100 percent above the rated voltage may eventually burn up the motor.

The current rating of a stepper is expressed in amps (or milliamps) per phase. The power supply driving the motor needs to deliver at least as much as the per-phase specification, preferably more if the motor is driving a heavy load. The four-step actuation sequence powers two phases at a time, which means the power supply must deliver at least twice as much current as the per-phase specification. If, for example, the current per phase is 0.25 amps, the power requirement at any one time is 0.50 amps.

Controlling a Stepper Motor

Steppers have been around for a long time. In the old days, stepper motors were actuated by a mechanical switch, a solenoid-driven device that pulsed each of the windings of the motor in the proper sequence. Now, stepper motors are invariably controlled by electronic means. Basic actuation can be accomplished via computer control by pulsing each of the four windings in turn. The computer can't directly power the motor, so transistors must be added to each winding, as shown in Fig. 19.6.

USING A STEPPER MOTOR CONTROLLER CHIP

In the absence of direct computer control, the easiest way to provide the proper sequence of actuation pulses is to use a custom stepper motor chip, such as the Allegro Microsystems

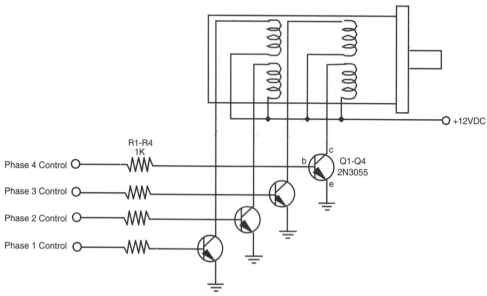

FIGURE 19.6 The basic hookup connection to drive a stepper motor from a computer or other electronic interface. The phasing sequence is provided by software or other means through a port following a four-bit binary sequence: 1010, 0110, 0101, 1001 (reverse the sequence to reverse the motor).

UCN5804. This chip is designed expressly for use with the common unipolar stepper motor and provides a four-step actuation sequence. Stepper motor translator chips tend to be modestly priced, at about $5 to $10, depending on their features and where you buy them.

Figure 19.7 (refer to the parts list in Table 19.1) shows a typical schematic of the UCN5804. Heavier duty motors (more than about 1A per phase) can be driven by adding power transistors to the four outputs of the chips, as shown in the manufacturer's application notes. Note the Direction pin. Pulling this pin high or low reverses the rotation of the motor.

Using logic gates to control stepper motors

Another approach to operating unipolar stepper motors is to use discrete gates and clock ICs. You can assemble a stepper motor translator circuit using just two IC packages. The circuit can be constructed using TTL or CMOS chips.

The TTL version is shown in Fig. 19.8 (refer to the parts list in Table 19.2). Four Exclusive OR gates from a single 7486 IC provide the steering logic. You set the direction by pulling pin 12 HIGH or LOW. The stepping actuation is controlled by a 7476, which contains two JK flip-flops. The Q and 'Q outputs of the flip-flops control the phasing of the motor. Stepping is accomplished by triggering the clock inputs of both flip-flops.

The 7476 can't directly power a stepper motor. You must use power transistors or MOSFETs to drive the windings of the motor. See the section titled "Translator Enhancements" for a complete power driving schematic as well as other options you can add to this circuit.

The CMOS version, shown in Fig. 19.9 (refer to the parts list in Table 19.3), is identical to the TTL version, except that a 4070 chip is used for the Exclusive OR gates and a 4027 is used for the flip-flops. The pinouts are slightly different, so follow the correct schematic for the type of chips you use. Note that another CMOS Exclusive OR package, the 4030, is also available. Don't use this chip; it behaved erratically in this, as well as other pulsed, circuits.

In both the TTL and CMOS circuits, the stepper motor itself can be operated from a supply voltage that is wholly different than the voltage supplied to the ICs.

TRANSLATOR ENHANCEMENTS

Four NPN power transistors, four resistors, and a handful of diodes are all the translator circuits described in the last section need to provide driving power. (You can also use this scheme to increase the driving power of the UCN5804, detailed earlier). The schematic for the circuit is shown in Fig. 19.10 (refer to the parts list in Table 19.4). Note that you can substitute the bipolar transistors and resistors with power MOSFETs. See Chapter 17, "Choosing the Right Motor for the Job," for more information on using power MOSFETs.

You can use just about any NPN power transistor that will handle the motor. The TIP31 is a good choice for applications that require up to one amp of current. Use the 2N3055 for heavier-duty motors. Mount the drive transistors on a suitable heat sink.

You must insert a bias resistor in series between the outputs of the translation circuit and the base of the transistors. Values between about 1K and 3K should work with most motors and most transistors. Experiment until you find the value that works without causing the flip-flop chips to overheat. You can also apply Ohm's law, figuring in the current draw of the motor and the gain of the transistor, to accurately find the correct value of the resistor. If this is new to you, see Appendix A, "Further Reading," for a list of books on electronic design and theory.

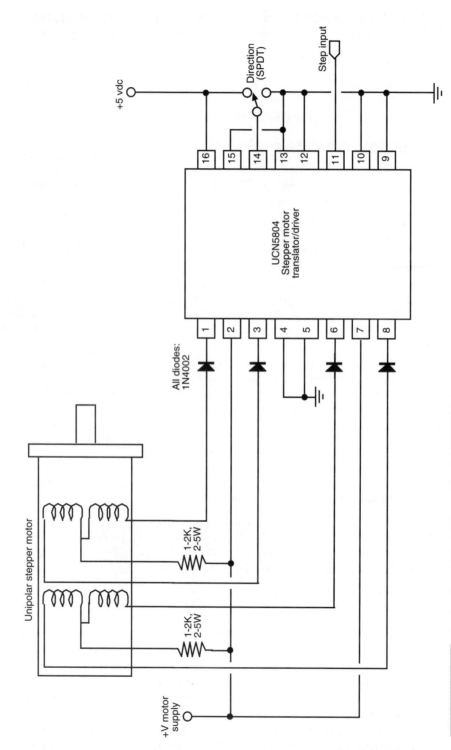

FIGURE 19.7 The basic wiring diagram for the UCN5804.

TABLE 19.1 PARTS LIST FOR UCN5804 STEPPER MOTOR TRANSLATOR/DRIVER.	
IC1	Allegro UCN5804 Stepper Motor Translator IC
R1, R2	1–2K resistor, 2–5 watts
D1–D4	1N4002 diode
M1	Unipolar stepper motor
Misc	SPDT switch, heat sinks for UCN5804 (as needed)

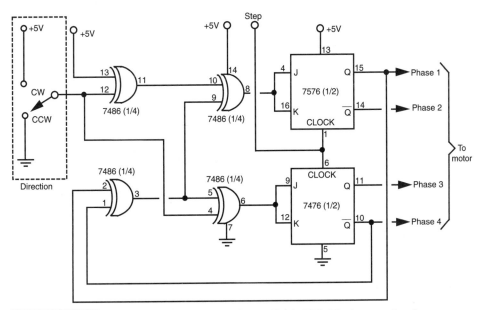

FIGURE 19.8 Using a pair of commonly available TTL ICs to construct your own stepper motor translator circuit.

TABLE 19.2 PARTS LIST FOR TTL STEPPER MOTOR TRANSLATOR	
IC1	7486 Quad Exclusive OR Gate IC
U2	7476 Dual "JK" flip-flop IC

It is sometimes helpful to see a visual representation of the stepping sequence. Adding an LED and current-limiting resistor in parallel with the outputs provides just such a visual indication. See Fig. 19.11 for a wiring diagram (refer to the parts list in Table 19.5). Note the special wiring to the flip-flop outputs. This provides a better visual indication of the stepping action than hooking up the LEDs in the same order as the motor phases.

Figure 19.12 shows two stepper motor translator boards. The small board controls up to two stepper motors and is designed using TTL chips. The LED option is used to provide a

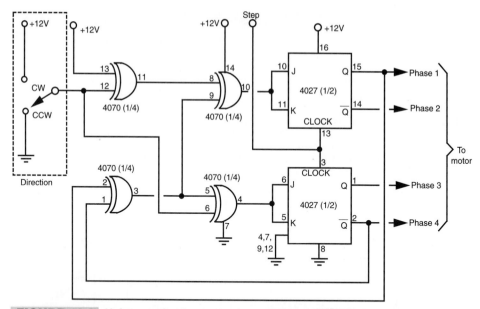

FIGURE 19.9 Using a pair of commonly available CMOS ICs to construct your own stepper motor translator circuit.

TABLE 19.3 PARTS LIST FOR CMOS STEPPER MOTOR TRANSLATOR.	
IC1	4070 Quad Exclusive OR Gate IC
IC2	4037 Dual "JK" flip-flop IC

visual reference of the step sequence. The large board uses CMOS chips and can accommodate up to four motors. The boards were wire-wrapped; the driving transistors are placed on a separate board and heat sink.

TRIGGERING THE TRANSLATOR CIRCUITS

You need a square wave generator to provide the triggering pulses for the motors. You can use the 555 timer wired as an astable multivibrator, or make use of a control line in your computer or microcontroller. When using the 555, remember to add the 0.1 μF capacitor across the power pins of the chip. The 555 puts a lot of noise into the power supply, and this noise regularly disturbs the counting logic in the Exclusive OR and flip-flop chips. If you are getting erratic results from your circuit, this is probably the cause.

USING BIPOLAR STEPPER MOTORS

As detailed earlier in the chapter, unipolar stepper motors contain four coils in which two of the coils are joined to make a center tap. This center tap is the "common"

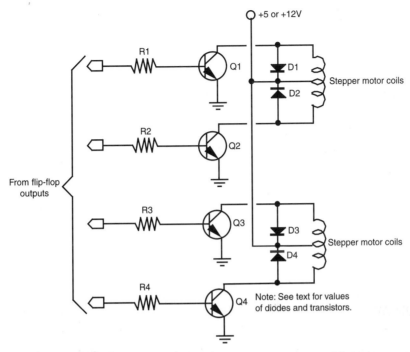

FIGURE 19.10 Add four transistors and resistors to provide a power
output stage for the TTL or CMOS stepper motor translator circuits.

TABLE 19.4	PARTS LIST FOR STEPPER MOTOR DRIVER.
Q1–Q4	Under 1 amp draw per phase: TIP32 NPN transistor
	1 to 3 amp draw per phase: TIP120 NPN Darlington transistor
R1–R4	1K resistor, 1 watt
D1–D4	1N4004 diode
Misc	Heat sinks for transistors

connection for the motor. Bipolar stepper motors contain two coils, do not use a common connection, and are generally less expensive because they are easier to manufacture. A bipolar stepper motor has four external connection points. An old method for
operating a bipolar stepper motor was to use relays to reverse the polarity of a DC voltage to two coils. This caused the motor to inch forward or backward, depending on the
phasing sequence.

Today, the more common method for operating a bipolar stepper motor is to use a specialty stepper motor translator, such as the SGS-Thompson L297D (the L297D can also
be used to drive unipolar stepper motors). To add more current driving capacity to the
L297D you can add a dual H-bridge driver, such as the L298N, which is available from

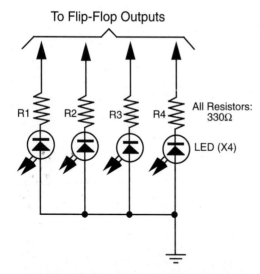

To Flip-Flop Outputs

R1   R2   R3   R4

All Resistors:
330Ω

LED (X4)

FIGURE 19.11 Add four LEDs and resistors to provide a visual indication of the stepping action.

TABLE 19.5 PARTS LIST FOR UCN5804 STEPPER MOTOR TRANSLATOR/DRIVER.	
R1–R4	300-ohm resistor
LED1–4	Light-emitting diode

All resistors have 5 or 10 percent tolerance, 1/4-watt.

the same company. The truth table for the typical driving sequence of a bipolar stepper motor is shown in Fig. 19.13 (see p. 292).

BUYING AND TESTING A STEPPER MOTOR

Spend some time with a stepper motor and you'll invariably come to admire its design and be able to think up all sorts of ways to make it work for you in your robot designs. But to use a stepper, you have to get one. That in itself is not always easy. Then after you have obtained it and taken it home, there's the question of figuring out where all the wires go! Let's take each problem one at a time.

SOURCES FOR STEPPER MOTORS

Despite their many advantages, stepper motors aren't nearly as common as the trusty DC motor, so they are harder to find. And when you do find them, they're expensive when new. The surplus market is by far the best source for stepper motors for hobby robotics. See Appendix B, "Sources," for a list of selected mail order surplus companies that regularly carry a variety of stepper motors. They carry most of the "name brand" steppers: Thompson-Airpax, Molon, Haydon, and Superior Electric. The cost of surplus steppers is often a quarter or fifth of the original list price.

The disadvantage of buying surplus is that you don't always get a hookup diagram or adequate specifications. Purchasing surplus stepper motors is largely a hit-or-miss

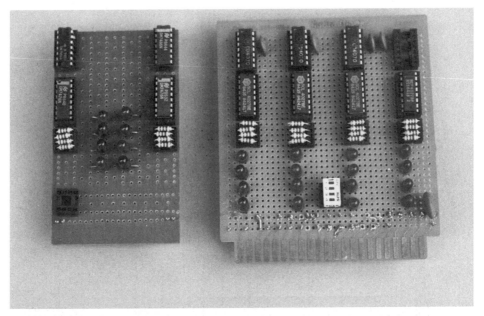

FIGURE 19.12 Two finished stepper motor translator boards, with indicator LEDs. The board on the left controls two stepper motors; the board on the right controls four stepper motors.

affair, but most outlets let you return the goods if they aren't what you need. If you like the motor, yet it still lacks a hookup diagram, read the following section on how to decode the wiring.

WIRING DIAGRAM

The internal wiring diagram of both a bipolar and unipolar stepper motor is shown in Fig. 19.14. The wiring in a bipolar stepper is actually easy to decode. You use a volt-ohm meter to do the job right. You can be fairly sure the motor is two-phase if it has only four wires leading to it. You can identify the phases by connecting the leads of the meter to each wire and noting the resistance. Wire pairs that give an open reading (infinite ohms) represent two different coils (phases). You can readily identify mating phases when there is a small resistance through the wire pair.

Unipolar steppers behave the same, but with a slight twist. Let's say, for argument's sake, that the motor has eight wires leading to it. Each winding, then, has a pair of wires. Connect your meter to each wire in turn to identify the mating pairs. As illustrated in Fig. 19.15, no reading (infinite ohms) signifies that the wires do not lead to the same winding; a reading indicates a winding.

If the motor has six wires, then four of the leads go to one side of the windings. The other two are commons and connect to the other side of the windings (see Fig. 19.16). Decoding this wiring scheme takes some patience, but it can be done. First, separate all those wires where you get an open reading. At the end of your test, there should be two three-wire sets that provide some reading among each of the leads.

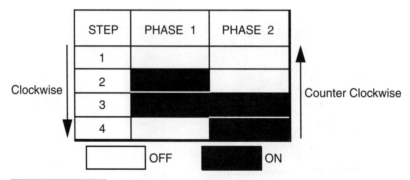

STEP	PHASE 1	PHASE 2
1		
2	■	
3	■	■
4		■

Clockwise ↓   Counter Clockwise ↑

☐ OFF   ■ ON

FIGURE 19.13 The phasing sequence for a bipolar stepper motor.

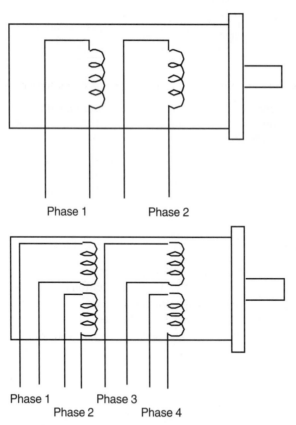

Phase 1   Phase 2

Phase 1   Phase 3
   Phase 2   Phase 4

FIGURE 19.14 Pictorial diagrams of the coils in a bipolar and unipolar stepper motor.

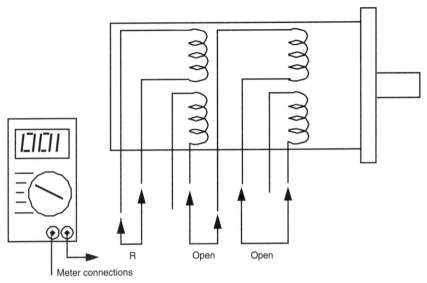

FIGURE 19.15 Connection points and possible readings on an 8-wire unipolar stepper motor.

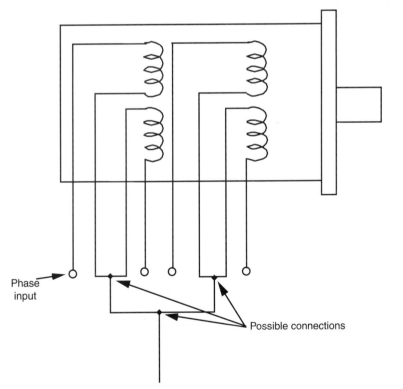

FIGURE 19.16 Common connections may reduce the wire count of the stepper motor to five or six, instead of eight.

Locate the common wire by following these steps. Take a measurement of each combination of the wires and note the results. You should end up with three measurements: wires 1 and 2, wires 2 and 3, and wires 1 and 3. The meter readings will be the same for two of the sets. For the third set, the resistance should be roughly doubled. These two wires are the main windings. The remaining wire is the common.

Decoding a five-wire motor is the most straightforward procedure. Measure each wire combination, noting the results of each. When you test the leads to one winding, the result will be a specified resistance (let's call it "R"). When you test the leads to two of the windings, the resistance will be double the value of "R," as shown in Fig. 19.17. Isolate this common wire with further testing and you've successfully decoded the wiring.

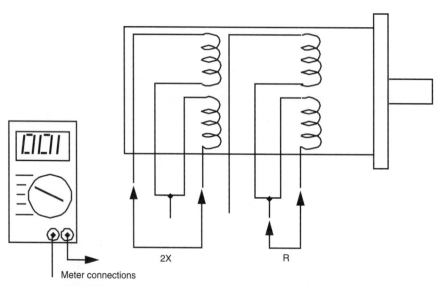

Meter connections

FIGURE 19.17 Connection points and possible readings on a 5- or 6-wire unipolar stepper motor.

From Here

To learn more about...	Read
Driving a robot	Chapter 16, "Robot Locomotion Principles"
Selecting a motor for your robot	Chapter 17, "Choosing the Right Motor for the Job"
Connecting motors to computers, microcontrollers, and other electronic circuitry	Chapter 29, "Interfacing with Computers and Microcontrollers"

20

WORKING WITH SERVO MOTORS

DC and stepper motors are inherently "open feedback" systems—you give them juice, and they spin. How much they spin is not always known, not even for a stepper motor, which turns by finite degrees based on the number of pulses it gets. Should something impede the rotation of the motor it may not turn at all, but there's no easy, built-in way that the control electronics would know that.

Servo motors, on the other hand, are designed for "closed feedback" systems. The output of the motor is coupled to a control circuit; as the motor turns, its speed and/or position are relayed to the control circuit. If the rotation of the motor is impeded for whatever reason, the feedback mechanism senses that the output of the motor is not yet in the desired location. The control circuit continues to correct the error until the motor finally reaches its proper point.

Servo motors come in various shapes and sizes. Some are smaller than a walnut, while others are large enough to take up their own seat in your car. They're used for everything from controlling computer-operated lathes to copy machines to model airplanes and cars. It's the last application that is of most interest to hobby robot builders: the same servo motors used with model airplanes and cars can readily be used with your robot.

These servo motors are designed to be operated via a radio-controlled link and so are commonly referred to as *radio-controlled* (or *R/C*) servos. But in fact the servo motor itself is not what is radio-controlled; it is merely connected to a radio receiver on the plane or car. The servo takes its signals from the receiver. This means you don't have to control your robot via radio signals just to use an R/C servo—unless you want to, of course. You can control a servo with your PC, a microcontroller such as the Basic Stamp, or even a simple circuit designed around the familiar 555 timer integrated circuit.

In this chapter we'll review what R/C servos are, and how they can be put to use in a robot. We will limit the discussion to R/C servos. While there are other types of servo motors, it is the R/C type that is commonly available and reasonably affordable. For simplicity's sake, when you see the term *servo* in the text that follows understand that it specifically means an R/C servo motor, even though there are other types.

How Servos Work

Figure 20.1 shows a typical standard-sized R/C servo motor, which is used with model flyable airplanes and model racing cars. The size and mounting of a standard servo is the same regardless of the manufacturer, which means that you have your pick of a variety of makers. There are other common sizes of servo motors besides that shown in Fig. 20.1, however. We'll discuss these in a bit.

Inside the servo is a motor, a series of gears to reduce the speed of the motor, a control board, and a potentiometer (see Fig. 20.2). The motor and potentiometer are connected to the control board, all three of which form a *closed feedback loop*. Both control board and motor are powered by a constant DC voltage (usually between 4.8 and 7.2 volts).

To turn the motor, a digital signal is sent to the control board. This activates the motor, which, through a series of gears, is connected to the potentiometer. The position of the potentiometer's shaft indicates the position of the output shaft of the servo. When the potentiometer has reached the desired position, the control board shuts down the motor.

As you can surmise, servo motors are designed for limited rotation rather than for continuous rotation like a DC or stepper motor. While it is possible to modify an R/C servo to rotate continuously (see later in this chapter), the primary use of the R/C servo is to achieve accurate rotational positioning over a range of 90° or 180°. While this may not sound like much, in actuality such control can be used to steer a robot, move legs up and down, rotate a sensor to scan the room, and more. The precise angular rotation of a servo in response to a specific digital signal has enormous uses in all fields of robotics.

Servos and Pulse Width Modulation

The motor shaft of an R/C servo is positioned by using a technique called *pulse width modulation* (PWM). In this system, the servo responds to the duration of a steady stream of digital pulses. Specifically, the control board responds to a digital signal whose pulses vary from about 1 millisecond (one thousandth of a second, or *ms*) to about 2 ms. These pulses are sent some 50 times per second. The exact length of the pulse, in fractions of a millisecond, determines the position of the servo. Note that it is not the number of pulses per second that controls the servo, but the *duration* of the pulses that matters. The servo requires about 30 to 60 of these pulses per second. This is referred to as the *refresh* rate; if the refresh rate is too low, the accuracy and holding power of the servo is reduced.

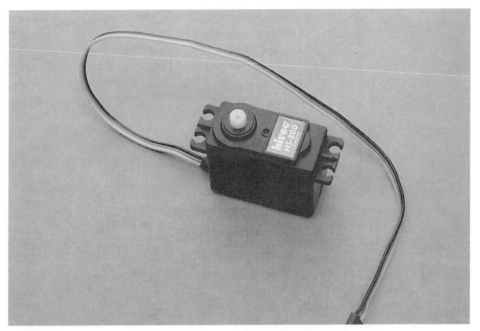

FIGURE 20.1 The typical radio-controlled (R/C) servo motor.

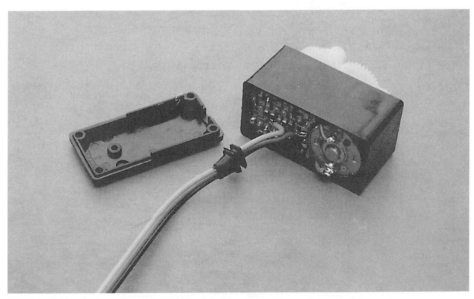

FIGURE 20.2 The internals of an R/C servo. The servo consists of a motor, a gear train, a potentiometer, and a control circuit.

At a duration of 1 ms, the servo is commanded to turn all the way in one direction (let's say counterclockwise, as shown in Fig. 20.3). At 2 ms, the servo is commanded to turn all the way in the other direction. Therefore, at 1.5 ms, the servo is commanded to turn to its center (or *neutral*) position. As mentioned earlier, the angular position of the servo is determined by the width (more precisely, the duration) of the pulse. This technique has gone by many names over the years. One you may have heard is *digital proportional*—the movement of the servo is proportional to the digital signal being fed into it.

The power delivered to the motor inside the servo is also proportional to the difference between where the output shaft is and where it's supposed to be. If the servo has only a little way to move to its new location, then the motor is driven at a fairly low speed. This ensures that the motor doesn't "overshoot" its intended position. But if the servo has a long way to move to its new location, then it's driven at full speed in order to get it there as fast as possible. As the output of the servo approaches its desired new position, the motor slows down. What seems like a complicated process actually happens in a very short period of time—the average servo can rotate a full 60° in a quarter to half second.

The actual length of the pulses varies between servo brands, and sometimes even between different models by the same manufacturer. The 1–2 ms range is typical but is by no means set in stone. When you are purchasing a servo brand for a model airplane or car, you should typically mate it with a radio receiver made by the same company to ensure compatibility. Since you're not likely to use a radio receiver with the R/C servos in your robot, you'll need to do some experimenting to find the optimum pulse width ranges for the servos you use. This is just part of what makes robot experimenting so fun!

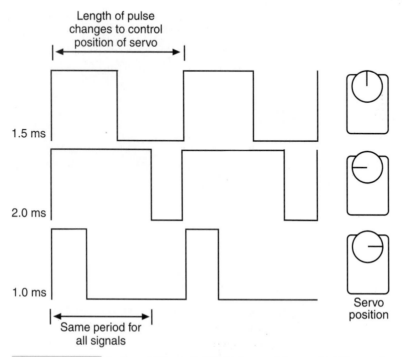

FIGURE 20.3 Pulse width modulation is used to control the position of the output shaft of the servo motor.

The Role of the Potentiometer

The potentiometer of the servo plays a key role in allowing the motor to set the position of its output shaft. The potentiometer is physically attached to the output shaft (and in some servo models, the potentiometer *is* the output shaft). In this way, the position of the potentiometer very accurately reflects the position of the output shaft of the servo. Recall that a potentiometer works by providing a varying voltage to a control circuit, as shown in Fig. 20.4. As the *wiper* inside the potentiometer moves, the voltage changes.

The control circuit in the servo correlates this voltage with the timing of the incoming digital pulses and generates an "error signal" if the voltage is wrong. This error signal is proportional to the difference between the position of the potentiometer and the timing of the incoming signal. To compensate, the control board applies the error signal to turn the motor. When the voltage from the potentiometer and the timing of the digital pulses match, the error signal is removed, and the motor stops.

Rotational Limits

Servos also vary by the amount of rotation they will perform for the 1–2 ms (or whatever) signal they are provided. Most standard servos are designed to rotate back and forth by 90° to 180°, given the full range of timing pulses. You'll find the majority of servos will be able to turn a full 180°, or very nearly so.

Should you attempt to command a servo beyond its mechanical limits, the output shaft of the motor will hit an internal stop. This causes the gears of the servo to grind or chatter. If left this way for more than a few seconds the gears of the motor may be permanently damaged.

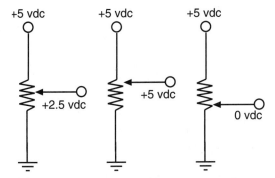

FIGURE 20.4 A potentiometer is often used as a variable voltage divider. As the potentiometer turns, its wiper travels the length of a resistive element. The output of the potentiometer is a varying voltage, from 0 to the +V of the circuit.

Therefore, when experimenting with servomotors exercise care to avoid pushing them beyond their natural limits.

Special-Purpose Servo Types and Sizes

While the standard-sized servo is the one most commonly used in both robotics and radio-controlled models, other R/C servo types, styles, and sizes exist as well.

- *Quarter-scale* (or *large-scale*) servos are about twice the size of standard servos and are significantly more powerful. Quarter-scale servos are designed to be used in large model airplanes, but they also make perfect power motors for a robot.
- *Mini-micro* servos are about half the size (and smaller!) of standard servos and are designed to be used in tight spaces in a model airplane or car. They aren't as strong as standard servos, however.
- *Sail winch* servos are designed with maximum strength in mind, and are primarily intended to move the jib and mainsail sheets on a model sailboat.
- *Landing-gear retraction* servos are made to retract the landing gear of medium- and large-sized model airplanes. The design of the landing gear often requires the servo to guarantee at least 170° degree rotation, if not more (i.e., up to and exceeding 360° of motion). It is not uncommon for retraction servos to have a slimmer profile than the standard variety because of the limited space on model airplanes.

Gear Trains and Power Drives

The motor inside an R/C servo turns at several thousand RPMs. This is too fast to be used directly on model airplanes and cars, or on robots. All servos employ a gear train that reduces the output of the motor to the equivalent of about 50–100 RPM. Servo gears can be made of plastic, nylon, or metal (usually brass or aluminum).

Metal gears last the longest, but they significantly raise the cost of the servo. Replacement gear sets are available for many servos, particularly the medium- to high-priced ones ($20+). Should one or more gears fail, the servo can be disassembled and the gears replaced. In some cases, you can "upgrade" the plastic gears in a less expensive servo to higher-quality metal ones.

Besides the drive gears, the output shaft of the servo receives the most wear and tear. On the least expensive servos this shaft is supported by a plastic "bearing," which obviously can wear out very quickly if the servo is used heavily. Actually, this piece is not a bearing at all but a *bushing*, a sleeve or collar that supports the shaft against the casing of the servo. Metal bushings, typically made from lubricant-impregnated brass, last longer but add to the cost of the servo. The best (and most expensive) servos come equipped with ball bearings, which provide longest life. Ball bearing "upgrades" are available for some servo models.

Typical Servo Specs

R/C servo motors enjoy some standardization. This sameness applies primarily to standard-sized servos, which measure approximately 1.6 inches by 0.8 inch by 1.4 inches. For other servo types the size varies somewhat between makers, as these are designed for specialized tasks.

Table 20.1 outlines typical specifications for several types of servos, including dimensions, weight, torque, and transit time. Of course, except for the size of standard servos, these specifications can vary between brand and model. A few of the terms used in the specs require extra discussion. As explained in Chapter 17, "Choosing the Right Motor for the Job," the *torque* of the motor is the amount of force it exerts. The standard torque unit of measure for R/C servos is expressed in ounce-inches—or the number of ounces the servo can lift when the weight is extended one inch from the shaft of the motor. Servos exhibit very high torque thanks to their speed reduction gear trains.

The *transit time* (also called *slew rate*) is the approximate time it takes for the servo to rotate the shaft $X°$ (usually specified as 60°). Small servos turn at about a quarter of a second per 60°, while larger servos tend to be a bit slower. The faster the transit time, the "faster acting" the servo will be.

You can calculate equivalent RPM by multiplying the 60° transit time by 6 (to get full 360° rotation), then dividing the result into 60. For example, if a servo motor has a 60° transit time of 0.20 seconds, that's one revolution in 1.2 seconds ($.2 \times 6 = 1.2$), or 50 RPM ($60 / 1.2 = 50$).

Bear in mind that there are variations on the standard themes for all R/C servo classes. For example, standard servos are available in more expensive high-speed and high-torque versions. Servo manufacturers list the specifications for each model, so you can compare and make the best choice based on your particular needs.

Many R/C servos are designed for use in special applications, and these applications can be adapted to robots. For example, a servo engineered to be used with a model sailboat will be water resistant and therefore useful on a robot that works in or around water.

TABLE 20.1	**TYPICAL SERVO SPECIFICATIONS**					
SERVO TYPE	**LENGTH**	**WIDTH**	**HEIGHT**	**WEIGHT**	**TORQUE**	**TRANSIT TIME**
Standard	1.6"	0.8"	1.4"	1.3 oz	42 oz-in	0.23 sec/60°
1/4-scale	2.3"	1.1"	2.0"	3.4 oz	130 oz-in	0.21 sec/60°
Mini-micro	0.85"	0.4"	0.8"	0.3 oz	15 oz-in	0.11 sec/60°
Low profile	1.6"	0.8"	1.0"	1.6 oz	60 oz-in.	0.16 sec/60°
Sail winch small	1.8"	1.0"	1.7"	2.9 oz	135 oz-in	0.16 sec/60° 1 sec/360°
Sail winch large	2.3"	1.1"	2.0	3.8 oz	195 oz-in	0.22 sec/60° 1.3 sec/360°

Connector Styles and Wiring

While many aspects of servos are standardized, there is much variety between manufacturers in the shape and electrical contacts of the connectors used to attach the servo to a receiver. While your robot probably won't use a radio receiver, you may still want to match up the servo with a properly mated connector on your controller board or computer. Or, you may decide the connector issue isn't worth the hassle, and just cut it off from the servo, hardwiring it to your electronics. This is an acceptable alternative, but hardwiring makes it more difficult to replace the servo should it ever fail.

CONNECTOR TYPE

There are three primary connector types found on R/C servos:

- "J" or Futaba style
- "A" or Airtronics style
- "S" or Hitec/JR style

Servos made by the principle servo manufacturers—Futaba, Airtronics, Hitec, and JR—employ the connector style popularized by that manufacturer. In addition, servos made by competing manufacturers are usually available in a variety of connector styles, and connector adapters are available.

PINOUT

The physical shape of the connector is just one consideration. The wiring of the connectors (called the *pinout*) is also critical. Fortunately, all but the "old-style" Airtronics servos (and the occasional oddball four-wire servo) use the same pinout, as shown in Fig. 20.5. With very few exceptions, R/C servo connectors use three wires, providing DC power, ground, and signal (or control). Table 20.2 lists the pinouts for several popular brands of servos.

COLOR CODING

Most servos use color coding to indicate the function of each connection wire, but the actual colors used for the wires vary between servo makers. Table 20.3 lists the most common colors used in several popular brands.

USING SNAP-OFF HEADERS FOR MATED CONNECTORS

The female connectors on most R/C servos are designed to mate with pins placed 0.100 inch apart. As luck would have it, this is the most common pin spacing used in electronics, and suitable *pin headers* are in ready supply. The "snap-off" variety of header is perhaps the most useful, as you can buy a long strip and literally snap off just the number of pins you want. For a servo, snap off three pins, then solder them to your circuit board, as shown in Fig. 20.6.

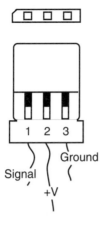

FIGURE 20.5 The standard pinout of servos is pin 1 for signal, pin 2 for +V, and pin 3 for ground. In this configuration damage will not usually occur if you accidentally reverse the connector.

TABLE 20.2 CONNECTOR PINOUTS OF POPULAR SERVO BRANDS

BRAND	PIN 1 (LEFT)	PIN 2 (CENTER)	PIN 3 (RIGHT)
Airtronics ("new" style)	Signal	+V	Gnd
Airtronics ("old" style)	Signal	Gnd	+V
Futaba	Signal	+V	Gnd
Hitec	Signal	+V	Gnd
JR	Signal	+V	Gnd

TABLE 20.3 COLOR CODING OF POPULAR SERVO BRANDS

SERVO	+V	GND	SIGNAL
Airtronics	Red	Black	White
	Red stripe Brown	Blue	
Cirrus	Red	Brown	Orange
Daehwah	Red	Black	White
Fleet	Red	Black	White
Futaba	Red	Black	White
Hitec	Red	Black	Yellow
JR	Red	Brown	Orange
KO	Red	Black	Blue
Kraft	Red (4.8 v) White (2.4 v)	Black	Orange Yellow
Sanwa	Red stripe	Black (in center)	Black

FIGURE 20.6 You can construct your own servo connectors using snap-off headers soldered to your robot control board.

You'll want to mark how the servo connector should attach to the header, as it's easy to reverse the connector and plug it in backward. Fortunately, this *probably* won't cause any damage to either the servo or electronics, since reversing the connector merely exchanges the signal and ground wires. This is *not* true of the "old-style" Airtronics connector: if you reverse the connector, the signal and +V lines are swapped. In this case, both servo and control electronics can be irreparably damaged.

Circuits for Controlling a Servo

Unlike a DC motor, which runs if you simply attach battery power to its leads, a servo motor requires proper interface electronics in order to rotate its output shaft. While the need for interface electronics may complicate to some degree your use of servos, the electronics are actually rather simple. And if you plan on operating your servos with a PC or microcontroller (such as the Basic Stamp), all you need for the job is a few lines of software.

A DC motor typically needs power transistors, MOSFETs, or relays if it is interfaced to a computer. A servo on the other hand can be directly coupled to a PC or microcontroller with no additional electronics. All of the power-handling needs are taken care of by the control board in the servo, saving you the hassle. This is one of the key benefits of using servos with computer-controlled robots.

CONTROLLING A SERVO VIA A 555 TIMER

You don't *need* a computer to control a servo. You can use the venerable 555 timer IC to provide the required pulses to a servo. Fig. 20.7 shows one common approach to using the 555 to control a servo.

In operation, the 555 produces a signal pulse of varying duty cycle, which controls the operation of the servo. Adjust the potentiometer to position the servo. Since the 555 can easily produce pulses of very short and very long duration, there is a good chance that the servo may be commanded to operate outside its normal position extremes. If the servo hits its stop and begins chattering *remove power immediately*! If you don't, the gears inside the servo will eventually strip out, and you'll need to either throw the servo away or replace its gears.

CONTROLLING A SERVO VIA A BASIC STAMP

The Basic Stamp II is a popular microcontroller used to interface with various robotic parts, including servos. The Stamp, which is discussed in more detail in Chapter 31, can directly control one or more servos. However, the more servos the more processing time is required to send pulses to each one (at least, not without resorting to some higher-level programming which we'll leave to the Stamp-specific books).

Fig. 20.8 shows the hookup diagram for connecting a standard servo to the Basic Stamp II. Note that the power to the servo *does not* come from the Basic Stamp II, or any prototyping board it is on. Servos require more current than the Stamp can provide. A pack of four AA batteries is sufficient to power the servo. For proper operation ensure that the grounds are connected between the Stamp and the battery pack. Use a 33–47 µF capacitor between the +V and ground of the AA pack to help kill any noise that may be induced into

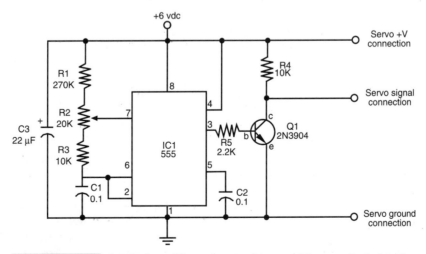

FIGURE 20.7 A 555 timer IC can be used to provide a control signal to a servo.

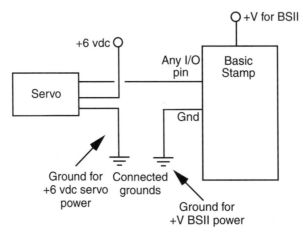

FIGURE 20.8 Hookup diagram for connecting a servo to a Basic Stamp II

the electronics when the servo turns on and off. See Chapter 31 for suitable code that you can use to command a servo using a Basic Stamp II.

USING A DEDICATED CONTROLLER

R/C receivers are designed with a maximum of eight servos in mind. The receiver gets a digital pulse train from the transmitter, beginning with a long sync pulse, followed by as many as eight servo pulses. Each pulse is meant for a given servo attached to the receiver: pulse 1 goes to servo 1, pulse 2 goes to servo 2, and so on. The eight pulses plus the sync pulse take about 20 ms. This means the pulse train can be repeated 50 times each second, which we earlier referred to as the refresh rate. As the refresh rate gets slower the servos aren't updated as quickly and can "throb" or lose position as a result.

Unless the control electronics you are using can simultaneously supply pulses to multiple servos at a time (multitasking), the control circuitry can no longer effectively send the refresh pulses (the continuous train of pulses) fast enough. For these applications, you can use a dedicated servo controller, which is available from a number of sources, including Scott Edwards Electronics and NetMedia (see Appendix B, "Sources," for addresses and Web sites). Dedicated servo controllers can operate five, eight, or even more servos autonomously, which reduces the program overhead of the microcontroller or computer you are using.

The main benefit of dedicated servo controllers is that a great number of servos can be commanded simultaneously, even if your computer, microcontroller, or other circuitry is not multitasking. For example, suppose your robot requires 24 servos. Say it's an eight-legged spider, and each leg has three servos on them; each servo controls a different "degree of freedom" of the leg. One approach would be to divide the work among three servo controllers, each capable of handling eight servos. Each controller would be responsible for a given degree of freedom. One might handle the rotation of all eight legs; another might handle the "flexion" of the legs; and the third might be for the rotation of the bottom leg segment.

Dedicated servo controllers must be used with a computer or microcontroller, as they need to be provided with real-time data in order to operate the servos. This data is commonly sent in a serial data format. A sequence of bytes sent from the computer or microcontroller is decoded by the servo controller, with each byte corresponding to a servo attached to it. Servo controllers typically come with application notes and sample programs for popular computers and microcontrollers, but to make sure things work it's very helpful to have a knowledge of programming and serial communications.

USING GREATER THAN 7.2 VOLTS

Servos are designed to be used with rechargeable model R/C battery packs, which put out from 4.8 to 7.2 volts, depending on the number of cells they have. Servos allow a fairly wide latitude in input voltage, and 6 volts from a four-pack of AAs provides more than enough juice. As the batteries drain, however, the voltage will drop, and you will notice your servos won't be as fast as they used to be. Somewhere below about 4.0 or 4.5 volts the servos will be too slow to do you much good, and they may not even function.

But what about going beyond the voltage of typical rechargeable batteries used for R/C models? Indeed, many servos can be operated in an intermittent fashion with up to about 12 volts, with few or no bad aftereffects. However, most servos will begin to overheat with more than 9 or 10 volts, and they may not like operating for long periods of time without a "cooling off" period.

Unless you need the extra torque or speed, it's best to keep the supply voltage to your servos at no more than 9 volts, and preferably between the rated 4.8- to 7.2-volts range. Of course, check the data sheet that comes with the servos you are using and note any special voltage requirements.

WORKING WITH AND AVOIDING THE "DEAD BAND"

References to the Grateful Dead notwithstanding, all servos exhibit what's known as a *dead band*. The dead band of a servo is the maximum time differential between the incoming control signal and the internal reference signal produced by the position of the potentiometer. If the time difference equates to less than the dead band—say, five or six microseconds—the servo will not bother trying to nudge the motor to correct for the error.

Without the dead band, the servo would constantly "hunt" back and forth to find the exact match between the incoming signal and its own internal reference signal. The dead band allows the servo to minimize this hunting so it will settle down to a position close to, though maybe not exactly, where it's supposed to be.

Dead band varies between servos and is often listed as part of the servo's specifications. A typical dead band is 5 microseconds (μs). If the servo has a full travel of 180° over a 1000 μs (1–2 ms) range, then the 5 μs dead band equates to one part in 200. You probably won't even notice the effects of dead band if your control circuitry has a resolution lower than the dead band.

However, if your control circuitry has a resolution higher than the dead band— which is the case with a microcontroller such as the Basic Stamp II or the Motorola MC68HC11—then small changes in the pulse width values may not produce any effect. For instance, if the controller has a resolution of 2 μs and if the servo has a dead band of 5 μs, then a change of just one or even two values—equal to a change of 2 or 4 μs in the pulse width—may not have an effect on the servo.

The bottom line: choose a servo that has a narrow dead band if you need accuracy and if your control circuitry or programming environment has sufficient resolution. Otherwise, ignore dead band since it probably won't matter one way or another. The trade-off here is that with a narrow dead band the servo will be more prone to hunt to its position and may even buzz after it has gotten there. (Hint: the way to minimize this is to stop the stream of pulses to the servo, assuming this is practical for your application.)

GOING BEYOND THE 1–2 MILLISECOND PULSE RANGE

You've already read that the "typical" servo responds to signals from 1 to 2 ms. While this is true in theory, in actual practice many servos can be fed higher and lower pulse values in order to maximize their rotational limits. The 1–2 ms range may indeed turn a servo one direction or another, but it may not turn it *all the way* in both directions. However, you won't know the absolute minimums and maximums for a given servo until you experiment with it. But take fair warning: Performing this experiment can be risky because operating a servo to its extremes can cause the mechanism to hit its internal stops. If left in this state for any period of time, the gears of the servo can become damaged.

If you just must have maximum rotation from your servo, connect it to your choice of control circuitry. Start by varying the pulse width in small increments below 1 ms (1000 µs), say in 10 µs chunks. After each additional increment, have your control program swing the servo back to its center or neutral position. When during your testing you hear the servo hit its internal stop (the servo will "chatter" as the gears slip), you've found the absolute lower-bound value for that servo. Repeat the process for the upper bound. It's not unusual for some servos to have a lower bound of perhaps 250 µs and an upper bound of over 2200 µs. Yet other servos may be so restricted that they cannot even operate over the "normal" 1–2 ms range.

Keep a notebook of the upper and lower operating bounds for each servo in your robot or parts storehouse. Since there can be mechanical differences between servos of the same brand and model, number your servos so you can tell them apart. When it comes time to program them, you can refer to your notes for the lower and upper bounds for that particular servo.

Modifying a Servo for Continuous Rotation

Many brands and models of R/C servos can be readily modified to allow them to rotate continuously, like a regular DC motor. Such modified servos can be used as drive motors for your robot. Modified servos can be easier to use than regular DC motors since they already have the power drive electronics built in, they come already geared down, and they are easy to mount on your robot.

BASIC MODIFICATION INSTRUCTIONS

Servo modification varies somewhat between makes and models, but the basic steps are the same:

1. Remove the case of the servo to expose the gear train, motor, and potentiometer. This is accomplished by removing the four screws on the back of the servo case and separating the top and bottom.
2. File or cut off the nub on the underside of the output gear that prevents full rotation. This typically means removing one or more gears, so you should be careful not to misplace any parts. If necessary, make a drawing of the gear layout so you can replace things in their proper location!
3. Remove the potentiometer and replace it with two 2.7K-ohm 1 percent tolerance ("precision") resistors, wired as shown in Fig. 20.9. This fools the servo into thinking it's always in the "center" position. An even better approach is to relocate the potentiometer to the outside of the servo case, so that you can make fine-tune adjustments to the center position. Alternatively, you can attach a new 5K- or 10K-ohm potentiometer to the circuit board outside the servo, as shown in Fig. 20.10.
4. Reassemble the case.

In the following two sections we provide more detailed modification instructions for two popular R/C servos, the Futaba S-148, and the Hitec HS-300. While there are certainly many more brands and models of servos to choose from, these two represent a good cross-section of the internal designs used with low- and medium-priced servos. With minor variations, the steps that follow can be applied to similarly designed servos.

STEPS FOR MODIFYING A FUTABA S-148 SERVO

The Futaba S-148 is among the most common servos used for hobby robotics. The S-148 uses a brass bushing on its output gear. See the previous section, "Basic Modification Instructions," for the generic steps for disassembling the servo.

1. Note the arrangement of all the gears, then remove them and set them aside. Try not to handle the gears too much, as this will remove the grease that was applied to the gears at the factory.

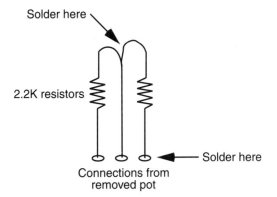

FIGURE 20.9 To modify a servo you must replace the internal potentiometer with two 2.7K resistors, wired as shown here.

FIGURE 20.10 For greater control and accuracy, use an external 5K or 10K pot to replace the one removed from the servo.

2. Locate and remove the two screws near the motor shaft. With these two screws removed you can separate the top of the case from the drive motor.
3. Press down on the metal output shaft (it is actually the shaft of the potentiometer) to remove the circuit board. You may need to work the circuit board loose by using a small screwdriver to pry it out by its four corners.
4. Snip the potentiometer off near where the leads connect to the circuit board.
5. If using fixed resistors, solder them in place as shown in Fig. 20.9. If using a 5K potentiometer, follow the additional steps provided in "Basic Modifications Instructions," earlier in the chapter.
6. Clip off the nub on the bottom of the output gear, as described in "Basic Modifications Instructions."

You may now reassemble the servo:

1. Insert the circuit board back into the top casing.
2. Attach the two small screws that secure the top casing to the motor.
3. Reassemble the gears in the proper sequence. The output gear will fit snugly over the brass bushing.
4. Reassemble the bottom casing, with screws.

The S-148 is representative of servos that are constructed with metal bushings or ball bearings for the output shaft. With minor variations, you can use these steps with other servos of similar design. For example, with only minor variations the same steps apply to the Hitec HS-422, another popular servo. Like the S-148, the Hitec HS-422 uses a brass bushing to support the output gear. The major difference between the S-148 and HS-422 is that the HS-422 lacks the two screws holding the top casing to the motor.

If you are modifying a servo with metal gears, you will not be able to easily clip off the mechanical stop that is located on the bottom of the output gear. For these you will need to use a file or small rotary grinder to remove the stop. You can use a Dremel or other motorized hobby tool to make short work of this task.

STEPS FOR MODIFYING A HITEC HS-300 SERVO

The HS-300 is an economical alternative to the S-148 and other servos with brass bushings or ball bearings. The output gear of the HS-300 is mounted directly to the potentiometer, and no bushing or bearing is used. This means that if you remove the potentiometer, you also remove the structure on which the output gear is attached. Therefore, the steps for modifying an HS-300 are different than those for the S-148. See "Basic Modification Instructions," earlier in the chapter for the generic steps to disassemble the servo.

1. Remove the center gear and the output gear. All the other gears can remain. If needed, place a small piece of electrical tape on the gears to hold them in place while you work (don't get any grease on the tape or it won't stick!).
2. Remove the control board from the bottom of the case.
3. Clip off the three wires leading to the potentiometer.
4. With a small flat-head screwdriver, pry off the three "fingers" holding the bottom of the potentiometer to its casing. Discard this part.
5. Using a small pair of needle-noise pliers, remove the small disc inside the potentiometer. Take care *not* to pull the shaft of the potentiometer out. It should remain held in place by a small retainer. Discard this part once it has been removed.
6. If using fixed resistors, solder them in place as shown in Fig. 20.9. If using a 5K potentiometer, follow the additional steps provided in "Basic Modifications Instructions," earlier in the chapter.
7. Clip off the nub on the bottom of the output gear, as described in "Basic Modifications Instructions."

Note:

An alternative to steps 4 and 5 is to ream or drill out the underside of the output gear so that it rotates freely around the potentiometer shaft. Be sure not to ream or drill out too much or you'll ruin the gear. At the same time, be sure you remove enough material so that the gear rotates freely, without any binding. You will still want to clip off the leads of the potentiometer so that it is no longer in circuit with the control board.

You may now reassemble the servo:

1. Insert the circuit board back into the top casing.
2. Replace the output gear onto the shaft of the modified potentiometer. Replace the center gear and make sure all the gears properly mesh.
3. Reassemble the bottom casing with screws.

The HS-300 is representative of servos that are constructed without metal bushings or ball bearings for the output shaft. With minor variations, you can use these steps with other servos of similar design.

APPLYING NEW GREASE

The gears in a servo are lubricated with a white or clear grease. As you remove and replace the gears during your modification surgery it's inevitable that some of the grease will come off on your fingers. If you feel too much of the grease has come off, you'll want to apply more. Most any viscous synthetic grease suitable for electronics equipment will work, though you can also splurge and buy a small tube of grease especially made for servo gears and other mechanical parts in model cars and airplanes.

When applying grease be sure to spread it around so that it gets onto all the mechanical parts of the servo that mesh or rub. However, avoid getting any of it inside the motor or on the electrical parts. Wipe off any excess.

While it may be tempting, don't apply petroleum-based oil to the gears, such as three-in-one oil or a spray lubricant like WD-40. Some oils may not be compatible with the plastics used in the servo, and spray lubricants aren't permanent enough.

TESTING THE MODIFIED SERVO

After reassembly but before connecting the servo to a control circuit, you'll want to test your handiwork to make sure the output shaft of the servo rotates smoothly. Do this by attaching a control disc or control horn to the output shaft of the servo. *Slowly* and carefully rotate the disc or horn and note any snags. Don't spin too quickly, as this will put undo stress on the gears.

If you notice any binding while you're turning the disc or horn, it could mean you didn't remove enough of the mechanical stop on the output gear. Disassemble the servo just enough to gain access to the output gear and clip or file off some more.

A CAUTION ON MODIFYING SERVOS

Modifying a servo typically entails removing or "gutting" the potentiometer and clipping off any mechanical stops or nubs on the output gear. For all practical purposes, this renders the servo unusable for its intended use, that is, to precisely control the angular position of its output shaft. So, before modifying a servo, be sure it's what you want to do. It'll be difficult to reverse the process.

Several sources of robotics parts provide premodified servos, which are a practical alternative if you don't care to do the modification yourself. The price is just a little higher than unmodified servos of the same brand and make.

SOFTWARE FOR RUNNING MODIFIED SERVOS

Even though a servo has been modified for continuous rotation, the same digital pulses are used to control the motor. Keep the following points in mind when using to run modified servos:

- If you've used fixed resistors in place of the original potentiometer inside the servo, sending a pulse of about 1.5 ms will stop the motor. Decreasing the pulse width will turn the motor in one direction; increasing the pulse width will turn the motor in the other direction. You will need to experiment with the exact pulse width to find the value that will cause the motor to stop.
- If you've used a replacement 5K potentiometer in place of the original that was inside the servo, you have the ability to set the precise "center point" that will cause the motor to stop. In your software, you can send a precise 1.5 ms pulse, then adjust the potentiometer until the servo stops. As with fixed resistors, values higher or lower than 1.5 ms will cause the motor to turn one way or another.

Specific software examples for running servos are provided in Part 5 of this book.

LIMITATIONS OF MODIFIED SERVOS

Modifying a servo for continual rotation carries with it a few limitations, exceptions, and "gotchas" that you'll want to keep in mind:

- The average servo is not engineered for lots and lots of continual use. The mechanics of the servo are likely to wear out after perhaps as little as 25 hours (that's elapsed time), depending on the amount of load on the servos. Models with metal gears and/or brass bushing or ball bearings will last longer.
- The control electronics of a servo are made for intermittent duty. Servos used to power a robot across the floor may be used minutes or even hours at a time, and they tend to be under additional mechanical stress because of the weight of the robot. Though this is not exactly common, it is possible to burn out the control circuitry in the servo by overdriving it.
- Standard-sized servos are not particularly strong in comparison to many other DC motors with gear heads. Don't expect a servo to move a 5- or 10-pound robot. If your robot is heavy, consider using either larger, higher-output servos (such as 1/4-scale or sail winch), or DC motors with built-in gear heads.
- Last and certainly not least, remember that modifying a servo voids its warranty. You'll want to test the servo before you modify it to ensure that it works.

MODIFYING BY REMOVING THE SERVO CONTROL BOARD

Another way to modify a servo for continuous rotation is to follow the steps outlined earlier and also remove the control circuit board. Your robot then connects directly to the servo motor. You'd use this approach if you don't want to bother with the pulse width schema. You get a nice, compact DC motor with gearbox attached.

However, since you've removed the control board, you will also need to provide adequate power output circuitry to drive the motor. Your PC or microcontroller will likely not be able to provide adequate current; in fact, trying to control a "gutted" servo motor directly will probably damage your PC or microcontroller.

A servo modified by removing its control board is essentially the same as an ordinary DC motor, and the control circuitry is exactly the same. See Chapter 18 for more information on working with DC motors.

Attaching Mechanical Linkages to Servos

One of the benefits of using R/C servos with robots is the variety of ways it offers you to connect stuff to the servos. In model airplane and car applications, servos are typically connected to a push/pull linkage of some type. For example, in a plane, a servo for controlling the rudder would connect to a push/pull linkage directly attached to the rudder. As the servo rotates, the linkage draws back and forth, as shown in Fig. 20.11. The rudder is attached to the body of the plane using a hinge, so when the linkage moves, the rudder flaps back and forth.

You can use the exact same hardware designed for model cars and airplanes with your servo-equipped robots. Visit the neighborhood hobby store and scout for possible parts you can use. Collect and read through Web sites and catalogs of companies that manufacture and sell servo linkages and other mechanics. Appendix B, "Sources," lists several such companies.

Attaching Wheels to Servos

Servos reengineered for full rotation are most often used for robot locomotion and are outfitted with wheels. Since servos are best suited for small- to medium-sized robots (under about three pounds), the wheels for the robot should ideally be between 2 and 5 inches in diameter. Larger-diameter wheels make the robot travel faster, but they can weigh more. You won't want to put extra large 7- or 10-inch wheels on your robot if each wheel weighs 1.5 pounds. There's your three-pound practical limit right there.

The general approach for attaching wheels to servos is to use the round control disc that comes with the servo (see Fig. 20.12). The underside of the disc fits snugly over the output shaft of the servo. You can glue or screw the wheel to the front of the disc. Here are some ideas:

- Large LEGO "balloon" tires have a recessed hub that exactly fits the control disc included with Hitec and many other servos. You can simply glue the disc into the rim of the tire.
- Lightweight foam tires, popular with model airplanes, can be glued or screwed to the control disc. The tires are available in a variety of diameters. If you wish, you can grind down the hub of the tire so it fits smoothly against the control disc.

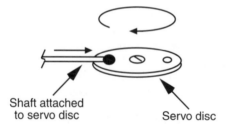

Shaft attached
to servo disc

Servo disc

FIGURE 20.11 Servos can be used to transform rotational motion to linear motion.

■ A gear glued or screwed into the control disc can be used as an ersatz wheel or as a gear that drives a wheel mounted on another shaft.

In all these cases, it's important to maintain access to the screw used to secure the control disc to the servo. When you are attaching a wheel or tire be sure not to block the screw hole. If necessary, insert the screw into the control disc first, then glue or otherwise attach the tire. Make sure the hub of the wheel is large enough to accept the diameter of your screwdriver, so you can tighten the screw over the output shaft of the servo.

Mounting Servos on the Body of the Robot

Servos should be securely mounted to the robot so the motors don't fall off while the robot is in motion. In my experience, the following methods do *not* work well, though they are commonly used:

■ *Duct tape or electrical tape.* The "goo" on the tape is elastic, and eventually the servo works itself lose. The tape can also leave a sticky residue.

FIGURE 20.12 Attaching a round control disc to the hub of a wheel.

■ *Hook-and-loop,* otherwise known as Velcro. Accurate alignment of the hook-and-loop halves can be tricky, meaning that every time you remove and replace the servos the wheels are at a slightly different angle with respect to the body of the robot. This makes it harder to program repeatable actions.

■ *Tie-wraps.* You must cinch the tie-wrap tightly in order to adequately hold the servo in place. Unless your robot is made of metal or strong plastic, you're bound to distort whatever part of the robot you've cinched the wrap against.

Over the years, I've found "hard mounting"—gluing, screwing, or bolting—the servos onto the robot body to be the best overall solution, and it greatly reduces the frustration level of hobby robotics.

ATTACHING SERVOS WITH GLUE

Gluing is a quick and easy way to mount servos on most any robot body material, including heavy cardboard and plastic. Use only a strong glue, such as two-part epoxy or hot-melt glue. I prefer hot-melt glue because it doesn't emit the fumes that epoxy does, and it sets much faster (about a minute in normal room temperatures versus a minimum of five minutes for fast-setting epoxy).

When gluing it is important that all surfaces be clean. Rough up the surfaces with a file or heavy-duty sandpaper for better adhesion. If you're gluing servos to LEGO parts, apply a generous amount so the extra adequately fills between the "nubs." LEGO plastic is hard and smooth, so be sure to rough it up first.

ATTACHING SERVOS WITH SCREWS OR BOLTS

A disadvantage of mounting servos with glue is that it's more or less permanent (and, according to Murphy's Law, more permanent than you'd like if you *want* to remove the servo, less permanent if you want the servo to stay in place!). For the greatest measure of flexibility, use screws or bolts to mount your servos to your robot body. All servos have mounting holes in their cases; it's simply a matter of finding or drilling matching holes in the body of your robot.

Servo mounts are included in many R/C radio transmitters and separately available servo sets. You can also buy them separately from the better-stocked hobby stores. The servo mount has space for one, two, or three servos. The mount has additional mounting holes that you can use to secure it to the side or bottom of your robot. Most servo mounts are made of plastic, so if you need to make additional mounting holes they are easy to drill.

You can also construct your own servo mounting brackets using 1/8-inch thick aluminum or plastic. A template is shown in Fig. 20.13. (Note: the template is not to scale, so don't trace it to make your mount. Use the dimensions to fashion your mount to the proper size.)

The first step in constructing your own servo mounting brackets is to cut and drill the aluminum or plastic, as shown in Fig. 20.13. Use a small hobby file to smooth off the edges and corners. The mounting hole centers provided in the template are designed to line up with the holes in LEGO Technic beams. This allows you to directly attach the servo mounts to LEGO pieces. Use 3/32 or 4/40 nuts and bolts, or 4/40 self-tapping screws, to attach the servo mount to the LEGO beam.

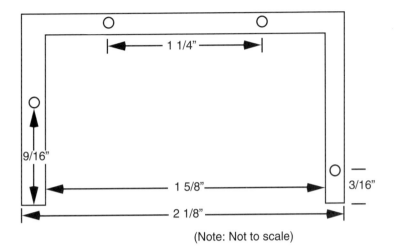

(Note: Not to scale)

FIGURE 20.13 Use this template to construct a servo mounting bracket. The template may not be reproduced in 1:1 size, so be sure to measure before cutting your metal or plastic.

Figure 20.14 shows a servo mounted on a bracket and attached to a LEGO beam. If necessary, the servos can be easily removed for repair or replacement.

FIGURE 20.14 A servo mounted on a homemade servo bracket.

From Here

To learn more about...	*Read*
Using batteries to power your robot	Chapter 15, "All about Batteries and Robot Power Supplies"
Fundamentals of robot locomotion	Chapter 16, "Robot Locomotion Principles"
Choosing the best motor for your robot	Chapter 17, "Choosing the Right Motor for the Job"
Ways to implement computers and microcontrollers to your robots	Chapter 28, "An Overview of Robot 'Brains'"
Interfacing servos and other motors to control circuitry	Chapter 29, "Interfacing with Computers and Microcontrollers"

PART 4

PRACTICAL ROBOTICS

PROJECTS

21

BUILD A ROVERBOT

Imagine a robot that can vacuum the floor for you, relieving you of that time-consuming household drudgery and freeing you to do other, more dignified tasks. Imagine a robot that patrols your house, inside or out, listening and watching for the slightest trouble and sounding the alarm if anything goes amiss. Imagine a robot that knows how to look for fire, and when it finds one, puts it out. Impossible? A dream?

Think again. The compact and versatile Roverbot introduced in this chapter can serve as the foundation for building any of these more advanced robots. You can easily add a small DC-operated vacuum cleaner to the robot, then set it free in your living room. Only the sophistication of the control circuit or computer running the robot limits its effectiveness at actually cleaning the rug.

You can attach light and sound sensors to the robot, providing it with eyes that help it detect potential problems. These sensors, as it turns out, can be the same kind used in household burglar alarm systems. Your only job is to connect them to the robot's other circuits. Similar sensors can be added so your Roverbot actively roams the house, barn, office, or other enclosed area looking for the heat, light, and smoke of fire. An electronically actuated fire extinguisher containing Halon is used to put out the fire.

The Roverbot described on the following pages represents the base model only. The other chapters in this book will show you how to add onto the basic framework to create a more sophisticated automaton. The Roverbot borrows from techniques described in Chapter 10, "Building a Metal Platform." If you haven't yet read that chapter, do so now. It will help you get more out of this one.

Building the Base

Construct the base of the Roverbot using shelving standards or extruded aluminum channel stock. The prototype Roverbot for this book used aluminum shelving standards because aluminum minimized the weight of the robot. The size of the machine didn't require the heavier-duty steel shelving standards.

The base measures 12 5/8 inches by 9 1/8 inches. These unusual dimensions make it possible to accommodate the galvanized nailing (mending) plates, which are discussed later in this chapter. Cut two pieces each of 12 5/8-inch stock, with 45° miter edges on both sides, as shown in Fig. 21.1 (refer to the parts list in Table 21.1). Do the same with the 9 1/8-inch stock. Assemble the pieces using 1 1/4-by-3/8-inch flat corner irons and 8/32 by 1/2-inch nuts and bolts. Be sure the dimensions are as precise as possible and that the cuts are straight and even. Because you are using the mending plates as a platform, it's doubly important with this design that you have a perfectly square frame. Don't bother to tighten the nuts and bolts at this point.

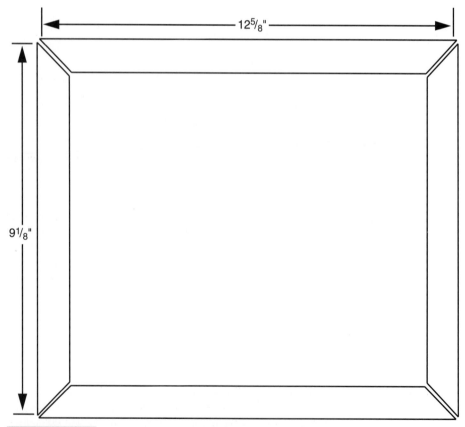

FIGURE 21.1 Cutting diagram for the Roverbot.

TABLE 21.1   PARTS LIST FOR ROVERBOT.

FRAME

2	12 5/8-inch length aluminum or steel shelving standard
2	9 1/8-inch length aluminum or steel shelving standard
3	4 3/16- by 9-inch galvanized nailing (mending) plate
4	1 1/4- by 3/8-inch flat corner iron

RISER

4	15-inch length aluminum or steel shelving standard
2	7-inch length aluminum or steel shelving standard
2	10 1/2-inch length aluminum or steel shelving standard
4	1- by 3/8-inch corner angle iron

MOTORS AND CASTER

2	Gear reduced output 6 or 12 volt DC motors
4	2 1/2- by 3/8-inch corner angle iron
2	5- to 7-inch diameter rubber wheels
2	1 1/4-inch swivel caster
Misc	Nuts, bolts, fender washers, tooth lock washers, etc. (see text)

POWER

2	6 or 12 volt, 1 or 2 amp hour batteries (voltage depending on motor)
2	Battery clamps

Attach one 4 3/16-inch-by-9-inch mending plate to the left third of the base. Temporarily undo the nuts and bolts on the corners to accommodate the plate. Drill new holes for the bolts in the plate if necessary. Repeat the process for the center and left mending plate. When the three plates are in place, tighten all the hardware. Make sure the plates are secure on the frame by drilling additional holes near the inside corners (don't bother if the corner already has a bolt securing it to the frame). Use 8/32 by 1/2-inch bolts and nuts to attach the plates into place. The finished frame should look something like the one depicted in Fig. 21.2. The underside should look like Fig. 21.3.

Motors

The Roverbot uses two drive motors for propulsion and steering. These motors, shown in Fig. 21.4, are attached in the center of the frame. The center of the robot was chosen to help distribute the weight evenly across the platform. The robot is less likely to tip over if you keep the center of gravity as close as possible to the center column of the robot.

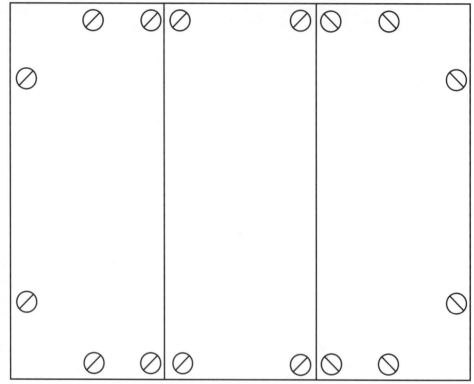

Top view

FIGURE 21.2 The top view of the Roverbot, with three galvanized mending plates added (holes in the plates not shown).

The 12-volt motors used in the prototype were found surplus, and you can use just about any other motor you find as a substitute. The motors used in the prototype Roverbot come with a built-in gearbox that reduces the speed to about 38 rpm. The shafts are 1/4 inch. Each shaft was threaded using a 1/4-inch 20 die to secure the 6-inch-diameter lawn mower wheels in place. You can skip the threading if the wheels you use have a setscrew or can be drilled to accept a setscrew. Either way, make sure that the wheels aren't too thick for the shaft. The wheels used in the prototype where 1 1/2 inches wide, perfect for the 2-inch-long motor shafts.

Mount the motors using two 2 1/2-inch-by-3/8-inch corner irons, as illustrated in Fig. 21.5. Cut about one inch off one leg of the iron so it will fit against the frame of the motor. Secure the irons to the motor using 8/32 by 1/2-inch bolts (yes, these motors have pretapped mounting holes!). Finally, secure the motors in the center of the platform using 8/32 by 1/2-inch bolts and matching nuts. Be sure that the shafts of the motors are perpendicular to the side of the frame. If either motor is on crooked, the robot will crab to one side when it rolls on the floor. There is generally enough play in the mounting holes on the frame to adjust the motors for proper alignment.

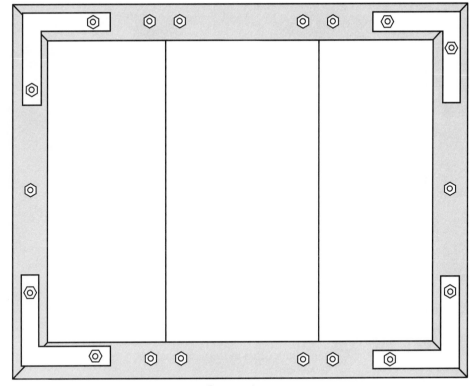

Bottom view

FIGURE 21.3 **Hardware detail for the frame of the Roverbot (bottom view).**

Now attach the wheels. Use reducing bushings if the hub of the wheel is too large for the shaft. If the shaft has been threaded, twist a 1/4-inch 20 nut onto it, all the way to the base. Install the wheel using the hardware shown in Fig. 21.6. Be sure to use the tooth lock washer. The wheels may loosen and work themselves free otherwise. Repeat the process for the other motor.

Support Casters

The ends of the Roverbot must be supported by swivel casters. Use a two-inch-diameter ball-bearing swivel caster, available at the hardware store. Attach the caster by marking holes for drilling on the bottom of the left and right mending plate. You can use the baseplate of the caster as a drilling guide. Attach the casters using 8/32 by 1/2-inch bolts and 8/32 nuts (see Fig. 21.7). You may need to add a few washers between the caster baseplate and the mending plate to bring the caster level with the drive wheels (the prototype used a 5/16-inch spacer). Do the same for the opposite caster.

If you use different motors or drive wheels, you'll probably need to choose a different size caster to match. Otherwise, the four wheels may not touch the ground all at once as

FIGURE 21.4 One of the drive motors, with wheel, attached to the base of the Roverbot.

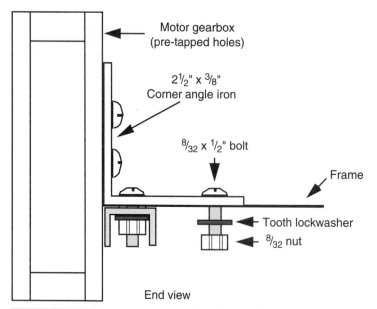

Motor gearbox
(pre-tapped holes)

$2\frac{1}{2}$" x $\frac{3}{8}$"
Corner angle iron

$\frac{8}{32}$ x $\frac{1}{2}$" bolt

Frame

Tooth lockwasher

$\frac{8}{32}$ nut

End view

FIGURE 21.5 Hardware detail for the motor mount. Cut the angle iron, if necessary, to accommodate the motor.

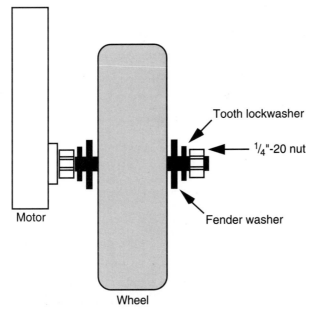

Tooth lockwasher

$\frac{1}{4}$"-20 nut

Fender washer

Motor

Wheel

FIGURE 21.6 Hardware detail for attaching the wheels to the motor shafts. The wheels can be secured by threading the shaft and using 1/4-inch 20 hardware, as shown, or secured to the shaft using a setscrew or collar.

they should. Before purchasing the casters, mount the motors and drive wheels, then measure the distance from the bottom of the mending plate to the ground. Buy casters to match. Again, add washers to increase the depth, if necessary.

Batteries

Each of the drive motors in the Roverbot consumes one-half amp (500 mA) of continuous current with a moderate load. The batteries chosen for the robot, then, need to easily deliver two amps for a reasonable length of time, say one or two hours of continuous use of the motors. A set of high-capacity Ni-Cads would fit the bill. But the Roverbot is designed so that subsystems can be added to it. Those subsystems haven't been planned yet, so it's impossible to know how much current they will consume. The best approach to take is to overspecify the batteries, allowing for more current than is probably necessary.

Six- and eight-amp-hour lead-acid batteries are somewhat common on the surplus market. As it happens, six or eight amps are about the capacity that would handle intermittent use of the drive motors. (The various electronic subsystems, such as an on-board computer and alarm sensors, should use their own battery.) These heavy-duty batteries are typically

A

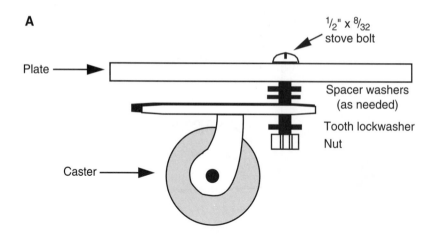

B

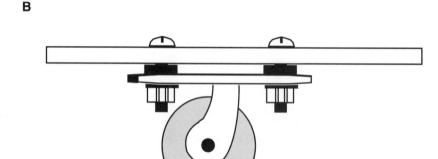

FIGURE 21.7 Adding the casters to the Roverbot. There is one caster on each end, and both must match the depth of the drive wheels (a little short is even better). *a.* Hardware detail; *b.* Caster mounted on mending plate.

available in six-volt packs, so two are required to supply the 12 volts needed by the motors. Supplementary power, for some of the linear ICs, like op amps, can come from separate batteries, such as a Ni-Cad pack. A set of "C" Ni-Cads don't take up much room, but it's a good idea to leave space for them now, instead of redesigning the robot later on to accommodate them.

The main batteries are rechargeable, so they don't need to be immediately accessible in order to be replaced. But you'll want to use a mounting system that allows you to remove the batteries should the need arise. The clamps shown in Fig. 21.8 allow such accessibility. The clamps are made from 1 1/4-inch wide galvanized mending plate, bent to match the contours of the battery. Rubber weather strip is used on the inside of the clamp to hold the battery firmly in place.

The batteries are positioned off to either side of the drive wheel axis, as shown in Fig. 21.9. This arrangement maintains the center of gravity to the inside center of the robot.

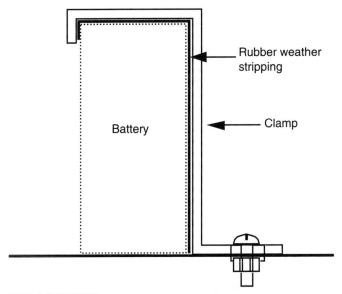

Rubber weather
stripping

Battery

Clamp

FIGURE 21.8 **A battery clamp made from a strip of galvanized plate, bent to the contours of the
battery. Line up the metal with weather
stripping for a positive grip.**

The gap also allows for the placement of one or two four-cell "C" battery packs, should
they be necessary.

Riser Frame

The riser frame extends the height of the robot by approximately 15 inches. Attached to
this frame will be the sundry circuit boards and support electronics, sensors, fire
extinguisher, vacuum cleaner motor, or anything else you care to add. The dimensions are
large enough to assure easy placement of at least a couple of full-size circuit boards, a
2 1/2-pound fire extinguisher, and a Black & Decker DustBuster. You can alter the dimensions of the frame, if desired, to accommodate other add-ons.

Make the riser by cutting four 15-inch lengths of channel stock. One end of each length
should be cut at 90°, the other end at 45°. Cut the mitered corners to make pairs, as shown
in Fig. 21.10. Make the crosspiece by cutting a length of channel stock to exactly seven
inches. Miter the ends as shown in the figure.

Connect the two sidepieces and crosspiece using a 1 1/2-inch-by-3/8-inch flat angle
iron. Secure the angle iron by drilling matching holes in the channel stock. Attach the
stock to the angle iron by using 8/32 by 1/2-inch bolts on the crosspieces and 8/32 by
1 1/2-inch bolts on the riser pieces. Don't tighten the screws yet. Repeat the process for
the other riser.

Construct two beams by cutting the angle stock to 10 1/2 inches, as illustrated in Fig.
21.11. Do not miter the ends. Secure the beams to the top corners of the risers by using

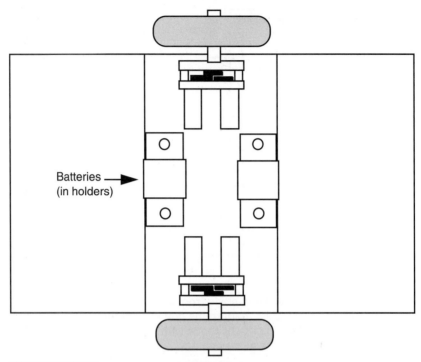

FIGURE 21.9 Top view of the Roverbot, showing the mounted motors and batteries. Note the even distribution of weight across the drive axis. This promotes stability and keeps the robot from tipping over. The wide wheelbase also helps.

1-inch-by-3/8-inch corner angle irons. Use 8/32 by 1/2-inch bolts to attach the iron to the beam. Connect the angle irons to the risers using the 8/32 by 1 1/2-inch bolts installed earlier. Add a spacer between the inside of the channel stock and the angle iron if necessary, as shown in Fig. 21.12. Use 8/32 nuts to tighten everything in place.

Attach the riser to the baseplate of the robot using 1-inch-by-3/8-inch corner angle irons. As usual, use 8/32 by 1/2-inch bolts and nuts to secure the riser into place. The finished Roverbot body and frame should look at least something like the one in Fig. 21.13.

Street Test

You can test the operation of the robot by connecting the motors and battery to a temporary control switch. See Chapter 8, "Building a Plastic Robot Platform," for a wiring diagram. With the components listed in Table 21.1, the robot should travel at a speed of about one foot per second. The actual speed will probably be under that because of the weight of the robot. Fully loaded, the Roverbot will probably travel at a moderate speed of about

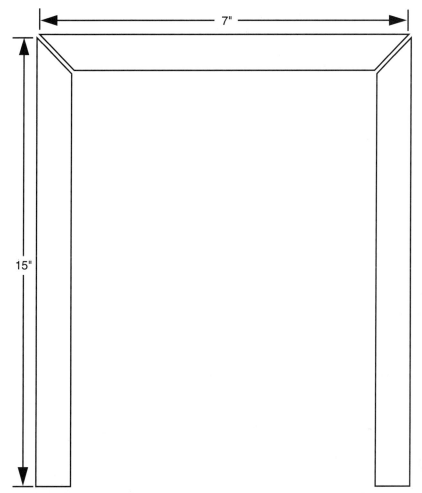

FIGURE 21.10 Cutting diagram for the Roverbot riser pieces (two sets).

eight or nine inches per second. That's just right for a robot that vacuums the floor, roams the house for fires, and protects against burglaries. If you need your Roverbot to go a bit faster, the easiest (and cheapest) solution is to use larger wheels. Using eight-inch wheels will make the robot travel at a top speed of 15 inches per second.

One problem with using larger wheels, however, is that they raise the center of gravity of the robot. Right now, the center of gravity is kept rather low, thanks to the low position of the two heaviest objects, the batteries and motors. Jacking up the robot using larger wheels puts the center of gravity higher, so there is a slightly greater chance of the robot tipping over. You can minimize any instability by making sure that subsystems are added to the robot from the bottom of the riser and that the heaviest parts are positioned closest to the base. You can also mount the motor on the bottom of the frame instead of on top.

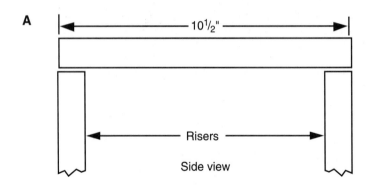

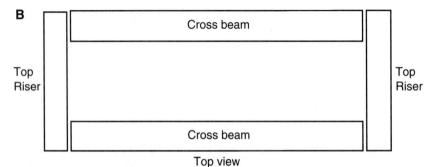

FIGURE 21.11 Construction details for the top of the riser. *a.* Side view showing the crosspiece joining the two riser sides; *b.* Top view showing the cross beams and the tops of the risers.

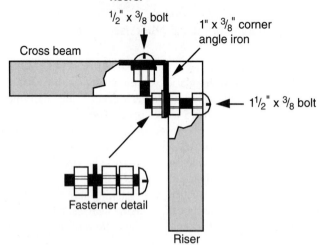

FIGURE 21.12 Hardware detail for attaching the risers to the cross beams.

FIGURE 21.13 The finished Roverbot (minus the batteries), ready for just about any enhancement you see fit.

From Here

To learn more about...	*Read*
Constructing robots using metal parts and pieces	Chapter 10, "Building a Metal Platform"
Powering your robot using batteries	Chapter 15, "All about Batteries and Robot Power Supplies"
Selecting a motor for your robot	Chapter 17, "Choosing the Right Motor for the Job"
Operating your robot with a computer or microcontroller	Chapter 28, "An Overview of Robot 'Brains'"

22

BUILD A HEAVY-DUTY SIX-LEGGED WALKING ROBOT

Let's be honest with each other. Do you like challenges? Do you like being faced with problems that demand decisive action on your part? Do you like spending many long hours tinkering in the garage or workshop? Do you like the idea of building the ultimate robot, one that will amaze you and your friends? If the answer is yes to all these questions, then maybe you're ready to build the Walkerbot, which we will describe in depth in this chapter.

This strange and unique contraption walks on six legs and turns corners with an ease and grace that belies its rather simple design. The basic Walkerbot frame and running gear can be used to make other types of robots as well. In Chapter 23, "Advanced Locomotion Systems," you'll see how to convert the Walkerbot to tracked or wheeled drive. The conversion is simple and straightforward. In fact, you can switch back and forth between drive systems.

The Walkerbot design described in this chapter is for the basic frame, motor, battery system, running gear, and legs. You can embellish the robot with additional components, such as arms, a head, computer control, you name it. The frame is oversized (in fact, it's too large to fit through some inside doors!), and there's plenty of room to add new subsystems. The only requirement is that that weight doesn't exceed the driving capacity of the motors and batteries and that the legs and axles don't bend. The prototype Walkerbot weighs about 50 pounds. It moves along swiftly and no structural problems have yet occurred. Another 10 or 15 pounds could be added without worry.

Frame

The completed Walkerbot frame measures 18 inches wide by 24 inches long by 12 inches deep. Construction is all aluminum, using a combination of 41/64-inch-by-1/2-inch-by-1/16-inch channel stock and 1-inch-by-1-inch-by-1/16-inch angle stock.

Build the bottom of the frame by cutting two 18-inch lengths of channel stock and two 24-inch lengths of channel stock, as shown in Fig. 22.1 (refer to the parts list in Table 22.1). Miter the ends. Attach the four pieces using 1 1/2-inch-by-3/8-inch flat angle irons and secure them with 3-by-1/2-inch bolts and nuts. For added strength, use four bolts on each corner.

In the prototype Walkerbot, I replaced many of the nuts and bolts with aluminum pop rivets in order to reduce the weight. Until the entire frame is assembled, however, use the bolts as temporary fasteners. Then, when the frame is assembled, square it up and replace the bolts and nuts with rivets one at a time. Construct the top of the frame in the same manner.

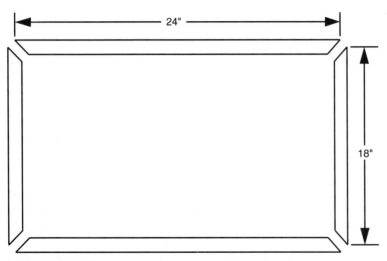

FIGURE 22.1 Cutting diagram for the frame of the Walkerbot (two sets).

TABLE 22-1	PARTS LIST FOR WALKERBOT FRAME.
4	24-inch lengths 41/64-inch-by-1/2-inch-by-1/16-inch aluminum channel stock
4	18-inch lengths 41/64-inch-by-1/2-inch-by-1/16-inch aluminum channel stock
4	12-inch lengths 1-inch-by-1-inch-by-1/16-inch aluminum angle stock
8	1 1/2-inch-by-3/8-inch flat angle iron
4	24-inch lengths 1-inch-by-1-inch-by-1/16-inch aluminum angle stock
2	17 5/8-inch lengths 1-inch-by-1-inch-by-1/16-inch aluminum angle stock
Misc	8/32 stove bolts, nuts, tooth lock washers, as needed

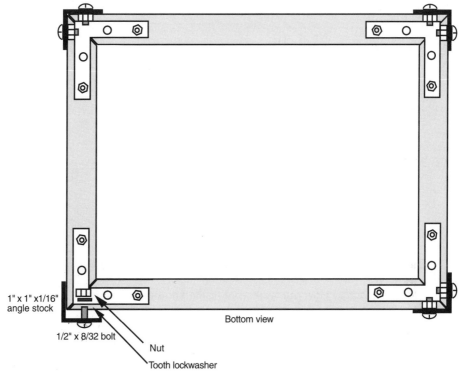

1" x 1" x1/16"
angle stock

1/2" x 8/32 bolt

Nut

Tooth lockwasher

Bottom view

FIGURE 22.2 **Hardware detail for securing the angle stock to the top and bottom frame pieces.**

Connect the two halves with four 12-inch lengths of angle stock, as shown in Fig. 22.2. Secure the angle stock to the frame pieces by drilling holes at the corners. Use 8/32 by 1/2-inch bolts and nuts initially; exchange for pop rivets after the frame is complete. The finished frame should look like the one diagrammed in Fig. 22.3.

Complete the basic frame by adding the running gear mounting rails. Cut four 24-inch lengths of 1-inch-by-1-inch-by-1/16-inch angle stock and two 17 5/8-inch lengths of the same angle stock. Drill 1/4-inch holes in four long pieces as shown in Fig. 22.4. The spacing between the sets of holes is important. If the spacing is incorrect, the U-bolts won't fit properly.

Refer to Fig. 22.5. When the holes are drilled, mount two of the long lengths of angle stock as shown. The holes should point up, with the side of the angle stock flush against the frame of the robot. Mount the two short lengths on the ends. Tuck the short lengths immediately under the two long pieces of angle stock you just secured. Use 8/32 by 1/2-inch bolts and nuts to secure the pieces together. Dimensions, drilling, and placement are critical with these components. Put the remaining two long lengths of drilled angle stock aside for the time being.

Legs

You're now ready to construct and attach the legs. This is probably the hardest part of the project, so take you time and measure everything twice to assure accuracy. Cut six 14-inch

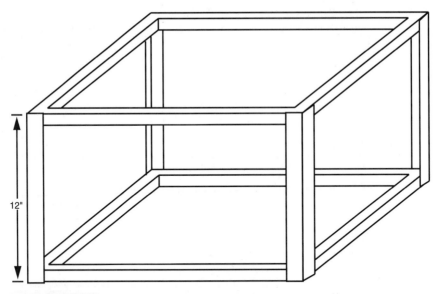

FIGURE 22.3 How the Walkerbot frame should look so far.

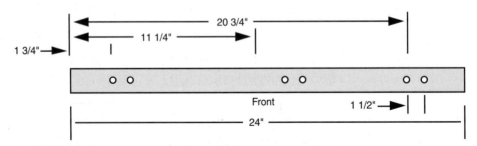

FIGURE 22.4 Cutting and drilling guide for the motor mount rails (four).

lengths of 57/64-inch-by-9/16-inch-by-1/16-inch aluminum channel stock. Do not miter the ends. Drill a hole with a #19 bit 1/2 inch from one end (the "top"); drill a 1/4-inch hole 4 3/4 inches from the top (see Fig. 22.6; refer to the parts list in Table 22.2). Make sure the holes are in the center of the channel stock.

With a 1/4-inch bit, drill out the center of six 4 5/8-inch-diameter circular electric receptacle plate covers. The plate cover should have a notched hole near the outside, which is used to secure it to the receptacle box. If the cover doesn't have the hole, drill one with a 1/4-inch bit 3/8 inch from the outside edge. The finished plate cover becomes a cam for operating the up and down movement of the legs.

Assemble four legs as follows: Attach the 14-inch-long leg piece to the cam using a 1/2-inch length of 1/2 Schedule 40 PVC pipe and hardware, as shown in Fig. 22.7. Be sure the ends of the pipe are filed clean and that the cut is as square as possible. The bolt should be tightened against the cam but should freely rotate within the leg hole.

Assemble the remaining two legs in a similar fashion, but use a 2-inch length of PVC pipe and a 3-inch stove bolt. These two legs will be placed in the center of the robot and

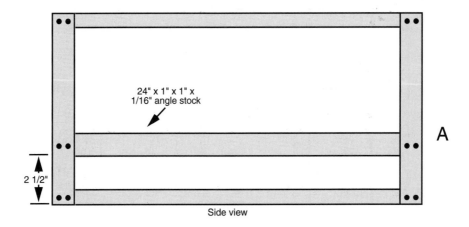

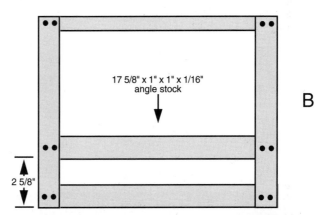

FIGURE 22.5 The motor mount rails secured to the robot. *a.* **The long rail mounts 2 1/2 inches from the bottom of the frame (the holes drilled earlier point up);** *b.* **The short end crosspiece rail mounts 2 5/8 inches from the bottom of the frame.**

will stick out from the others. This allows the legs to cross one another without interfering with the gait of the robot. The "bearings" used in the prototype were 1/2-inch-diameter closet door rollers.

Now refer to Fig. 22.8. Thread a 5-inch-by-1/4-inch 20 carriage bolt through the center of the cam, using the hardware shown. Next, install the wheel bearings to the shafts, 1-inch from the cam. The 1 1/4-inch-diameter bearings are the kind commonly used in lawn mowers and are readily available. The bearings used in the prototype had 1/2-inch hubs. A 1/2-inch-to-1/4-inch reducing bushing was used to make the bearings compatible with the diameter of the shaft.

Install 3 1/2-inch-diameter 30 tooth #25 chain sprocket (another size will also do, as long as all the leg mechanism sprockets in the robot are the same size). Like the bearings,

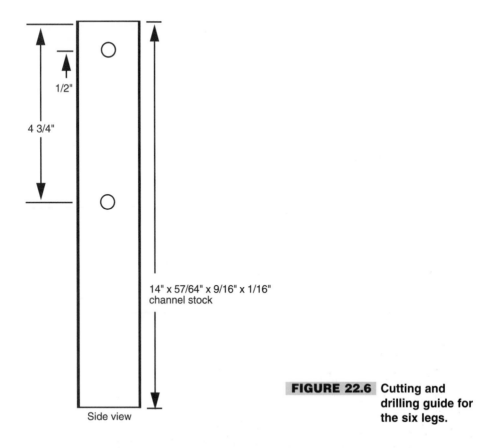

1/2"

4 3/4"

14" x 57/64" x 9/16" x 1/16"
channel stock

Side view

FIGURE 22.6 Cutting and
drilling guide for
the six legs.

TABLE 22.2	PARTS LIST FOR WALKERBOT LEGS
6	14-inch lengths 57/64-inch-by-9/16-inch-by-1/16-inch aluminum channel stock
6	6-inch lengths 41/64-inch-by-1/2-inch-by-1/16-inch aluminum channel stock
6	Roller bearings
6	Steel electrical covers (4 5/8-inch diameter)
6	5-inch hex-head carriage bolt
6	2-inch-by-3/8-inch flat mending iron
6	1 1/4-inch 45° "Ell" Schedule 40 PCV pipe fitting
Misc	10/24 and 8/32 stove bolts, nuts, tooth lock washers, locking nuts, flat washers, as needed. 1/2-inch Schedule 40 PVC cut to length (see text)

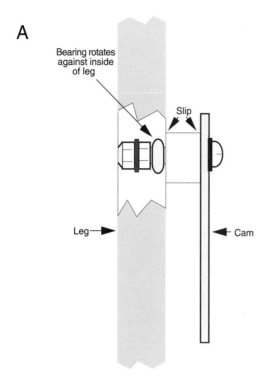

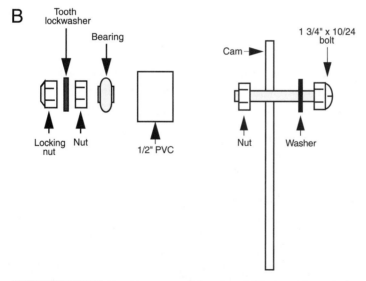

FIGURE 22.7 Hardware detail for the leg cam. *a.* Complete cam and leg; *b.* Exploded view. Note that two of the legs use a 2-inch piece of PVC and a 3-inch bolt.

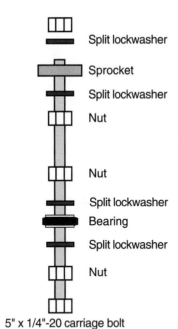

Split lockwasher

Sprocket

Split lockwasher

Nut

Nut

Split lockwasher

Bearing

Split lockwasher

Nut

5" x 1/4"-20 carriage bolt

FIGURE 22.8   Hardware detail of the leg shafts.

a reducing bushing was used to make the 1/2-inch I.D. hubs of the sprockets fit on the shaft. The exact positioning of the sprockets on the shaft is not important at this time, but follow the spacing diagram shown in Fig. 22.9 as a guide. You'll have to "fine-tune" the sprockets on the shaft as a final alignment procedure anyway.

Once all the legs are complete, install them on the robot using U-bolts. The 1 1/2-inch-wide-by-2 1/2-inch-long-by-1/4-inch 20 thread U-bolts fit over the bearings perfectly. Secure the U-bolts using the 1/4-inch 20 nuts supplied.

Refer to Fig. 22.10 for the next step. Cut six 6-inch lengths of 41/64-inch-by-1/2-inch-by-1/16-inch aluminum channel stock. With a #19 bit, drill holes 3/8 inch from the top and bottom of the rail. With a nibbler tool, cut a 3 1/2-inch slot in the center of each rail. The slot should start 1/2 inch from one end.

Alternatively, you can use a router, motorized rasp, or other tool to cut the slot. In any case, make sure the slot is perfectly straight. Once cut, polish the edges with a piece of 300 grit wet-dry Emory paper, used wet. Use your fingers to find any rough edges. There can be none. This is a difficult task to do properly, and you may want to take this portion to a sheet metal shop and have them do it for you (it'll save you an hour or two of blister-producing nibbling!). An alternative method, which requires no slot cutting, is shown in Fig. 22.11. Be sure to mount the double rails exactly parallel to one another.

Mount the rails using 8/32 by 2-inch bolts and 8/32 nuts. Make sure the rails are directly above the shaft of each leg or the legs may not operate properly. You'll have to drill through both walls of the channel in the top of the frame.

The rails serve to keep the legs aligned for the up-and-down pistonlike stroke of the legs. Attach the legs to the rails using 3/8-inch-by-1 1/2-inch bolts. Use nuts and locking nuts fasteners as shown in Fig. 22.12. This finished leg mechanism should look like the one depicted in Fig. 22.13. Use grease or light oil to lubricate the slot. Be sure that there

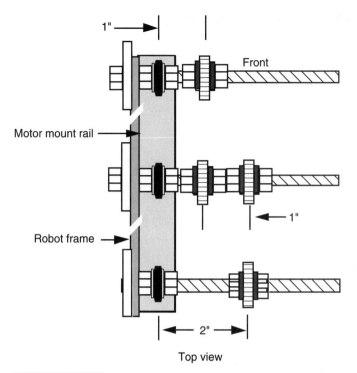

FIGURE 22.9 The leg shafts attached to the motor mount rails (left side shown only).

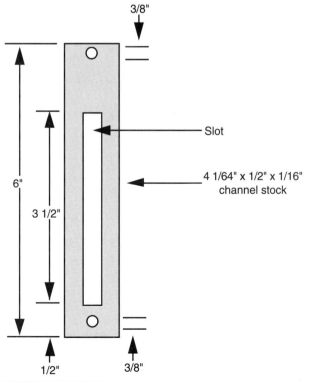

FIGURE 22.10 Cutting and drilling guide for the cam sliders (six required).

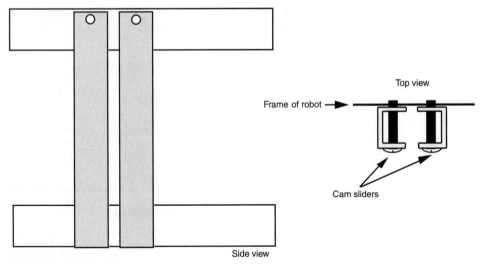

Top view

Frame of robot →

Cam sliders

Side view

FIGURE 22.11 An alternative approach to the slotted cam sliders.

A   B

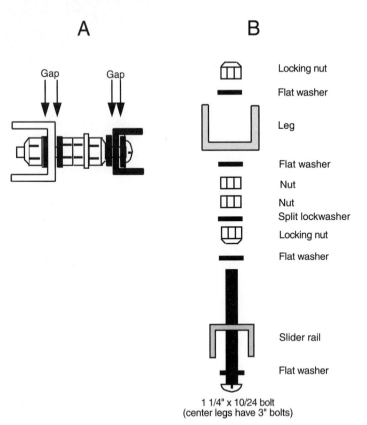

Gap   Gap

Locking nut

Flat washer

Leg

Flat washer

Nut

Nut
Split lockwasher

Locking nut

Flat washer

Slider rail

Flat washer

1 1/4" x 10/24 bolt
(center legs have 3" bolts)

FIGURE 22.12 Cam slider hardware detail. *a.* **Complete
assembly;** *b.* **Exploded view. Note that the
center legs have 2-inch bolts.**

FIGURE 22.13 The slider cam and hardware. The slot must be smooth and free of burrs, or the leg will snag.

is sufficient play between the slot and the bolt stem. The play cannot be excessive, however, or the leg may bind as the bolt moves up and down inside the slot. Adjust the sliding bolt on all six legs for proper clearance.

Drill small pilot holes in the side of six 45° 1 1/4-inch PVC pipe elbows. These serve as the feet of the legs. Paint the feet at the point if you wish. Using #10 wood screws, attach a 2-inch-by-3/8-inch flat mending iron to each of the elbow feet. Drill 1/4-inch holes 1 1/4 inches from the bottom of the leg. Secure the feet onto the legs using 1/2-inch-by-1/4-inch 20 machine bolts, nuts, and lock washers. Apply a 3-inch length of rubber weather strip to the bottom of each foot for better traction. The leg should look like the one in Fig. 22.14. The legs should look like the one in Fig. 22.15. A close-up of the cam mechanism is shown in Fig. 22.16.

Motors

The motors used in the prototype Walkerbot were surplus finds originally intended as the driving motors in a child's motorized bike or go-cart. The motors have a fairly high torque at 12 volts DC and a speed of about 600 rpm. A one-step reduction gear was added to bring the speed down to about 230 rpm. The output speed is further reduced to about 138 rpm by using a drive sprocket. For a walking machine, that's about right, although it could stand to be a bit slower. Electronic speed reduction can be used to slow the motor output down to about 100

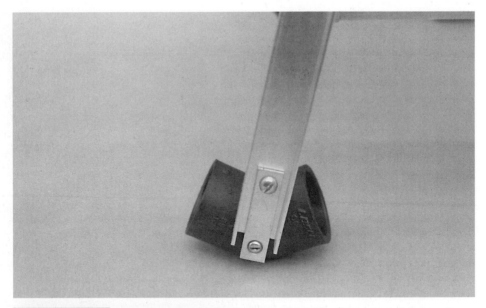

FIGURE 22.14   PVC plumbing fittings used as feet. The feet use a flat mending iron. Add pads or rubber to the bottom of the feet as desired.

FIGURE 22.15   One of six legs, completed (shown already attached to the robot).

FIGURE 22.16 A close-up detail of the leg cam.

rpm. You can use other motors and other driving techniques as long as the motors have a (pre-reduced) torque of at least 6 lb-ft. and a speed that can be reduced to 140 rpm or so.

Mount the motors inside two 6 1/2-inch-by-1 1/2-inch mending plate Ts. Drill a large hole, if necessary, for the shaft of the motor to stick through, as shown in Fig. 22.17 (refer to the parts list in Table 22.3). The motors used in the prototype came with a 12-pitch 12-tooth nylon gear. The gear was not removed for assembly, so the hole had to be large enough for it to pass through. The 30-tooth 12-pitch metal gear and 18-tooth 1/4-inch chain sprocket were also sandwiched between the mending plates.

The 1/4-inch shaft of the driven gear and sprocket is free running. You can install a bearing on each plate, if you wish, or have the shaft freely rotate in oversize holes. The sprocket and gear have 1/2-inch I.D. hubs, so reducing bushings were used. The sprocket and gear are held in place with compression. Don't forget the split washers. They provide the necessary compression to keep things from working loose.

Before attaching the two mending plates together, thread a 28 1/2-inch length of #25 roller chain over the sprocket. The exact length can be one or two links off; you can correct for any variance later on. Assemble the two plates using 8/32 by 3-inch bolts and 8/32 nuts and lock washers. Separate the plates using 2-inch spacers.

Attach the two 17 7/8-inch lengths of angle bracket on the robot, as shown in Fig. 22.18. The stock mounts directly under the two end pieces. Use 1/2-inch-by-8/32 bolts and nuts to secure the crosspieces in place. Secure the leg shafts using 1 1/4-inch bearings and U-bolts.

Mount the motor to the newly added inner mounting rails using 3-inch-by-1/2-inch mending plate Ts. Fasten the plates onto the motor mount, as shown in Fig. 22.19, with

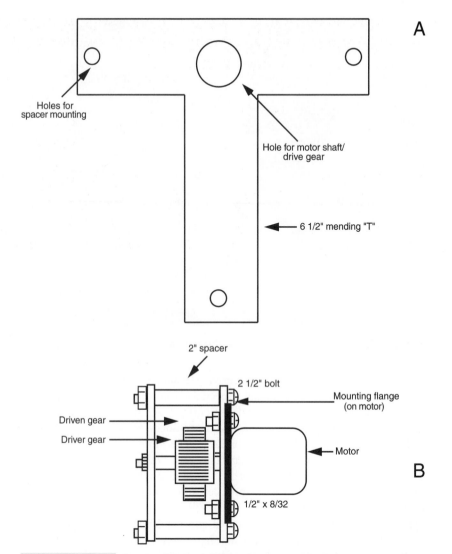

FIGURE 22.17 Motor mount details. *a*. Drilling guide for the mending T; *b*. The motor and drive gear-sprocket mounted with two mending Ts.

8/32 by 1/2-inch bolts and nuts. Position the shaft of the motor approximately 7 inches from the back of the robot (you can make any end of the shaft the back; it doesn't matter). Thread the roller chain over the center sprocket and the end sprocket. Position the motor until the roller chain is taut. Mark holes and drill. Secure the motor and mount to the frame using 8/32 by 1/2-inch bolts and nuts. Repeat the process for the opposite motor. The final assembly should look like Fig. 22.20.

Thread a 28 1/2-inch length of #25 roller chain around sprockets of the center and front legs. Attach an idler sprocket 7 1/2 inches from the front of the robot in line with the leg mounts. Use a diameter as close to 2 inches as possible for the idler; otherwise, you may

TABLE 22-3	PARTS LIST FOR WALKERBOT MOUNT-DRIVE SYSTEM
4	6 1/2-inch galvanized mending plate T
4	3-inch galvanized mending plate T
2	Heavy-duty gear-reduction DC motors
12	3 1/2-inch-diameter 30-tooth #15 chain sprocket
4	28 1/2-inch-length #25 roller chain
12	2 1/2-inch-by-1 1/2-inch-by-1/4-inch 20 U-bolts, with nuts and tooth lock washers
12	1 1/2-inch O.D. 1/4-inch-to-1/2-inch ID bearing
Misc	Reducing bushings (see text)

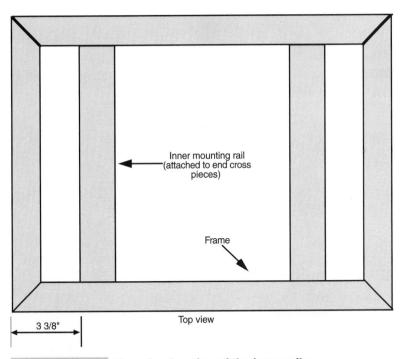

FIGURE 22.18 Mounting location of the inner rails.

need to shorten or lengthen the roller chain. Thread the roller chain around the sprocket, and find a position along the rail until the roller chain is taut (but not overly tight). Make a mark using the center of the sprocket as a guide and drill a 1/4-inch hole in the rail. Attach the sprocket to the robot. Figs. 22.21 through 22.23 show the motor mount, idler sprocket, and roller chain locations.

FIGURE 22.19 One of the drive motors mounted on the robot using smaller galvanized mending Ts.

FIGURE 22.20 Drive motor attached to the Walkerbot, with drive chain joining the motor to the leg shafts.

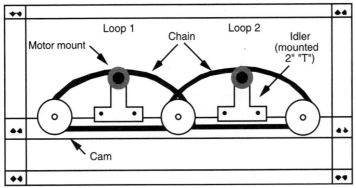

FIGURE 22.21 Mounting locations for idler sprocket.

FIGURE 22.22 A view of the mounted motor, with chain drive.

Batteries

The Walkerbot is not a lightweight robot, and its walking design requires at least 30 percent more power than a wheeled robot. The batteries for the Walkerbot are not trivial. You have a number of alternatives. One workable approach is to use two 6-volt motorcycle batteries, each rated at about 30 ampere-hours (AH). The two batteries together equal a slimmed-down version of a car battery in size and weight.

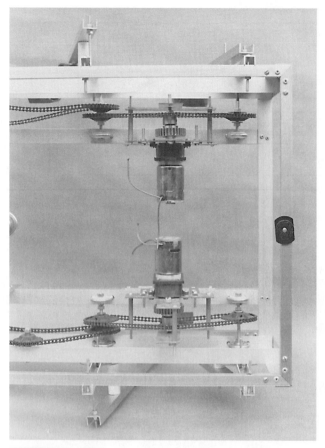

FIGURE 22.23 Left and right motors attached to the robot.

You can also use a 12-volt motorcycle or dune buggy battery, rated at more than 20 AH. The prototype Walkerbot used 12-AH 6-volt gel cell batteries. The amp-hour capacity is a bit on the low side, considering the two-amp draw from each motor, and the planned heavy use of electronics and support circuits. In tests, the 12-AH batteries provided about two hours of use before requiring a recharge.

There is plenty of room to mount the batteries. A good spot is slightly behind the center legs. By offsetting the batteries a bit in relation to the drive motors, you restore the center of gravity to the center of the robot. Of course, other components you add to the robot can throw the center of gravity off. Add one or two articulated arms to the robot, and the weight suddenly shifts toward the front. For flexibility, why not mount the batteries on a sliding rail, which will allow you to shift their position forward or back depending on the other weight you add to the Walkerbot.

The complete Walkerbot, minus the batteries, is shown in Fig. 22.24. Some additional hardware and holes are apparent on this version. Pay no attention to them. These were either my mistakes (!), or were made for components removed for the illustration.

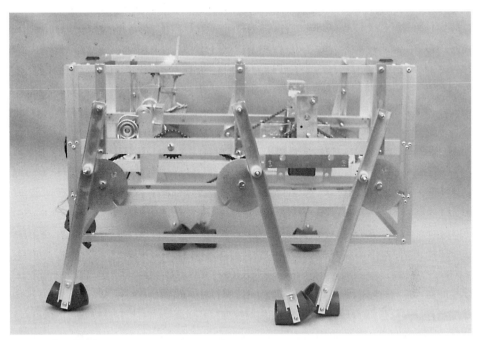

FIGURE 22.24 The completed Walkerbot.

Testing and Alignment

You can test the operation of the Walkerbot by temporarily installing a wired control box. The box consists of two DPDT switches wired to control the forward and backward motion of the two legs. See Chapter 8, "Robots of Plastic," for more details and a wiring diagram.

But before you test the Walkerbot, you need to align its legs. The legs on each side should be positioned so that either the center leg touches the ground or the front and back leg touch the ground. When the two sets of legs are working in tandem, the walking gait should be as shown in Fig. 22.25. This gait is the same as an insect's and provides a great deal of stability. To turn, one set of legs stops (or reverses) while the other set continues. During this time, the "tripod" arrangement of the gait will be lost, but the robot will still be supported by at least three legs.

An easy way to align the legs is to loosen the chain sprockets (so you can move the legs independently) and position the middle leg all the way forward and the front and back legs all the way back. Retighten the sprockets, and look out for misalignment of the roller chain and sprockets. If a chain bends to mesh with a sprocket, it is likely to pop off when the robot is in motion.

During testing, be on the lookout for things that rub, squeak, and work loose. Keep your wrench handy and adjust gaps and tighten bolts as necessary. Add a dab of oil to those parts that seem to be binding. You may find that a sprocket or gear doesn't stay tightened on a shaft. Look for ways to better secure the component to the shaft, such as by using a setscrew or another split lock washer. It may take several hours of "tuning up" to get the robot working at top efficiency.

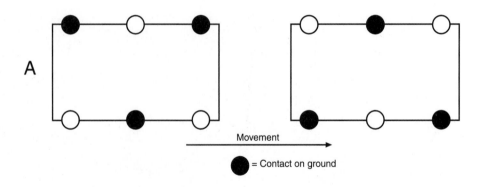

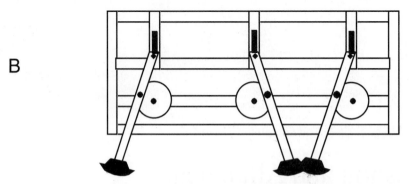

FIGURE 22.25 The walking gait. *a.* **The alternating tripod walking style of the Walkerbot, shared with thousands of crawling insects;** *b.* **The positioning of the legs for proper walking (front and back legs in synchrony; middle leg 180 degrees out of sync). The middle leg doesn't hit either the front or back leg because it is further from the body of the robot.**

Once the robot is aligned, run it through its paces by having it walk over level ground, step over small rocks and ditches, and navigate tight corners. Keep an eye on your watch to see how long the batteries provide power. You may need to upgrade the batteries if they cannot provide more than an hour of fun and games.

The Walkerbot is ideally suited for expansion. Fig. 22.26 shows an arm attached to the front side of the robot. You can add a second arm on the other side for more complete dexterity. Attach a dome on the top of the robot, and you've added a "head" on which you can attach a video camera, ultrasonic "ears" and "eyes," and lots more. Additional panels can be added to the front and back ends; attach them using hook-and-loop (such as Velcro) strips. That way, you can easily remove the panels should you need quick access to the inside of the robot.

FIGURE 22.26 An arm attached to the front side of the Walkerbot.

From Here

To learn more about...	*Read*
Working with metal	Chapter 10, "Building a Metal Platform"
Robot locomotion styles, including wheels, treads, and legs	Chapter 16, "Robot Locomotion Principles"
Using DC motors	Chapter 18, "Working with DC Motors"
Additional locomotion systems based on the Walkerbot frame	Chapter 23, "Advanced Locomotion Systems"
Constructing an arm for the Walkerbot	Chapter 25, "Build a Revolute Coordinate Arm"

23

ADVANCED LOCOMOTION SYSTEMS

Two drive wheels aren't the only way to move a robot across the living room or workshop floor. If you read Chapter 22, you learned how to build a six-legged walking robot. Here, in this chapter, you'll learn the basics of applying some other unique drive systems to propel your robot designs, including a stair-climbing robot, an outdoor tracked robot, and even a six-wheeled "Buggybot."

Track o' My Robot

There is something exciting about seeing a tank climb embankments, bounding over huge boulders as if they were tiny dirt clods. A robot with tracked drive is a perfect contender for an automaton that's designed for outdoor use. Where a wheeled or legged robot can't go, the tracked robot can roll in with relative ease. Experimental tracked robots, using metal tracks just like tanks, have been designed for the U.S. Navy and U.S. Army and are even used by many police and fire departments. Everyone has liked what they've seen so far; development of tracked autonomous vehicles continues still.

Using a heavy metal track for your personal robot is decidedly a bad idea. The track is too heavy and much too hard to fabricate. For a homebrew robot, a rubber track is more than adequate. You can use a large timing belt, even an automotive fan belt, for the track.

Another alternative I've used with some success is rubber wetsuit material. Most diving shops have long strips of the rubber lying around that they'll sell or give to you. You can mend the rubber using a special waterproof adhesive. You can glue the strip together to make a band, then glue small rubber cleats on to the band. Fig. 23.1 shows the basic idea.

The drive train for a tracked robot must be specifically engineered or modified for the task of driving a track. The Walkerbot described in Chapter 22 makes a good base for a tracked robot. Remove the legs and install three small drive pulleys, as diagrammed in Fig. 23.2. The track fits inside the groove of the pulleys, so it won't easily slip out. You must add rollers to the bottom of the carriage against which the track passes. Unless the track is

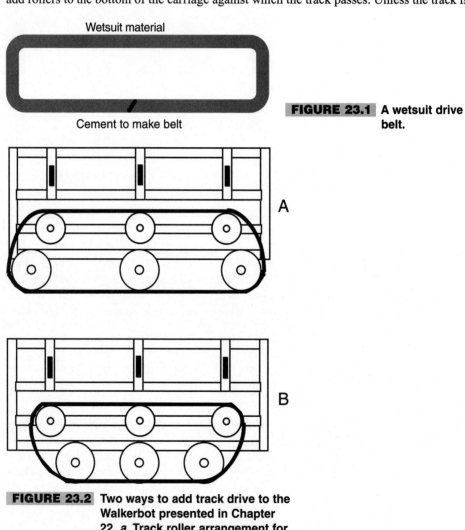

Wetsuit material

Cement to make belt

FIGURE 23.1 A wetsuit drive belt.

FIGURE 23.2 Two ways to add track drive to the Walkerbot presented in Chapter 22. *a.* Track roller arrangement for good traction and stability but relatively poor turning radius. *b.* Track roller arrangement for good turning radius, but hindered traction and stability.

thick there can be no groove in the rollers. Otherwise, the track would ride *inside* the rollers, instead of outside. Wide rubber tires make good rollers.

With this design, the track may pop off the rollers and drive wheels under certain circumstances. To help minimize the chances of throwing the track every few minutes, add a guide roller to the bottom of the carriage, as diagrammed in Fig. 23.3. The track rides inside a groove (or flange) in the small guide roller, and prevents the track from popping out of place.

To propel the robot, you activate both motors so the tracks move in the same direction and at the same speed. To steer, you simply stop or reverse one side. For example, to turn left, stop the left track. To make a hard left turn, reverse the left track. The Walkerbot has six driven wheels. The three wheels on each side are linked together, so they all provide power to the track. But you don't need a three-wheel drive system. In fact, you can usually get by with just one driver wheel on each side of the robot.

Steering Wheel Systems

Using dual motors to effect propulsion and steering is just one method for getting your robot around. Another approach is to use a pivoting wheel to steer the robot. The same wheel can provide power, or power can come from two wheels in the rear (the latter is much more common). The arrangement is not unlike golf carts, where the two rear wheels provide power and a single wheel in the front provides steering. See Fig. 23.4 for a diagram of a typical steering-wheel robot. Fig. 23.5 shows a detail of the steering mechanism.

The advantage of a steering-wheel robot is that you need only one powerful drive motor. The motor can power both rear wheels at once. The steering wheel motor needn't be as powerful since all it has to do is swivel the wheel back and forth a few degrees. The biggest disadvantage of steering-wheel systems is the steering! You must build stops into the steering mechanisms (either mechanical or electronic) to prevent the wheel from turning more than 50° or 60° to either side. Angles greater than about 60° cause the robot to suddenly steer in the other direction. They may even cause the robot to lurch to a sudden stop because the front wheel is at a right angle to the rear wheels.

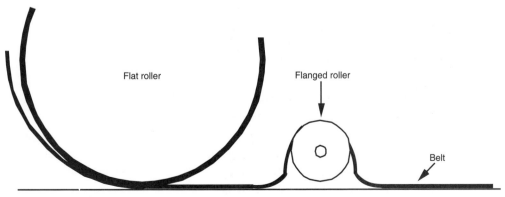

Flat roller   Flanged roller

Belt

FIGURE 23.3 A close-up view of the flanged roller used to prevent the track from popping off the drive.

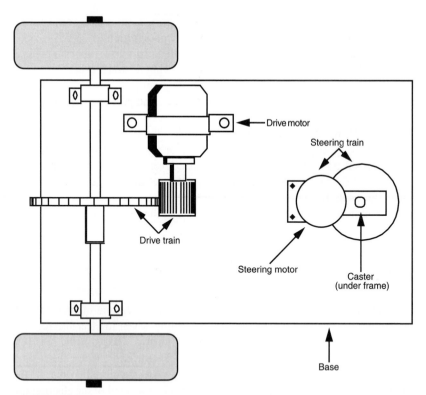

FIGURE 23.4 A basic arrangement for a robot using one drive motor and steering wheel.

The servomechanism that controls the steering wheel must "know" when the wheel is pointing forward. The wheel must return to this exact spot when the robot is commanded to forge straight ahead. Not all servomechanisms are this accurate. The motor may stop one or more degrees off the center point, and the robot may never actually travel in a straight line. A good steering motor, and a more sophisticated servomechanism, can reduce this limitation.

A number of robot designs with steering-wheel mechanisms have been described in other robot books and on various Web pages. Check out Appendix A, "Further Reading," and Appendix C, "Robot Information on the Internet," for more information.

Six-Wheeled Robot Cart

You can also modify the Walkerbot described in Chapter 22 into a six-wheeled rugged terrain cart, or Buggybot. Simply remove the legs and attach wheels, as diagrammed in Fig. 23.6. The larger the wheels the better, as long as they aren't over about nine inches (the centerline diameter between each drive shaft).

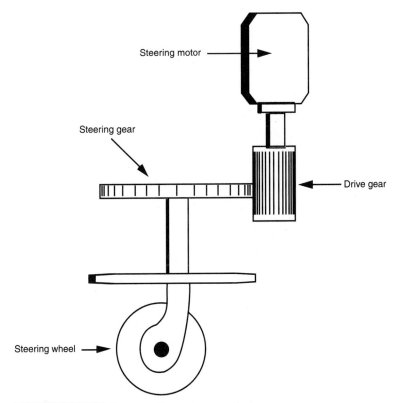

Steering motor

Steering gear

Drive gear

Steering wheel

FIGURE 23.5 The steering gear up close.

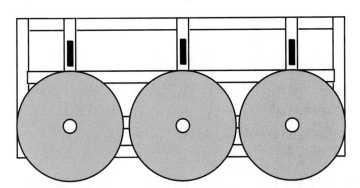

FIGURE 23.6 Converting the Walkerbot (from Chapter 22) into a six-wheeled, all-terrain Buggybot.

Pneumatic wheels are the best choice because they provide more "bounce" and handle rough ground better than hard rubber tires. Most hardware stores carry a full assortment of pneumatic tires. Most are designed for things like wheelbarrows and hand dollies. Cost can be high, so you may want to check out the surplus or used industrial supply houses. Just be sure the tire doesn't have a flat!

Steering is accomplished as with a two-wheeled robot. The series of three wheels on each side act as a kind of track tread, so the vehicle behaves much like a tracked vehicle. The maneuverability isn't as good as with a two-wheeled robot, but you can still turn the robot in a radius a little longer than its length. Sharp turns require you to reverse one set of wheels while applying forward motion to the other.

Tri-Star Wheels

In the science fiction film *Damnation Alley,* starring Jan-Michael Vincent and George Peppard, the earth has been decimated by an atomic war, and the heroes must trek across the county amid radioactive storms, marauders, and other postwar denizens. To help them get there in one piece, they use an incredible 35-foot-long steel vehicle called the "Landmaster." This thing crashes through walls, hops over large ditches as if they were nothing but potholes, even swims across the water (see the movie production still in Fig. 23.7). The drive system used by the Landmaster is an unusual tri-star arrangement that's perfectly suited for robotic tasks such as climbing stairs, maneuvering through rough terrain, and, yes, even going through water.

The Landmaster was built by Dean Jeffries Automotive Styling, of Hollywood, California. The full-size vehicle can still be seen parked in the lot of his workshop off Cahuenga Boulevard. Jeffries modeled the wheel arrangement of the Landmaster after a design patented (but never actively used) by Lockheed Aircraft for an all-terrain vehicle. Each of the four "wheels" on the vehicle is actually a set of three smaller wheels, clustered in a triangle, as shown in Fig. 23.8.

All three tires continually rotate, driven by a central shaft, but in normal operation only two of them are touching the ground. When the vehicle encounters an obstacle or hole, the wheel gang flips and rotates the wheels into a new position, as diagrammed in Fig. 23.9. You have to either see the movie or build your own robot based on this design to believe it.

All four wheel gangs are powered by a central motor. Steering is accomplished by bending the midsection of the robot. In the Landmaster vehicle, this is accomplished by using hydraulic rams, but for a homebrew robot you'd probably use a servo or stepping motor. Fig. 23.10 shows how such a robot might be constructed. One large motor, which rests in one of the robot halves, drives the four wheel gangs. Since the robot is made in two parts and the midsection bends, the wheels in the other half of the robot are driven via a flexible cable, the same kind used with handheld electric drills. To balance the robot, the batteries are placed in the second half. A wiring harness connects the two halves together.

FIGURE 23.7 The "Landmaster," from the motion picture *Damnation Alley*. Photo courtesy 20th Century Fox.

FIGURE 23.8 A close-up of the tri-star wheels used in the Landmaster.

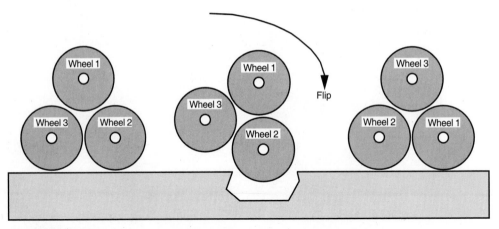

FIGURE 23.9 How the wheel gang "flips" when it encounters a hole or obstacle. The same basic motion can be used to climb stairs.

Building Robots with Shape-memory Alloy

A metal that has a memory? You bet. As early as 1938, scientists observed that certain metal alloys, once bent into odd shapes, returned to the original form when heated. This property was considered little more than a laboratory curiosity because the metal alloys were weak, were difficult and expensive to manufacture, and broke apart after just a couple of heating/cooling cycles.

Research into metals with memory took off in 1961, when William Beuhler and his team of researchers at the U.S. Naval Ordnance Laboratory developed a titanium-nickel alloy that repeatedly displayed the memory effect. Beuhler and his cohorts developed the first commercially viable shape-memory alloy, or SMA. They called the stuff Nitinol, a fancy-sounding name derived from *Ni*ckel *Ti*tanium *Na*val *O*rdnance *L*aboratory.

Since its introduction, Nitinol has been used in a number of commercial products—but not many. For example, several Nitinol engines have been developed that operate with only hot and cold water. In operation, the metal contracts when exposed to hot water and relaxes when exposed to cold water. Combined with various assemblies of springs and cams, the contraction and relaxation (similar to a human muscle) causes the engine to move.

Other commercial applications of Nitinol include pipe fittings that automatically seal when cooled, large antenna arrays that can be bent (using hot water) into most any shape desired, sunglass frames that spring back to their original shape after being bent, and a novel antiscald device that shuts off water flow in a shower should the water temperature exceed a certain limit.

Regular Nitinol contracts and relaxes in heat (in air, water, or other liquid). That limits the effectiveness of the metal in many applications where local heat can't be applied. Researchers have attempted to heat the Nitinol metal using electrical current in an

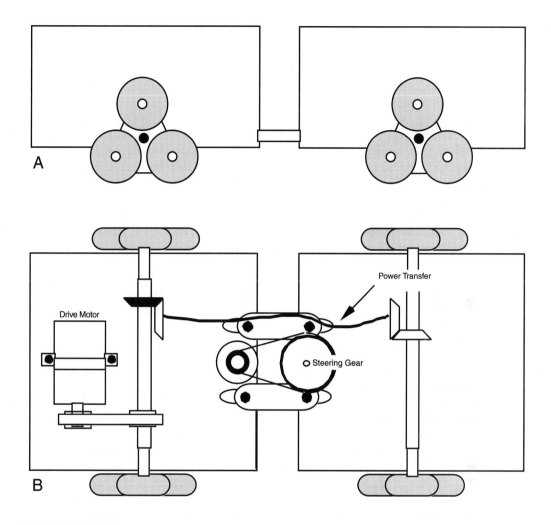

FIGURE 23.10 A preliminary design for a robot based on the tri-star wheel arrangement. *a.* Side view of the robot; *b.* Top view of the robot showing the one drive motor and the central steering motor.

effort to exactly control the contraction and relaxation. But because of the molecular construction of Nitinol, hot spots develop along the length of the metal, causing early fatigue and breakage.

In 1985, a Japanese company, Toki Corp., unveiled a new type of shape-memory alloy specially designed to be activated by electrical current. Toki's unique SMA material, trade-named BioMetal, offers all of the versatility of the original Nitinol, with the added bene-fit of near instant electrical actuation. BioMetal and materials like it—Muscle Wire from Mondo-Tronics or Flexinol from Dynalloy—have many uses in robotics, including novel locomotive actuation. From here on out we'll refer to this family of materials generically as shape-memory alloy, or simply SMA.

BASICS OF SMA

At its most basic level, SMA is a strand of nickel titanium alloy wire. Though the material may be very thin (a typical thickness is 0.15 mm—slightly wider than a strand of human hair) it is exceptionally strong. In fact, the tensile strength of SMA rivals that of stainless steel: the breaking point of the slender wire is a whopping six pounds. Even under this much weight, SMA stretches little. In addition to its strength, SMA also shares the corrosion-resistance of stainless steel.

Shape memory alloys change their internal crystal structure when exposed to certain higher-than-normal temperatures (this includes the induced temperatures caused by passing an electrical current through the wire). The structure changes again when the alloy is allowed to cool. More specifically, during manufacture the SMA wire is heated to a very high temperature, which embosses or "memorizes" a certain crystal structure. The wire is then cooled and stretched to its practical limits. When the wire is reheated, it contracts because it is returning to the memorized state.

Although most SMA strands are straight, the material can also be manufactured in spring form, usually as an expansion spring. In its normal state, the spring exerts minimum tension, but when current is applied the spring stiffens, exerting greater tension. Used in this fashion, SMAs become an "active spring" that can adjust itself to a particular load, pressure, or weight.

Shape memory alloys have an electrical resistance of about one ohm per inch. That's more than ordinary hookup wire, so SMAs will heat up more rapidly when an electrical current is passed through them. The more current passes through, the hotter the wire becomes and the more contracted the strand. Under normal conditions, a two- to three-inch length of SMA is actuated with a current of about 450 milliamps. That creates an internally generated temperature of about 100–130°C; 90°C is required to achieve the shape-memory change. Most SMAs can be manufactured to change shape at most any temperature, but 90°C is the standard value for off-the-shelf material.

Excessive current should be avoided. Why? Extra current causes the wire to overheat, which can greatly degrade its shape-memory characteristics. For best results, current should be as low as necessary to achieve the contraction desired. Shape memory alloys will contract by 2 to 4 percent of their length, depending on the amount of current applied. The maximum contraction of typical SMA material is 8 percent, but that requires heavy current that can, over a period of just a few seconds, damage the wire.

USING SMA

Shape memory alloys need little support paraphernalia. Besides the wire itself, you need some type of terminating system, a bias force, and an actuating circuit. We'll discuss each of these in the following sections.

Terminating system   The terminators attach the ends of the SMA wires to the support structure or mechanism you are moving. Because SMAs expand as they contract, using glue or other adhesive will not secure the wire to the mechanism. Ordinary soldering is not recommended as the extreme heat of the soldering can permanently damage the wire. The best approach is to use a crimp-on terminator. These and other crimp terminators are available from companies that sell shape memory alloy wire (either in the experimenter's kit or separately).

You can make your own crimp-on connectors using 18-gauge or smaller solderless crimp connectors (the smaller the better). Although these connectors are rather large for the thin 0.15 mm SMA, you can achieve a fairly secure termination by folding the wire in the connector and pressing firmly with a suitable crimp tool. Be sure to completely flatten the connector. If necessary, place the connector in a vise to flatten it all the way.

Bias force   Apply current to the ends of an SMA wire and it just contracts in air. To be useful, the wire must be attached to one end of the moving mechanism and biased (as shown in Fig. 23.11) at the other end. Besides offering physical support, the bias offers the counteracting force that returns the SMA wire to its limber condition once current is removed from the strand.

Actuating circuit   SMAs can be actuated with a 1.5-volt penlight battery. Because the circuit through the SMA wire is almost a dead short, the battery delivers almost its maximum current capacity. But the average 1.5-volt alkaline penlight battery has a maximum current output of only a few hundred milliamps, so the current is limited through the wire. You can connect a simple on/off switch in line with the battery, as detailed in Fig. 23.12, to contract or relax the SMA wire.

The problem with this setup is that it wastes battery power, and if the power switch is left on for too long, it can do some damage to the SMA strand. A more sophisticated approach uses a 555 timer IC that automatically shuts off the current after a short time. The schematic in Fig. 23.13 shows one way of connecting a 555 timer IC to control a length of SMA. Table 23.1 provides a parts list for the 555 SMA circuit.

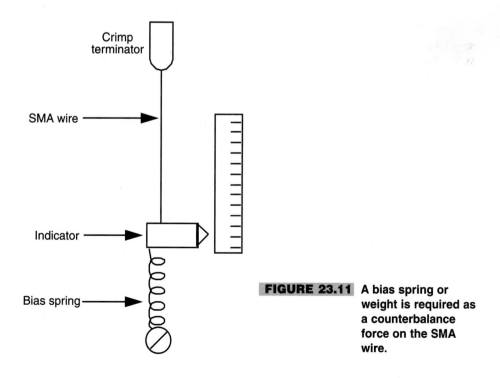

FIGURE 23.11 A bias spring or weight is required as a counterbalance force on the SMA wire.

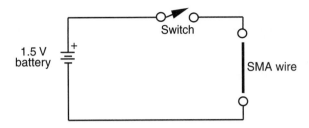

FIGURE 23.12 A simple switch in series with a 1.5-volt penlight battery forms a simple SMA driving circuit. The low current delivered by the penlight battery prevents damage to the SMA wire.

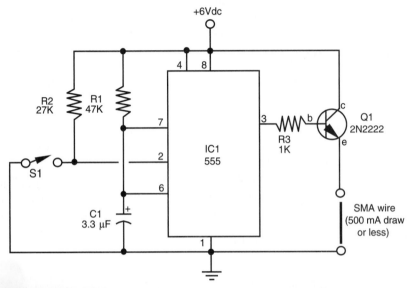

FIGURE 23.13 A 555 timer IC is at the heart of an ideal driving circuit for SMA wire. The 555 removes the current shortly after you release activating switch S1.

TABLE 23.1	PARTS LIST FOR 555 SMA DRIVER.
IC1	555 timer
Q1	2N2222 NPN transistor
R1	47K resistor
R2	27K resistor
R3	1K resistor
C1	3.3 µF polarized electrolytic capacitor
Misc	Momentary SPST switch, SMA wire

All resistors are 5 to 10 percent tolerance, 1/4 watt. All capacitors are 10 to 20 percent tolerance, rated at 35 volts or more.

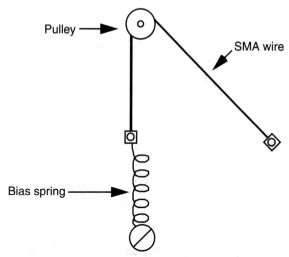

FIGURE 23.14 Concept of using SMA wire
with a mechanical pulley.

In operation, when you press momentary switch S1 it activates the wire and it contracts. Release S1 immediately, and the SMA stays contracted for an extra fraction of a second, then releases as the 555 timer shuts off. Since the total ON time of the 555 depends on how long you hold S1 down, plus the 1/10-of-a-second delay, you should depress the switch only momentarily.

SHAPE MEMORY ALLOY MECHANISMS

With the SMA properly terminated and actuated, it's up to you and your own imagination to think of ways to use it in your robots. Fig. 23.14 shows a typical application using an SMA wire in a pulley configuration. Apply current to the wire and the pulley turns, giving you rotational motion. A large-diameter pulley will turn very little when the SMA tenses up, but a small-diameter one will turn an appreciable distance.

Fig. 23.15 shows a length of SMA wire used in a lever arrangement. Here, the metal strand is attached to one end of a bell crank. On the opposite end is a bias spring. Applying juice to the wire causes the bell crank to move. The spot where you attach the drive arm dictates the amount of movement you will obtain when the SMA contracts.

SMA wire is tiny stuff, and you will find that the miniature hardware designed for model R/C airplanes is most useful for constructing mechanisms. Most any well-stocked hobby store will stock a full variety of bell cranks, levers, pulleys, wheels, gears, springs, and other odds and ends to make your work with SMA more enjoyable.

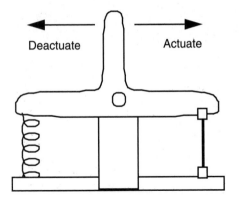

Deactuate   Actuate

FIGURE 23.15 The bell crank changes the contraction of the SMA for sideways movement of the lever. The spring enables the bell crank to return to its original position after the current is removed from the wire.

From Here

To learn more about...	*Read*
Selecting a motor for driving a robot	Chapter 17, "Choosing the Right Motor for the Job"
Using DC motors for robot locomotion	Chapter 18, "Working with DC Motors"
Building Walkerbot, a large six-legged walking robot	Chapter 22, "Build a Heavy-duty, Six-legged Walking Robot"

24

AN OVERVIEW OF ARM SYSTEMS

Robots without arms are limited to rolling or walking about, perhaps noting things that occur around them, but little else. The robot can't, as the slogan goes, "reach out and touch someone," and it certainly can't manipulate its world.

The more sophisticated robots in science, industry, and research and development have at least one arm to grasp, reorient, or move objects. Arms extend the reach of robots and make them more like humans. For all the extra capabilities arms provide a robot, it's interesting that they aren't at all difficult to build. Your arm designs can be used for factory-style, stationary "pick-and-place" robots, or they can be attached to a mobile robot as an appendage.

This chapter deals with the concept and design theory of robotic arms. Specific arm projects are presented in Chapters 25 and 26. Incidentally, when we speak of arms, we will usually mean just the arm mechanism minus the hand (also called the gripper). Chapter 27, "Experimenting with Gripper Designs," talks about how to construct robotic hands and how you can add them to arms to make a complete, functioning appendage.

The Human Arm

Take a close look at your own arms for a moment. You'll quickly notice a number of important points. First, your arms are amazingly adept mechanisms, no doubt about it. They are capable of being maneuvered into just about any position you want. Your arm has two major joints: the shoulder and the elbow (the wrist, as far as robotics is concerned, is usually considered part of the gripper mechanism). Your shoulder can move in two planes, both up and down and back and forth. Move your shoulder muscles up, and your entire arm is raised away from your body. Move your shoulder muscles forward, and your entire arm moves forward. The elbow joint is capable of moving in two planes as well: back and forth and up and down.

The joints in your arm, and your ability to move them, are called *degrees of freedom*. Your shoulder provides two degrees of freedom in itself: shoulder rotation and shoulder flexion. The elbow joint adds a third and fourth degree of freedom: elbow flexion and elbow rotation.

Robotic arms also have degrees of freedom. But instead of muscles, tendons, ball and socket joints, and bones, robot arms are made from metal, plastic, wood, motors, solenoids, gears, pulleys, and a variety of other mechanical components. Some robot arms provide but one degree of freedom; others provide three, four, and even five separate degrees of freedom.

Arm Types

Robot arms are classified by the shape of the area that the end of the arm (where the gripper is) can reach. This accessible area is called the *work envelope.* For simplicity's sake, the work envelope does not take into consideration motion by the robot's body, just the arm mechanics.

The human arm has a nearly spherical work envelope. We can reach just about anything, as long as it is within arm's length, within the inside of about three-quarters of a sphere. Imagine being inside a hollowed-out orange. You stand by one edge. When you reach out, you can touch the inside walls of about three-quarters of the orange peel.

In a robot, such a robot arm would be said to have *revolute coordinates.* The three other main robot arm designs are *polar coordinate, cylindrical coordinate,* and *Cartesian coordinate.* You'll note that there are three degrees of freedom in all four basic types of arm designs. Let's take a closer look at each one.

REVOLUTE COORDINATE

Revolute coordinate arms, such as the one depicted in Fig. 24.1, are modeled after the human arm, so they have many of the same capabilities. The typical robotic design is somewhat different, however, because of the complexity of the human shoulder joint.

The shoulder joint of the robotic arm is really two different mechanisms. Shoulder rotation is accomplished by spinning the arm at its base, almost as if the arm were mounted on a record player turntable. Shoulder flexion is accomplished by tilting the upper arm member backward and forward. Elbow flexion works just it does in the human arm. It moves the forearm up and down. Revolute coordinate arms are a favorite design choice for hobby robots. They provide a great deal of flexibility, and, besides, they actually *look* like arms. For details on how to construct a revolute coordinate arm, see Chapter 25.

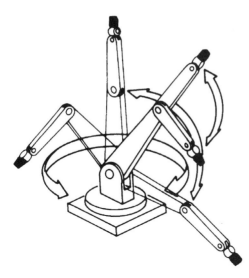

FIGURE 24.1 A revolute coordinate arm.

POLAR COORDINATE

The work envelope of the polar coordinate arm is the shape of a half sphere. Next to the revolute coordinate design, polar coordinate arms are the most flexible in terms of the ability to grasp a variety of objects scattered about the robot. Fig. 24.2 shows a polar coordinate arm and its various degrees of freedom.

A turntable rotates the entire arm, just as it does in a revolute coordinate arm. This function is akin to shoulder rotation. The polar coordinate arm lacks a means for flexing or bending its shoulder, however. The second degree of freedom is the elbow joint, which moves the forearm up and down. The third degree of freedom is accomplished by varying the reach of the forearm. An "inner" forearm extends or retracts to bring the gripper closer to or farther away from the robot. Without the inner forearm, the arm would only be able to grasp objects laid out in a finite two-dimensional circle in front of it. Not very helpful.

The polar coordinate arm is often used in factory robots and finds its greatest application as a stationary device. It can, however, be mounted to a mobile robot for increased flexibility. Chap. 26 shows you how to build a rather useful stationary polar coordinate arm.

CYLINDRICAL COORDINATE

The cylindrical coordinate arm looks a little like a robotic forklift. Its work envelope resembles a thick cylinder, hence its name. Shoulder rotation is accomplished by a revolving base, as in revolute and polar coordinate arms. The forearm is attached to an elevatorlike lift mechanism, as depicted in Fig. 24.3. The forearm moves up and down this column to grasp objects at various heights. To allow the arm to reach objects in three-dimensional space, the forearm is outfitted with an extension mechanism, similar to the one found in a polar coordinate arm.

CARTESIAN COORDINATE

The work envelope of a Cartesian coordinate arm (Fig. 24.4) resembles a box. It is the arm most unlike the human arm and least resembles the other three arm types. It has no

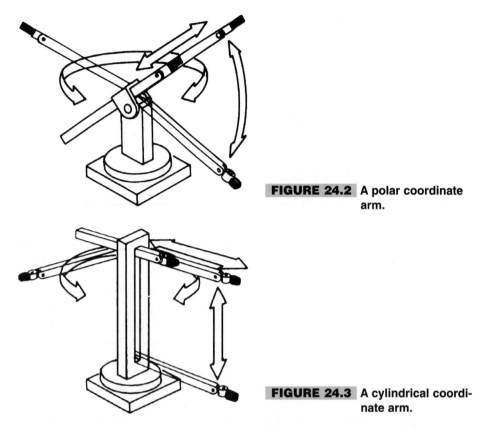

FIGURE 24.2 A polar coordinate arm.

FIGURE 24.3 A cylindrical coordinate arm.

rotating parts. The base consists of a conveyer belt-like track. The track moves the elevator column (like the one in a cylindrical coordinate arm) back and forth. The forearm moves up and down the column and has an inner arm that extends the reach closer to or farther away from the robot.

Activation Techniques

There are three general ways to move the joints in a robot arm:

- Electrical
- Hydraulic
- Pneumatic

Electrical actuation is done with motors, solenoids, and other electromechanical devices. It is the most common and easiest to implement. The motors for elbow flexion, as well as the motors for the gripper mechanism, can be placed in or near the base. Cables, chains, or belts connect the motors to the joints they serve.

Electrical activation doesn't always have to be via an electromechanical device such as a motor or solenoid. Other types of electrically induced activation are possible using a

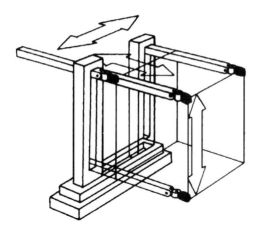

FIGURE 24.4 A Cartesian coordinate arm.

variety of technologies. One of particular interest to hobby robot builders is shape memory alloy, or SMA, as discussed in Chapter 23. SMA material goes by a number of trade names, such as BioMetal, Dynalloy, Nitinol, and Muscle Wire.

The construction of SMA material differs from manufacturer to manufacturer, but the activation technique is about the same: when heat is applied to the metal, it contracts to a predefined state. Heat can be applied directly, through a flame or with hot water, or by passing an electrical current through the material. The electrical activation is the technique most commonly used in robotics.

A disadvantage of SMA is its slow expansion rate: the metal must cool before it relaxes and returns to its preheated shape and size. The larger the piece of metal is, the longer it takes to cool, so the slower the "muscle" returns to its noncontracted state. As a result, most of the shape memory alloy material you'll see available, such as Muscle Wire from Mondo-Tronics, is hair-thin. Don't let the small diameter of the wire fool you, however: Muscle Wire and many other SMA materials can bear considerable weight—several pounds in both the contracted and noncontracted state. See Chapter 23 for a full discussion of shape memory alloy.

Hydraulic actuation uses oil-reservoir pressure cylinders, similar to the kind used in earth-moving equipment and automobile brake systems. The fluid is noncorrosive and inhibits rust: both are the immediate ruin of any hydraulic system. Though water can be used in a hydraulic system, if the parts are made of metal they will no doubt eventually suffer from rust, corrosion, or damage by water deposits. For a simple homebrew robot, however, a water-based hydraulic system using plastic parts is a viable alternative.

Pneumatic actuation is similar to hydraulic, except that pressurized air is used instead of oil or fluid (the air often has a small amount of oil mixed in it for lubrication purposes). Both hydraulic and pneumatic systems provide greater power than electrical actuation, but they are more difficult to use. In addition to the actuation cylinders themselves, such as the one shown in Fig. 24.5, a pump is required to pressurize the air or oil, and values are used to control the retraction or extension of the cylinders. For the best results, you need a holding tank to stabilize the pressurization.

An interesting variation on pneumatic actuation is the "Air Muscle," an ingenious combination of a small rubber tube and black plastic mesh. The rubber tube acts as an expandable bladder, and the plastic mesh forces the tube to inflate in a controllable manner. Air Muscle is available premade in various sizes; it is activated by pumping air into the tube.

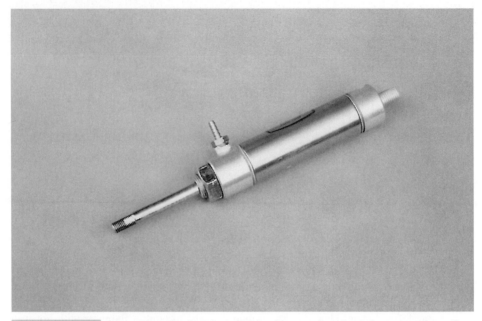

FIGURE 24.5 One of hundreds of available sizes and styles of pneumatic cylinders. This one has a bore of about 1/2 inch and a stroke of three inches.

When filled with air, the tube expands its width but contracts its length (by 25 percent). The result is that the tube and mesh act as a kind of mechanical muscle. The Air Muscle is said to be more efficient than the standard pneumatic cylinder, and according to its makers it has about a 400:1 power-to-weight ratio.

From Here

To learn more about...	*Read*
Building a robotic revolute coordinate arm	Chapter 25, "Build a Revolute Coordinate Arm"
Building a robotic stationary polar coordinate arm	Chapter 26, "Build a Polar Coordinate Arm"
Creating hands for robot arms	Chapter 27, "Experimenting with Gripper Designs"
Endowing robot arms and hands with the sense of touch	Chapter 35, "Adding the Sense of Touch"

25

BUILD A REVOLUTE COORDINATE
ARM

The revolute coordinate arm design provides a great deal of flexibility, yet requires few components. The arm described in this chapter enjoys only two degrees of freedom. You'll find, however, that even with two degrees of freedom, the arm can do many things. It can be used by itself as a stationary pick-and-place robot, or it can be attached to a mobile platform. The construction details given here are for a left hand; to build a right hand, simply make it a mirror image of the left.

You'll note that the arm lacks a hand—a gripper. You can use just about any type of gripper. In fact, you can design the forearm so it accepts many different grippers interchangeably. See Chapter 27, "Experimenting with Gripper Designs," for more information on robot hands.

Design Overview

The design of the revolute coordinate arm is modeled after the human arm. A shaft-mounted shoulder joint provides shoulder rotation (degree of freedom #1). A simple swing-arm rotating joint provides the elbow flexion (degree of freedom #2).

You could add a third degree of freedom, shoulder flexion, by providing another joint immediately after the shoulder. In tests, however, I found that this basic two-degree-of-freedom arm is quite sufficient for most tasks. It is best used, however, on a mobile platform where the robot can move closer to or farther away from the object it's grasping. That's cheating, in a way, but it's a lot simpler than adding another joint.

Shoulder Joint and Upper Arm

The shoulder joint is a shaft that connects to a bearing mounted on the arm base or in the robot. Attached to the shaft is the drive motor for moving the shoulder up and down. The motor is connected by a single-stage gear system, as shown in Fig. 25.1 (refer to the parts list in Table 25.1). In the prototype arm for this book, the output of the motor was approximately 22 rpm, or roughly one-third of a revolution per second.

For a shoulder joint, 22 rpm is a little on the fast side. I chose a gear ratio of 3:1 to decrease the speed by a factor of three (and increase the torque of the motor roughly by a factor of three). With the gear system, the shoulder joint moves at about one revolution every eight seconds. That may seem slow, but remember that the shoulder joint swings in an arc of a little less than 50°, or roughly one-seventh of a complete circle. Thus, the shoulder will go from one extreme to the other in under two seconds.

Refer to Fig. 25.2. The upper arm is constructed from a 10-inch length of 57/64-inch-by-9/16-inch-by-1/16-inch aluminum channel stock and a matching 10-inch length of 41/64-inch-by-1/2-inch-by-1/16-inch aluminum channel stock. Sandwich the two stocks together to make a bar. Drill a 1/4-inch hole 1/2 inch from the end of the channel stock pieces. Cut a piece of 1/4-inch 20 all-thread rod to a length of seven inches (this measurement depends largely on the shoulder motor arrangement, but seven inches gives you room to make changes). Thread a 1/4-inch 20 nut, flat washer, and locking washer onto one end of the rod. Leave a little extra—about 1/8 inch to 1/4 inch—on the outside of the nut. You'll need the room in a bit.

Drill a 1/4-inch hole in the center of a 3 3/4-inch-diameter metal electrical receptacle cover plate. Insert the rod through it and the hole of the larger channel aluminum. Thread two 1/4-inch 20 nuts onto the rod to act as spacers, then attach the smaller channel aluminum. Lock the pieces together using a flat washer, tooth washer, and 1/4-inch 20 nut. The shoulder is now complete.

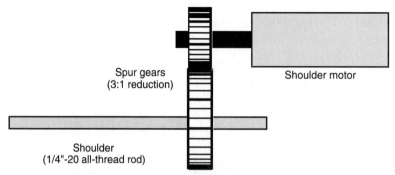

Spur gears
(3:1 reduction)

Shoulder motor

Shoulder
(1/4"-20 all-thread rod)

FIGURE 25.1 The gear transfer system used to actuate the shoulder of the revolute arm. You can also use a motor with a built-in reduction gear if the output of the motor is not slow enough for the arm.

Table 25.1	PARTS LIST FOR REVOLUTE ARM.
1	10-inch length 57/64-inch-by-9/16-inch-by-1/16-inch aluminum channel stock
1	10-inch length 41/64-inch-by-1/2-inch-by-1/16-inch aluminum channel stock
1	8-inch length 57/64-inch-by-9/16-inch-by-1/16-inch aluminum channel stock
1	8-inch length 41/64-inch-by-1/2-inch-by-1/16-inch aluminum channel stock
1	7-inch length 1/4-inch 20 all-thread rod
2	1 3/4-inch-by-10/24 stove bolt
2	1 1/2-inch-by-3/8-inch flat corner iron
1	3-inch-by-3/4-inch mending plate "T" (for motor mounting)
2	1/2-inch aluminum spacer
1	1/4-inch aluminum spacer
2	3/4-inch-diameter, 5-lugs-per-inch timing belt sprocket
1	20 1/2-inch-length timing belt (5 lugs per inch)
2	Stepper motors (see text)
1	3:1 gear reduction system (such as one 20-tooth 24-pitch spur gear and one 60-tooth 24-pitch spur gear)
Misc	6/32, 10/24, and 1/4-inch 20 nuts, washers, tooth lock washers, fishing tackle weights

Elbow and Forearm

The forearm attaches to the end of the upper arm. The joint there serves as the elbow. The forearm is constructed much like the upper arm: cut the small and large pieces of channel aluminum to eight inches instead of ten inches. Construct the elbow joint as shown in Figs. 25.3 and 25.4, using two 1 1/2-inch-by-3/8-inch flat corner angles, 1/2-inch spacers, and 10/24 hardware. The 3/4-inch timing belt sprocket (5 lugs per inch) is used to convey power from the elbow motor, which is mounted at the shoulder. The completed joint is shown in Fig. 25.5.

You can actually use just about any size of timing belt or sprocket. When using the size of sprockets specified in Table 25.1, the timing belt is 20 1/2 inches. If you use another size sprocket for the elbow or the motor, you may need to choose another length. You can adjust for some slack by mounting the elbow joint closer or farther to the end of the upper arm.

You may also use #25 roller chain to power the elbow. Use a sprocket on the elbow and a sprocket on the motor shaft. Connect the two with a #25 roller chain. You'll need to experiment based on the size of sprockets you use to come up with the exact length for the roller chain.

When the elbow and forearm are complete, mount the motor on the shoulder, directly on the plate cover. The motor we chose for the prototype revolute coordinate arm was a

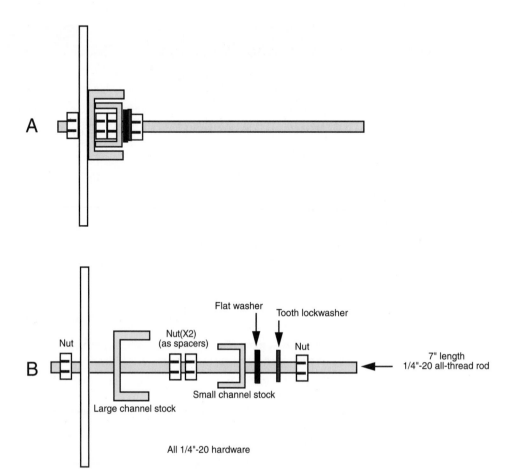

FIGURE 25.2 Shoulder shaft detail. *a.* **Completed shaft;** *b.* **Exploded view.**

one-amp medium-duty stepper motor. Predrilled holes on the face of the motor made it easier to mount the arm. A 3-inch-by-3/4-inch mending plate T was used to secure the motor to the plate, as illustrated in Fig. 25.6. New holes were drilled in the plate to match the holes in the motor (1 7/8-inch spacing), and the "T" was bent at the cross.

Unscrew the nut holding the cover plate and upper arm to the shaft, place the "T" on it, and retighten. Make sure the motor is perpendicular to the arm. Then, using the other hole in the "T" as a guide, drill a hole through the cover plate. Secure the T in place with an 8/32 by 1/2-inch bolt and nut. The finished arm, with a gripper attached, is shown in Fig. 25.7.

Refinements

As it is, the arm is unbalanced, and the shoulder motor must work harder to position the arm. You can help to rebalance the arm by relocating the shoulder rotation shaft and by adding counterweights or springs. Before you do anything hasty, however, you may want to attach a gripper to the end of the forearm. Any attempts to balance the arm now will be

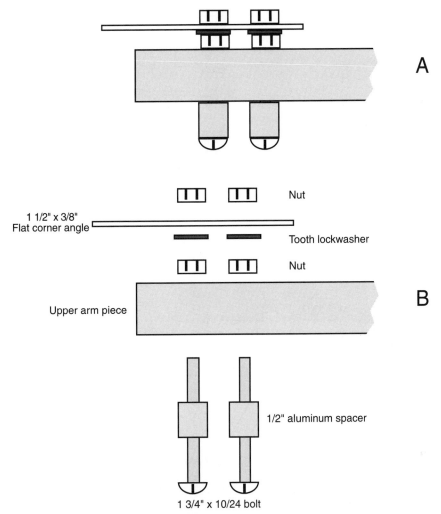

FIGURE 25.3 Upper arm elbow joint detail. *a.* Complete joint; *b.* Exploded view.

severely thwarted when you add the gripper.

The center of gravity for the whole arm, with the elbow drive motor included, is approximately midway along the length of the upper arm (at least this is true of the prototype arm; your arm may be different). Remove the long shaft from the present shoulder joint, and replace it with a short 1 1/2-inch- or 2-inch-long 1/4-inch 20 bolt. Drill a new 1/4-inch hole through the upper arm at the approximate center of gravity, and thread the shoulder shaft through it. Attach it as before, using 1/4-inch 20 nuts, flat washers, and toothed lock washers.

The forearm is also out of balance, and you can correct it in a similar manner, by attaching the shoulder joint nearer to the center of the arm. This has the unfortunate side effect, however, of shortening the reach of the forearm. One solution is to make the arm longer to compensate. In effect, you'll be keeping the elbow joint where it is, just adding extra length

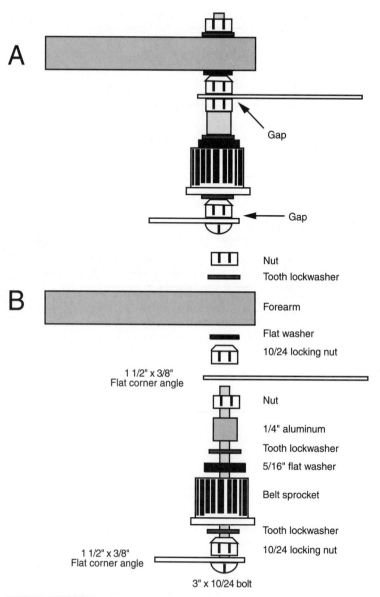

A

Gap

Gap

Nut

Tooth lockwasher

B

Forearm

Flat washer

10/24 locking nut

1 1/2" x 3/8"
Flat corner angle

Nut

1/4" aluminum

Tooth lockwasher

5/16" flat washer

Belt sprocket

Tooth lockwasher

10/24 locking nut

1 1/2" x 3/8"
Flat corner angle

3" x 10/24 bolt

FIGURE 25.4 Forearm elbow joint detail. *a.* Complete joint; *b.*
Exploded view.

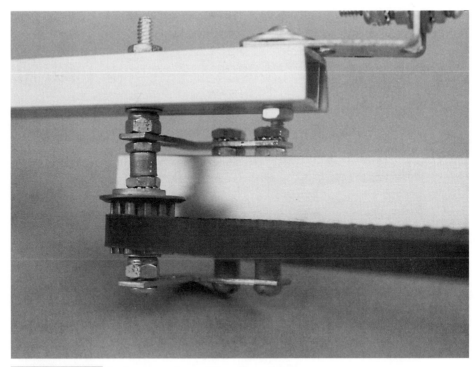

FIGURE 25.5 A close-up view of the elbow joint.

behind it.

This may interfere with the operation of the arm or robot, however, so you may want to opt for counterweights attached to the end of the arm. I successfully used two four-ounce fishing tackle weights attached to the arm with a 2-inch-by-3/4-inch corner angle bracket (see Fig. 25.8).

Position Control

The stepper motors used for the shoulder and elbow joints of the prototype provide a natural control over the position of the arm. Under electronic control, the motors can be commanded to rotate a specific number of steps, which in turn moves the upper arm and forearm a specified amount.

You should supplement the open-loop servo system with limit switches. These switches provide an indication when the arm joints have moved to their extreme positions. The most common limit switches are small leaf switches. You can also construct optical switches using photo-interrupters. A small patch of plastic or metal interrupts the flow of light between an LED and phototransistor, thus signaling the limit of movement. You can build these interrupters by mounting an infrared LED and phototransistor on a small perforated board or purchase ready-made modules (they are common surplus finds).

When using continuous DC motors, you need to provide some type of feedback to

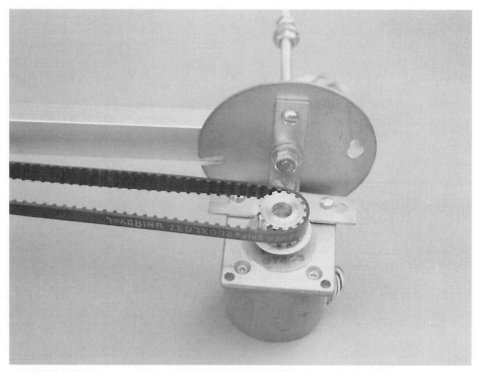

FIGURE 25.6 The motor mounted on the shoulder.

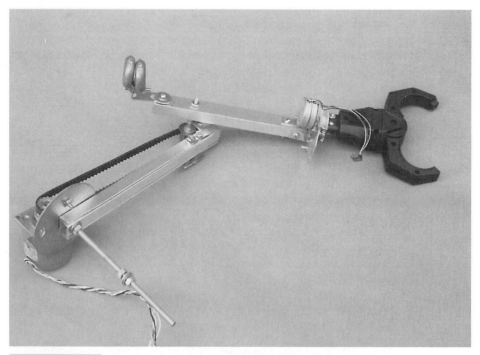

FIGURE 25.7 The completed arm, with gripper (hand) attached.

FIGURE 25.8 Counterbalance weights attached to the end of the forearm help redistribute the weight. You can also use springs, which will help reduce the overall weight of the arm.

report the position of the arm. Otherwise, the control electronics (almost always a computer) will never know where the arm is or how far it has moved. There are several ways you can provide this feedback. The most popular methods are a potentiometer and an incremental shaft encoder.

POTENTIOMETER

Attach the shaft of a potentiometer to the shoulder or elbow joint or motor (see Fig. 25.9), and the varying resistance of the pot serves as an indication of the position of the arm. Just about any pot will do, as long as it has a travel rotation the same as or greater than the travel rotation of the joints in the arm. Otherwise, the arm will go past the internal stops of the potentiometer. Travel rotation is usually not a problem in arm systems, where joints seldom move more than 40° or 50°. If your arm design moves more than about 270°, use a multiturn pot. A three-turn pot should suffice.

Another method is to use a slider-pot. You operate a slider-pot by moving the wiper up and down, rather than by turning a shaft. Slider-pots are ideal when you want to measure linear distance, like the amount of travel (distance) of a chain or belt. Fig. 25.10 shows a slider-pot mounted to a cleat in the timing belt used to operate the elbow joint.

The value of the pot is a function of the control electronics you have hooked up to it, but 10K to 100K potentiometers usually work well with most any circuit. The potentiome-

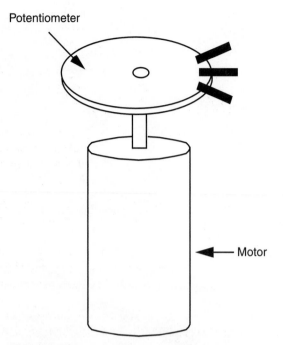

Potentiometer

Motor

FIGURE 25.9 Using a potentiometer as a
position feedback device.
Mount the potentiometer on a
drive motor or on a joint of
the arm.

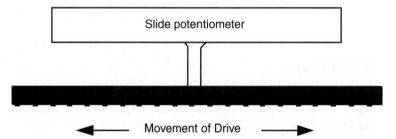

Slide potentiometer

Movement of Drive

FIGURE 25.10 Using a slide potentiometer to register position
feedback. The wiper of the pot can be linked to
any mechanical device, like a chain or belt, that
moves laterally.

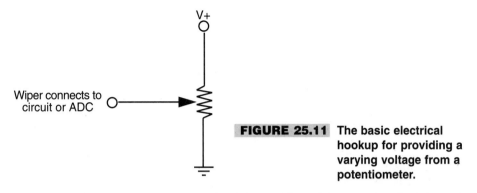

FIGURE 25.11 The basic electrical hookup for providing a varying voltage from a potentiometer.

ter may provide a relative measurement of the position of the arm, but the information is in analog form, as a resistance or voltage, neither of which can be directly interpreted by a computer.

By connecting the pot as shown in Fig. 25.11, you gain an output that is a voltage between 0 and the positive supply voltage (usually 5 or 12 volts). The wiper of the pot can be connected to the input of an analog-to-digital converter (ADC), which translates voltage levels into bytes.

Now, before you go off screaming about the complexity of ADCs, you should really try one first. The latest chips are relatively inexpensive (under $5) and require a very minimum number of external components to operate. The best part about ADC chips is that most have provisions for connecting eight or more analog signals. You select which signal input you want to convert into digital data. That means you can use one $5 ADC for all of the joints in a two-arm robot system.

To be useful, the ADC should be connected to a microprocessor or computer. You can also use your personal computer as the controlling electronics for your robot. Read Chapter 29, "Interfacing with Computers and Microcontrollers," for more information about ADCs and computer control.

Also note that some microcontrollers have their own ADCs built in. For example, the BasicX-24 from NetMedia sports eight ADC inputs; the OOPic microcontroller offers a pair of ADC inputs. Neither of these microcontrollers requires any external components to be connected to the ADC inputs. See Chaps. 32 and 33, respectively, for more information on the BasicX and OOPic microcontroller chips.

INCREMENTAL SHAFT ENCODER

The incremental shaft encoder was first introduced in Chapter 18, "Working with DC Motors." The shaft encoder is a disc that has many small holes or slots near its outside circumference. You attach the disc to a motor shaft or the shoulder or elbow joint. See Chapter 18 for more information on using shaft encoders. To review, shaft encoders are typically composed of a circuit connected to the phototransistor (the latter of which is baffled to block off ambient light). The phototransistor counts the number of on/off flashes and then converts that number into distance traveled. For example, one on/off flash may equal a 2° movement of the joint. Two flashes may equal a 4° movement, and so forth.

The advantage of the incremental shaft encoder is that its output is inherently digital. You can use a computer, or even a simple counter circuit, to simply count the number of on/off flashes. The result, when the movement ends, is the new position of the arm.

From Here

To learn more about...	Read
Using DC motors and shaft encoders	Chapter 18, "Working with DC Motors"
Using stepper motors to drive robot parts	Chapter 19, "Working with Stepper Motors"
Different robotic arm systems and assemblies	Chapter 24, "An Overview of Arm Systems"
Attaching hands to robotic arms	Chapter 27, "Experimenting with Gripper Designs"
Interfacing feedback sensors to computers and microcontrollers	Chapter 29, "Interfacing with Computers and Microcontrollers"

26

BUILD A POLAR COORDINATE ARM

Polar coordinate arms are ideal for use in a stand-alone robotic manipulator. They are fairly inexpensive and easy to build, and they can be adapted to a number of useful applications, especially robotic training. The design described in this chapter is a three-degree-of-freedom polar coordinate arm that is mounted on a stationary base. You can, if you wish, attach the arm to a mobile base or, for an even more outrageous project, add wheels or track to the base itself and make a giant rolling arm.

The arm design presented here has no gripper, or hand, mechanism. You can attach any number of different grippers to the end of the arm. Choose the gripper based on the application you have in mind. Read more about robotic grippers in Chapter 27, "Experimenting with Gripper Designs."

Constructing the Base

The base measures 10 inches by 12 inches by 4 inches. The prototype for this book was made from aluminum shelving standards. Refer to Fig.26.1. You can also use 41/64-inch-by-1/2-inch-by-1/16-inch aluminum channel stock, which is recommended. Construct the base by cutting four 10-inch and four 12-inch lengths. Cut each end at a 45° angle. Cut four 2 1/2-inch riser pieces. Do not miter the ends of these lengths. Assemble the top and bottom frames using 1 1/2-inch-by-3/8-inch flat corner angles. Secure the stock to the corner angles with 8/32 by 1/2-inch bolts and 8/32 nuts.

Refer to Fig. 26.2. Attach a 1-inch-by-1/2-inch corner angle bracket using 8/32 by 1/2-inch bolts and nuts to each one of the short riser pieces. Attach 2-inch-by-1/2-inch flat mending plates to the top of the riser pieces. Connect the top and bottom frames with the risers spaced 2 3/4 inches from the corners. Use 8/32 by 1/2-inch bolts and 8/32 nuts.

Shoulder Rotation Mechanism

The shoulder rotation mechanism consists of a motor, a turntable, and a roller chain gear system. Start by adding a cross brace to the top of the base. Cut a 10 5/8-inch length of 57/64-inch-by-9/16-inch-by-1/16-inch aluminum channel stock. Mount it lengthwise in the center base using two 2 1/2-inch-by-1/2-inch flat mending iron Ts. Use 8/32 by 1/2-inch bolts and 8/32 nuts to secure the Ts and cross brace into place.

Drill a 3/8-inch hole in the center of the cross brace. Position one 3-inch-diameter ball-bearing turntable (lazy Susan) over the hole. Using the mounting holes on the baseplate of the turntable as a guide, mark corresponding mounting holes in the cross brace. Drill for 6/32 bolts (#28 bit) and attach the turntable using two 6/32 by 1/2-inch bolts and 6/32 nuts (see Fig. 26.3).

Construct the center shaft of the arm with a 3-inch-by-10/24 pan-head stove bolt. Place a 1/2-inch-diameter bearing on either side of the channel stock. Be sure the center (rotating part) of the bearings rest over the hole, or they won't turn properly, and that the head of the bolt is positioned over the inner wheel of the bearing. Add a 1/4-inch spacer and lock the assembly into place with a 10/24 nut.

On to the drive mechanics. The drive sprocket (35 teeth, 3-inch diameter, #25 roller chain) is sandwiched between two plastic spacers, as shown in Fig. 26.3. These spacers are actually closet pole holders. They already have holes drilled in the center; so you can just

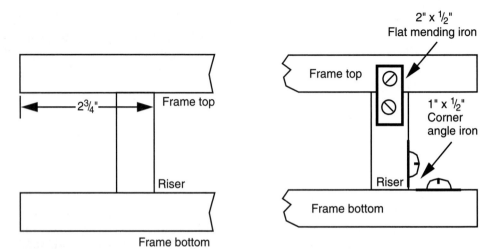

FIGURE 26.1 Cutting and assembly for the polar arm base risers. *a.* Riser placement; *b.* Hardware assembly detail.

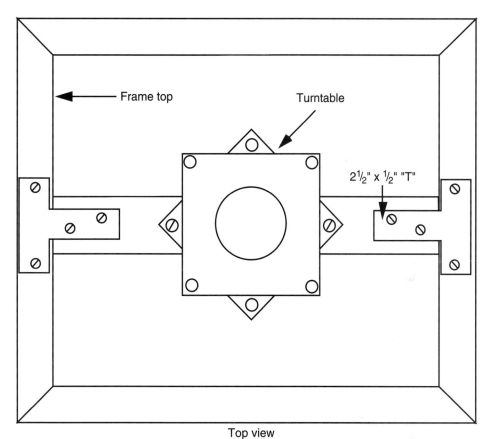

Top view

FIGURE 26.2 Cutting and assembly detail for crosspiece and turntable.

TABLE 26-1 PARTS LIST FOR POLAR COORDINATE ARM.	
Frame/BASE	
4	12-inch lengths 41/64-inch-by-1/2-inch-by-1/16-inch aluminum channel stock
4	10-inch lengths 41/64-inch-by-1/2-inch-by-1/16-inch aluminum channel stock
4	2 1/2-inch lengths 41/64-inch-by-1/2-inch-by-1/16-inch aluminum channel stock
4	2-inch-by-1/2-inch flat mending iron
4	1-inch-by-1/2-inch corner angle iron
8	1 1/2-inch-by-3/8-inch flat corner angle iron
Misc	8/32 stove bolts, nuts, tooth lock washers
Shoulder Base	
1	10 5/8-inch length 57/64-inch-by-9/16-inch-by-1/16-inch aluminum channel stock
1	9-inch length 57/64-inch-by-9/16-inch-by-1/16-inch aluminum channel stock

TABLE 26.1   PARTS LIST FOR POLAR COORDINATE ARM. (*Continued*)

Shoulder Base

2	7-inch length 57/64-inch-by-9/16-inch-by-1/16-inch aluminum channel stock
2	2 1/2-inch mending plate "T"
2	1 1/4-inch-by-5/8-inch corner angle iron
1	3-inch-diameter ball-bearing turntable
1	3-inch-by-10/24 stove bolt, nuts, flat washers, tooth lock washers
2	Plastic closet pole holders
2	1/2-inch bearings
1	1/2-inch aluminum spacer
1	3-inch-diameter 35-tooth chain sprocket (#25 chain)
1	Stepper motor
1	1 3/4-inch-diameter 20-tooth chain sprocket (#25 chain)
1	17-inch-long (nominal) #25 roller chain
Misc	1/2-inch-by-8/32 stove bolts, nuts, tooth lock washers

Elbow

2	6-inch lengths 57/64-inch-by-9/16-inch-by-1/16-inch aluminum channel stock
2	3 1/2-inch lengths 1-inch-by-1-inch-by-1/16-inch aluminum angle stock
1	2 1/2-inch length 1-inch-by-1-inch-by-1/16-inch aluminum angle stock
1	7 1/2-inch length 1/4-inch 20 all-thread rod, nuts, locking nuts, flat washers, tooth lock washers
2	1 3/4-inch-diameter 20-tooth chain sprocket (#25 chain)
1	17-inch-long #25 roller chain
1	Stepper motor
Misc	1/2-inch-by-8/32 stove bolts, nuts, tooth lock washers

Forearm

1	16-inch-long (nominal) drawer rail
2	1-inch-to-1 1/2-inch-diameter spur gears, with setscrew recessed in hub
1	3 1/2-inch length 1-inch-by-1-inch-by-1/16-inch aluminum angle stock
1	18-inch length (approx.) 1/16-inch diameter steel aircraft cable
2	14–16 gauge wire lug
1	Stepper motor
Misc	8/32 stove bolts, nuts, tooth lock washers

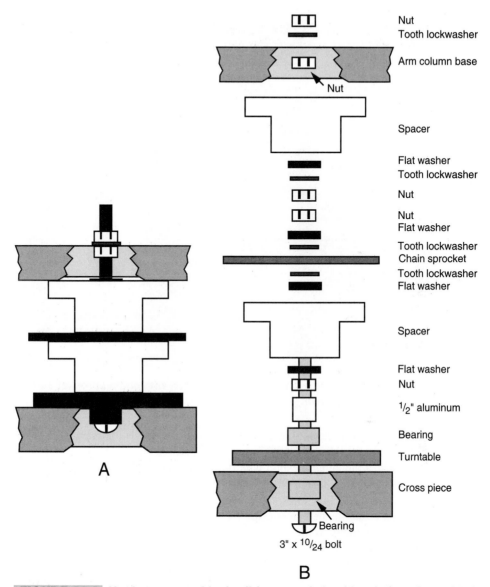

Nut
Tooth lockwasher
Arm column base
Nut
Spacer
Flat washer
Tooth lockwasher
Nut
Nut
Flat washer
Tooth lockwasher
Chain sprocket
Tooth lockwasher
Flat washer
Spacer
Flat washer
Nut
$^1/_2$" aluminum
Bearing
Turntable
Cross piece
Bearing
3" x $^{10}/_{24}$ bolt

A

B

FIGURE 26.3 Hardware assembly detail for central shoulder shaft. *a.* Assembled shaft; *b.* Exploded view.

plop them onto the shaft. The drive sprocket should be approximately one inch above the cross brace. Use a 10/24 bolt and tooth lock washer and a flat washer to clamp the drive mechanism into place.

Attach a 20-tooth 1 3/4-inch diameter #25 chain sprocket to the shaft of the motor, as shown in Fig. 26.4. The prototype arm used a medium-duty stepper motor with a 1/4-inch shaft. The 1/2-inch I.D. hub of the sprocket was reduced to 1/4-inch with reducing bushings. If you use a similar motor (they are common on the new and surplus market) and the same-size sprockets, the roller chain length should be a nominal 17 inches. You can use a

FIGURE 26.4 The mounted shoulder motor, with chain sprocket and roller chain.

slightly longer or shorter length of roller chain because you can position the motor anywhere along the length of the frame to compensate.

Choose a mounting location with the roller chain in place. Move the motor along the edge of the frame until the chain is taut (but not overly tight), and mark the mounting location. At the mark, attach the motor to the frame using two 1 1/2-inch-by-3/8-inch flat corner braces. Elevate the braces using 1/2-inch spacers. Use 6/32 by 1/2-inch bolts and nuts to secure the motor to the braces and 8/32 by 1/2-inch bolts and nuts to secure the braces to the frame.

Construct the arm column using two 7-inch lengths of 57/64-inch-by-9/16-inch-by-1/16-inch aluminum channel stock and one 9-inch length of the same. Mount a 7-inch length flush to one end of the 9-inch member. Use 1 1/4-inch-by-5/8-inch corner angle brackets and 8/32 by 1/2-inch bolts and nuts to secure the pieces in place. Mount the other 7-inch length 3 inches from the opposite end of the 9-inch member (see Fig. 26.5 for details). Likewise, secure it using an angle bracket. Drill a 1/4-inch hole in the center of the 9-inch piece, and mount the assembly on the shoulder rotation shaft, as shown in the figure.

Building the Elbow Mechanism

The elbow mechanism consists of a platform driven by a stepper motor. To distribute the weight, mount the motor on the 9-inch shoulder member. Refer to Fig. 26.6. Construct the

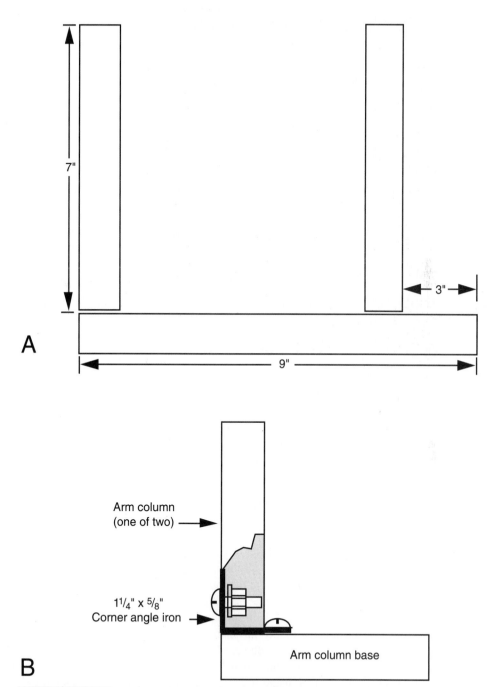

FIGURE 26.5 Cutting and assembly for the arm column. *a.* Dimensions of pieces; *b.* Hardware assembly detail.

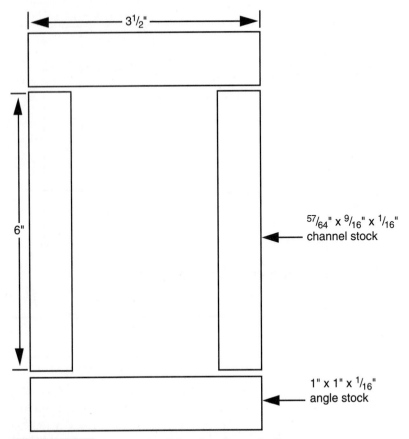

$^{57}/_{64}$" x $^{9}/_{16}$" x $^{1}/_{16}$"
channel stock

1" x 1" x $^{1}/_{16}$"
angle stock

FIGURE 26.6 Cutting detail for the elbow shaft.

elbow platform by cutting two 6-inch lengths of 57/64-inch-by-9/16-inch-by-1/16-inch aluminum channel stock. Couple them together with two 3 1/2-inch lengths of 1-inch-by-1-inch-by-1/16-inch aluminum angle stock. Connect the pieces using 1/2-inch-by-8/32 bolts and 8/32 nuts.

Drill a 1/4-inch hole in the center of each angle stock for the elbow rotation shaft. Cut a 7 1/2-inch length of 1/4-inch 20 all-thread rod and attach a locking nut to one end. Drill a 1/4-inch hole 1/2-inch down from the top of each 7-inch arm column piece. Thread the elbow rotation shaft through the pieces, using the hardware noted in Fig. 26.7. Secure a 20-tooth, 1 1/2-inch diameter #25 chain sprocket to the end.

Mount a matching 20-tooth #25 chain sprocket on the shaft of the elbow stepper motor (it is the same type as the one used for shoulder rotation). Attach a 17-inch length of #25 roller chain between the two sprockets, and mount the motor to the end of a 9-inch shoulder cross brace using a 2 1/2-inch length of 1-inch-by-1-inch-by-1/16-inch aluminum angle stock. Use 6/32 by 1/2-inch bolts and nuts to secure the motor to the angle stock. The angle stock is riveted to the cross brace (the head of a machine bolt is too thick). The motor, as attached, should look like the one in Fig. 26.8.

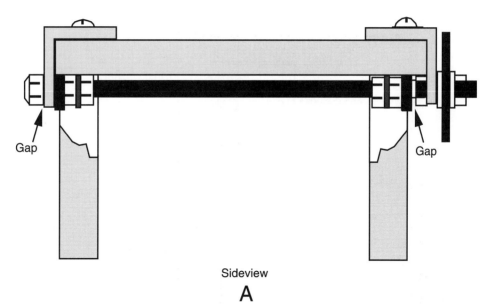

Gap   Gap

Sideview

A

FIGURE 26.7 Assembly detail for the elbow shaft. *a.* Assembled shaft. (See p. 398 for an exploded view.)

Building the Forearm

The retractable forearm is a rather simple mechanism, but you must exercise some patience when constructing it. The forearm uses commonly available parts, and to make your job easier you may want to stick with the parts specified in this chapter. No sense in both of us sweating this one out.

The retractable forearm is constructed using a metal drawer rail. The rail, shown in Fig. 26.9, is composed of two pieces: an 11-inch "base" and a 16 3/4-inch long retracting rail. The rail rides within the base on a set of ball bearings. If you add a little bit of grease, the rail slides smoothly along the length of the rail without trouble. The drawer rail used in the prototype required no modification, but some rails have stops and locks that you may want to defeat. Usually, this involves nothing more than filing down a piece of metal or drilling out the offending stop.

Drill mounting holes in the rail to match the bolts already in place on the elbow platform (you may need to remove the inner rail to get to some portions of the base). Unfasten the bolts on the side opposite the sprocket, and attach the rail. Retighten the bolts.

Mount the rail motor directly in the center of the elbow platform. Cut another 3 1/2-inch length of 1-inch-by-1-inch-by-1/16-inch aluminum angle stock and attach it to the platform using 8/32 by 1/2-inch bolts and nuts. Secure the motor using the mounting technique that is best suited for it. The stepper motor used in the prototype arm already had threaded mounting holes on the shaft end. These were used to secure the motor in place.

Attach two 1 1/2-inch-diameter gears to the motor shaft. Position the gears so the hubs face each other. The idea is to create a spool-like shaft for the forearm cable (see Fig. 26.10). Alternatively, you can use sprockets or fashion a real spool out of metal or wood.

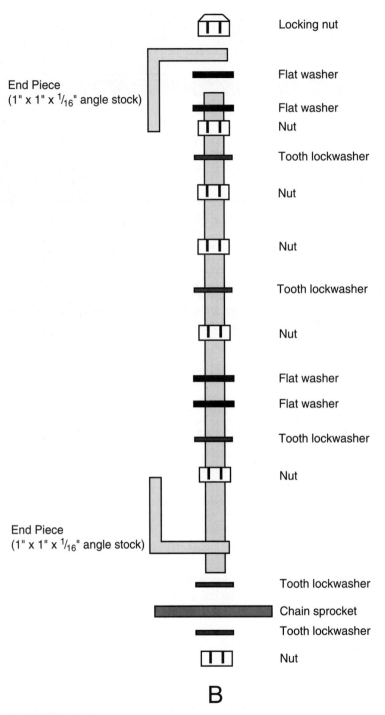

FIGURE 26.7 *b.* Exploded view.

FIGURE 26.8 Mounted elbow motor with chain sprocket and roller chain.

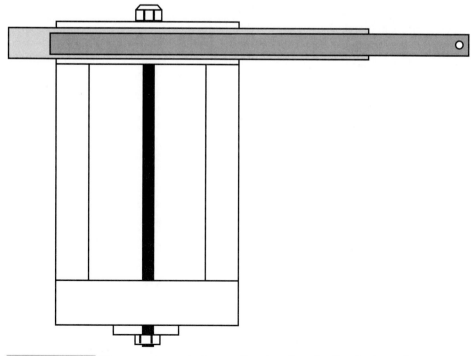

FIGURE 26.9 The recommended mounting location for the drawer rail.

FIGURE 26.10 The rail motor mounted in place.

The main design consideration is that the inside of the spool must be flush. Setscrews that bulge out will tangle with the cable.

Cut a length of 1/16-inch-round steel aircraft cable to a length of approximately 18 inches. On both ends, clamp a 14-to-16-gauge wire lug using a pair of pliers or clamping tool (see Fig. 26.11). Secure one lug to the back end of the rail using 6/32 by 1/2-inch bolts and nuts (there may already be a hole for the hardware; if not, drill your own).

Loop the cable once around the spool shaft and pull it tight to the other end. Remove as much slack as possible and make a mounting mark using the wire lug as a guide. Drill the hole and secure the lug using 6/32 by 1/2-inch bolts and 6/32 nuts. The assembly should look similar to Fig. 26.12.

You may find that when you using metal or plastic gears or sprockets for the spool, the cable slops around and doesn't have much traction. One solution is to line the spool shaft with a couple of layers of masking tape. This approach has proved satisfactory for the prototype arm, even after several years of use. Alternatively, you can rough up the shaft using coarse sandpaper.

The finished polar coordinate arm is shown in Fig. 26.13. Note that the shoulder is able to rotate continuously in a 360° circle and will keep on rotating indefinitely like a wheel. In actual use, the wires to the motors will prevent the shoulder from rotating more than one complete turn.

When the forearm is completely extended, its reach is not quite to the ground (but it is to the arm's own base). This reach is just about right as as most gripper designs will add at least five or six inches to the length of the arm. The robot should be able to pick up small

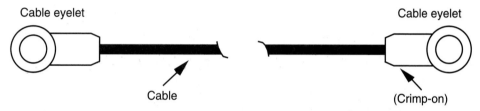

FIGURE 26.11 A length of aircraft cable terminated with crimp-on electrical lugs. The 14-to-16-gauge wire seems to work with 1/16-inch-diameter steel cable.

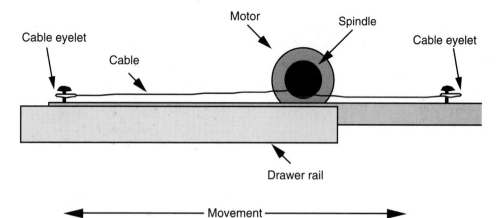

FIGURE 26.12 Threading detail for the drawer cable assembly.

FIGURE 26.13 The finished polar coordinate arm.

objects placed a half-foot or more from the base. You can increase the reach by making the base thinner or by using a longer drawer rail.

Going Further

There is room to improve this basic design for a polar coordinate arm. One improvement you can make quite easily is to add crosspieces to support the turntable used for shoulder rotation. As it is, there is a great deal of side-to-side slop, and additional braces would largely eliminate it.

Arm systems need a great deal of position control if the robot is to manipulate objects without direct intervention from you, its human master. See Chapter 25, "Build a Revolute Coordinate Arm," for more complete details on adding position control to the joints of arms.

From Here

To learn more about...	*Read*
Using DC motors and shaft encoders	Chapter 18, "Working with DC Motors"
Using stepper motors to drive robot parts	Chapter 19, "Working with Stepper Motors"
Different robotic arm systems and assemblies	Chapter 24, "An Overview of Arm Systems"
Attaching hands to robotic arms	Chapter 27, "Experimenting with Gripper Designs"
Interfacing feedback sensors to computers	Chapter 29, "Interfacing with Computers and and microcontrollers"

27

EXPERIMENTING WITH
GRIPPER DESIGNS

The arm systems detailed in Chapters 25 through 26 aren't much good without hands. In the robotics world, hands are usually called *grippers* (also *end effectors*) because the word more closely describes their function. Few robotic hands can manipulate objects with the fine motor control of a human hand; they simply grasp or grip the object, hence the name gripper. Never sticklers for semantics, we'll use the terms *hands* and *grippers* interchangeably.

Gripper designs are numerous, and no single design is ideal for all applications. Each gripper technique has unique advantages over the others, and you must fit the gripper to the application at, er, hand. This chapter outlines a number of useful gripper designs you can use for your robots. Most are fairly easy to build; some even make use of inexpensive plastic toys. The gripper designs encompass just the finger or grasping mechanisms. The last section of this chapter details how to add wrist rotation to any of the gripper designs.

The Clapper

The "clapper" gripper is a popular design, favored because of its easy construction and simple mechanics. You can build the clapper using metal, plastic, or wood, or a combination of all three. The details given in Table 27.1 are for a metal and plastic clapper.

TABLE 27.1	PARTS LIST FOR THE CLAPPER.
2	1 1/2-inch-by-2 1/2-inch-by-1/16-inch thick acrylic plastic sheet
2	1-inch-by-3/8-inch corner angle bracket
1	1 1/2-inch-by-1-inch brass or aluminum hinge
1	Small 6- or 12-vdc spring-loaded solenoid
8	1/2-inch-by-6/32 stove bolts, nuts

The clapper consists of a wrist joint (which, for the time being, we'll assume is permanently attached to the forearm of the robot). Connected to the wrist are two plastic plates. The bottom plate is secured to the wrist; the top plate is hinged. A small spring-loaded solenoid is positioned inside, between the two plates. When the solenoid is not activated, the spring pushes the two flaps out, and the gripper is open. When the solenoid is activated, the plunger pulls in, and the gripper closes. The amount of movement at the end of the gripper is minimal—about 1/2 inch with most solenoids. However, that is enough for general gripping tasks.

Cut two 1/16-inch-thick acrylic plastic pieces to 1 1/2 inches by 2 1/3 inches. Attach the lower flap to two 1-inch-by-3/8-inch corner angle brackets. Place the brackets approximately 1/8 inch from either side of the flap. Secure the pieces using 6/32 by 1/2-inch bolts and 6/32 nuts. Cut a 1 1/2-inch length of 1 1/2-inch-by-1/8-inch aluminum bar stock. Mount the two brackets to the bottom of the stock as shown in the figure. Attach the top flap to a 1 1/2-inch-by-1-inch (approximately) brass or aluminum miniature hinge. Drill out the holes in the hinge with a #28 drill to accept 6/32 bolts. Secure the hinge using 6/32 bolts and nuts.

The choice of solenoid is important because it must be small enough to fit within the two flaps and it must have a flat bottom to facilitate mounting. It must also operate with the voltage used in your robot, usually 6 or 12 volts. Some solenoids have mounting flanges opposite the plunger. If yours does, use the flange to secure the solenoid to the bottom flap. Otherwise, mount the solenoid in the center of the bottom flap, approximately 1/2 inch from the back end (nearest the brackets), with a large glob of household cement. Let it stand to dry.

Align the top flap over the solenoid. Make a mark at the point where the plunger contacts the plastic. Drill a hole just large enough for the plunger; you want a tight fit. Insert the plunger through the hole and push down so that the plunger starts to peek through. Align the top and bottom flaps so they are parallel to one another.

Using the mounting holes in the hinges as a guide, mark corresponding holes in the aluminum bar. Drill holes and mount the hinge using 1/2-inch-by-6/32 bolts and nuts. The finished clapper should look like Fig. 27.1.

Test the operation of the clapper by activating the solenoid. If the plunger works loose, apply some household cement to keep it in place. You may want to add a short piece of rubber weather stripping to the inside ends of the clappers so they can grasp objects easier. You can also use stick-on rubber feet squares, available at most hardware and electronics stores.

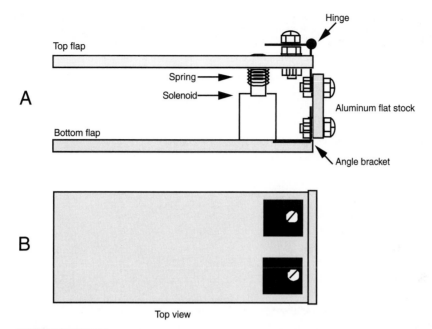

FIGURE 27.1 The clapper gripper. *a.* Assembly detail; *b.* Top view.

Two-Pincher Gripper

The two-pincher gripper consists of two movable fingers, somewhat like the claw of a lobster. The steps for constructing a basic and two advanced models are described in this section.

BASIC MODEL

For ease of construction, the basic two-pincher gripper is made from extra Erector Set parts (the components from a similar construction kit toy may also be used). Cut two metal girders to 4 1/2 inches (since this is a standard Erector Set size, you may not have to do any cutting). Cut a length of angle girder to 3 1/2 inches, as shown in Fig. 27.2 (refer to the parts list in Table 27.2). Use 6/32 by 1/2-inch bolts and nuts to make two pivoting joints. Cut two 3-inch lengths and mount them (see Fig. 27.3). Nibble the corner off both pieces to prevent the two from touching one another. Nibble or cut through two or three holes on one end to make a slot. As illustrated in Fig. 27.4, use 6/32 by 1/2-inch bolts and nuts to make pivoting joints in the fingers.

The basic gripper is finished. You can actuate it in a number of ways. One way is to mount a small eyelet between the two pivot joints on the angle girder. Thread two small cables or wire through the eyelet and attach the cables. Connect the other end of the cables to a solenoid or a motor shaft. Use a light compression spring to force the fingers apart when the solenoid or motor is not actuated.

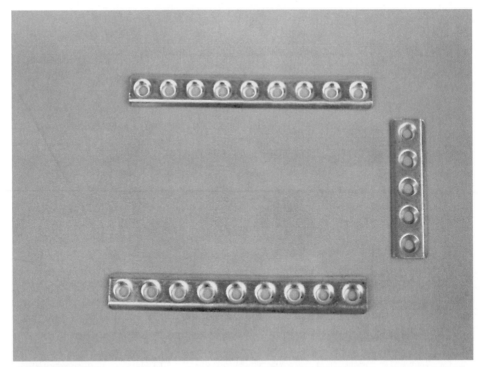

FIGURE 27.2 An assortment of girders from an Erector Set toy construction kit.

TABLE 27.2 PARTS LIST FOR TWO-FINGER ERECTOR SET GRIPPER.	
2	4 1/2-inch Erector Set girder
1	3 1/2-inch-length Erector Set girder
4	1/2-inch-by-6/32 stove bolts, fender washer, tooth lock washer, nuts
Misc	14- to 16-gauge insulated wire ring lugs, aircraft cable, rubber tabs, 1/2 by 1/2-inch corner angle brackets (galvanized or from Erector Set)

You can add pads to the fingers by using the corner braces included in most Erector Set kits and then attaching weather stripping or rubber feet to the brace. The finished gripper should look like the one depicted in Fig. 27.5.

ADVANCED MODEL NUMBER 1

You can use a readily available plastic toy and convert it into a useful two-pincher gripper for your robot arm. The toy is a plastic "extension arm" with the pincher claw on one end and a hand gripper on the other (see Fig. 27.6). To close the pincher, you pull on the hand gripper. The contraption is inexpensive—usually under $10—and it is available at many toy stores.

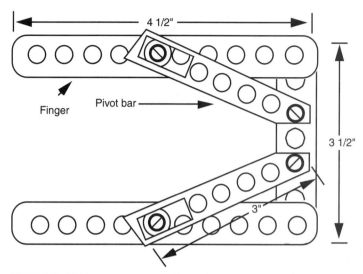

FIGURE 27.3 Construction detail of the basic two-pincher gripper, made with Erector set parts.

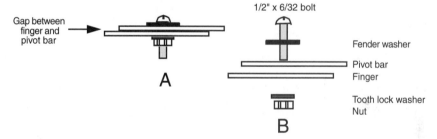

FIGURE 27.4 Hardware assembly detail of the pivot bar and fingers of the two-pincher gripper. *a.* Assembled sliding joint; *b.* Exploded view.

Chop off the gripper three inches below the wrist. You'll cut through an aluminum cable. Now cut off another 1 1/2 inches of tubing—just the arm, but not the cable. File off the arm tube until it's straight, then fashion a 1 1/2-inch length of 3/4-inch-diameter dowel to fit into the rectangular arm. Drill a hole for the cable to go through. The cable is off-centered because it attaches to the pull mechanism in the gripper, so allow for this in the hole. Place the cable through the hole, push the dowel at least 1/2 inch into the arm, and then drill two small mounting holes to keep the dowel in place (see Fig. 27.7). Use 6/32 by 3/4-inch bolts and nuts to secure the pieces.

You can now use the dowel to mount the gripper on an arm assembly. You can use a small 3/4-inch U-bolt or flatten one end of the dowel and attach it directly to the arm. The gripper opens and closes with only a 7/16-inch pull. Attach the end of the cable to a heavy-duty solenoid that has a stroke of at least 7/16 inch. You can also attach the gripper cable to a 1/8-inch round aircraft cable. Use a crimp-on connector designed for 14- to 16-gauge electrical wire to connect them end to end, as shown in Fig. 27.8. Attach the aircraft cable

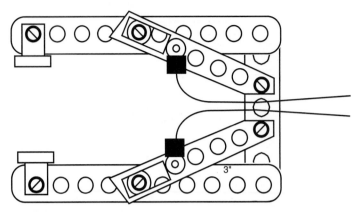

FIGURE 27.5 The finished two-pincher gripper, with fingertip pads and actuating cables.

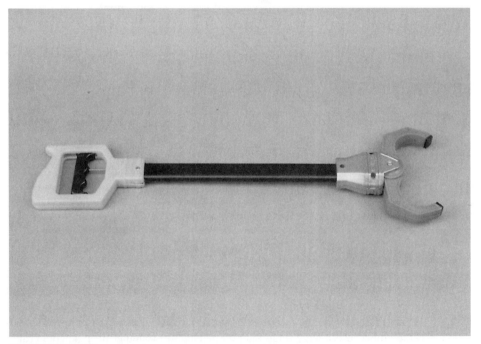

FIGURE 27.6 A commercially available plastic two-pincher robot arm and claw toy. The gripper can be salvaged for use in your own designs.

to a motor or rotary solenoid shaft and activate the motor or solenoid to pull the gripper closed. The spring built into the toy arm opens the gripper when power is removed from the solenoid or motor.

ADVANCED MODEL NUMBER 2

This gripper design uses a novel worm gear approach, without requiring a hard-to-find (and expensive) worm gear. The worm is a length of 1/4-inch 20 bolt; the gears are

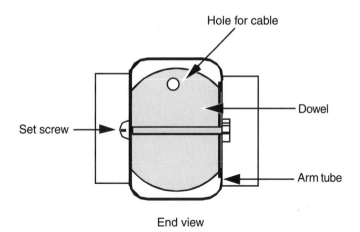

Hole for cable

Dowel

Set screw

Arm tube

End view

FIGURE 27.7 Assembly detail for the claw gripper and wooden dowel. Drill a hole for the actuating cable to pass through.

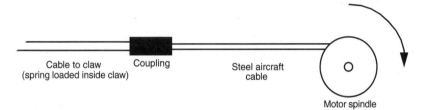

Cable to claw
(spring loaded inside claw)

Coupling

Steel aircraft
cable

Motor spindle

FIGURE 27.8 One method for actuating the gripper: Attach the solid aluminum cable from the claw to a length of flexible steel aircraft cable. Anchor the cable to a motor or rotary solenoid. Actuate the motor or solenoid and the gripper closes. The spring in the gripper opens the claw when power to the motor or solenoid is removed.

standard 1-inch-diameter 64-pitch aluminum spur gears (hobby stores have these for about $1 apiece). Turning the bolt opens and closes the two fingers of the gripper. Refer to the parts list in Table 27.3.

Construct the gripper by cutting two 3-inch lengths of 41/64-inch-by-1/2-inch-by-1/16-inch aluminum channel stock. Using a 3-inch flat mending "T" plate as a base, attach the fingers and gears to the "T" as shown in Fig. 27.9. The distance of the holes is critical and depends entirely on the diameter of the gears you have. You may have to experiment with different spacing if you use another gear diameter. Be sure the fingers rotate freely on the base but that the play is not excessive. Too much play will cause the gear mechanism to bind or skip.

Secure the shaft using a 1 1/2-inch-by-1/2-inch corner angle bracket. Mount it to the stem of the "T" using an 8/32 by 1-inch bolt and nut. Add a #10 flat washer between the "T" and the bracket to increase the height of the bolt shaft. Mount a 3 1/2-inch-long 1/4-inch 20 machine bolt through the bracket. Use double nuts or locking nuts to form a free-spinning shaft. Reduce the play as much as possible without locking the bolt to the bracket. Align the finger gears to the bolt so they open and close at the same angle.

TABLE 27.3	PARTS LIST FOR WORM DRIVE GRIPPER.
2	3-inch lengths 41/64-inch-by-1/2-inch-by-1/16-inch aluminum channel
2	1-inch-diameter 64-pitch plastic or aluminum spur gear
1	2-inch flat mending "T"
1	1 1/2-inch-by-1/2-inch corner angle iron
1	3 1/2-inch-by-1/4-inch 20 stove bolt
2	1/4-inch 20 locking nuts, nuts, washers, tooth lock washers
2	1/2-inch-by-8/32 stove bolts, nuts, washers
1	1-inch-diameter 48-pitch spur gear (to mate with gear on driving motor shaft)

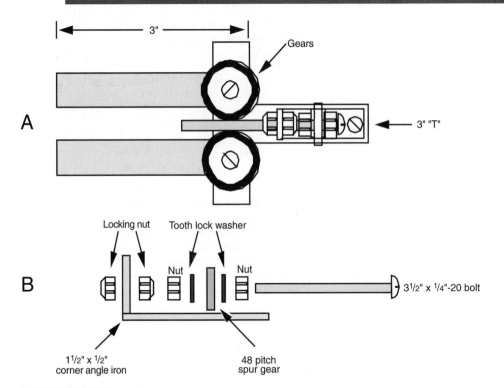

FIGURE 27.9 A two-pincher gripper based on a homemade work drive system. *a.* Assembled gripper; *b.* Worm shaft assembly detail.

To actuate the fingers, attach a motor to the base of the bolt shaft. The prototype gripper used a 1/2-inch-diameter 48-pitch spur gear and a matching 1-inch 48-pitch spur gear on the drive motor. Operate the motor in one direction and the fingers close. Operate the motor in the other direction and the fingers open. Apply small rubber feet pads to the inside ends of the grippers to facilitate grasping objects. The finished gripper is shown in Fig. 27.10.

Figs. 27.11 through 27.14 show another approach to constructing two-pincher grippers. By adding a second rail to the fingers and allowing a pivot for both, the fingertips remain

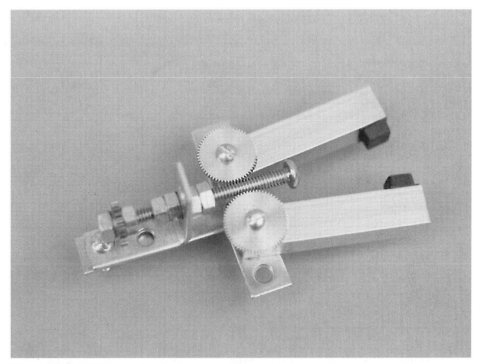

FIGURE 27.10 The finished two-pincher worm drive gripper.

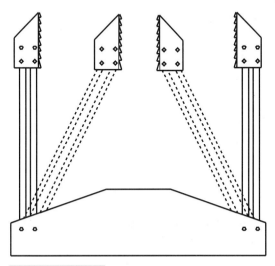

FIGURE 27.11 Adding a second rail to the fingers and allowing the points to freely pivot causes the fingertips to remain parallel to one another.

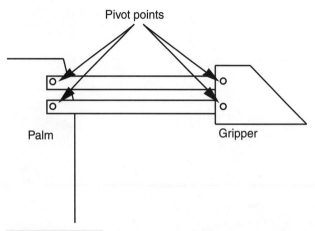

FIGURE 27.12 Close-up detail of the dual-rail finger system. Note the pivot points.

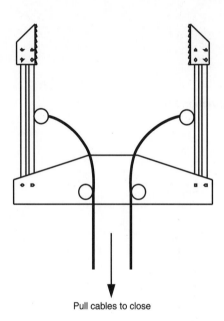

FIGURE 27.13 A way to actuate the gripper. Attach cables to the fingers and pull the cables with a motor or solenoid. Fit a torsion spring along the fingers and palm to open the fingers when power is removed from the motor or solenoid.

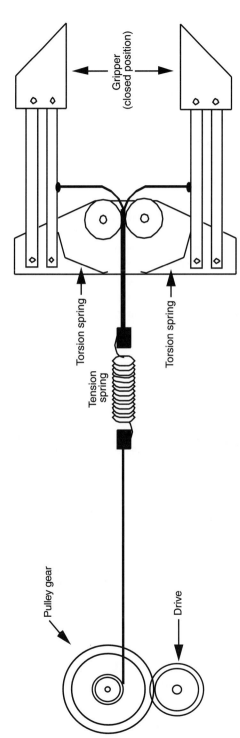

FIGURE 27.14 Actuation detail of a basic two-pincher gripper using a motor. The tension spring prevents undo pressure on the object being grasped. Note the torsion springs in the palm of the gripper.

413

parallel to one another as the fingers open and close. You can employ several actuation techniques with such a gripper. Fig. 27.15 shows the gripping mechanism of the still-popular Radio Shack/Tomy Armatron. Note that it uses double rails to effect parallel closure of the fingers. You can model your own gripper using the design of the Armatron or amputate an Armatron and use its gripper for your own robot.

Flexible Finger Grippers

Clapper and two-pincher grippers are not like human fingers. One thing they lack is a *compliant* grip: the capacity to contour the grasp to match the object. The digits in our fingers can wrap around just about any oddly shaped object, which is one of the reasons we are able to use tools successfully.

You can approximate the compliant grip by making articulated fingers for your robot. At least one toy is available that uses this technique; you can use it as a design base. The plastic toy arm described earlier is available with a handlike gripper instead of a claw gripper. Pulling on the handgrip causes the four fingers to close around an object, as shown in Fig. 27.16. The opposing thumb is not articulated, but you can make a thumb that moves in a compliant gripper of your own design.

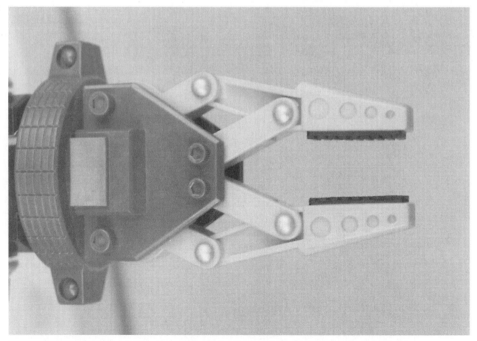

FIGURE 27.15 A close-up view of the Armatron toy gripper. Note the use of the dual-rail finger system to keep the fingertips parallel. The gripper is moderately adaptable to your own designs.

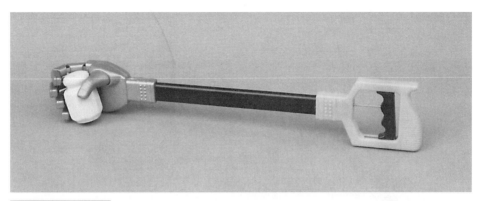

FIGURE 27.16 Commercially available plastic robotic arm and hand toy. The gripper can be salvaged for use in your own designs. The opposing thumb is not articulated, but the fingers have a semi-compliant grip.

Make the fingers from hollow tube stock cut at the knuckles. The mitered cuts allow the fingers to fold inward. The fingers are hinged by the remaining plastic on the topside of the tube. Inside the tube fingers is semiflexible plastic, which is attached to the fingertips. Pulling on the handgrip exerts inward force on the fingertips. The result? the fingers collapse at the cut joints.

You can use the ready-made plastic hand for your projects. Mount it as detailed in the previous section on the two-pincher claw arm. You can make your own fingers from a variety of materials. One approach is to use the plastic pieces from some of the toy construction kits. Cut notches into the plastic to make the joints. Attach a length of 20- or 22-gauge stove wire to the fingertip and keep it pressed against the finger using nylon wire ties. Do not make the ties too tight, or the wire won't be able to move. An experimental plastic finger is shown in Fig. 27.17.

You can mount three of four such fingers on a plastic or metal "palm" and connect all the cables from the fingers to a central pull rod. The pull rod is activated by a solenoid or motor. Note that it takes a considerable pull to close the fingers, so the actuating solenoid or motor should be fairly powerful.

The finger opens again when the wire is pushed back out as well as by the natural spring action of the plastic. This springiness may not last forever, and it may vary if you use other materials. One way to guarantee that the fingers open is to attach an expansion spring, or a strip of flexible spring metal, to the tip and base of the finger, on the back side. The spring should give under the inward force of the solenoid or motor, but adequately return the finger to the open position when power is cut.

Wrist Rotation

The human wrist has three degrees of freedom: it can twist on the forearm, it can rock up and down, and it can rock from side to side. You can add some or all of these degrees of freedom to a robotic hand. A basic schematic of a three-degree-of-freedom wrist is shown in Fig. 27.18.

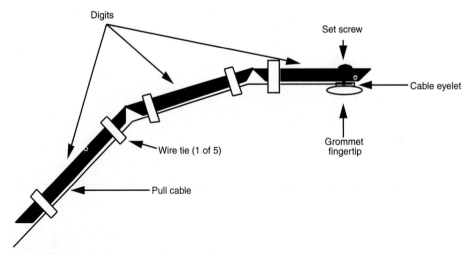

FIGURE 27.17   A design for an experimental compliant finger. Make the finger spring-loaded by attaching a spring to the back of the finger (a strip of lightweight spring metal also works).

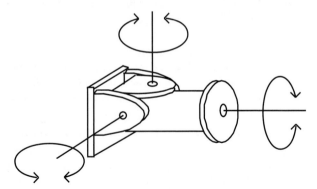

FIGURE 27.18   The three basic degrees of freedom in a human or robotic wrist (wrist rotation in the human arm is actually accomplished by rotating the bones in the forearm).

With most arm designs, you'll just want to rotate the gripper at the wrist. Wrist rotation is usually performed by a motor attached at the end of the arm or at the base. When the motor is connected at the base (for weight considerations), a cable or chain joins the motor shaft to the wrist. The gripper and motor shaft are outfitted with mating spur gears. You can also use chains (miniature or #25) or timing belts to link the gripper to the drive motor. Fig. 27.19 shows the wrist rotation scheme used to add a gripper to the revolute coordinate arm described in Chapter 25.

You can also use a worm gear on the motor shaft. Remember that worm gears introduce a great deal of gear reduction, so take this into account when planning your robot. The wrist should not turn too quickly or too slowly.

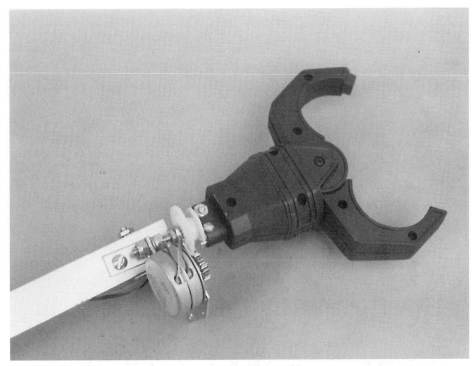

FIGURE 27.19 A two-pincher gripper (from the plastic toy robotic arm detailed earlier in the chapter), attached to the revolute arm described in Chapter 25. A small stepper motor and gear system provide wrist rotation.

Another approach is to use a rotary solenoid. These special-purpose solenoids have a plate that turns 30° to 50° in one direction when power is applied. The plate is spring-loaded, so it returns to its normal position when the power is removed. Mount the solenoid on the arm and attach the plate to the wrist of the gripper.

From Here

To learn more about...	*Read*
Using DC motors and shaft encoders	Chapter 18, "Working with DC Motors"
Using stepper motors to drive robot parts	Chapter 19, "Working with Stepper Motors"
Different robotic arm systems and assemblies	Chapter 24, "An Overview of Arm Systems"
Building a robotic revolute coordinate arm	Chapter 25, "Build a Revolute Coordinate Arm"
Building a robotic stationary polar coordinate arm	Chapter 26, "Build a Polar Coordinate Arm"
Interfacing feedback sensors to computers and microcontrollers	Chapter 29, "Interfacing with Computers and Microcontrollers"

PART 5

COMPUTERS AND ELECTRONIC CONTROL

AN OVERVIEW OF ROBOT "BRAINS"

"Brain, brain, what is brain?" If you're a Trekker, you know this is a line from one of the original Star Trek episodes of the 1960s, entitled "Spock's Brain." The quality of the story notwithstanding, the episode was about how Spock's brain was surgically removed by a race of women who needed it to run their air conditioning system. Dr. McCoy rigged up a gizmo to operate Spock's brainless body by remote control. Clearly, without his brain Spock wasn't much good to anyone, least of all to Dr. McCoy, who never got the hang of the buttons he needed to push to start Spock walking.

"Brains" are what differentiate robots from simple automated machines—brainless Spocks who might as easily crash into walls as move in a straight line. The brains of a robot process outside influences, like sonar sensors or bumper switches; then based on the programming or wiring, they determine the proper course of action. Without a brain of some type, a robot is really nothing more than just a motorized toy, repeating the same actions over and over again, oblivious to anything around it.

A computer of one type or another is the most common brain found on a robot. A robot control computer is seldom like the PC on your desk, though robots can certainly be operated by most any personal computer. And of course not all robot brains are computerized. A simple assortment of electronic components—a few transistors, resistors, and capacitors—are all that's really needed to make a rather intelligent robot. Hey, it worked for Mr. Spock!

In this chapter we'll review the different kinds of "brains" found on the typical hobby or amateur robot, including the latest microcontrollers—computers that are specially made to interact with (control) hardware. Endowing your robot with smarts is a big topic, so additional material is provided in Chapters 29 through 33, including individual discussions on using several popular microcontrollers, such as the Basic Stamp II.

Brains from Discrete Components

You can use the wiring from discrete components (transistors, resistors, etc.) to control a robot. This book contains numerous examples of this type of brain, such as the line-tracing robot circuits in Chapter 38, "Navigating through Space." The line-tracing functionality is provided by just a few common integrated logic circuits and a small assortment of transistors and resistors. Light falling on either or both of two photodetectors causes motor relays to turn on or off. The light is reflected from a piece of tape placed on the ground.

Fig. 28.1 shows another common form of robot brain made from discrete component parts. This brain makes the robot reverse direction when it sees a bright light. The circuit is simple, as is the functionality of the robot: light shining on the photodetector turns on a relay. Variations of this circuit could make the robot stop when it sees a bright light. By using two sensors, each connected to separate motors (much like the line-tracers of Chapter 38), you could make the robot follow a bright light source as it moves. By simply reversing the sensor connections to the motors, you can make the robot behave in the oppo-

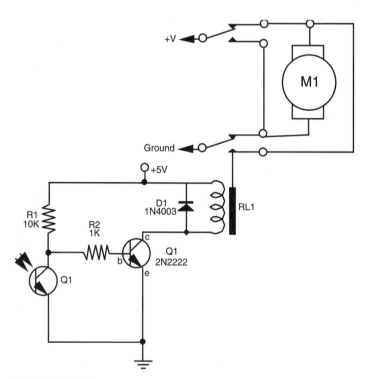

FIGURE 28.1 Only a few electronic components are needed to control a robot using the stimulus of a sensor.

site manner as shown in Fig. 28.1, such as steering away from the light source, instead of driving toward it. See Fig. 28.2 for an example.

You could add additional simple circuitry to extend the functionality of robots that use discrete components for brains. For instance, you could use a 555 timer as a time delay: trigger the timer and it runs for five or six seconds, then stops. You could wire the 555 to a relay so it applies juice only for a specific amount of time. In a two-motor robot, using two 555 timers with different time delays could make the thing steer around walls and other obstacles.

Brains from Computers and Microcontrollers

Perhaps the biggest downside of making robot brains from discrete components is that because the brains are hardwired as circuitry, changing the behavior of the machine requires additional work. You either need to change the wires around or add and remove components. Using an experimenter's breadboard (Chapter 3) makes it easier to try out different designs simply by plugging components and wiring into the board. But this soon gets tiresome and can lead to errors because parts can work loose from the board.

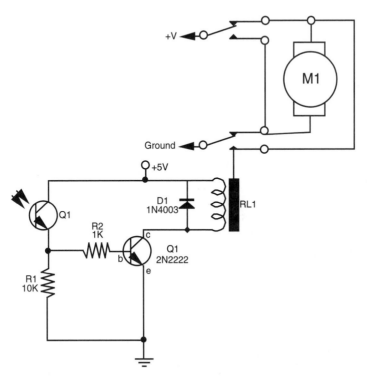

FIGURE 28.2 By connecting the sensors and control electronics differently, a robot can be made to "behave" in different ways.

You can "rewire" a robot controlled by a computer simply by changing the *software* running on the computer. For example, if your robot has two light sensors and two motors, you don't need to do much more than change a few lines of programming code to make the robot come toward a light source, rather than move away from it. No changes in hardware are required. In fact, this exact program functionality was demonstrated in Chapter 14, which discussed how to use the LEGO Mindstorms RCX robot with the Not Quite C programming language.

Types of Computers for Robots

An almost endless variety of computers can be used as a robot's brain. The three most common are as follows:

- *Microcontroller.* These are programmed either in assembly language or a high-level language such as Basic or C. The LEGO Mindstorms RCX is a good example of a robot run from a microcontroller.
- *Single-board computer.* These are also programmed either in assembly language or a high-level language, but they generally offer more processing power than a microcontroller.
- *Personal computer.* Examples include an IBM PC compatible or an Apple Macintosh, or even an older model such as the venerable Commodore 64.

MICROCONTROLLERS

Microcontrollers are fast becoming a favorite method for endowing a robot with smarts. Microcontrollers are inexpensive, have simple power requirements (usually just +5 volts), and most can be programmed using software on your PC. Once programmed, the microcontroller is disconnected from the PC and operates on its own.

Microcontrollers are available in two basic flavors: *low-level programmable* and *embedded-language programmable.* These loosely defined terms relate to the programming of the controller. Both kinds of microcontroller are *fully programmable,* but one contains a kind of operating system that allows it to be programmed with a higher-level language, such as Basic.

Microcontrollers are available in 4-, 8-, 16-, and 32-bit versions (plus a few others, used for special purposes). While PCs have long since "graduated" to 16-bit and higher architectures, most applications for microcontrollers do not require more than 8 bits; hence, the 8-bit controller is still very popular.

Low-level programmable Microcontrollers are, in effect, programmable integrated circuits in which you define how the innards of the chip are connected and how the various connections interact with one another. Following the cues of your program, the microcontroller accepts input, analyzes it in one way or another, and outputs some value. This is fundamentally the same as any computer, except that a microcontroller is primarily designed to operate things (motors, relays, lamps, etc.) rather than interact with people through a keyboard and display monitor.

The traditional way to program a microcontroller is with assembly language, using your PC as a host development system. Assembler appears somewhat arcane to newcomers. However, because microcontrollers use a limited set of instructions, with adequate study it is not overly difficult to master.

The exact format and contents of an assembly-language microcontroller program vary between manufacturers. The popular PIC microcontrollers from Microchip follow one language convention. Microcontrollers from Intel, Atmel, Motorola, NEC, Texas Instruments, Philips, Hitachi, Holtek, and other companies may follow a different convention. While the basic functionality of microcontrollers from these different companies is similar, learning to use each one involves a learning curve. As a result, microcontroller developers tend to fixate on one brand, and even one model, since learning a new language syntax can entail a lot of extra work.

Assembly language is a common method for programming microcontrollers, but it is by no means the only method. A number of compilers are available that convert the syntax of a higher-level language—such as Basic, C, or Pascal—into a language the controller can use. In one approach, the compiler transforms your Basic, C, or other program into the machine code required by the microcontroller. Once compiled the program is downloaded from the PC to the controller

Popular microcontrollers commonly used in robot control include those listed in the following table.

PART NAME	MANUFACTURER
PIC16F84*	Microchip
68HC05	Motorola
68HC11	Motorola, Toshiba
8051	Intel and various**
AVR	Atmel
H8/3292***	Hitachi
Z8	Zilog
80186, 80188 80386 EX	Intel

Notes: *PIC16F84 is just one of several dozen microcontrollers in the PICMicro line of microcontrollers from Microchip. The PIC controllers vary by internal architecture (e.g., 8- or 16-bit), number of inputs, and special I/O features such as built-in analog-to-digital converters.

**The 8051 has become an industry-standard microcontroller design and is available from a number of companies, which include (as of this writing) Intel, Atmel, Philips, Dallas Semiconductor, and several others. As such, the functionality and capabilities of the 8051 systems can vary.

*** The "H8" is the microcontroller used in the popular LEGO Mindstorms RCX robot.

Embedded-language programmable   In this popular microcontroller "flavor," the microcontroller contains a high-level language interpreter that is permanently stored on the chip. For lack of a better term, we'll refer to these as embedded-language programmable. With this system, the compiler on your computer converts your program into an interme- diate "tokenized" language. The interpreter in the microcontroller finishes the job of trans- lating the tokens to the low-level machine code needed by the chip.

Among the most popular embedded-language programmable microcontrollers for hobby robots is the Basic Stamp. Over the past few years, a number of competitors to the Basic Stamp have appeared, including the OOPic from Savage Industries and the BasicX from NetMedia. These use Basic or a Basic-like syntax to save you from having to pro- gram the microcontroller in assembler. Basic Stamp, BasicX, and OOPic are discussed in much more detail in Chapter 31, 32, and 33, respectively.

Standard and semistandard variants of the Basic programming language permeate microcontrollers. For example, a number of microcontrollers use Basic-52 (as found on the Micromint 80C52, for example), a fast and efficient version of Basic that fits in about 8K of memory space. Basic-52 provides additional command statements to support direct interfacing with the hardware of the chip. This includes interfacing with the chip's real- time clock, hardware interrupts, assembly language routines (when speed is required), and more.

Another popular flavor of Basic, currently available for the 8051 and Atmel AVR micro- controllers, is BASCOM, from MCS Electronics, based in Holland. BASCOM is a develop- ment environment in which you write code in Basic, then compile the result in machine-read- able code, which is then sent to the microcontroller. Users of BASCOM enjoy the easier Basic development language, while still being able to take advantage of all the microcontroller's hardware, including timers and interrupts.

A microprocessor with built-in I/O   A key benefit of microcontrollers is that they combine a microprocessor component with various inputs/outputs (I/O) that are typically needed to interface with the real world. For example, the 8051 controller sports the fol- lowing features, many of which are fairly standard among microcontrollers:

- Central processing unit (CPU)
- Hardware interrupts
- Built-in timer or counter
- Programmable full-duplex serial port
- 32 I/O lines (four 8-bit ports)
- RAM and ROM/EPROM in some models

Some microcontrollers will have greater or fewer I/O lines, and not all have hardware interrupt inputs. Some will have special-purpose I/O (see the section "Of Inputs and Outputs" later in this chapter) for such things as voltage comparison or analog-to-digital conversion. Just as there is no one car that's perfect for everyone, each microcontroller's design will make it more suitable for one application than for another.

Microcontrollers and program or data storage   One potential downside to microcontrollers is that they have somewhat limited memory space for programs. The typ- ical low-cost microcontroller may have only a few thousand bytes of program storage. While this may seem terribly confining, in reality most microcontrollers are programmed to do a single job. This one job may not require more than a few dozen lines of program code. If a human-readable display is used, it's typically limited to a small 2-by-16 character LCD, not entire screens of color graphics and text.

By using *external addressing,* advanced microcontrollers may handle more storage—8K or 32K are not uncommon, and a few can support well over a megabyte. Compared to what you may be used to on your personal computer, this may still not be a lot of space. Fortunately, most robot control programs don't take up nearly as much room as the average Windows application! However, keep the program storage limitations in mind when you're planning which brain to get for your robot.

Harvard versus Princeton   Some microcontrollers—and computers for that matter—stuff programs and data into one lump area and have a single data bus for fetching both program instructions and data. These are said to use the "Princeton," or more commonly *Von-Neumann,* architecture. This is the architecture common to the IBM PC compatible and many desktop computers, but it is not as commonly found in microcontrollers. Rather, most microcontrollers use the *Harvard* architecture, where programs are stored in one place and data in another. Two busses are used: one for program instructions and one for data.

The difference is not trivial. A microprocessor using the Harvard architecture can run faster because it can keep track of its current program location while handling all of the data needs. When using the Von Neumann architecture, the processor must constantly switch between going to a data location and a program location on the same bus.

Because of the clear delineation in program and data space in the Harvard architecture, such microcontrollers have two separate memory areas: ROM (read-only memory) for program space and RAM (random access memory) for holding data used while the program runs. For this reason you will often see two data storage specifications for microcontrollers. The data storage space is typically quite small—perhaps 256 bytes or less. The program storage space can be 1K and over, depending on the controller. And as mentioned earlier, some microcontrollers also support external addressing, which allows you to expand the amount of memory available to the controller.

Erasing and starting over   Since microcontrollers are meant to be programmed (and sometimes reprogrammed many times over), the ROM is often designed to be erasable using any of several techniques. One of the oldest techniques, still used, is to erase the contents of the ROM program area using ultraviolet (UV) light. The microcontroller has a clear plastic or glass "window" that exposes the semiconductor die within. By leaving the controller out in full sunlight for several hours or exposing it to a special UV light source made for the job, the old contents of the ROM are erased and it is made ready to accept a new program. These controllers are said to use EPROM, or erasable programmable read-only memory.

A more convenient method uses electrically erasable ROM (called EEPROM), or even static RAM memory with a built-in 5- to 10-year battery. With EEPROM, an electrical signal erases the old contents of the ROM so that new bits can be written to it. EEPROM tends to be slow, and there is a limit to the number of times the ROM can be erased (something in the 100,000+ range). Both battery-backed static RAM as well as the latest Flash memory are faster than EEPROM. Flash memory can only be erased and rewritten about a thousand times; battery-backed static RAM can be erased an indefinite number of times.

In-field programming and reprogramming A key benefit of microcontrollers with EEPROM or Flash memory is that they can be programmed and reprogrammed "in the field" (or "in circuit"). This has enormous potential for use in your programmable robot. With in-field programming there is no need to remove the microcontroller chip from its circuit in your robot to reprogram it. Instead, you merely connect a cable from your PC and download the new program. Of course, this requires that the microcontroller have an on-board connector so it can be attached to your PC cable. Most ready-made controllers for robotics (the Basic Stamp, the BasicX-24, etc.) come with, or have options for purchasing, a "carrier board" that has the proper cable connections.

One-time programmable Not all microcontrollers are meant to be reprogrammed. In fact, the reprogrammable controllers (with EEPROM or Flash memory) are among the most expensive of the lot. Less costly alternatives are made to be programmed only once and are intended for permanent installations. These one-time programmable (OTP) microcontrollers are popular in consumer goods and automotive applications. In quantity, an 8-bit OTP microcontroller might cost just a dollar, or even less.

For hobby robotics applications, the OTP is useful for dedicated processes, such as controlling servos or triggering and detecting a sonar ping from an ultrasonic distance measurement system. You'll find that a number of the ready-made hobby robotic solutions on the market today have, at their heart, an OTP microcontroller. The microcontroller takes the place of more complex circuitry that uses individual integrated circuits.

An OTP microcontroller requires a special "burner" programmer module that accepts the download from your PC. The burner is not complicated for the low- and medium-end microcontrollers that are programmed via a serial connection. A number of mail order and electronics firms sell inexpensive programmers (under $30) for use with both one-time and in-field programmable microcontrollers.

SINGLE-BOARD COMPUTERS

Single-board computers (SBCs) are a lot like "junior PCs," but on a single circuit board. In fact, many SBCs are IBM PC-compatible and use Intel microprocessors that are capable of running any Intel-based program, including the MS-DOS operating system. SBCs are full-blown computers in every way, except that all the necessary components are on one board. This includes the CPU, input/output, and timers. Because of their architecture, SBCs can support many kilobytes, and even megabytes, of program and data storage. Whether you need a lot of storage depends on your application, but it's nice to know the SBC can support it if you do.

Like microcontrollers, an SBC can be programmed in either assembly language or in a high-level language such as Basic or C. SBCs that are based on Intel microprocessors can often run MS-DOS and programs designed to be used on a PC-compatible machine (some can even run Windows). The DOS or Windows operating system is typically loaded in read-only memory (ROM) so it doesn't take up program storage space. In an SBC that supports DOS, for example, you can write programs on your personal computer and test them out. When they're perfected, you can download them (via a cable) to the SBC, where they will reside. The program will remain until erased if the SBC is equipped with Flash memory or EEPROM.

SBC form factors   Single-board computers come in a variety of shapes and forms. A standard form factor that is supported by several hundred manufacturers is the PC/104, which measures just four inches square. This is an ideal size for most any medium- or large-sized robot project—and some small ones, too! PC/104 gets its name from "Personal Computer" (originally of IBM fame) and the number of pins (104) used to connect two or more PC/104-compatible boards together.

SBC kits   To handle different kinds of jobs, SBCs are available in larger or smaller sizes than the four-by-four-inch PC/104. And while most SBCs are available in ready-made form, they are also popular as kits. For example, the BotBoard series of single-board computers, designed by robot enthusiast Marvin Green, combine a Motorola 68HC11 microcontroller with outboard interfacing electronics (the HC11 has its own interfacing capability as well, though many robot engineers like to add more). The Miniboard and HandyBoard, designed by instructors at MIT, are other single-board computers based on the HC11; both are provided in kit and ready-made form from various sources.

PERSONAL COMPUTERS

Having your personal computer control your robot is a good use of available resources because you already have the computer to do the job (you *do* have a computer, right?!). Of course, it also means that your automaton is constantly tethered to your PC, either with a wire or with a radio frequency or infrared link. (Chapter 30 discusses how to use the standard IBM PC-compatible parallel port to interface with robot circuitry.)

Just because the average PC is deskbound doesn't mean you can't mount it on your robot and use it in a portable environment. Whether you'd *want* to is another matter. Certain PCs are more suited for conversion to mobile robot use than others. Here are the qualities to look for if you plan on using your PC as the brains in an untethered robot:

- *Small size.* In this case, small means that the computer can fit in or on your robot. A computer small enough for one robot may be a King Kong to another. Generally speaking, however, a computer larger than about 12 by 12 inches is too big for any reasonably sized 'bot.
- *Standard power supply requirements.* Some computers need only a few power supply voltages, most often +5 and sometimes +12. A few, like the IBM PC compatible, require negative reference voltages of -12 and -5. (However, some IBM PC-compatible motherboards will still function if the -12 and -5 voltages are absent.)
- *Accessibility to the microprocessor system bus or an input/output port.* The computer won't do you much good if you can't access the data, address, and control lines. The IBM PC architecture provides for ready expansion using "daughter" cards that connect to the motherboard. It also supports a variety of standard I/O ports, including parallel and serial.
- *Uni- or bidirectional parallel port.* If the computer lacks access to the system bus, or if you elect not to use that bus, you should have a built-in parallel port. This allows you to use 8-bit data to control the functionality of your robot. The Commodore 64, no longer made but still available in the used market, supports a fully bidirectional parallel port.
- *Programmability.* You must be able to program the computer using either assembly language or a higher-level language such as Basic, C, Logo, or Pascal.

■ *Mass storage capability.* You need a way to store the programs you write for your robot, or every time the power is removed from the computer you'll have to rekey the program back in. (Recall that microcontrollers and SBCs equipped with Flash or EEPROM memory retain their programs even when power is removed.) Floppy disks or small, low-power hard disk drives are possible contenders here.

■ *Availability of technical details.* You can't tinker with a computer unless you have a full technical reference manual. The reference manual should include full schematics or, at the very least, a pinout of all the ports and expansion slots. Some manufacturers do not publish technical details on their computers, but the information is usually available from independent book publishers. Visit the library or a bookstore to find a reference manual for your computer.

IBM PC compatible motherboard The IBM PC or PC compatible may seem an unlikely computer for robot control, but it offers many worthwhile advantages: expansion slots, large software base, and readily available technical information. Another advantage is that these machines are plentiful on the used market—$20 at some thrift stores. As software for PCs has become more and more sophisticated, older models have to be junked to make room for faster processors and larger memories.

You don't want to put the entire PC on your mobile robot; it would be too heavy. Instead, remove the motherboard from inside the PC, and install that on your 'bot. How successful you are doing this will depend on the design of the motherboard you are using. The supply requirements of older PC-compatible motherboards are rather hefty: you need one or more large batteries to provide power and tight voltage regulation.

Later models of motherboards (those made after about 1990) used large-scale integration chips that dramatically cut down on the number of individual integrated circuits. This reduces the power consumption of the motherboard. Favor these "newer" motherboards (sometimes referred to as "green" motherboards, for their energy-saving qualities), as they will save you the pain and expense of providing extra battery power.

As mentioned earlier, PC motherboards "require" four different operating voltages:

■ +5 volts, for the main circuit board logic. The +5 vdc is high-current demand; some early PC motherboards may draw an amp or more.

■ +12 and −12 volts, used for powering disk drives and, in the case of −12 volts, for RS-232 serial communications. For serial communications the current demand for the ±12 volts is low: 100 mA or less. Additional current capacity is needed for the +12 volt source if you use a floppy or hard disk.

■ −5 volts, used as a reference voltage in some applications. Current demand is low at 100 mA or less.

Note that some motherboards may run fine with just +5 vdc, especially if you do not use their serial ports. (If your 'bot motherboard uses a hard disk drive or floppy disk drive, it may need +12 volts for its drive motors. You should account for this in your power supply requirements.) Others will not even turn on unless all four voltages are present. Only testing will determine this.

You can build a suitable power supply for an IBM PC-compatible motherboard using linear or switching voltage regulators and voltage inverters. Maxim makes several easy-to-use and affordable integrated circuits for these applications. You can also sometimes find power supplies for battery-powered laptop computers that provide the four voltage taps. The power supply will likely have modest current ratings, so use it only with a newer, lower-amperage motherboard.

If you're buying a motherboard, new or used, make sure it comes with a BIOS chip (Basic Input/Output System). The motherboard can't work without it. The motherboard should also have its CPU and some memory; otherwise, you'll need to purchase these separately, at additional cost. Take note of the I/O ports that are part of the motherboard and those that must be added separately. Most PC-compatible motherboards will have one or more expansion slots, so you can add your own I/O ports. Keep in mind that every expansion board you add will increase the current consumption of the motherboard.

The keyboard is separate and connects to the motherboard by way of a small connector. The BIOS in some motherboards allow you to turn the keyboard detection off. You *want* this feature. Without it, the motherboard won't boot the operating system, and you'll need to either keep a keyboard connected to the motherboard or rig up some kind of "dummy" keyboard adapter. The same goes for the video display. Make sure you can operate the motherboard without the display.

IBM PC compatible laptop   Motherboards from desktop IBM PC-compatible computers can be a pain to use because of the power supply requirements. An old laptop, though harder to find and more expensive on the used market, makes for an ideal robot-control computer. A perfect PC-compatible laptop is the older monochrome liquid crystal display variety. Battery consumption was reasonable, and the switch to all-color displays a few years back made the used black-and-white models more affordable. Check online auction sites such as eBay, a used computer store, or your local classified ads.

You can probably use the laptop as is, without removing its parts and mounting them on the robot. That way, you can still access the keyboard and display. Use the parallel and/or serial ports on the laptop to connect to the robot. While these ports don't provide the same flexibility as a direct connection to the motherboard's system bus, they should function admirably for most robotic applications.

You can use the unit's standard batteries to power the laptop or use external batteries to provide operating juice. Note that most laptops use rechargeable batteries, where the per-cell voltage is lower than in traditional batteries (1.2 volts versus 1.5 volts per cell). Note the voltage of a freshly charged battery, and duplicate that using any external power source you may provide. If the laptop is equipped with a DC input for a car lighter adapter or AC adapter, use it rather than directly wiring an external power source to the battery terminals.

Commodore 64   The Commodore 64 was first introduced in 1982 and became one of best-selling computers ever. Some 20 to 25 million people owned a Commodore 64, partly because of the wide variety of software and peripherals for it, but mostly because it was so inexpensive. In its day, the Commodore 64 was routinely discounted to under $150. Add a TV set and you had a computer you could use immediately. Some programs (mostly games) came in cartridge form; the cartridges plugged into the back of the computer.

The Commodore 64 is no longer manufactured, but it can still be found at swap meets and other used markets. The keyboard enclosure of the Commodore 64 holds the computer. The power supply for the Commodore 64 is external to the computer. You'll want to replace it with regulated power from your robot, using a 6- to 12-volt battery source. The power supply also provides an odd 9-volt AC source, which you'll need to supply using an adapter circuit.

The Commodore 64 does not offer direct access to the microprocessor system bus, but it does have a unique "user port" that can be used for most robotic applications. The user port is primarily a parallel printer port with a twist: it allows bidirectional data transfer between the computer and the external device. Using the Basic programming language built into the Commodore 64, you can make some of the eight data line inputs and some outputs. There can always be two-way communication between the computer and your robot. The computer also has a serial port.

Of Inputs and Outputs

The architecture of robots requires inputs, for such things as mode setting or sensors, as well as outputs, for things like motor control or speech. The basic input and output of a computer or microcontroller is a two-state binary voltage level (off and on), usually 0 and 5 volts. For example, to place an output of a computer or microcontroller to HIGH, the voltage on that output is brought, under software control, to +5 volts.

In addition to standard LOW/HIGH inputs and outputs, there are several other forms of I/O found on single-board computers and microcontrollers. The more common are listed in the following sections, organized by type. Several of these are discussed in more detail in Chapter 29.

SERIAL COMMUNICATIONS

The most common types of serial communications include the following:

I2C (inter-integrated circuit). This is a two-wire serial network protocol used by Philips to allow integrated circuits to communicate with one another. With I2C you can install two or more microcontrollers in a robot and have them communicate with one another. One I2C-equipped microcontroller may be the "master," while the others are used for special tasks, such as interrogating sensors or operating the motors.

Microwire. This is a serial synchronous communications protocol used in National Semiconductor products and popular for use with the PICMicro line of microcontrollers from Microchip Technologies. Most Microwire-compatible components are used for interfacing with microcontroller or microprocessor support electronics, such as memory and analog-to-digital converters.

SCI (serial communications interface). This is an enhanced version of the UART, which is detailed later in this list.

SPI (serial peripheral interface). This is a standard used by Motorola and others to communicate between devices. Like Microwire, SPI is most often used to interface with microcontroller or microprocessor support electronics, especially outboard EEPROM memory.

Synchronous serial port. In this technology, data is transmitted one bit at a time, using two wires. One wire contains the transmitted data, and the other wire contains a clock signal. The clock serves as a timing reference for the transmitted data. Note that this is different from asynchronous serial communication (discussed next), which does not use a separate clock signal.

UART (universal asynchronous receiver transmitter). This is used for serial communications between devices, such as your PC and the robot's computer or microcontroller. Asynchronous means that there is no separate synchronizing system for the data. Instead, the data itself is embedded with special bits (called *start* and *stop* bits) to ensure proper flow. The USART (*Universal Synchronous/Asynchronous Receiver Transmitter*) can be used in either asynchronous or synchronous mode, thus making possible faster throughput of data.

DATA CONVERSION

There are two principle types of data conversion:

ADC. Analog-to-digital conversion transforms analog (linear) voltage changes to binary (digital). ADCs can be outboard, contained in a single integrated circuit, or included as part of a microcontroller. Multiple inputs on an ADC chip allow a single IC to be used with several inputs (4-, 8-, and 16-input ADCs are common).

DAC. Digital-to-analog conversion transforms binary (digital) signals to analog (linear) voltage levels. DACs are not as commonly employed in robots; rather, they are commonly found on such devices as compact disc players.

PULSE AND FREQUENCY MANAGEMENT

The three major types of pulse and frequency management are the following:

Input capture. This is an input to a timer that determines the frequency of an incoming digital signal. With this information, for example, a robot could differentiate between inputs, such as two different locator beacons in a room. Input capture is similar in concept to a tunable radio.

PMW. Pulse width modulator is a digital output that has a square wave of varying duty cycle (e.g., the "on" time for the waveform is longer or shorter than the "off" time). PMW is often used with a simple resistor and capacitor to approximate digital-to-analog conversion, to create sound output, and to control the speed of a DC motor.

Pulse accumulator. This is an automatic counter that counts the number of pulses received on an input over X period of time. The pulse accumulator is part of the architecture of the microprocessor or microcontroller and can be programmed autonomously. That is, the accumulator can be collecting data even when the rest of the microprocessor or microcontroller is busy running some other program.

SPECIAL FUNCTIONS

Hardware interrupts. Interrupts are special input that provides a means to get the attention of a microprocessor or microcontroller. When the interrupt is triggered, the microprocessor can temporarily suspend normal program execution and run a special subprogram.

Comparator. This is an input that can compare a voltage level against a reference. The value of the input is then lower (0) or higher (1) than the reference. Comparators are most often used as simple analog-to-digital converters where HIGH and LOW are represented by something other than the normal voltage levels (which can vary, depending on the kind of logic circuit used). For example, a comparator may trigger HIGH at 2.7 volts. Normally, a digital circuit will treat any voltage over about 0.5 or 1 volt as HIGH; anything else is considered LOW.

Analog/mixed-signal (A/MS). These are inputs (and often outputs) that can handle analog or digital signals, under software guidance. Many microcontrollers are designed to handle both analog and digital signals on the same chip and even mix and match analog/digital on the same pins of the device.

External reset. This is an input that resets the computer or microcontroller so it clears any data in RAM and restarts its program (the program stored in EEPROM or elsewhere is not erased).

Switch debouncer. This cleans up the signal transition when a mechanical switch (push button, mercury, magnetic reed, etc.) opens or closes. Without a debouncer, the control electronics may see numerous signal transitions and could interpret each one as a separate switch state. With the debouncer the control electronics sees just a single transition.

Input pullup. Pullup resistors (5–10K) are required for many kinds of inputs to control electronics. If the source of the input is not actively generating a signal, the input could "float" and therefore confuse the robot's brain. The pullup resistors, which can be built into a microcontroller and activated via software, prevent this floating from occurring.

From Here

To learn more about...	*Read*
Connecting computers and other control circuits to the outside world	Chapter 29, "Interfacing with Computers and Microcontrollers"
Creating a robot controlled by the parallel port on an IBM PC-compatible computer	Chapter 30, "Computer Control via PC Parallel Port"
Using the Basic Stamp microcontroller	Chapter 31, "Using the Basic Stamp"
Using the BasicX microcontroller	Chapter 32, "Using the BasicX Microcontroller"
Using the OOPic microcontroller	Chapter 33, "Using the OOPic Microcontroller"

INTERFACING WITH COMPUTERS AND MICROCONTROLLERS

The brains of a robot don't operate in a vacuum. They need to be connected to motors to make the robot move and to sensors to make the robot perceive its surroundings. In most cases, these outside devices cannot be directly connected to the computer or microcontroller of a robot. Instead, it is usually necessary to condition these inputs so they can be adequately used by the robot's brain.

In this chapter we discuss the most common and practical methods for interfacing real-world devices to computers and microcontrollers. For your convenience, some of the material presented in this chapter is replicated, in context, in other chapters of the book.

Sensors as Inputs

By far the most common use for inputs in robotics is sensors. There are a variety of sensors, from the super simple to the amazingly complex. All share a single goal: providing the robot with data it can use to make intelligent decisions. A temperature sensor, for example, might help a robot determine if it's too hot to continue a certain operation. Or an "energy watch" robot might record the temperature as it strolls throughout the house, looking for locations where the temperature varies widely (indicating a possible energy leak).

TYPES OF SENSORS

Broadly speaking, there are two types of sensors, shown in Fig. 29.1:

■ Digital sensors provide on/off or true/false results. A switch is a good example of a digital sensor: either the switch is open or it's closed.

■ Analog sensors provide a range of values, usually a voltage. In many cases, the sensor itself provides a varying resistance or current, which is then converted by an external circuit into a voltage. For example, when exposed to light the resistance of a CdS (cadmium sulfide) cell changes dramatically. In a simple circuit with a fixed resistor, this resistance is used to output a voltage.

In both digital and analog sensors, the result is a voltage level that can be fed to a computer, microcontroller, or other electronic device. In the case of a digital sensor, the robot electronics are only interested in whether the voltage is a logical LOW (usually 0 volts) or a logical HIGH (usually 5 volts). As such, digital sensors can often be directly connected to a robot control computer without any additional interfacing electronics.

In the case of an analog sensor, you need additional robot electronics to convert the varying voltage levels into a form that a control computer can use. This typically involves using an analog-to-digital converter, which is discussed later in this chapter.

EXAMPLES OF SENSORS

One of the joys of building robots is figuring out new ways of having them react to changes in the environment. This is readily done with the wide variety of affordable sensors now available. New sensors are constantly being introduced, and it pays to stay abreast of the latest developments. Not all new sensors are affordable for the hobby robot builder, of course—you'll just have to dream about getting that $10,000 vision system. But there are plenty of other sensors that cost much, much less; many are just a few dollars.

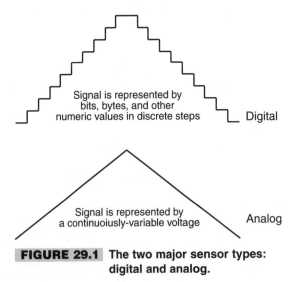

FIGURE 29.1 The two major sensor types: digital and analog.

Part 6 of this book discusses many different types of sensors commonly available today that are suitable for robotic work. Here is just a short laundry list to whet your appetite:

- *Sonar range finder.* Reflected sound waves are used to judge distances. Its effective use is typically from about a foot to 30–40 feet.
- *Sonar proximity or movement.* Reflected sound waves are used to determine if the robot is close to an object (this is called "proximity detection"). Movement (a person, dog, whatever) changes the reflected sound waves and can likewise be detected. Its range is from 0 inches to 20–30 feet.
- *Infrared range finder or proximity.* Reflected infrared light is used to determine distance and proximity. The detected range is typically from 0 inches to two or three feet.
- *Light sensors.* Various light sensors detect the presence or absence of light. Light sensors can detect patterns when used in groups (called "arrays"). A sensor with an array of thousands of light-sensitive elements, like a CCD video camera, can be used to construct eyes for a robot.
- *Pyroelectric infrared.* A pyroelectric infrared sensor detects changes in heat patterns and is often used in motion detectors. The detected range is from 0 to 30 feet and beyond.
- *Speech input or recognition.* Your own voice and speech patterns can be used to command the robot.
- *Sound.* Sound sources can be detected by the robot. You can tune the robot to listen to only certain sound wavelengths or to those sounds above a certain volume level.
- *Contact switches.* Used as "touch sensors," when activated these switches indicate that the robot has made contact with some object.
- *Accelerometer.* Used to detect changes in speed and/or the pull of the earth's gravity, accelerometers can be used to determine the traveling speed of a robot or whether it's tilted dangerously from center.
- *Gas or smoke.* Gas and smoke sensors detect dangerous levels of noxious or toxic fumes and smoke.
- *Temperature.* A temperature sensor can detect ambient or applied heat. Ambient heat is the heat of the room or air; applied heat is some heat (or cold) source directly applied to the sensor.

Motors and Other Outputs

A robot uses outputs to take some physical action. Most often, one or more motors are attached to the outputs of a robot to allow the machine to move. On a mobile robot, the motors serve to drive wheels, which scoot the 'bot around the floor. On a stationary robot, the motors are attached to arm and gripper mechanisms, allowing the robot to grasp and manipulate objects.

Motors aren't the only ways to provide motility to a robot. Your robot may use solenoids to "hop" around a table or pumps and valves to power pneumatic or hydraulic pressure systems. No matter what system the robot uses, the basic concepts are the same: the robot's control circuitry (e.g., a computer) provides a voltage to the output, which turns the motor, solenoid, or pump on. When voltage is removed, the motor, et al, stops.

OTHER COMMON TYPES OF OUTPUTS

Other types of outputs are used for the following purposes:

■ *Sound.* The robot may use sound to warn you of some impending danger ("Danger, Will Robinson, danger!") or to scare away intruders. If you've built an R2-D2 like robot (from *Star Wars* fame), your robot might use chirps and bleeps to communicate with you. Hopefully, you'll know what "bebop, pureeep!" means.

■ *Voice.* Either synthesized or recorded, a voice lets your robot communicate in more human terms.

■ *Visual indication.* Using light-emitting diodes (LEDs), numeric displays, or liquid crystal displays (LCDs), visual indicators help the robot communicate with you in direct ways.

CONSIDERING POWER-HANDLING REQUIREMENTS

Outputs typically drive heavy loads: motors, solenoids, pumps, and even high-volume sound demand lots of current. The typical robotic control computer cannot provide more than 15–22 mA (milliamps) of current on any output. That's enough to power one or two LEDs, but not much else.

To use an output to drive a load, you need to add a power element that provides adequate current. This can be as simple as one transistor, or it can be a ready-made power driver circuit capable of running large, multi-horsepower motors. One common power driver is the H-bridge, so called because the transistors used inside it are in a "H" pattern around the motor or other load (see Chapter 18, "Working with DC Motors," for more information on H-bridges). The H-bridge can connect directly to the control computer of the robot and provides adequate voltage and current to the load.

Input and Output Architectures

The architecture of robots requires inputs, for such things as mode settings or sensors, as well as outputs, such as motor control or speech. As we've already seen, the basic input and output of a computer or microcontroller are a two-state binary voltage level (off and on), usually between 0 and 5 volts. Two types of interfaces are used to transfer these HIGH/LOW digital signals to the robot's control computer: parallel and serial.

PARALLEL INTERFACING

In a parallel interface, multiple bits of data are transferred at one time using (typically) eight separate wires. Parallel interfaces enjoy high speed because more information can be shuttled about in less time. A typical parallel interface is the computer port on your personal computer. It sends data an entire byte (eight bits) at a time. When printing text, each byte represents a different character, like an *A* or an *8*. Such characters can be represented by different combinations of the eight-bit data.

SERIAL INTERFACING

The downside to parallel interfaces is that they consume input/output lines on the robot computer or microcontroller. There are only a limited number of wires (*I/O lines*) on the control computer; typically 16 or even fewer. If the robot uses two 8-bit parallel ports, that leaves no I/O lines for anything else.

Serial interfaces, on the other hand, conserve I/O lines because they send data on a single wire. They do this by separating a byte of information into its constituent bits, then sending each bit down the wire at a time, in single-file fashion. There are a variety of serial interface schemes, using one, two, three, or four I/O lines. Additional I/O lines are used for such things as timing and coordinating between the data sender and the data recipient.

A number of the sensors you may use with your robot have serial interfaces, and on the surface it may appear they are a tad harder to interface than parallel connections. But, in fact, they aren't if you use the right combination of hardware and software. Before you can use the serial data from the sensor, you have to "clock out" all of the bits and assemble them into 8- or 16-bit data, which is used to represent some meaningful value (such as distance between the sensor and some object, for example). The task of reconstructing serial data is made easier when you use a computer or microcontroller because software on the control computer does all the work for you. The Basic Stamp II, for example, provides a single command that does just this job. You can read more about computers and other electronic control for your robot in the remaining chapters of this section of the book.

Interfacing Outputs

As mentioned previously, most output circuits require more voltage and current than the control electronics (computer, microprocessor, microcontroller) of your robot can provide. Therefore, you need some type of power driver to convert the 0–5 volt (off/on) signals provided by the control circuitry into the current and/or voltage levels required by the output.

Figs. 29.2 through 29.7 show various approaches for doing this, including relay, transistor, power MOSFET, discrete component H-bridge, single-package H-bridge, and buffer circuits. All have their advantages and disadvantages, and they are described in context throughout this book. See especially Chapter 18, "Working with DC Motors," and Chapter 19, "Working with Stepper Motors," for more information on these power drive techniques.

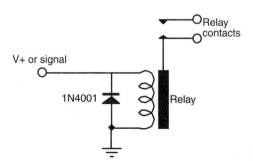

FIGURE 29.2 Relay interface.

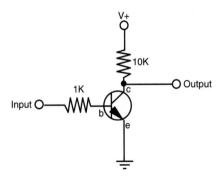

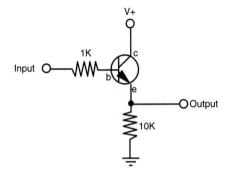

FIGURE 29.3 Bipolar transistor interface.

Interfacing Digital Inputs

The following sections describe common ways to connect digital inputs to the control electronics (microprocessor, computer, or microcontroller) of your robot.

BASIC INTERFACE CONCEPTS

Switches and other strictly digital (on/off) sensors can be readily connected to control electronics. Figs. 29.8 through 29.10 show a variety of techniques, including direct connection of a switch sensor, interface via a switch debouncer, and interface via a buffer. The buffer is recommended to help you isolate the source of the input from the control electronics.

INTERFACING FROM DIFFERENT VOLTAGE LEVELS

Some digital input devices may operate a voltage that differs from the control electronics. Erratic behavior and even damage to the input device or control electronics could result if you connected components with disparate voltage sources together. So-called *logic translation* circuits are needed for these kinds of interfaces. Several integrated circuits provide these functions in off-the-shelf solutions. You can create most of the interfaces you need using standard CMOS and TTL logic chips.

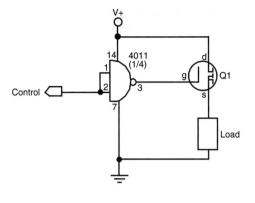

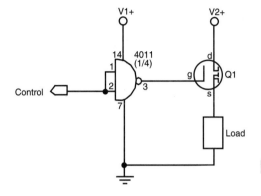

FIGURE 29.4 Power MOSFET interface.

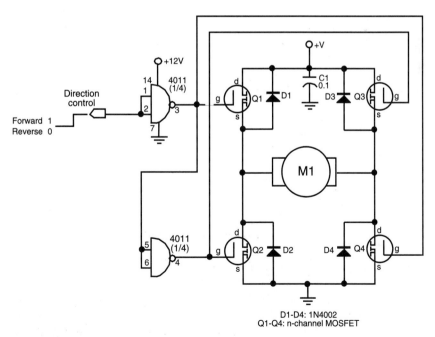

D1-D4: 1N4002
Q1-Q4: n-channel MOSFET

FIGURE 29.5 Discrete component H-bridge interface.

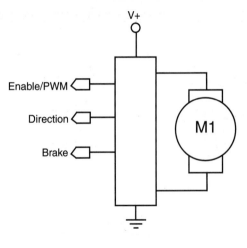

FIGURE 29.6 Packaged H-bridge interface.

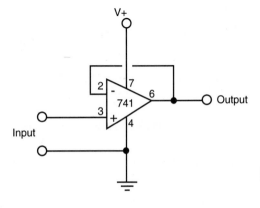

FIGURE 29.7 Non-inverting buffer follower interface.

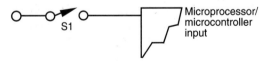

FIGURE 29.8 Direct connection of switch/digital input.

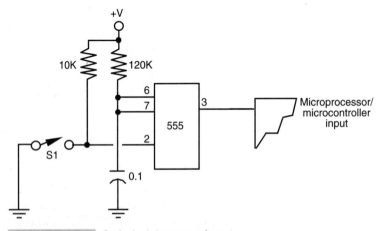

FIGURE 29.9 Switch debouncer input.

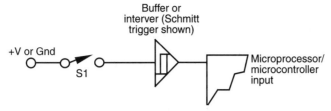

FIGURE 29.10 Buffer input.

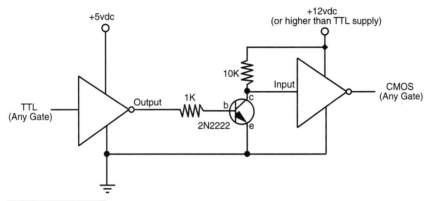

FIGURE 29.11 TTL-to-CMOS translation interface.

Fig. 29.11 shows how to interface TTL (5 volt) to CMOS circuits that use different power sources (use this circuit even if both circuits run under +5 vdc). Fig. 29.12 shows the same concept, but for translating CMOS circuits to TTL circuits that use different power sources.

USING OPTO-ISOLATORS

Note that in both circuits the ground connection is shared. You may wish to keep the power supplies of the inputs and control electronics totally separate. This is most easily done using opto-isolators, which are readily available in IC-like packages. Fig. 29.13 shows the basic concept of the opto-isolator: the source controls a light-emitting diode. The input of the control electronics is connected to a photodetector of the opto-isolator.

Note that since each "side" of the opto-isolator is governed by its own power supply, you can use these devices for simple level shifting, for example, changing a +5 vdc signal to +12 vdc, or vice versa.

ZENER DIODE INPUT PROTECTION

If a signal source may exceed the operating voltage level of the control electronics, you can use a zener diode to "clamp" the voltage to the input. Zener diodes act like valves that turn on only when a certain voltage level is applied to them. As shown in Fig. 29.14, by putting a zener diode across the +V and ground of an input, you can basically shunt any excess voltage and prevent it from reaching the control electronics.

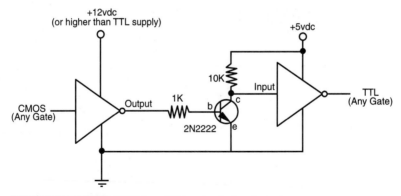

FIGURE 29.12 CMOS-to-TTL translation interface.

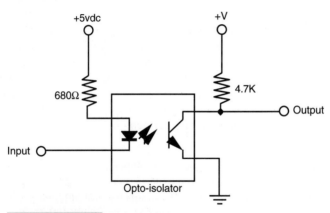

FIGURE 29.13 Opto-isolator.

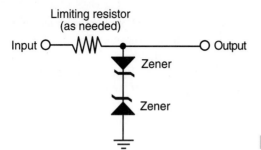

FIGURE 29.14 Zener diode shunt.

Zener diodes are available in different voltages; the 4.7- or 5.1-volt zeners are ideal for interfacing to inputs. Use the resistor to limit the current through the zener. The wattage rating of the zener diode you use depends on the maximum voltage presented to the input as well as the current drawn by the input. For most applications where the source signal is no more than 12–15 volts, a quarter-watt zener should easily suffice. Use a higher wattage resistor for higher current draws.

Interfacing Analog Input

In most cases, the varying nature of analog inputs means they can't be directly connected to the control circuitry of your robot. If you want to *quantify* the values from the input you need to use some form of analog-to-digital conversion (see the section "Using Analog-to-Digital Conversion" later in this chapter for more information).

Additionally, you may need to condition the analog input so its value can be reliably measured. This may include amplifying and buffering the input, as detailed later in this section.

VOLTAGE COMPARATOR

The voltage comparator takes a linear, analog voltage and outputs a simple on/off (LOW/HIGH) signal to the control electronics of your robot. The comparator is handy when you're not interested in knowing the many possible levels of the input, but you want to know when the level exceeds a certain threshold.

Fig. 29.15 shows the voltage comparator circuit. The potentiometer is used to determine the "trip point" of the comparator. To set the potentiometer, apply the voltage level you want to use as the trip point to the input of the comparator. Adjust the potentiometer so the output of the comparator just changes state. Note that the pullup resistor is used on the output of the comparator chip (LM339) used in the circuit. The LM339 uses an open collector output, which means that it can pull the output LOW, but it cannot pull it HIGH. The pullup resistor allows the output of the LM339 to pull HIGH.

SIGNAL AMPLIFICATION

Many analog inputs provide on and off signals but not at a voltage high enough to be useful to the control electronics of your robot. In these instances you must amplify the signal, which can be done by using a transistor or an operational amplifier. The op-amp method is the easiest in most cases, and the LM741 is probably the most commonly used op-amp. Fig. 29.16 shows the basic op-amp as an amplifier. R1 sets the input impedance of the amplifier; R2 sets the gain.

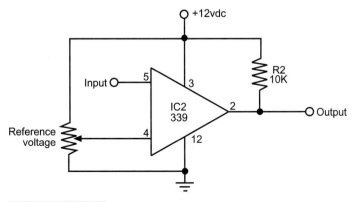

FIGURE 29.15 Voltage comparator input.

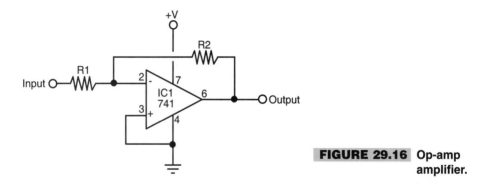

FIGURE 29.16 Op-amp amplifier.

SIGNAL BUFFERING

The control electronics of your robot may "load down" the input sources that you use. This is usually caused by a low impedance on the input of the control electronics. When this happens, the electrical characteristics of the sources change, and erratic results can occur. By *buffering* the input you can control the amount of loading and reduce or eliminate any unwanted side effects.

The op-amp, as shown in Fig. 29.17, is but one common way of providing high-impedance buffering for inputs to control electronics. R1 sets the input impedance. Note that there is no R2, as in Fig. 29.18. In this case, the op-amp is being used in *unity gain* mode, where it does not amplify the signal.

OTHER SIGNAL TECHNIQUES FOR OP-AMPS

There are many other ways to use op-amps for input signal conditioning, and they are too numerous to mention here. A good source for simple, understandable circuits is the *Engineer's Mini-Notebook: Op-Amp Circuits*, by Forrest M. Mims III, which is available through Radio Shack. No robotics lab should be without Forrest's books.

COMMON INPUT INTERFACES

Figs. 29.18 and 29.19 show common interfaces for analog inputs. These can be connected to analog-to-digital converters (ADC), comparators, buffers, and the like. The most common interfaces are as follows:

- *CdS* (cadmium-sulfide) cells are, in essence, variable resistors. By putting a CdS cell in series with another resistor between the +V and ground of the circuit, a varying voltage is provided that can be read directly into an ADC or comparator. No amplification is typically necessary.
- A *potentiometer* forms a voltage divider when connected as shown in Fig. 29-19. The voltage varies from ground and +V. No amplification is necessary.
- The output of a *phototransistor* is a varying current that can be converted to a voltage by using a resistor. (The higher the resistance is, the higher the sensitivity of the device.) The output of a phototransistor is typically ground to close to +V, and therefore no further amplification is necessary.

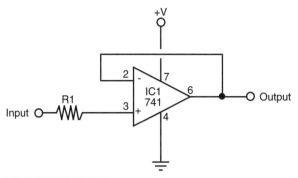

FIGURE 29.17 Op-amp buffer.

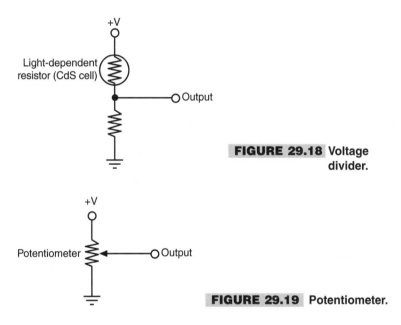

FIGURE 29.18 Voltage
divider.

FIGURE 29.19 Potentiometer.

■ Like a phototransistor, the output of a *photodiode* is a varying current. This output can
  also be converted into a voltage by using a resistor (see Fig. 29.18). (The higher the
  resistance, the higher the sensitivity of the device.) This output tends to be fairly weak—
  on the order of millivolts instead of volts. Therefore, amplification is usually required.

Using Analog-to-Digital Conversion

Computers are binary devices: their *digital* data is composed of strings of 0s and 1s, strung
together to construct meaningful information. But the real world is analog, where data can
be most any value, with literally millions of values between "none" and "lots"!

Analog-to-digital conversion is a system that takes analog information and translates it
into a digital, or more precisely *binary*, format suitable for your robot. Many of the sensors

you will connect to the robot are analog in nature. These include temperature sensors, microphones and other audio transducers, variable output tactile feedback (touch) sensors, position potentiometers (the angle of an elbow joint, for example), light detectors, and more. With analog-to-digital conversion you can connect any of them to your robot.

HOW ANALOG-TO-DIGITAL CONVERSION WORKS

Analog-to-digital conversion (ADC) works by converting analog values into their binary equivalents. In most cases, low analog values (like a weak light striking a photodetector) might have a low binary equivalent, such as "1" or "2." But a high analog value might have a high binary equivalent, such as "255" or even higher. The ADC circuit will convert small changes in analog values into slightly different binary numbers. The *smaller* the change in the analog signal required to produce a different binary number, the *higher* the "resolution" of the ADC circuit. The resolution of the conversion depends on both the voltage span (0–5 volts is most common) and the number of bits used for the binary value.

Suppose the signal spans 10 volts and 8 bits (or a byte) are used to represent various levels of that voltage. There are 256 possible combinations of 8 bits, which means the span of 10 volts will be represented by 256 different values. Given 10 volts and 8 bits of conversion, the ADC system will have a resolution of 0.039 volts (39 millivolts) per step. Obviously, the resolution of the conversion will be finer the smaller the span or the higher the number of bits. With a 10-bit conversion, for instance, there are 1024 possible combination of bits, or roughly 0.009 volts (9 millivolts) per step.

INSIDE THE SUCCESSIVE APPROXIMATION ADC

There are a number of ways to construct an analog-to-digital converter, including successive approximation, single slope, delta-sigma, and flash. Perhaps the most commonly used is the successive approximation approach, which is a form of systematized "20 questions." The ADC arrives at the digital equivalent of any input voltage within the expected range by successively dividing the voltage ranges by two, narrowing the possible result each time. Comparator circuits within the ADC determine if the input value is higher or lower than a built-in reference value. If higher, the ADC "branches" toward one set of binary values; if lower, the ADC branches to another set.

While this sounds like a roundabout way, the entire process takes just a few microseconds. One disadvantage of successive approximation (and some other ADC schemes) is that the result may be inaccurate if the input value changes before the conversion is complete. For this reason, most modern analog-to-digital converters employ a built-in "sample and hold" circuit (usually a precision capacitor and resistor) that temporarily stores the value until conversion is complete.

ANALOG-TO-DIGITAL CONVERSION ICS

You can construct analog-to-digital converter circuits using discrete logic chips—basically a string of comparators strung together. But an easier approach is a special-purpose ADC integrated circuit. These chips come in a variety of forms besides conversion method (e.g., successive approximation, discussed in the last section):

- *Single or multiplexed input.* Single-input ADC chips, such as the ADC0804, can accept only one analog input. Multiplexed-input ADC chips, like the ADC0809 or the ADC0817, can accept more than one analog input (usually 4, 8, or 16). The control circuitry on the ADC chip allows you to select the input you wish to convert.
- *Bit resolution.* The basic ADC chip has an 8-bit resolution (the ADC08xx ICs discussed earlier are all 8 bits). Finer resolution can be achieved with 10- and 12-bit chips. A few 16-bit analog-to-digital ICs are available, but these are not widely used in robotics. One of the most popular 12-bit ADC chips is the LTC1298, which can transform an input voltage (usually 0–5 volts) into 4096 steps.
- *Parallel or serial output.* ADCs with parallel outputs provide separate data lines for each bit. (10- and 12-bit converters may still only have eight data lines; the converted data must be read in two passes.) Serial output ADCs have a single output, and the data is sent 1 bit at a time. Serial output ADCs are handy when used with microcontrollers and single-board computers, where input/output lines can be scarce. In the most common scheme, a program running on the microcontroller or computer "clocks in" the data bits one by one in order to reassemble the converted value. The ADC08xx chips have parallel outputs; the 12-bit LTC1298 has a serial output.

INTEGRATED MICROCONTROLLER ADCS

Many microcontrollers and single-board computers come equipped with one or more analog-to-digital converters built in. This saves you the time, trouble, and expense of connecting a stand-alone ADC chip to your robot. You need not worry whether the ADC chip provides data in serial or parallel form since all the data manipulation is done internally. You just tell the system to fetch an analog input, and it tells you the resulting digital value.

On the downside, the ADCs on most microcontrollers are typically more limited than the stand-alone variety. For example, with most stand-alone ADCs you can set a particular span of voltages, say from 2 volts to 4.5 volts, rather than the usual 0 to 5 volts. The full bit range (8, 10, 12 bits, etc.) then applies to this narrow span. The result is better overall resolution since the same number of bits is used with a smaller voltage range. Most ADCs built into microcontrollers and computers have no way to set the span, which makes limited-range conversions less accurate. Additionally, you're stuck with the ADC resolution that is built into the microcontroller or computer. If the chip uses 8-bit resolution and you need 10 or 12, you'll have to add an outboard converter.

SAMPLE CIRCUITS

Fig. 29.20 shows a basic circuit for using the ADC0809, which provides eight analog inputs and an 8-bit conversion resolution. The input you want to test is selected using a 3-bit control sequence—000 for input 1, 001 for input 2, and so on. Note the ~500 kHz time base, which can come from a ceramic resonator or other clock source or from a resistor/capacitor (RC) time constant. If you need precise analog-to-digital conversion, you should use a more accurate clock than an RC circuit.

Fig. 29.21 shows the pinout diagram for the popular ADC0804, an 8-bit successive approximation analog-to-digital conversion IC with one analog input.

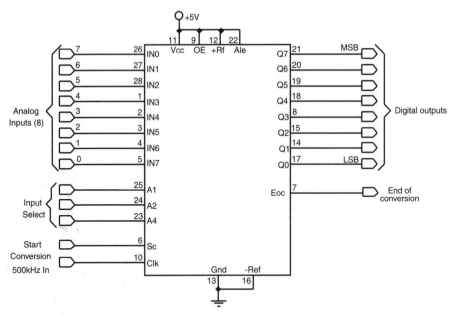

FIGURE 29.20 Basic hookup circuit for the ADC0809 analog-to-digital converter.

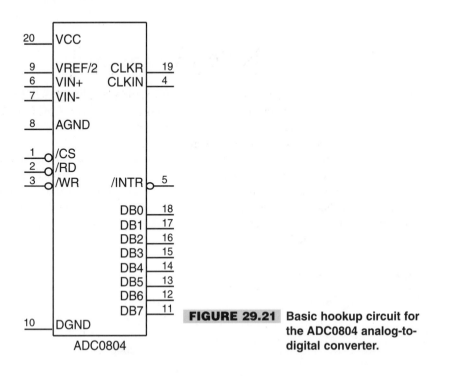

FIGURE 29.21 Basic hookup circuit for the ADC0804 analog-to-digital converter.

Using Digital-to-Analog Conversion

Digital-to-analog conversion (DAC) is the inverse of analog-to-digital conversion. With a DAC, a digital signal is converted to a varying analog voltage. DACs are common in some types of products, such as audio compact discs, where the digital signal impressed upon the disc is converted into a melody pleasing to the human ear.

At least in the robotics world, however, DACs are not as commonly used as ADCs, and when they are, simpler "approximation" circuits are all that's usually necessary. A common technique is to use a capacitor and resistor to form a traditional RC time-constant circuit. A digital device sends periodic pulses through the RC circuit. The capacitor discharges at a more or less specified rate. The more pulses there are during a specific period of time, the higher the voltage that will get stored in the capacitor.

The speed of DC motors is commonly set using a kind of digital-to-analog conversion. Rather than vary the voltage to a motor directly, the most common approach is to use pulse width modulation (PWM), in which a circuit applies a continuous train of pulses to the motor. The longer the pulses are "on," the faster the motor will go. This works because motors tend to "integrate" out the pulses to an average voltage level; no separate digital-to-analog conversion is required. See Chapter 18 for additional information on PWM with DC motors.

You can accomplish digital-to-analog conversion using integrated circuits specially designed for the task. The DAC08, for example, is an inexpensive eight-bit digital-to-analog converter IC that converts an eight-bit digital signal into an analog voltage.

Expanding Available I/O Lines

A bane of the microcontroller- and computer-controlled robot is the shortage of input/output pins. It always seems that your robot needs one more I/O pin than the computer or microcontroller has. As a result, you think you either need to drop a feature or two from the robot or else add a second computer or microcontroller.

Fortunately, there are alternatives. Perhaps the easiest is to use a data demultiplexer, a handy device that allows you to turn a few I/O lines into many. Demultiplexers are available in a variety of types; a common component offers three input lines and eight output lines. You can individually activate any one of the eight output lines by applying a binary control signal on the three inputs. The following table shows which input control signals correspond to which selected outputs.

INPUT CONTROL	SELECTED OUTPUT
000	1
001	2
010	3
011	4

INPUT CONTROL	SELECTED OUTPUT (*CONTINUED*)
100	5
101	6
110	7
111	8

The demultiplexer includes the venerable 74138 chip, which is designed to bring the selected line LOW, while all the others stay HIGH. One caveat regarding demultiplexers is that only one output can be active at any one time. As soon as you change the input control, the old selected output is deselected, and the new one is selected in its place.

One way around this is to use an addressable latch such as the 74259; another way is to use a serial-to-parallel shift register, such as the 74595. The 74595 chip uses three inputs (and optionally a fourth, but for our purposes it can be ignored) and provides eight outputs. You set the outputs you want to activate by sending the 74595 an eight-bit serial word. For example,

SERIAL WORD	SELECTED OUTPUT(S)
00000001	1
00001001	1 and 4
01000110	2, 3, and 7

…and so on. Fig. 29.22 shows how to interface to the 74595. In operation, software on your robot's computer or microcontroller sends eight clock pulses to the Clock line. At each clock pulse, the Data line is sent one bit of the serial word you want to use. When all eight pulses have been received, the Latch line is activated. The outputs of the 74595 remain active until you change them (or power to the chip is removed, of course).

If this seems like a lot of effort to expend just to turn three I/O lines into eight, many microcontrollers (and some computers) used for robotics include a "Shiftout" command that does all the work for you. This is the case, for example, with the Basic Stamp II (but not the Basic Stamp I), the BasicX-24, and several others. To use the Shiftout command, you indicate the data you want to send and the I/O pins of the microcontroller that are connected to the 74595. You then send a short pulse to the Latch line, and you're done! A key benefit of the 74595 is that you can "cascade" them to expand the I/O options even more.

There are still other ways to expand I/O lines, including serial peripheral interface (SPI), the Dallas 1-Wire protocol, and others. We briefly introduced several of the more commonly used systems earlier in this chapter. If your computer or microcontroller supports one or more of these systems you may wish to investigate using these systems in case you find you are running out of I/O lines for your robot.

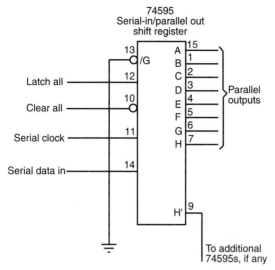

FIGURE 29.22 The 74595 serial-in/parallel-out (SIPO) shift register lets you expand the data lines and select multiple lines at the same time.

Bitwise Port Programming

Controlling a robot typically involves manipulating one or more input/output lines ("bits") on a port attached to a computer or built into a microcontroller. A common layout for an I/O port is eight bits, comprising eight individual connection pins. This is the same general layout as the parallel port found on IBM PC-compatible computers, which provides eight data lines for sending characters to a printer or other device (along with a few additional input and output lines used for control and status).

The design of the typical microcontroller or computer, as well as the usual program tools for it, doesn't make it easy to directly manipulate the individual bits of a port. Rather, you must manipulate the whole port all at once and, in doing so, hopefully alter only the desired bits. The alternative is to send a whole value—from 0 to 255 for an eight-bit port, and 0 to 15 for a four-bit port—to the port at the same time. This value corresponds to the bits you want to control. For example, given an eight-bit port, the number 54 in binary is 00110110.

Fortunately, with a little bit of programming it's not hard to convert numeric values into their corresponding bits, and vice versa. Each programming language provides a different mechanism for these procedures, and what follows in the next few sections are *simple* approaches using Visual Basic. Other languages, such as C, offer more robust bit-handling operators that you can take advantage of. The sample code that follows is meant more to teach you the fundamentals than to be applied directly with a robot. Take the ideas and adapt them to your particular case.

DETERMINING BITS FROM A STRING PRESENTED IN BINARY FORMAT

Binary digits expressed as strings are a convenient way to represent a nibble or a byte of information. For example, the string "0010" is a nibble that represents the number 2:

Bit Weight	8	4	2	1	
Value	0	0	1	0	= 2

Another example: The number 11 is expressed as a four-bit binary string as "1011":

Bit Weight	8	4	2	1	
Value	1	0	1	1	= 11

There are a number of programming approaches for determining the individual bits of a binary-format string. For instance, you may want to determine which bits are 0 and which are 1 given any four-bit value from 0 to 15. You could create a lookup table that matches all 16 strings to their binary equivalents, but there are several other methods you may want to use instead, depending on the commands and statements provided in the programming language you are using.

One approach is to use the *Mid* statement, as shown here, to "parse" through a string and return the value of each character. This code is Visual Basic-compatible and places the bit value (0 or 1) into a four-element array, Bit(0) to Bit (3). While this example shows only 4-bit binary strings being used, the same technique can be used with 8- or 16-bit strings:

```
Sub BitArray()
Dim Count As Integer, BitString As String
Dim Bits(4) As Integer    ' Use Bits(8) or Bits(16) as appropriate
BitString = "0110"        ' Sample binary-format string
For Count = 0 To Len(BitString) - 1
Bits(Count) = Val(Mid(BitString, Count + 1, 1))
Next Count
MsgBox Bits(3)            ' Example: peek into element 3
End Sub
```

The working part of this code is the *For* loop. It, along with the *Mid* statement, separates each character of the string—in this case, "0110"—into the four elements of the array. The *Val* statement converts the isolated string character into a numeric digit.

CONVERTING A VALUE INTO A BINARY-FORMAT STRING

Numeric values can be readily converted into strings once made into a string of 0s and 1s. Here's one approach:

```
Sub MakeBitString()
Dim Temp As String, Count As Integer, X As Integer
X = 9              ' Example value
Temp = ""
```

```
For Count = 0 To 3
  If (X And (2 ^ Count)) = 0 Then
    Temp = "0" & Temp
  Else
    Temp = "1" & Temp
  End If
Next Count
MsgBox Temp
End Sub
```

X is the value you want to convert into a string, in this case 9. The message box displays the binary equivalent in string format (which is "1001").

DETERMINING BITS FROM A DECIMAL VALUE

You can determine each bit of a decimal value by using the bitwise *And* operator, first introduced in Chapter 7, "Programming Concepts: The Fundamentals." The use of the *And* operator is straightforward:

X = Value *And* BitWeight

where *Value* is the value you want to test (0 to 15 for a four-bit number), and *BitWeight* is the binary digit you wish to determine. The bit weights for a value between 0 and 15 are 1, 2, 4, and 8. For example:

```
X = 7 And 4
```

This tests to see if the third bit (bit weight of 4) is 0 or 1 for the number 7. Expressed in binary form, 7 is 0111:

Bit Weight	8	4	2	1	
Value	0	1	1	1	= 7

So, for this example, *X* would contain a nonzero result (4 in this case). But suppose you want to test for bit weight 8:

```
X = 7 And 8
```

Now, *X* contains zero because the fourth bit is not part of the value 7.

The following is a quick demonstration routine, suitable for use in Visual Basic and other compatible programming environments. Bits are counted 0 to 3. The message box displays the result of the *And*'ing expression. Change *X* to different values; change the *If* expression to experiment with different bit weights:

X And 1—bit 0

X And 2—bit 1

X And 4—bit 2

X And 8—bit 3

```
Sub DefBit()
Dim X As Integer
X = 8                        ' Experiment with values 0 thru 15
If (X And 8) <> 0 Then       ' Bit weight = 8
 MsgBox "Bit 3 is 1"
Else
 MsgBox "Bit 3 is 0"
End If
End Sub
```

SUMMING BITS INTO A DECIMAL VALUE

You will have plenty of occasions to convert a set of binary digits into a decimal value. This can be done with simple addition and multiplication, as shown in the Visual Basic-compatible code that follows. Here, values for four bits have been specified and named D0 through D3. The message box displays the numeric equivalent of the bits you specify. For example, the bits

1001

result in 9.

```
Sub SumBits()
Dim X As Integer
Dim D0 As Integer, D1 As Integer, D2 As Integer, D3 As Integer
D0 = 1
D1 = 0
D2 = 0
D3 = 1
X = (D3 * 8) + (D2 * 4) + (D1 * 2) + D0
MsgBox X
End Sub
```

MASKING VALUES BY *OR*'ING

It is not uncommon to manipulate the individual bits of a computer or microcontroller port. Quite often, however, it is not possible or practical to address each individual bit. Rather, you must control the bits in sets of four or eight. As we've already seen, in binary notation bits have different bit weights, so a number like 12 is actually composed of these bits:

Bit Weight	8	4	2	1	
Value		1	1	0	0	= 12

Imagine you have a four-bit port on your computer or microcontroller. You control the setting of each bit by using a value from 0 to 15, with 0 representing the bits 0000, and 15 representing the bits 1111. You need a way to control each bit separately, without changing the state of the other bits. This is done by using an *Or* mask, and it's painfully simple:

X = CurrentValue *Or* BitToTurnOn

where *CurrentValue* is the current binary value of the four bits (which you can often determine by querying the value of the port, such as with the *Inp* statement in QBasic), and *BitToTurnOn* is the bit(s) you want to turn on (make 1). For instance, suppose *CurrentValue* is 7 (0111) and you want to turn on bit 3 (bit weight 8) as well:

```
X = 7 Or 8
```

X is 15, or 1111.

Note:

In many cases, but not all, the result of *Or*ing will be the same as if you just added the numbers. This isn't always true, however, so don't get into the habit of just adding the numbers.

The following is Visual Basic-compatible code that demonstrates the use of *Or* masking to turn on specific bits, without changing the others. Assume *X* is the *CurrentValue* you previously obtained from the port, and *Mask* is the *BitToTurnOn* value from above:

```
Sub MaskValues()
Dim X As Integer, Mask As Integer, Result As Integer
X = 3          ' 0011
Mask = 2       ' 0010
Result = X Or Mask
MsgBox Result
End Sub
```

With the values specified ($X = 3$, *Mask*= 2), the result is still 3 because the bit for a "2" is already set 1. For practice, change *X* to another value, say 4 (binary 0100). The result is now 6, which represents binary (0110).

You can readily turn off a bit by using the *Xor* operator, as shown here:

```
Result = (X Or Mask) Xor Mask
```

The following table shows some sample results from using *Xor* masking:

X	MASK	RESULT
2	1	3 (binary 0011)
7	3	4 (binary 0100)
10	2	8 (binary 1000)
3	8	11 (binary 1011)

To understand how *Xor* works consult this truth table:

VALUE 1	VALUE 2	OUTPUT
0	0	1
0	1	0
1	0	1
1	1	1

From Here

To learn more about...	*Read*
Motor specifications	Chapter 17, "Choosing the Right Motor for the Job"
Interfacing circuitry to DC motor loads	Chapter 18, "Working with DC Motors"
Computers and microcontrollers for robotic control	Chapter 28, "An Overview of Robot 'Brains'"
Input and output using an IBM PC-compatible parallel port	Chapter 30, "Computer Control via PC Parallel Port"
Interfacing sensors	Part 6, "Sensors and Navigation"

30

COMPUTER CONTROL VIA PC
PRINTER PORT

In the "Wizard of Oz," the Scarecrow laments "If I only had a brain." He imagines the wondrous things he could do and how important he'd be if he had more than straw filling his noggin! In a way, your robot is just like the Scarecrow. Without a computer to control it, your robot can only be so smart. Hardwiring functions into the robot is a suitable alternative to computer control, and you should always look to simpler approaches than immediately connecting all the parts of your robot to a super Cray-2 computer.

Yet there are plenty of applications that cry out for computer control; some tasks, like image and voice recognition, require a computer. One of the easiest ways to connect a robot to a computer is to use an IBM PC compatible. You can readily wire up your robot to the PC's parallel port. The parallel port is intended primarily for connecting the computer to printers, plotters, and some other computer peripherals. With a few ICs and some rudimentary programming, it can also be used to directly control your robots. If your computer has several parallel ports, you can use them together to make a very sophisticated control system.

Despite the many advantages of the computer's parallel port, using it involves some disadvantages too. You are limited to controlling only a handful of functions on your robot because the parallel port has only so many input and output lines—although with some creative design work you can effectively increase that number. Also, most parallel ports are designed primarily to get data *out* of the computer, not into it, though many parallel ports are bidirectional under low-level software control. The average parallel port also has a number of input-specific lines for directly communicating with a printer or other peripheral, though the number of input lines is small.

This chapter deals primarily with how to use the parallel port on an IBM PC compatible. Why the PC? It's a common computer—hundreds of millions of them are in use today. While the PC comes in many styles, shapes, and sizes, they all do basically the same thing and provide the same specifications for both software and parallel port output. If you don't want to use your main PC for your robot work, you can probably find a secondhand machine for under $100.

Note:

> The text that follows pertains to the parallel port on an IBM PC compatible used in standard mode, and not in enhanced, bidirectional mode. On some computers, you may need to modify the system BIOS settings to turn off enhanced and/or bidirectional settings. Otherwise, the port may not behave the way you want it to. You can modify the BIOS settings by restarting your computer and following the "Setup" instructions shown on the screen as the PC boots.

The Fundamental Approach

In the original design of the IBM PC, system input and output—such as the parallel port, serial port, and video display—were handled by "daughter boards" that were plugged into the computer's main motherboard. This design practice continues, though today the average PC compatible comes with features such as parallel and serial port, video display, modem, and even a sound card already built into the motherboard. Whether these features are built into the motherboard or added by plugging in a daughter board, all of are input/output (I/O) ports of one type or another.

The PC accesses its various I/O ports by using an *address* code. Each device or board in the computer has an address that is unique to itself, just as you have a home address that no one else in the world shares with you. Very old IBM PCs and compatibles used a monochrome display adapter board, which included its own parallel port. The printer port on this board used a starting address of 956. This address is in *decimal,* or base-10 numbering form. You may also see PC system addresses specified in *hexadecimal,* or base-16, form. In hex, the starting address is 3BCH (the address is really 3BC; the *H* means that the number is in hex). By convention, the parallel port contained on an I/O expansion board, or built into the motherboard, has a decimal address of 888 (or 378H hex) or 632 (278H).

Parallel ports in the PC are given the *logical* names LPT1:, LPT2:, and LPT3:. Every time the system is powered up or reset, the ROM BIOS (Basic Input/Output System) chip on the computer motherboard automatically looks for parallel ports at these I/O addresses, 3BCH, 378H, and 278H, in that order. (It skips 3BCH if you don't have a monochrome card or printer port installed, which you probably don't unless your machine is *ancient!*). The logical names are assigned to these ports as they are found.

Table 30.1 shows the port addresses for the parallel ports in the PC. Applications software often use the logical port names instead of the actual addresses, but in attaching a robot to the computer we'll need to rely on the actual address—hence the need to go into these details.

TABLE 30.1 ADDRESSES OF PARALLEL PORTS.

ADAPTER	DATA	STATUS	CONTROL
Parallel port on monochrome display card	3BCH, 956D	3BDH, 957D	3BEH, 958D
PC/XT/AT printer adapter	378H, 888D	379H, 889D	37AH, 890D
Secondary LPTx card (as LPT2:)	278H, 632D	279H, 633D	27AH, 634D
"H" Suffix = Hex			
"D" Suffix 5 Decimal			

The PC parallel port is a 25-pin connector, which is referred to as a DB-25 connector. Cables and mating connectors are in abundant supply, which makes it easy for you to wire up your own peripherals. You can buy connectors that crimp onto 25-conductor ribbon cable or connectors that are designed for direct soldering. Fig. 30.1 shows the pinout designations for the connector (shown with the end of the connector facing you). Note that only a little more than half of the pins are in use. The others are either not connected inside the computer or are grounded to the chassis. Table 30.2 shows the meaning of the pins.

Notice that not one address is given, but three. The so-called starting address is used for *data output register*. The data output register is comprised of eight binary weighted bits, something on the order of 01101000 (see Fig. 30.2). There are 256 possible combinations of the eight bits. In a printer application, this means that the computer can send specific code for up to 256 different characters. The data output pins are numbered 2 through 9. The bit positions and their weights are shown in Table 30.3.

The other two registers of the parallel port, have different addresses (base address of the port, plus either one or two). These registers are for *status* and *control*. The most commonly used status and control bits (for a printing application, anyway) are shown in Fig. 30.3 on page 464. The function of the status and control bits is shown in Table 30.4 on page 465.

To a printer, one of the most important control pins is pin number 1. This is the STROBE line, which is used to tell the peripheral (printer, robot) that the parallel data on lines 2 through 9 is ready to be read. The STROBE line is used because all the data may not arrive at their outputs at the same time. It is also used to signal a change in state. The output lines are latched, meaning that whatever data you place on them stays there until you change it or turn off the computer. During printing, the STROBE line toggles HIGH to LOW and then HIGH again. You don't have to use the STROBE line when commanding your robot, but it's a good idea if you do.

Other control lines you may find on parallel printer ports include the following (some of these lines aren't always implemented):

■ Auto form feed
■ Select/deselect printer

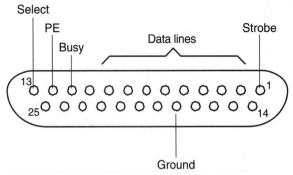

FIGURE 30.1 Pinout of the DB25 parallel port connector, as used on IBM PC-compatible computers.

TABLE 30.2   PARALLEL PORT PINOUT FUNCTIONS.

PIN	FUNCTION (PRINTER APPLICATION)
1	Strobe
2	Data bit 0
3	Data bit 1
4	Data bit 2
5	Data bit 3
6	Data bit 4
7	Data bit 5
8	Data bit 6
9	Data bit 7
10	Acknowledge
11	Busy
12	OE (out of paper, or empty)
13	Printer online
14	Auto line feed after carriage return
15	Printer error
16	Initialize printer
17	Select/deselect printer
18–25	Unused or grounded

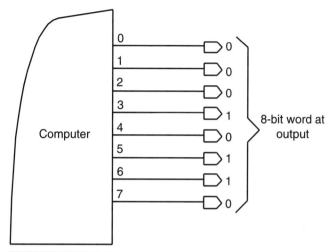

FIGURE 30.2 The parallel port outputs eight bits at a time.

TABLE 30.3 BIT POSITION WEIGHTS.		
BIT POSITION		**WEIGHT**
D7	=	128
D6	=	64
D5	=	32
D4	=	16
D3	=	8
D2	=	4
D1	=	2
D0	=	1

■ Initialize printer
■ Printer interrupt

Traditionally, the status lines are the only ones that feed back into the computer (as mentioned earlier, most parallel printer ports are now bidirectional, but this is not a feature we'll get into this time around). There are five status lines, and not all parallel ports support every one. They are as follows:

■ Printer error
■ Printer not selected
■ Paper error

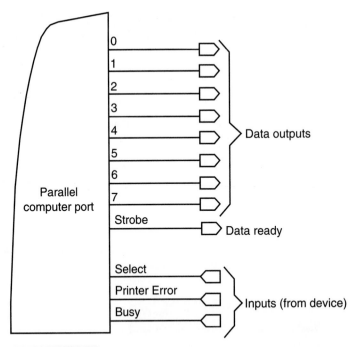

FIGURE 30.3 The minimum parallel port: eight data outputs, a STROBE (Data Ready) line, and inputs from the printer, including Select, Printer Error, and Busy.

- Acknowledge
- Busy

The acknowledge and busy lines are commonly used for the same thing in a printer application. However, depending on the design of the port in your computer, you can use the two separately in your own programs. (One helpful tidbit: for a printing application when the BUSY line is LOW, the ACK line is HIGH.)

Robot Experimenter's Interface

It's not generally a good idea to connect robot parts directly to a parallel port because wiring mistakes in the robot could damage the circuitry in your PC. Moreover, the parallel port in your PC may not have the drive current needed to directly operate relays, solenoids, and power transistors. By using an interface, discussed later in this chapter, you can help protect the circuitry inside your computer and provide more drive current for operating robotic control devices. This interface, called the Robot Experimenter's Interface for lack of a better name, lets your PC control up to 12 robotic functions (such as motors) and read the values of up to four robotic switches or other digital sensing devices.

TABLE 30.4 PARALLEL PORT STATUS AND CONTROL BITS.

CONTROL BITS

BIT	FUNCTION
0	LOW = normal; HIGH = output of byte of data
1	LOW = normal; HIGH = auto linefeed after carriage return
2	LOW = initialize printer; HIGH = normal
3	LOW = deselect printer; HIGH = select printer
4	LOW = printer interrupt disables; HIGH = enabled
5–7	Unused

STATUS BITS

BIT	FUNCTION
0–2	Unused
3	LOW = printer error; HIGH = no error
4	LOW = printer not on line; HIGH = printer on line
5	LOW = printer has paper; HIGH = out of paper
6	LOW = printer acknowledges data sent; HIGH = normal
7	LOW = printer busy: HIGH = out of paperH

Caution:

Any time you mess around with a computer there is a risk of damaging it, and this goes for the circuit presented next. This is not a project you should consider if you're new to electronics and aren't sure what you're doing.

CONSTRUCTING THE INTERFACE

The schematic diagram for the Robot Experimenter's Interface is shown in Fig. 30.4. You can build it in under an hour, and it requires very few components. The interface uses a solderless experimenter's breadboard so you can create circuits right on the interface. The input and output buffering is provided by the 74367 hex buffer driver. Three such chips are used to provide 18 buffered lines, which is more than enough.

You may wish to build the interface in an enclosure that is large enough to hold the breadboard and the wire-wrapping socket. Make or buy a cable using a male DB-25 connector and a four- or five-foot length of 25-connector ribbon cable. Solder the data output, status, and control line conductors to the proper pins of the 74367 ICs. Route the outputs to the bottom of the wire-wrap socket. A finished interface should look something like the one in Fig. 30.5. Using the interface requires you to provide a +5 vdc source. *Do not* try to power the interface from the parallel port! Use a length of 22 AWG solid

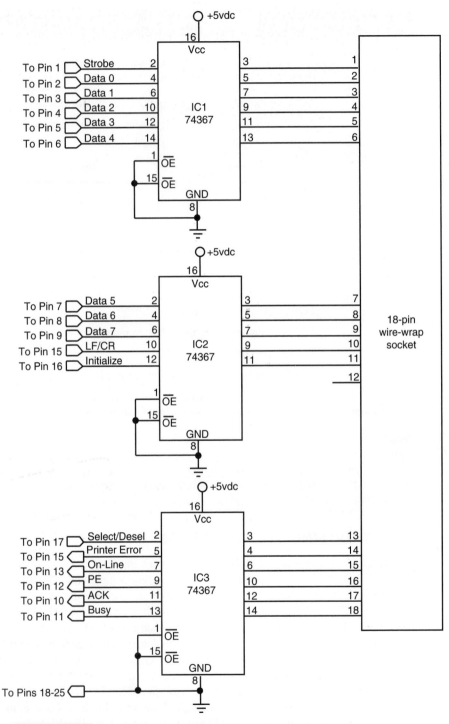

FIGURE 30.4 Schematic for the Robot Experimenter's Interface.

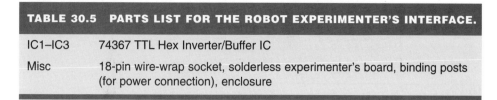

TABLE 30.5	PARTS LIST FOR THE ROBOT EXPERIMENTER'S INTERFACE.
IC1–IC3	74367 TTL Hex Inverter/Buffer IC
Misc	18-pin wire-wrap socket, solderless experimenter's board, binding posts (for power connection), enclosure

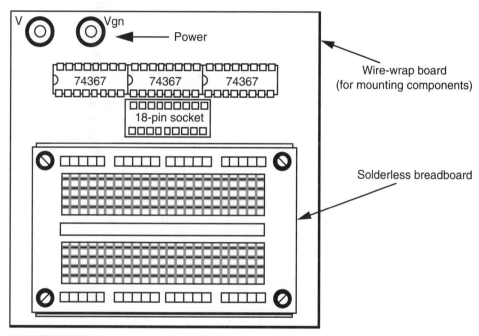

FIGURE 30.5 The completed Robot Experimenter's Interface. Mount the bread-board, wire-wrap socket, ICs, and power terminals on a perf board, and secure the board into a project case.

conductor wire to connect the signals at the wire-wrap socket to whatever points on the breadboard you desire.

TESTING THE INTERFACE

The first order of business is to connect the Robot Experimenter's Interface as shown in Fig. 30.6. Connect the cable to the parallel port of your computer (some of the LEDs will light). Use a DOS-based Basic interpreter program to manipulate the three registers (data, control, and status) of the parallel port. Most older PCs will have a Basic interpreter either built into the BIOS (as was the case with the original IBM PC) or provided as a separate .com or .exe executable file. If your PC has MS-DOS 5.0 or later, look for QBasic, an updated version of the venerable Microsoft Basic from the late 1970s. All of the program examples in this chapter assume you're using QBasic, or a similar updated Basic variant.

Note that if you're using Microsoft Windows 95 or later, QBasic probably isn't installed on your computer, but it is provided on the Windows CD-ROM. Look for the *qbasic.exe*

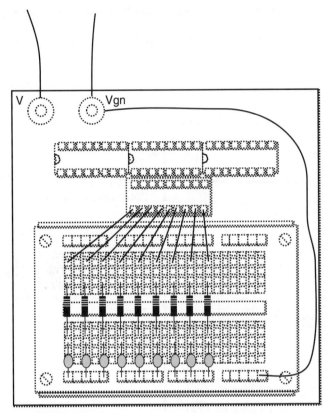

FIGURE 30.6 Component arrangement for testing the Robot Experimenter's Interface.

program file in the *OtherOldmsdos* directory, or visit Microsoft's Web page at www.microsoft.com for additional information.

Type the program shown in Listing 30.1. The program assumes you're using the standard LPT1: port, which has an address of 888 decimal (378 hex). If you're using a different parallel port, change the *BaseAddress* as required. Refer to Table 30.1 earlier in this chapter. You're now ready to run the program (in QBasic, press Shift+F5).

LISTING 30.1.

```
BaseAddress = 888          ' Base address of parallel port
DataPort = BaseAddress     ' Address of data register

FOR Count = 0 TO 255
    OUT DataPort, Count
FOR x = 1 TO 500: NEXT x
NEXT Count
```

The LEDs connected to each of the data lines should flash on and off very rapidly. Some of the LEDs will flash more than the others; this is normal. When the program finishes all of the LEDs should stay lit. If the LEDs do not flash, recheck your wiring and make sure the program has been typed correctly. The LEDs that are on represent a logic 1 state; those that are off represent a logic 0 state.

Fig. 30.7 is a blank dotted-line version of the Robot Experimenter's Interface. Feel free to use it to sketch out your own designs.

Using the Port to Operate a Robot: The Basics

The 74367 used in the Robot Experimenter's Interface cannot sink or source more than about 20 mA of current per output, and as you would expect you can't operate a motor directly from it. However, it can drive a low-power relay, transistor, or H-bridge. See Chapter 18, "Working with DC Motors," for some popular ways to bridge the low-level output of the interface to control a real-world robot.

The simplest way to operate your robot via computer is to connect each of the data output lines to a suitable transistor, small relay, or H-bridge input. You can control the on/off state of up to eight motors or other devices using just the eight data lines of the parallel port (you can actually control even more devices; more on this in a bit).

Let's say that you only have three motors connected to the interface and that you are using lines 0, 1, and 2 (pins 2, 3, and 4, respectively). To turn on motor 1, you must

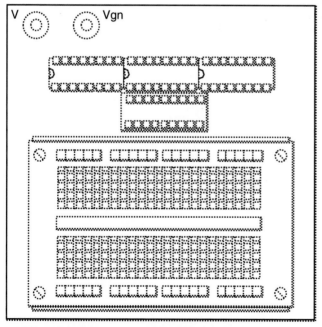

FIGURE 30.7 A blank Robot Experimenter's Interface layout. Feel free to copy it and use it to make your own designs.

activate the bit for line 0, that is, make it HIGH. To do this, output a *bit pattern* number to the port using the BASIC OUT command. The OUT command is used to send data to an I/O port. The command is used with two parameters: *port address* and *value.* The two are separated by a comma. For port address, use the base address of the parallel port; for data, use the value you want to send to the port. Here's an example:

```
OUT 888, 10
```

(Note: In the test code you used variables, *BaseAddress* and *DataPort,* instead of "hard-wired" literal values for the port address. It's a better practice to use variables because that makes it easier to change your program. For right now, however, I'll use literal values such as *888* for short examples, but revert back to using variables in the larger ready-to-go program code.)

The base address is 888, and the value is decimal 10. Table 30.6 shows the first 16 binary numbers and the bit pattern that constitutes them.

For most robotic applications where you use the parallel port to control motors, you'll need two data lines for each motor: one to turn the motor on and off and another to con-

TABLE 30.6   DECIMAL AND BINARY EQUIVALENTS (0–15 ONLY).

DECIMAL	BINARY
0	0000
1	0001
2	0010
3	0011
4	0100
5	0101
6	0110
7	0111
8	1000
9	1001
10	1010
11	1011
12	1100
13	1101
14	1110
15	1111

trol its direction. You can use the four bits in Table 30.6 to control the on/off state of the motors as well as their direction. For this you might use data lines 0, 1, 2, and 3 (pins 2, 3, 4, and 5, respectively, of the interface).

Table 30.7 lists all the possible bit patterns for data lines 0–3. You can connect the motor relays to the pins in any order, but the table assumes the following:

- Data line 0 (bit 1) controls the On/Off relay for motor 1
- Data line 1 (bit 2) controls the On/Off relay for motor 2
- Data line 2 (bit 3) controls the Direction relay for motor 1
- Data line 3 (bit 4) controls the Direction relay for motor 2

You should get into the habit of initializing the port at the beginning of the program by outputting decimal 0. That prevents the motors from energizing at random. The line of code for this is as follows:

TABLE 30.7 DATA BITS FOR CONTROLLING TWO MOTORS.

| BINARY | DECIMAL VALUE | MOTOR1 | | MOTOR 2 | |
		CONTROL (BIT 1)	DIRECTION (BIT 3)	CONTROL (BIT 2)	DIRECTION (BIT 4)
0000	0	Off	Forward	Off	Forward
0001	1	On	Forward	Off	Forward
0010	2	Off	Forward	On	Forward
0011	3	On	Forward	On	Forward
0100	4	Off	Reverse	Off	Forward
0101	5	On	Reverse	Off	Forward
0110	6	Off	Reverse	On	Forward
0111	7	On	Reverse	On	Forward
1000	8	Off	Forward	Off	Reverse
1001	9	On	Forward	Off	Reverse
1010	10	Off	Forward	On	Reverse
1011	11	On	Forward	On	Reverse
1100	12	Off	Reverse	Off	Reverse
1101	13	On	Reverse	Off	Reverse
1110	14	Off	Reverse	On	Reverse
1111	15	On	Reverse	On	Reverse

```
OUT 888, 0
```

To activate just motor 1, choose a decimal number where only the first bit changes. There is only one number that meets that criterion: it is decimal 1, or 0001 (we will ignore bits 5 through 8 for this discussion since they are not in use). So type:

```
OUT 888, 1
```

Run this program; motor 1 turns on. To turn it off, send a decimal 0 to the port, as described earlier. You use the same technique to turn on motor 2 or both motors 1 and 2 at the same time. To turn on both motors at the same time, for example, look for the binary bit pattern where the first and second bits are 1 (it's decimal 3), and output this value to the port.

Controlling a Two-wheel Robot

Controlling the common two-wheeled robot is a simple matter of sending the right bit patterns to the parallel port. Note that binary 0000 (decimal 0) turns off both motors, so the robot stops. Changing the binary bit pattern activates the right or left motor and controls its direction. Table 30.8 lists the most common bit patterns you will use.

When writing the control program for your robot you may find it necessary to insert short pauses between each state change (motor 1 forward and reverse, for example). You can create simple pauses in Basic with "do nothing" *FOR...NEXT* loops as shown in the testing program in Listing 30.2. The program first resets all bits to 0, then sleeps (waits) one second. The program then goes through a timed routine turning on different motors and reversing their direction: forward, reverse.

Note that do-nothing *FOR...NEXT* loops are processor-speed dependent. Adjust the value of one or both loops to control the actual delay for your computer. You may also wish to use the *SLEEP* statement, which inserts a delay for the number of seconds you specify. Other versions of Basic provide for additional time-delay commands. Most Basic programming environments, such as Microsoft QBasic (QuickBasic), allow you to terminate

TABLE 30.8 COMMON BIT PATTERNS FOR CONTROLLING TWO MOTORS.		
BINARY	**DECIMAL**	**FUNCTION**
0000	0	All stop
0011	3	Forward
1111	15	Reverse
0010	2	Right turn
0001	1	Left turn
0111	7	Hard right turn (clockwise spin)
1011	11	Hard left turn (counterclockwise spin)

the program at any time by pressing Ctrl+Break (break is the Pause/Break key, usually located near the numeric keypad).

LISTING 30.2.

```
DECLARE SUB DelaySub ()
BaseAddress = 888              ' Base address of parallel port
DataPort = BaseAddress         ' Address of data register

OUT DataPort, 0
SLEEP 1
OUT DataPort, 3
DelaySub
OUT DataPort, 15
DelaySub
OUT DataPort, 2
DelaySub
OUT DataPort, 1
SLEEP      2
OUT DataPort, 0

SUB DelaySub
' adjust delay as necessary
FOR DELAY = 1 TO 100000: NEXT DELAY
END SUB
```

Controlling More Than Eight Data Lines

As shown in the previous examples each motor requires two bits. Therefore, one parallel port can control the action and direction of four motors. However, you can actually control more motors (or other devices) by using a number of simple schemes and without resorting to using additional parallel ports.

The most straightforward method for expanding a single parallel port is to make use of some or all of the data lines of the control register. You send bits to these control lines in exactly the same way as you send bits to the data output lines, except that you use a different address. For the standard LPT1: port at decimal 888, the decimal address for the control lines is 890. Only the first five bits of the address are used in the port, which means the decimal numbers you use will be between 0 and 31.

Let's say you are using bit 2 of the control address (in a printer application, bit 2 is used to initialize the printer). You turn that bit on—and no others—by entering the following program line:

```
OUT 890, 4
```

Note that you can output a binary pattern to address 890, and it will not affect the data output lines.

USING EXPANDED IO

Another way of increasing the number of controlled devices is to use a data demultiplexer. There are several types in both the TTL and CMOS IC families. A popular data demultiplexer (or "demux") is the 74154. This chip takes four binary weighted input lines (1, 2, 4, 8) and provides 16 outputs. Only one output can be on at a time. See the schematic in Fig. 30.8 to see how to hook it up. The IC is shown connected to the first four data output lines of the parallel port. You can actually connect it to any four, and you don't even have to use all four lines. With just three lines, the demux allows you to control up to eight devices.

To select the device connected to the number 3 output of the demux, for example, you apply a binary 3 (0011) to its input lines. Write the line as follows:

```
OUT 888, 3
```

A limitation of the demux is that you can't control more than one device connected to it at any one time. You can't, for example, attach both drive motors to the demux outputs and have

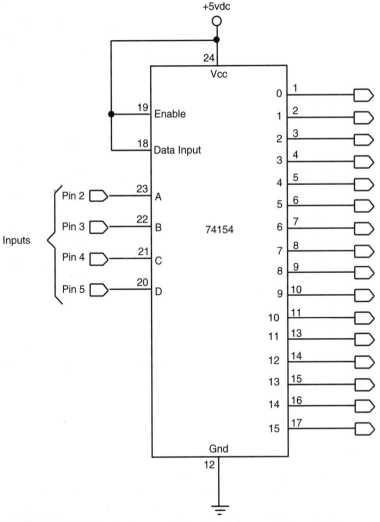

FIGURE 30.8 Basic wiring diagram for the 74154 demultiplexer chip.

them on at the same time. There will be many times, however, when your robot will only be doing one thing (such as triggering an ultrasonic ranger). In these cases, the demux is perfect.

EXTERNAL ADDRESSING

As mentioned earlier in this chapter, all sorts of data and control lines are inside the computer, on the microprocessor bus. There is also a set of special-purpose lines, the address lines, that are used to pass data to specific devices and expansion boards. For example, you address the data output lines of the parallel port by sending out the address 888.

The address for the parallel port triggers just the parallel port, but with some ingenuity (and no extra components) you can wire up a "subaddress" scheme so the one parallel port can fully control a very large number of devices. This is the third and most sophisticated way to sap all the power out of the parallel port.

You can disable the 74367 hex buffer IC, which is used to link the port to the outside world. In the Robot Experimenter's Interface, the ENABLE lines of the chip, pins 1 and 15, are held LOW by tying them to the ground, so data is passed from the input to the output. When the ENABLE pins are brought HIGH, the outputs are driven to a high-impedance state and no longer pass digital data. In this way, the 74367 acts as a kind of valve. The two ENABLE lines control different input/output pairs, as shown in Fig. 30.9. The high-impedance disabled state is engineered so that many 74367 chips can be paralleled on the same data lines, without loading the rest of the circuit.

You can use the ENABLE pins of the 74367 and a few of the unused control lines in the parallel port to make yourself an electronic data selector switch. In operation, you output a binary word onto the data output lines. You then send the word to the desired device by addressing it with the control lines.

Here is an example: Let's say that you have connected three subaddress ports to the parallel printer port, as shown in Fig. 30.10. Control lines 1, 2, and 3 are connected to the ENABLE pins of the 74367. The inputs of the three 74367s are connected together. The outputs of each feed to the specific device.

To turn on bits 0 and 1 on device 2, enter the following lines into Basic and run the program:

```
OUT 888, 3
OUT 890, 2
```

The first line of the program outputs a decimal 3 to the data output register. That places the binary bit pattern 00000011 on the parallel port data output lines. The next line enables device 2 because it turns on the second 74367.

Inputting Data

Recall that a parallel port has a third and final input register for providing status. Most parallel ports support four or five status lines, which you can use to input data back into the computer. The Robot Experimenter's Interface uses the four status lines you are likely to find in any parallel port. To read data from the port, you use the Basic *INP* statement (*INP* for input). The input command is used as follows:

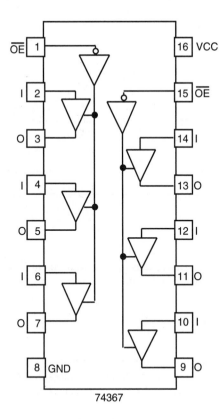

74367

Inputs		Outputs
$\overline{OE}$	I	O
L	L	L
L	H	H
H	X	HI-Z

Truth table -- 74367

FIGURE 30.9 The internal configuration of the 74367 chip. Note the two independent ENABLE lines, on pins 1 and 15.

```
Y=INP(x)
```

In place of x you put the decimal address of the port you want to read. In the case of the main printer port at starting address 888, the address of the status register is 889. The Y is a variable used to store the return value for future use in the program. For testing, you can PRINT the value of Y, which shows the decimal equivalent of the binary bit pattern on the screen.

Listing 30.3 is a sample program that displays the current values of the four inputs connected to the Robot Experimenter's Interface. The values are shown as 0 ("false") and -1 ("true"). Bear in mind that the Busy and Online lines are *active-low*; therefore their logic is the reverse of the others. The code in the test program "compensates" for the active-low condition by reversing the logic in the *If* expressions.

Also note the less-than-straightforward method for determining if pins 15 and 12 are triggered exclusively. These extra *If* tests are needed because the parallel port (most, anyway) will automatically bring pin 12 HIGH if pin 15 is brought HIGH. Weirdness is also

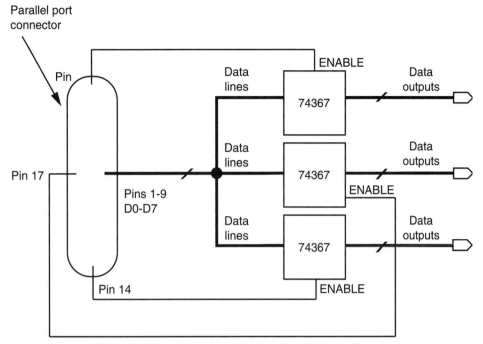

FIGURE 30.10 Block diagram for a selectable parallel port, using three 74367 ICs to independently control three separate devices.

encountered if pin 15 is brought HIGH while trying to read the values of pins 10 and 11. The port reads pins 10 and 11 as LOW, even though they may be HIGH on the interface. Again, this is the action of pin 15 (printer error), and for this reason, it's usually a good idea to limit its use or to ensure that the values of other inputs are ignored whenever pin 15 is HIGH.

LISTING 30.3.

```
DIM BaseAddress AS INTEGER, StatusPort AS INTEGER
DIM DataPort AS INTEGER, ControlPort AS INTEGER
DIM x AS INTEGER, Count
AS INTEGER
BaseAddress = 888
DataPort = BaseAddress
StatusPort = BaseAddress + 1
ControlPort = BaseAddress + 2

WHILE (1)
    x = INP(StatusPort) + 1
    IF (x AND 64) = 64 THEN
        PRINT "Pin 10: 1"
    ELSE
        PRINT "Pin 10: 0"
    END IF
     IF (x AND 128) <> 128 THEN
        PRINT "Pin 11: 1"
     ELSE
```

```
      PRINT "Pin 11: 0"
END IF
IF ((x AND 16) = 0) AND ((x AND 8) = 0) THEN
    PRINT "Pin 12: 1"
ELSE
    PRINT "Pin 12: 0"
END IF
IF ((x AND 32) = 32) AND (x AND 8) = 8 AND (x AND 16) = 16 THEN
    PRINT "Pin 15: 1"
ELSE
    IF ((x AND 32) = 0) AND (x AND 8) = 0 THEN
        PRINT "Pin 15: 1"
    ELSE
        PRINT "Pin 15: 0"
    END IF
END IF
PRINT "": PRINT ""
FOR Count = 1 TO 10000: NEXT Count
CLS
WEND
```

Before moving on, notice the use of the *DIM* keyword in the program shown in Listing 30.3. The *DIM* (for "dimension") keyword tells Basic what kind of variables are used in the program. While using *DIM* is not absolutely mandatory (in QBasic and later), you'll find that adopting it in your programs will not only help reduce errors and bugs. Most of all, it will make your programs run *much* faster. Without the *DIM* keyword, the Basic interpreter creates an all-purpose "variant" variable type that can hold numbers of different sizes, as well as strings. Every use of the variable requires Basic to rethink the best way to store the variable contents, and this takes time.

A Practical Application of the Parallel Port Input Lines

You can use the status bits for the robot's various sensors, like whiskers, line-tracing detectors, heat and flame detectors, and so forth. The simple on/off nature of these sensors makes them ideal for use with the parallel port. Listing 30.4 shows a simple demonstrator program that turns two drive motors forward until either switch located on the front of the robot is activated. Upon activation of either switch, the robot will back up for one second, spin on its axis for two seconds, then go forward again.

The demonstrator program is an amalgam of techniques discussed previously in this chapter. The program assumes you have a two-wheel robot of the type described earlier in the chapter, with the motors controlled according to the definitions in Table 30.8. Whisker or bumper switches are attached to pins 10 and 11.

LISTING 30.4.

```
DECLARE SUB GetAway ()
DIM BaseAddress AS INTEGER, StatusPort AS INTEGER
DIM SHARED DataPort AS INTEGER
```

```
DIM ControlPort AS INTEGER
DIM x AS INTEGER, Count AS INTEGER

BaseAddress = 888
DataPort = BaseAddress
StatusPort = BaseAddress + 1
ControlPort = BaseAddress + 2

CLS
PRINT "Press Ctrl+Break to end program..."

WHILE (1)
     OUT DataPort, 3            ' drive forward
     x = INP(StatusPort) + 1    ' read sensors
     IF (x AND 64) = 64 THEN    ' if sensor 1 active
        GetAway
     END IF
     IF (x AND 128) <> 128 THEN ' if sensor 2 active
        GetAway
     END IF
      FOR Count = 1 TO 500: NEXT Count
WEND
```

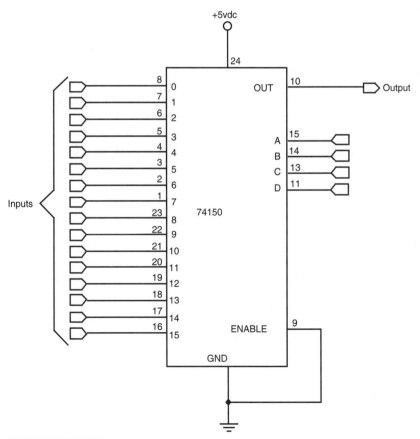

FIGURE 30.11 Basic wiring diagram for the 74150 multiplexer chip.

```
SUB GetAway
OUT DataPort, 15      ' back up
SLEEP 1               ' wait one second
OUT DataPort, 7       ' hard left turn
SLEEP 2               ' wait two seconds
END SUB
```

Expanding the Number of Inputs

Normally, you can have up to five sensors attached to the parallel port (though many ports only support three or four inputs, depending on their specific design). However, by using the ENABLE pins of the buffers in the 74367 chips, it is possible to select the input from a wide number of sensors. For example, using just four control lines with a 74150 data selector means you can route up to 16 sensors to the parallel port. See Fig. 30.11, above, for a pinout diagram of the 74150.

From Here

To learn more about...	*Read*
Computers and microcontrollers for robots	Chapter 28, "An Overview of Robot 'Brains'"
Connecting computers and microcontrollers to "real-world" devices such as motors and sensors	Chapter 29, "Interfacing with Computers and Microcontrollers"
Using remote control to activate your robot	Chapter 34, "Remote Control Systems"
Using sensors to aid in robot navigation	Part 6, "Sensors and Navigation"

31

USING THE BASIC STAMP

Since its inception, the Basic Stamp, from Parallax, Inc., has provided the "on-board brains" for countless robotics projects. This thumbprint-sized microcontroller uses Basic-language commands for instructions and is popular among robot enthusiasts, electronics and computer science instructors, and even design engineers looking for an inexpensive alternative to microprocessor-based systems. The original Basic Stamp has been greatly enhanced, and new models sport faster speeds, more memory capacity, easier software programming, and additional data lines for interfacing with motors, switches, and other robot parts.

In this chapter, you'll learn the fundamentals of the Basic Stamp and how to use it in your robotics projects. You will also want to read Chapters 32 and 33, which provide full coverage of the BasicX and the OOPic, two other microcontrollers that use an embedded high-level language for programming.

Inside the Basic Stamp

The Basic Stamp is really an off-the-shelf PIC from Microchip Technologies ("PIC" stands for "programmable integrated circuit," though other definitions are also commonly cited, including "peripheral interface controller" and "programmable interface controller"). Embedded in this PIC is a proprietary Basic-like language interpreter called *PBasic*. The chip stores commands downloaded from a PC or other development environment. When you run the program, the language interpreter built inside the Stamp converts the instructions into code the chip can use. Common instructions involve such things as assigning a given data line as an input or output or toggling an output line from high to low in typical computer-control fashion.

The net result is that the Basic Stamp acts like a programmable electronic circuit, with the added benefit of intelligent control—but *without* the complexity and circuitry overhead of a dedicated microprocessor. Instead of building a logic circuit out of numerous inverters, AND gates, flip-flops, and other hardware, you can use just the Basic Stamp module to provide the same functionality and doing everything in software. (To be truthful, the Stamp often requires that at least some external components interface with real-world devices.) Nor do you need to construct a microprocessor-based board for your robot followed by the contortions of programming the thing in some arcane machine language.

Because the Stamp accepts input from the outside world, you can write programs that interact with that input. For instance, it's a slam dunk to activate an output line—say, one connected to a motor—when some other input (like a switch) changes logic states. You could use this scheme, for instance, to program your robot to reverse its motors if a bumper switch is activated. Since this is done under program control and not as hardwired circuitry, it's easier to change and enhance your robot as you experiment with it.

As of this writing there are several versions of the Basic Stamp, including the original Basic Stamp Rev D, the Basic Stamp I ("BSI"), the Basic Stamp II ("BSII"), and the Basic Stamp II-SX. Though in their day they were useful, the Rev D and BSI products are of limited use in most robotics applications, which leaves the BSII and BSII-SX as the serious contenders. The BSII and BSII-SX share many of the same features, though the latter is faster. In this chapter, I'll concentrate on the BSII, but in most cases the specifications and command sets apply to the BSII-SX as well. You should expect continued development of the Basic Stamp, with new and updated versions. Be sure to check the Parallax Web site at *www.parallaxinc.com* for news.

The microcontroller of the Basic Stamp uses two kinds of memory: PROM (programmable read-only memory) and RAM. The PROM memory is used to store the PBasic interpreter; the RAM is used to store data while a PBasic program is running. Memory for the programs that you download from your computer is housed in a separate chip (but is still part of the Basic Stamp itself; see the description of the BSII module in the next section). This memory is EEPROM, for "electrically erasable programmable read-only memory" (the "read-only" part is a misnomer, since it can be written to as well).

In operation, your PBasic program is written on a PC, then downloaded—via a serial connection—to the Basic Stamp, where it is stored in EEPROM, as shown in Fig. 31.1. The program in the EEPROM is in the form of "tokens"; special instructions that are read, one at a time, by the PBasic interpreter stored in the Basic Stamp's PROM memory. During program execution, temporary data is kept in RAM. Note that the EEPROM memory of the Basic Stamp is nonvolatile—remove the power and its contents remain. The same is not true of the RAM. Remove the power from the Basic Stamp and any data stored in the RAM is gone. Also note that the PBasic interpreter, which is stored in the PROM memory of the microcontroller, is not replaceable.

As a modern microcontroller, the Basic Stamp II is a little tight when it comes to available memory space. The chip sports only 2K of EEPROM and just 32 *bytes* of

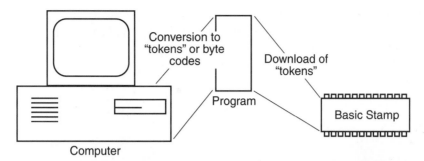

Computer

FIGURE 31.1 **Programs are downloaded from your PC to the Basic Stamp, where they are stored in "tokenized" format in EEPROM. The PBasic interpreter executes these tokens one by one.**

RAM. Of those 32 bytes, 6 are reserved for storing the settings information of the input/output pins of the Basic Stamp, leaving only 26 bytes for data. For many robotics applications, the 2K EEPROM (program storage) and 26-byte RAM (for data storage) are sufficient. However, for complex designs you may need to use a second Basic Stamp or select a microcontroller—such as the Basic Stamp II-SX—that provides more memory.

Stamp Alone or Developer's Kit

The Basic Stamp is available directly from its manufacturer or from a variety of dealers the world over. The prices from most sources are about the same. In addition to the BSI, BSII, and BSII-SX variations mentioned earlier, you'll find that the Basic Stamp is available in several different premade kits as well as a stand-alone product.

- *BSII Module.* The Basic Stamp module (see Fig. 31.2) contains the actual microcontroller chip as well as other support circuitry. All are mounted on a small printed circuit board that is the same general shape as a 24-pin IC. In fact, the BSII is designed to plug into a 24-pin IC socket. The BSII module contains the microcontroller that holds the PBasic interpreter, a 5-volt regulator, a resonator (required for the microcontroller), and a serial EEPROM chip.
- *BSII Starter Kit.* The starter kit is ideal for those just, well, starting out. It includes a BSII module, a carrier board, a programming cable, a power adapter, and software on CD-ROM. The carrier board, shown in Fig. 31.3, has a 24-pin socket for the BSII module, a connector for the programming cable, a power adapter jack, and a prototype area for designing your own interface circuitry.
- *Basic Stamp Activity Board.* The Activity Board, which is typically sold without a BSII module, offers you a convenient way to experiment with the Basic Stamp. It contains four LEDs, four switches, a modular jack for experimenting with X-10 remote control modules, a speaker, and two sockets so you easily interface such things as serial analog-to-digital converters (ADCs).

FIGURE 31.2 The Basic Stamp II module, containing microcontroller, voltage regulator, resonator, and EEPROM.

FIGURE 31.3 The Basic Stamp carrier board, ideal for experimenting with the BSII. Sockets are provided for both the BSI and BSII.

- *Growbot and BOE Bot.* The Growbot and BOE Bot products are small mobile robot kits that are designed to use the Basic Stamp microcontroller. The robots are similar (the BOE Bot is a little larger and heavier) and are able to accommodate more experiments. A BSII module is generally not included with either robot kit.
- *Basic Stamp Bug II.* Another robot kit, the Basic Stamp Bug II, is a six-legged walking robot. The Bug is meant to be controlled with a BSI microcontroller, though you could refit it to use the BSII. The Basic Stamp module is extra.

Physical Layout of the BSII

The Basic Stamp II is a 24-pin device; 16 of the pins are input/output (I/O) lines that you can use to connect with your robot. For example, you can use I/O pins to operate a radio-controlled (R/C) servo. Or you can use a stepper motor or a regular DC motor, when you use them with the appropriate power interface circuitry. As outputs, each pin can source (that is, output 5 volts) 20mA of current or sink (output 0 volts) about 25 mA. However, the entire BSII should not source or sink more than about 80-100mA for all pins. You can readily operate a series of LEDs, without needing external buffer circuitry to increase the power-handling capability.

Or you can connect the BSII to a Polaroid sonar range-finding module (see Chapter 38, "Navigating through Space"), various bumper switches, and other sensors. The "direction" of each I/O pin can be individually set, so some pins can be used for outputs and others for inputs. You can dynamically configure the direction of I/O pins during program execution. This allows you to use one pin as both an input and an output, should this be called for.

Fig. 31.4 shows the pin layout of the BSII. The Basic Stamp II supports three ports, referred to as A, B, and C. Port A is used for internal connections, namely, the serial lines to the outboard EEPROM chip, as well as the RS-232 serial connections to and from the PC that is used for programming. This leaves two full 8-bit ports, B and C, for use as I/O lines. Through PBasic commands, you can control all eight bits of the each port together or each pin individually.

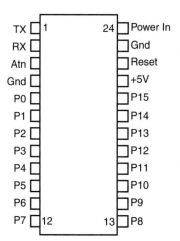

FIGURE 31.4 The layout of the pins on the Basic Stamp II module.

Understanding and Using PBasic

As you've read earlier in this chapter, at the heart of the Basic Stamp is PBasic, which is the language used to program the Basic Stamp device. PBasic has undergone changes during the life of the Basic Stamp products, and the syntax and the commands between PBasic for the Basic Stamp I (known as PBasic1) and PBasic for the Basic Stamp II (PBasic2) are different. What follows is strictly PBasic2 for the Basic Stamp II.

PBasic programs for the Basic Stamp are developed in the Basic Stamp Editor, an application that comes with the Starter Kit (and is also available for free download at the Parallax Web site). The Editor lets you write, edit, save, and open Basic Stamp programs. It also allows you to compile and download your finished programs to a Basic Stamp. The download step requires that your Basic Stamp be connected to a carrier board or other circuit board attached to a download cable. The download cable is connected to your PC via one of its serial ports. Fig. 31.5 shows the Basic Stamp Editor.

Like any language, PBasic is composed of a series of *statements* that are strung together in a logical syntax. A statement forms an instruction that the BSII is to carry out. For example, one statement may tell the chip to fetch a value on one of its I/O pins, while another may tell it to wait a certain period of time. The majority of PBasic statements can be categorized into three broadly defined groups: variable and pin or port definitions, flow control, and special function. We'll cover each of these next.

VARIABLE AND PIN/PORT DEFINITIONS

As with any programming language, PBasic uses variables to store bits and pieces of information during program execution. Variables can be of several different sizes; you should always strive to choose the smallest variable size that will accommodate the data you wish

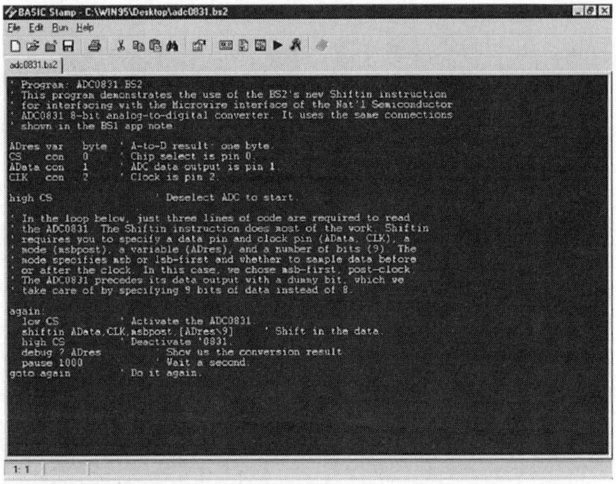

FIGURE 31.5 The Basic Stamp Editor is used to create, compile, and download programs for the Basic Stamp.

to store. In this way you will conserve precious RAM space (remember, you only have 26 bytes of RAM to work with!).

PBasic provides the following four variable types:

- Bit—1 bit (one eighth of a byte)
- Nibble—4 bits (4 bits)
- Byte—1 byte (8 bits)
- Word—2 bytes (16 bits)

Variables must be declared in a PBasic program before they can be used. This is done using the *var* statement, as follows:

```
VarName        var        VarType
```

where *VarName* is the name (or *symbol*) of the variable, and *VarType* is one of the variable types just listed. Here's an example:

```
Red               var                bit
Blue              var                byte
```

Red is a bit, and *Blue* is a byte. Note that capitalization does not matter in a PBasic program. The following has the same result:

```
Red               Var                Bit
BLUE              VAR                BYTE
```

Once declared, variables can be used throughout a program. The most rudimentary use for variables is with the = (equals) assignment operator, as in

Red = 1

Blue = 12

Variables can also be assigned as the result of a math expression (2 + 2) or as the value of an input pin. For example, suppose an input pin is connected to a mechanical switch. Ordinarily, the switch is open, and the value at the pin is LOW (0). Suppose a variable, called *Switch,* stores the current value of the pin. The *Switch* variable would contain 0 as long as the switch is opened. If the switch is closed, the *Switch* variable then stores 1 (or logical HIGH). More about I/O pins in a bit.

Variables store values that are expected to change as the program runs. PBasic also supports constants, which are used as a convenience for the programmer. Constants are declared much as variables are, using the *con* statement:

```
MyConstant             con             5
```

MyConstant is the name of the constant, and its value is 5. Constants do not consume any RAM and are typically used to make it easier to modify the program later on.

The Basic Stamp treats its 16 I/O pins like additional memory. The instantaneous value of an I/O pin functions exactly like a one-bit variable: the value is either 0 or 1. If the I/O pin is an input, then the value of that input will be 0 or 1, depending on the condition of the circuit on the outside of the Basic Stamp. The mechanical switch is a good example of this: depending on whether the switch is opened or closed, the value of the input pin is 0 (open) or 1 (closed).

When I/O pins are used as outputs, their logical state is changed using the *high, low,* and *toggle* statements. In each case, the number of the pin (0 through 15) is given to tell the Basic Stamp which I/O you want to change:

■ High brings the I/O pin HIGH (1)
■ Low brings the I/O pin LOW (0)
■ Toggle changes the state of the I/O pin from 0 to 1, or vice versa, depending on its previous value.

Here's an example (using the traditional ' character for comments):

```
high 1          ' put I/O pin 1 (RB1) high
low 12          ' put I/O pin 12 (RC4) low
toggle 5        ' change I/O pin = (RB5) opposite to its
                  previous value
```

There are many ways to determine the current value of an I/O pin that is used as an input. Most are used with special functions, which are outlined later in this chapter. You can also directly reference the value of an input by using the *Inx* statement, where *x* is a number from 0 to 15. For instance, to read the value of pin 3 and put it into a variable, you'd use the following:

```
SomeVar = In3
```

FLOW CONTROL

Flow control statements tell your program what to do next. A commonly used flow control statement is *if,* which is used in conditional expressions that execute one part of the program if condition A exists and another part of the program if condition B exists. Two other flow control statements are *goto* and *gosub,* which are used to unilaterally jump from one part of a program, and the *for* statement, which is used to repeat a block of code a specific number of times. Let's look first at the *if* statement.

The *if* statement, which is always used in conjunction with the *then* statement, conditionally branches execution depending on the outcome of an expression. The syntax is as follows:

if Expression *then* Label

Expression is the condition that must resolve to a True or False statement, and *Label* is an identified label elsewhere in the program that the execution is to jump to. An example of a typical *Expression* is checking the value of a variable or input pin against an expected value:

```
if MyVar=1 then Flash
```

If the contents of *MyVar* is equal to 1 (the expression is True), then the program is expected to jump to the *Flash* label. This label is identified by the label name, followed by a colon, as in:

```
Flash:
...rest of the code goes here
```

The *if* expression can use a number of logical operators:

=	equal to
<>	not equal to
>	greater than
<	less than
>=	greater than or equal to
<=	less than or equal to

The *if* statement is a little funky compared to other modern programming languages, in that the result of the expression branches execution to a label. Note that there is no explicit *else* keyword in PBasic, that is, an action to be taken if the expression is False. As cited in the Basic Stamp manual, you must write *if* statements that have a True and False component along these lines:

```
if aNumber < 100 then isLess
debug "greater than or equal to 100"
stop

isLess:
debug "less than 100"
stop
```

Notice how this bit of programming works: Should *aNumber* be less than 100, then the program jumps to the *isLess* label, and the *debug* statement (which prints text in the debug window of the Basic Stamp programming environment on your PC) prints "less than 100." However, if *aNumber* is 100 or higher, the jump to *isLess* is ignored, and the program simply executes the next line, which is yet another debug statement ("greater than or equal to 100"). Note the introduction of another flow control statement: *stop*. The *stop* statement stops program execution.

The *goto* and *gosub* statements are used with labels to divert execution to another part of the program. *Goto* is most often used to create endless loops, as shown here:

```
high 1
RepeatCode:
    pause 100
    toggle 1
    goto RepeatCode
```

In this program, I/O pin 1 is set to 1 (HIGH). The program then pauses for 100 milliseconds (one-tenth of a second) and then toggles I/O pin 1 to its opposite state. The *goto* statement makes the program jump back to the *RepeatCode* label. With each trip through the code, I/O pin 1 is toggled HIGH or LOW. If the pin is connected to an LED, for example, it would flash on and off rapidly 10 times each second.

Gosub is similar to *goto,* except that when the code at the label is done, the program returns to the statement immediately after *gosub.* Here's an example:

```
high 1
low 2
gosub FlashLED
'... some other code here
stop

FlashLED:
    toggle 1
    toggle 2
    pause 100
    return
```

The program begins by setting I/O pin 1 to HIGH and I/O pin 2 to LOW. It then "calls" the *FlashLED* routine, using the *gosub* statement. The code in the *FlashLED* routine toggles I/O pins 1 and 2 from their previous state, waits one-tenth of a second (100 milliseconds), and then returns with the *return* flow control statement. Note the *stop* statement used before the *FlashLED* label. It prevents the code from re-executing the *FlashLED* routine when it is not intended.

The *for* statement is used with the *to* and *next* statements. All form a controlled counter that is used to repeat the code a set number of times. The syntax for the *for* statement is:

for Variable = StartValue *to* EndValue [more statements] *next*

Variable is a variable that is used to contain the current count of the *for* loop. *StartValue* is the initial value applied to *Variable*. Conversely, *EndValue* marks the maximum value that will be applied to *Variable*. The loop breaks out—and the rest of the program continues to execute—when the *Variable* exceeds *EndValue*. For example, if you use the following,

```
for VarName = 1 to 10
```

the loop starts with 1 in *VarName* and counts to 10. The *for* loop is repeated 10 times. You don't have to start with 1, and you can use an optional *step* keyword to tell the *for* loop that you want to count by 2s, 3s, or some other value:

```
for VarName = 5 to 7         ' counts from 5 to 7
for VarName = 1 to 100 step 10    ' counts from 1 to 100, but steps by 10
```

For loops are used to execute whatever programming lies between the *for* and *next* statements. Here's a simple example:

```
high 1
for VarName = 1 to 10
    toggle 1
    pause 100
next
```

This program repeats the *for* loop a total of 10 times. At each iteration through the loop, I/O pin 1 is toggled (HIGH to LOW, and back again).

The Basic Stamp supports additional flow control statements, all of which are detailed in the Basic Stamp manual. These include *Branch* and *End*.

SPECIAL FUNCTIONS

The PBasic language supports several dozen special functions that are used to control some activity of the chip, including ones to sound tones through an I/O pin or to wait for a change of state on an input. I'll briefly review here the special functions most useful for robotics. You'll want to study these statements more fully in the Basic Stamp manual (available for free download from Parallax and also included in the Starter Kit as a printed book).

- *button*. The *button* statement momentarily checks the value of an input and then branches to another part of the program if the button is in a LOW (0) or HIGH (1) state. The *button* statement lets you choose which I/O pin to examine, the "target state" you are looking for (either 0 or 1), and the delay and rate parameters that can be used for such things as switch debouncing. The *button* statement doesn't stop program execution, which allows you to monitor a number of I/O pins at once.

- *debug*. The Basic Stamp Editor has a built-in terminal that displays the result of bytes sent from the Basic Stamp back to the PC. The *debug* statement "echoes" numbers or text to the screen and is highly useful during testing. For example, you can have the *debug* statement display the current state of an I/O pin, so you can visually determine whether or not the program is working properly.

- *freqout*. The *freqout* statement is used to generate tones primarily intended for audio reproduction. You can set the I/O pin, duration, and frequency (in Hertz) using the *freqout* statement. An interesting feature of *freqout* is that you can apply a second frequency, which intermixes with the first. For example, you can combine a straight middle A (440 Hz) with a middle C (523 Hz) to create a kind of chord. Don't expect a symphonic sound, but it works for simple tunes. When *freqout* is used to drive a speaker you should connect capacitors (and resistors, as required) to build a filter.

- *input*.The *input* statement makes the specified I/O pin an input. As an input, the value of the pin can be read in the program. Many of the special function statements, such as *button* and *pulsin,* automatically set an I/O pin as an input, so the *input* statement is not needed for these. See the next section, "Interfacing Switches and Other Digital Inputs," for additional information on the *input* statement.

- *pause*. The *pause* statement is used to delay execution by a set amount of time. To use *pause* you specify the number of milliseconds (thousandths of a second) to wait. For example, *pause 1000* pauses for one second.

- *pulsin*. The *pulsin* statement measures the width of a single pulse with a resolution of two microseconds (2 µs). You can specify which I/O pin to use, whether you're looking for a 0-to-1 or 1-to-0 transition, as well as the variable you want to store the result in. *Pulsin* is handy for measuring time delays in circuits, such as the return "ping" of an ultrasonic sonar.

- *pulsout*. *Pulsout* is the inverse of *pulsin:* with *pulsout* you can create a finely measured pulse with a duration of between 2 µs and 131 milliseconds (ms). The *pulsout* statement is ideal when you need to provide highly accurate waveforms.

- *rctime*. The *rctime* statement measures the time it takes for an RC (resistor/capacitor) network to discharge to an opposite logical state. The *rctime* statement is often used to indirectly measure the capacitance or resistance of a circuit, or simply as a kind of simplified analog-to-digital circuit. Fig. 31.6 shows a sample circuit.

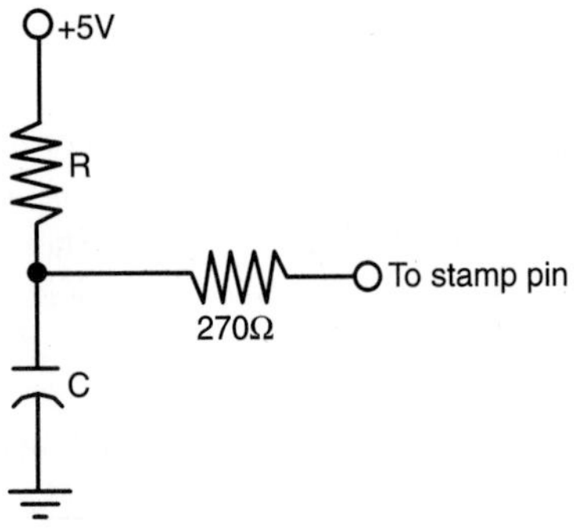

FIGURE 31.6 The *rctime* **statement is used to measure the time it takes for a capacitor to discharge (the timing is accurate to 2 μs intervals). With this information you can indirectly measure capacitance or resistance.**

- *serin* and *serout*. *Serin* and *serout* are used to send and receive asynchronous serial communications. They represent one method for communicating with other devices, even other Basic Stamps, all connected together. Both commands require that you set the particulars of the serial communications, such as data (baud) rate, and the number of data bits for each received word. One application of *serout* is to interface a liquid crystal display (LCD) to the Basic Stamp. You use *serout* to send commands and text to the LCD.
- *shiftin* and *shiftout* The *serin* and *serout* statements are used in one-wire asynchronous serial communications. The *shiftin* and *shiftout* statements are used in two- or three-wire synchronous serial communications. The main difference is that with *shiftin/shiftout* a separate pin is used for clocking the data between its source and destination. If you're only sending or receiving data, you can use just two pins: one for data and one for clock. If you're both sending and receiving, your best bet is to use three pins: data in, data out, and clock. These statements are useful when communicating with a variety of external hardware, including serial-to-parallel shift registers and serial analog-to-digital converters.

Interfacing Switches and Other Digital Inputs

You can easily connect switches, either for control or for "bump" or other contact sensors, to the Basic Stamp using either of the approaches shown in Fig. 31.7. You can use the *button* statement, described briefly earlier in the chapter, to determine the current value of the switch. You can watch for a transition from 0 (LOW) to 1 (HIGH), or vice versa. The *button* statement also includes a built-in debounce feature, so the Basic Stamp will ignore the typical "noise" that occurs when a mechanical switch closes. Without debounce, if you press a switch it may cause a number of button trigger events in the code—as many as 10–20, depending on the switch and whether there's other code in the program.

To use the *button* statement you must define a variable in which to store the switch closure result and then set up a repeating loop to test if or when the button is closed. The full syntax of *button* is as follows:

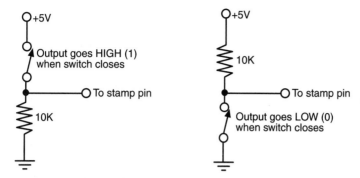

FIGURE 31.7 **You can connect a switch to a Basic Stamp pin so that it triggers a 0 or a 1 state when it closes.**

button Pin, Downstate, Delay, Rate, ByteVariable, TargetState, Label

■ *Pin* is the pin you wish to test, from 0 to 15.
■ *Downstate* specifies the logical state when the button is pressed. Use 0 for an active-low switch connection, 1 for an active-high switch connection.
■ *Delay* specifies how long the button must be pressed before the auto-repeat feature starts. Valid values are 0 to 255, with 255 denoting that you wish to use the debounce feature.
■ *Rate* specifies the number of cycles between auto-repeat. Valid values are 0 to 255.
■ *ByteVariable* is a temporary variable used to store the value of the button. You need to clear it to 0 (zero) before the *button* statement is called the first time.
■ *TargetState* specifies which state the button should be in (0 or 1) for a branch to occur.
■ *Label* is the name of the label to branch to when the button is in its target state.

In the following example, the code notes when the button closes and produces a 0 (LOW) on input pin 1. The following code uses the debounce feature of the *button* statement in order to trigger just once when the button is pressed.

```
Bttn          var         byte       ' defines workspace variable for button
Bttn = 0                             ' initializes
Loop:
    button 1,0,255,250,Bttn,0,NotPressed
    debug "Button pressed! "
NotPressed: goto Loop                ' repeat forever
```

Note that the *button* statement does not pause the program. The loop is required to keep the Basic Stamp continually scanning pin 1 in anticipation of the switch changing state from 1 to 0. You can use this to your advantage to watch for any number of buttons on different pins.

You can use the *button* statement for any kind of input, not just switches. *Button* can be used with any digital input, whether the input changes from 0 to 1 or 1 to 0. You can also use the *Inx* statement (*x* is a number from 0 to 15 that denotes a pin) to watch for a change in pin states. You should not use this method with mechanical switches since there is no built-in debounce software to eliminate multiple triggers. Like *button,* you can put multiple *Inx* statements in a loop in order to scan one or more pins at a time. You can place other code in the loop as well should you need the Basic Stamp to process other parts of the program. The syntax is relatively simple:

```
IOPin3          var        bit
IOPin5          var        bit
Loop:
    IOPin3 = In3
    IOPin5 = In5
    . . .some other code here
    goto Loop
```

You can add additional code to determine what should happen if the *IOPin3* and/or *IOPin5* variables are a given state. For instance, if you want to execute some code when both pins are 1 (HIGH), you might use the following:

```
if IOPin3 = 1 And IOPin5 = 1 then
```

Interfacing DC Motors to the Basic Stamp

The Basic Stamp is ideal for controlling a DC motor that is connected to an H-bridge circuit (see Chapter 18, "Working with DC Motors"). In the typical H-bridge for a single motor, the Basic Stamp controls the on/off operation of the motor using one pin and the direction using another pin. By using the *high* and *low* statements, you can control the motor easily, turning it on and off and reversing its direction.

The Basic Stamp code for controlling a DC motor is relatively straightforward: use the *high* or *low* statements and indicate which I/O pins you wish to use. For example, suppose your motor H-bridge is connected to pins 0 and 1, with pin 0 used for on/off control and pin 1 used for direction. Note that when pin 0 is low (0), the motor is off and therefore the setting of pin 1 doesn't matter.

PIN	LOW	HIGH
0 (on/off)	Motor off	Motor on
1 (direction)	Forward	Reverse

Here is an example:

```
low 1           ' set direction to forward
high 0          ' turn on motor
pause 2000      ' wait two seconds
low 0           ' turn off motor
pause 100       ' wait 1/10 second
```

```
high 1                    ' set direction to reverse
high 0                    ' turn on motor
pause 2000                ' wait two seconds
low 0                     ' turn off motor
```

By using labeled routines and the *gosub* statement you can define common actions and develop more compact programs:

```
gosub motorOnFwd
gosub waitLong
gosub motorOff
pause 1
gosub motorOnRev
gosub waitShort
gosub motorOff
end
motorOnFwd:
low 1
high 0
return
motorOnRev:
high 1
high 0
return
motorOff:
low 0
return
waitLong:
pause 5000
return
waitShort:
pause 1000
return
```

Interfacing RC Servo motors to the Basic Stamp

Servo motors for radio-controlled (R/C) cars and airplanes can be easily connected to and controlled with a Basic Stamp. In fact, the code required for operating a servo motor is surprisingly simple, which is one of the aspects of the Basic Stamp that has so endeared it to robot experimenters.

Hobby servo motors contain their own interface circuitry, so you don't need an H-bridge or power driver. You may connect any I/O pin of the Basic Stamp directly to the signal input of the servo (see Chapter 20, "Working with Servo Motors," for more information on servomotors, how they work, and their electronic connections). Keep in mind that the Basic Stamp cannot provide operating power to the servo motor; you must use a separate battery or power supply for it. Otherwise, you run the risk of damaging the Basic Stamp or, at the least, having a program malfunction as the Basic Stamp continually resets itself because of the power sag.

Fig. 31.8 shows a good approach for connecting a common RC servo to the Basic Stamp by using a separate battery supply for the servo. Note that the ground connections of the power supplies for both the Basic Stamp and the servo are connected and that a 1 µF tantalum capacitor has been added across the power supply connections for both the Basic Stamp and the servomotor. This helps eliminate the problems caused by the electrical noise that can be generated when the servo turns on and off. If noise poses a problem for you, you can try adding an additional 100 µF electrolytic capacitor along with the 1 µF tantalum capacitors.

To operate the servo you need only a few lines of code, and you can easily control more than one servo at a time. The trick is to use the *pulsout* statement, which sends a pulse of a specific duration to an I/O pin. Servos need to be "refreshed" with a pulse from 30 to 60 times each second to maintain their position. By adding a delay (using the *pause* statement) and a loop your Basic Stamp can move and maintain any position of the servo. The following examples show the basic program, using pin 0 as the control signal line to the servo. It sets the servo in its approximate center position:

```
low 0
Loop:
  pulsout 0,750
  pause 20
  goto Loop
```

Here's how the program works: pin 0 is set as an output and set to lo2, with the *low 0* statement. The repeating loop is defined as the code between the *Loop:* label and the *goto Loop* statement. The *pulsout* statement sends a 1500-microsecond pulse (LOW-HIGH-LOW) to pin 0. The value of 750 is used because *pulsout* has a minimum resolution of 2 µs; 750 times 2 is 1500 µs. As mentioned in Chapter 20, a servo motor has a typical oper-

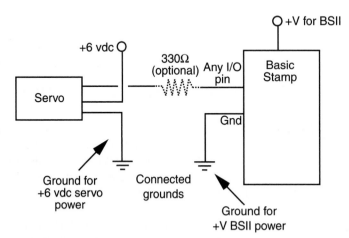

FIGURE 31.8 A typical wiring diagram for connecting the Basic Stamp to a hobby servo. A separate power supply, delivering 4.8 to 7.2 volts (6 volts are typical), is provided for the servo. The capacitors prevent any electrical noise from the servo from interfering with the operation of the Basic Stamp.

ating range of 1000 to 2000 μs (or 1-2 milliseconds, the same thing). Pulses within this range control the angular position of the servo output shaft. A pulse of 1500 (μ will position the output shaft of the servo in its approximate center position. I say "approximate" because the actual mechanical center of the shaft can vary from one servo to the other.

The *pause* statement pauses execution for 20 milliseconds (ms). When the loop is run, it will repeat about 50 times each second (20 ms * 50 = 1000 ms, or one second). Note that you can insert additional *pulsout* statements if you need to control other servos. The Basic Stamp can adequately control seven or eight servos, but in doing so, it won't have much leftover processing time for anything else. For this reason, dedicated servo control chips are typically used with the Basic Stamp to enable it to control multiple servos in a "set-and-forget" fashion. We will talk more about these in the next section.

To change the angular position of the servo motor, merely alter the timing of the *pulsout* statement:

```
pulsout 0,1000          ' 2000 usec pulse, approx. 180 degrees position
pulsout 0,500           ' 1000 usec pulse, approx. 0 degrees position
```

These values assume a strict 1000-2000 μs operating range for a full 0 to 180° rotation. This is actually *not* typical. You will likely find that your servo will have a full 180° rotation with higher and lower values than the nominal 1000–2000 μs. You can only determine this through experimentation. One servo may have a full 180° rotation with values of 500 to 2300 μs, for example, while another may need pulses in the 800 to 2500 μs range. If you need full 180° rotation of your servo you'll need to calibrate your Basic Stamp programs for each servo that you use. This is not a limitation of the Basic Stamp but of the manufacturing variations of the typical hobby servo.

Enhancement Add-in Products for the Stamp

One of the great advantages of the Basic Stamp is the multitude of program examples and hardware add-ons available, both from Parallax and independent companies. The Basic Stamp manual (available in printed form and as a free download from the Parallax Web site) is replete with examples and application notes. Unfortunately, in at least the edition that was current as of this writing, most of the application notes are for the older Basic Stamp I, not the newer and more capable Basic Stamp II. In some cases, you can use the sample programs in the application notes as is, but you should anticipate needing to "convert" them for use on the BSII. The manual contains application notes for controlling servo motors, LEDs, stepper motors, and other components common to the typical small robot.

The Web is also an excellent source of sample programs. See Appendix C, "Robot Information on the Internet," for a selected list of information about the Basic Stamp. Note that this list is quite limited. Use your favorite Web search engine—such as AltaVista (*www.altavista.com*) or Google (*www.google.com*)—to find additional resources.

Add-on hardware, including motor drivers, LCD displays, and keyboard interfaces, is also popular for the Basic Stamp. This hardware serves two important purposes:

■ The external hardware "offloads" the processing requirements from the Basic Stamp. Rather than have the Basic Stamp control multiple servos, for example, a servo motor controller chip can do the job, freeing the Basic Stamp to do other things. In most cases, the external hardware provides for "set-and-forget" functionality: the Basic Stamp sends a command to the hardware, then goes about its business with other things. The external hardware does all the rest.

■ The external hardware often reduces the degree of I/O pin usage the Basic Stamp needs. This is an important consideration given that the Basic Stamp has only 16 I/O lines. That may seem like a lot, but it's amazing how fast those lines "disappear" when you're designing and building a robot! You typically connect the external hardware to the Basic Stamp by way of a one- or two-wire serial connection. The Basic Stamp's *serout* and *serin* statements are used to send and receive data.

Typical of the external hardware that supports the Basic Stamp (and most other microcontrollers, for that matter) is the serial LCD display shown in Fig. 31.9. This display, manufactured by Scott Edwards Electronics (see Appendix B for contact information), connects to the Basic Stamp via a simple serial connection. You then issue simple commands to display text on the display. The display is available in a number of character-by-line formats (16 characters by 2 lines is common), including an all-graphic LCD panel.

Serially programmed servo modules are also available for the Basic Stamp (and other microcontrollers). These modules control one or more hobby servo motors. All are "set-and-forget," so once you command the module, it does the rest—including continually sending pulses to the servo motor so it can keep its position. Most servo modules control a number of servos, typically five or eight.

Dedicated modules for controlling other kinds of motors also exist. For example, the Motor Mind B (Fig. 31.10) from Solutions Cubed (see Appendix B for contact information) is a fully featured DC motor controller and power driver. The Motor Mind can operate a motor with up to 2 amps continuous current and up to 30 volts. You can use an optional tachometer input for built-in speed control (the tachometer frequency can span from 0 to 65,528 Hz). You can set the speed of the motor in 254 discrete steps and, of course, alter the direction of the motor.

See Appendix B for other useful hardware that you can interfaced to the Basic Stamp and other microcontrollers.

From Here

To learn more about...	*Read*
Using DC motors for robot locomotion	Chapter 18, "Working with DC Motors"
Using servo motors for robot control and locomotion	Chapter 20, "Working with Servo Motors"
Choices and alternatives for robot computers and microcontrollers	Chapter 28, "An Overview of Robot 'Brains'"
Attaching real-world hardware to computers and microcontrollers	Chapter 29, "Interfacing with Computers and Microcontrollers"

FIGURE 31.9 This LCD display connects to the Basic Stamp via a serial line and is programmed using simple PBasic statements.

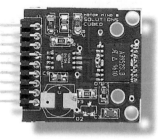

FIGURE 31.10 The Motor Mind B, shown with its heat sink removed, is an all-in-one intelligent motor controller that is often used with Basic Stamp (it can also be used with other microcontrollers and computers as well). Because it offers "set-and-forget" motor control and speed, Motor Mind B permits the Basic Stamp to carry on other tasks without taking up precious processing time.

32

USING THE BASICX
MICROCONTROLLER

Microcontrollers are fast becoming a favorite method for endowing robots with smarts. In fact, they're a robot builder's dream come true. Microcontrollers are single-chip computers complete with their own input/output ports and even memory. The typical cost of a microcontroller is from $5 to $15 and most can be programmed using the software on your PC. Once programmed, the microcontroller is disconnected from the PC and operates on its own. Microcontrollers are power misers too. Nearly all have simple power requirements (usually just 3.3 or 5 volts) and require just a few milliamps for their own operation, even when running at speeds of 5 or 10 megahertz.

Microcontrollers are available in two basic flavors: *low-level programmable* and *embedded-language programmable*. As we noted in Chapter 28, these loosely defined terms relate to the programming of the controller. Both kinds of microcontroller are fully programmable, but one contains a kind of built-in operating system that allows it to be programmed with a higher-level language, such as Basic.

Let's talk about low-level microcontrollers first. You program these with assembly language or C, using your PC as a host development system. Assembly language seems somewhat arcane to newcomers, but the language offers full control over the internal workings of the microcontroller. Unfortunately, there's no standard when it comes to assembly languages.

Popular alternatives to these low-level programmable microcontrollers are products that have a built-in programming interface, such as the Basic Stamp from Parallax or the OOPic from Savage Industries. These controllers support a high-level programming language—typically Basic—that is permanently embedded within the chip. Using your PC as a development platform, you write software for the microcontroller using a custom program editor. The software is then compiled to a series of *tokens* or *bytecodes* and then downloaded to the microcontroller.

Joining the ranks of powerful embedded-language programmable microcontrollers is the BasicX-24, by NetMedia, a company founded by the creator of the popular LANtastic networking software (which sold some 10 million copies). The BasicX-24 is actually a member of a family of microcontrollers from NetMedia that also includes the less expensive (but network-capable) BasicX-01. However, all things considered, the BasicX-24 is perhaps the most versatile, so this chapter will focus on it exclusively.

Inside the BasicX-24 Microcontroller

A selling point of the BasicX-24 (which we'll refer to as the BX-24 from here on) is that it is pin-for-pin compatible with Parallax's Basic Stamp II. That is, the functions of all 24 pins of the BX-24 replicate the functions of the Basic Stamp II, including power and ground connections. It's important to note, however, that the BX-24 is not a Stamp "clone." The two microcontrollers don't share the same programming languages, so programs written for one will not work on the other. Additionally, the BX-24 has several additional features not found in the Basic Stamp II, such as built-in analog-to-digital conversion and 32K of EEPROM memory.

Fig. 32.1 shows the BX-24 "chip," which (like the Basic Stamp) is actually several integrated circuits on a small circuit board. The layout of the pins on the BX-24 is identical to that of any standard-sized 24-pin IC, so it will plug into a regular 24-pin socket. Additional plated-through holes are provided on either end of the BX-24 board, making it just slightly longer than the Basic Stamp II. These holes provide connections to additional input/output lines provided on the BX-24. I'll get to those in a bit.

The BX-24 directly supports 16 input/output (I/O) lines, the same number as the Basic Stamp II. For each I/O line, or pin, you can change the direction from an input or an output. When an I/O line is an output, you can individually control the value of the pin, either 0 (logic LOW) or 1 (logic HIGH). When an I/O line is an input, you can read a digital or analog value from a TTL-compatible device connected to the BX-24. Eight of the 16 I/O lines can be used for analog connections. The BX-24 incorporates its own built-in 10-bit analog-to-digital converter (ADC). Under software control, you can indicate which of the 8 input lines is to be read.

Three of the plated-through holes of the BX-24 serve as optional I/O and are programmatically referred to as pins 25, 26, and 27. This makes a total of 19 input/output pins. The remaining plated-through holes provide a way to connect to the chip's serial peripheral interface, or SPI, lines. I do not recommend that you connect to these lines unless you're familiar with SPI interfaces, especially since the BX-24's EEPROM is controlled by these same I/O lines.

A nice feature of the BX-24 is its two LEDs: one red and one green. The green LED is normally used to indicate that the chip is powered on, but you can individually control both LEDs from your own programs. You might use the LEDs as status indicators, for example. The LEDs share two of the additional plated-through hole connectors on the BX-24.

FIGURE 32.1 The BasicX-24 consists of surface-mount integrated circuits on a small circuit board. The BX-24 circuit board has the same dimensions as a standard 24-pin IC.

The BX-24 board comes with its own five-volt voltage regulator, which provides enough operating current for all the components on the board, plus several LEDs or logic ICs. If you plan on using the BX-24 to operate a robot, you'll want to provide a separate power supply of adequate current rating to the other components of the robot. You should not rely on the BX-24's on-board regulator for this task.

Pinout Diagram for the BX-24

Fig. 32.2 shows the pinout diagram of the BX-24 as well as the functions of its 24 pins. Of main interest are the following:

- *Pin 24.* This is the unregulated power input. Apply an unregulated DC voltage of 5.5 to 15 volts here. The onboard regulator will provide a stable 5 vdc input for the BX-24 circuitry.
- *Pin 23, 4.* This is the ground. You can use either or both of these pins when connecting to other circuitry.
- *Pin 21.* This is for +5 vdc input. Instead of using pin 24 for power, you may directly apply *regulated* 5 vdc to this pin. Or, if power is applied through pin 24, pin 21 serves as a convenient source of regulated 5 vdc power. The voltage regulator on the BX-24

can supply approximately 70 mA of total additional current, either through this pin or through the I/O pins described next.

■ *Pins 5 through 11.* This is I/O Port C, one of two eight-bit ports on the BX-24. Pin 12 serves "double duty" as an input capture pin, which can be used for very accurate timing. Pin 11 serves double duty as an external interrupt. With the appropriate programming, the BX-24 can be commanded to automatically run certain code when this line goes HIGH.

■ *Pins 13 through 20.* This is Port D, the second of two eight-bit ports on the BX-24. All of the pins in this port serve double duty as analog-to-digital conversion inputs. That is, in addition to on/off (1 and 0) digital inputs and outputs, these pins can accept analog inputs. The range of the analog inputs spans 0 to 5 volts.

■ Pins 25 and 26 are additional I/O lines that are available if you solder connections directly to the BX-24 chip (as such as they are not strictly "pins," but we'll treat them as if they were). Access to pins 25 and 26 are provided via plated-through holes, which can be connected to wires or pin headers. These two pins share I/O with the on-board red and green LEDs.

■ Pin 27 serves as the output capture I/O line. As with pins 25 and pin 26, this pin is available if you solder directly to the BX-24 chip.

Programming the BX-24

To program the BX-24 you need to purchase the BasicX-24 developer's kit, which contains one BX-24, a programming cable, a power supply, a "carrier board" (see Fig. 32.3), and programming software on CD-ROM. You plug the BX-24 into the carrier board, which has a 24-pin socket and empty solder pads that you can use to add your own circuitry. The programming cable connects between the carrier board and a serial port on your PC. The power supply is the "wall wart" variety and provides about 12–16 vdc.

The BX-24 uses a proprietary programming environment, consisting of an editor and a download console, which also serves double duty as a terminal for data sent from the

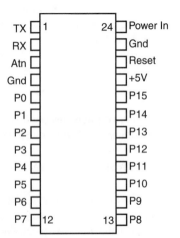

FIGURE 32.2 Pinout diagram of the BasicX-24 chip. Note that several of the pins serve double duty (as explained in the text).

FIGURE 32.3 The easiest way to experiment with the BX-24 is to use the carrier board that is included as part of the BX-24 developer's kit. The carrier board includes a DB-9 connector for hooking the system up to a PC for programming.

microcontroller. The program editor, shown in Fig. 32.4, supports the BasicX language, which is a subset of Microsoft Visual Basic. Don't expect all Visual Basic commands to be available in BasicX, however. BasicX supports the same general syntax as Visual Basic, and many of the same data types (bytes, integers, strings, and so forth).

If you're familiar with Visual Basic then you should feel right at home with BasicX. The BasicX language supports the usual control structures, such as *If...End If, While...Wend, For—Next,* and *Select...Case.* Your BasicX programs can be subroutines, and you can call those subroutines from anywhere in the program.

Depending on how you've used Visual Basic, however, you may discover that BasicX is far less forgiving of certain programming habits. BasicX uses a "strict" data-typing syntax that requires you to use the *Dim* statement—or one of its variations, such as *Const*—to define each variable before it is used. With the *Dim* statement you must also indicate the variable type, such as *Byte* or *String.*

Modern versions of Visual Basic support a special type of variable called the *variant.* Variants can hold most any kind of data, which allows you to freely "mix and match" data types, such as adding an integer to a string (i.e., adding the number one to the name

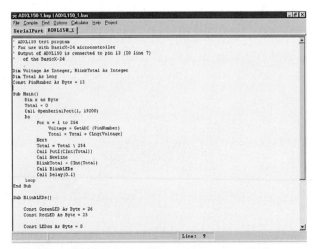

FIGURE 32.4 Use the BasicX program editor to create, edit, compile, and (optionally) download programs for the BX-24.

"Smith" to get "Smith1"). Apart from the danger that you will introduce bugs by mixing data types, variants consume a lot of memory. They also tend to slow down execution speed, since it must determine the type of variable each time it is accessed.

Visual Basic provides the variant feature because memory is abundant on PC systems, and—at least with the latest machines—processor speed is fairly fast. Conversely, memory in a microcontroller must be carefully rationed. The BX-24 supports 400 bytes (that's bytes, not megabytes or even kilobytes) of RAM memory to store data. For a microcontroller, that's actually a copious amount of memory! (By the way, if you're wondering, your programs are stored separately in a 32K block of EEPROM, which is enough for some 8000 instructions. You'll be hard-pressed to create programs that large for your robot.)

When using BasicX, you must be constantly aware of the data type being stored in each variable. If you need to manipulate two variables that contain different types of data, you must remember to use the various data conversion commands that BasicX supports. This is perhaps one of the most frustrating aspects of BasicX programming for newcomers.

A particularly nice feature of the BasicX editor is that it allows you to build "projects" consisting of multiple files. This allows you, for instance, to build a library of commonly used programming functions that you may regularly use in your robotics work. When building a new program for the BX-24, you create a new project and then include any constituent files. This saves you from having to manually cut and paste commonly used code to make one big program file.

Advanced programmers will appreciate the ability to work with real arrays in the BX-24 environment. You can create arrays of any data type except strings or other arrays. You can then reference the elements of the array using an index number. This feature makes it handy to manipulate such things as data streams, where you want to store a series of bytes in one compact package.

Before you can send your programs to the BX-24 chip they must be compiled. This is done in the BasicX editor by choosing the *Compile* command from the *Compile* menu.

Compiling can take a while on slower machines, so be patient. Syntax errors are flagged, and if they are found, compiling stops. When you have successfully compiled the program it can then be downloaded to the BX-24 chip. This can be done from the BasicX editor or from the download console. After the program has been successfully compiled, it can be redownloaded any number of times. It does not need to be recompiled before each download.

Multitasking with the BX-24

One of the more valuable uses subroutines provide is the ability to create multitasking programs. Multitasking is a built-in feature of the BasicX operating system. In most instances, the multitasking is "preemptive," meaning that the BasicX operating system forces the BX-24 microcontroller to "time-slice" between each multitasked subroutine. Each slice is given 1/512 of a second, more than enough to complete over a hundred instructions before moving on to the next subroutine. (The BX-24 processes some 65,000 instructions per second, or approximately 127 instructions per time-slice.) A few of the commands supported in the BasicX system suspend multitasking because they are sensitive to timing. These include such commands as *InputCapture* (explained later in this chapter), which accurately measures the duration of signals received by the BX-24.

While multitasking is a powerful feature of the BX-24, it's not always easy to implement. For each subroutine that you wish to multitask you must manually calculate the amount of RAM needed to hold data for that subroutine while the system switches. This calculation is necessary so sufficient "stack space" is allocated to hold the data as the BX-24 services each task. If you underestimate the RAM requirements, your program won't work properly; if you overestimate the requirements, you waste precious memory.

BasicX Functions for Robotics

The BX-24 is a general-purpose microcontroller, so many of its built-in features are geared toward any typical personal or commercial microcontroller application. Still, a number of features of the BasicX programming language lend themselves for use in robotics. These features are implemented as functions added to the BasicX language. To use a feature, you merely include it in your program along with any necessary command parameters.

Note: Several of these functions require you to use version 1.45 or later of the BasicX compiler. If you're already a BX-24 owner, you'll also need to make sure that your chip has the latest BasicX operating system firmware embedded into it. Check the BasicX site (www.basicx.com) for details.

REAL-TIME CLOCK

The BX-24 contains its own real-time clock (RTC), accurate to within several seconds per day. You must set the correct time whenever you power up the BX-24, but once the time is set, you can use the RTC to measure events. For example, you can write a robot program that accurately marks the time it takes to travel from one room to another. The RTC is also handy for data logging, which allows your robot to roam around the house or yard and store data from its sensors. Coupled with the BX-24's ability to optionally store data in EEPROM, the data log will survive even if power is removed to the chip.

GETADC AND PUTDAC

As mentioned earlier, the BX-24 has its own eight-channel, 10-bit ADC. With the *GetADC* function, you can read a voltage level on any of eight I/O pins and correlate that voltage level with a binary number (from 0 to 1023). Conversely, you can use the *PutDAC* function to output a pulse train that will mimic a variable voltage.

SHIFTIN AND SHIFTOUT

With *ShiftIn* you can receive a series of bits on a single I/O pin and convert them to a single byte in a variable. *ShiftOut* does the inverse, converting a byte into a series of bits. Both functions allow you to specify an I/O pin to be used as the data source and another I/O pin for the clock. The BasicX software automatically triggers the clock pin for each bit received or sent. The *ShiftIn* and *ShiftOut* functions are particularly handy when you are using serially based components, which allow you to interface with devices using only two I/O lines.

OPENCOM

The BX-24 supports as many serial ports as you have available I/O pins. With *OpenCom* you can establish serial communications with other BX-24 chips or any other device that supports serial data transfer. One common use for *OpenCom* is to establish a link from the BX-24 chip back to the download window of your PC. This window can serve as a terminal for debugging and other monitoring tasks.

PULSEIN AND PULSEOUT

The *PulseIn* function waits for the level at a given I/O pin to change state. One practical application of this feature is to activate some function on your robot when a critical button is pressed. *PulseOut* sends a pulse of a certain duration (in 1.085 microsecond units) out a given I/O pin. *PulseOut* is one of the most commonly used functions and is used to blink LEDs, trigger sonar pings, and command servo motors to move to a new location. Note that both *PulseIn* and *PulseOut* turn off the task-switching feature of the BX-24. Several other BasicX functions behave in the same way because they literally "take over" the chip. Because these functions hog processor time, both can also cause errors in the real-time clock.

INPUTCAPTURE

Somewhat akin to *PulseIn*, *InputCapture* watches for signal transition on a specific I/O pin of the BX-24. *InputCapture* can time the duration of these transitions, thereby giving you a "snapshot" of a digital pulse train, including how long each pulse lasted. One application of *InputCapture* is watching for and decoding the serial signals from an infrared remote control.

PLAYSOUND

The *PlaySound* function outputs a waveform that, when connected to an amplifier via a decoupling capacitor, allows you to play previously sampled sound that has been stored in the EEPROM. You can play back sounds at various sampling rates and control the number of times the sound is repeated. The repeat function is a handy way to stretch a relatively short sound sample into a longer one—for example, the "chug-chug" of a machine motor or a series of blips.

ADDITIONAL USEFUL FUNCTIONS FOR ROBOTICS

In addition to BX-24's built-in functions, you can access many of the internal hardware registers of the BX-24 chip. The BX-24 is based on the Atmel AT90S8535 microcontroller (download the data sheet for the '8535 to learn more about the internals of this powerful chip). By controlling the hardware registers of the BX-24 you can program features that the BasicX language itself does not directly support. For example, by setting a few registers for Timer1 (one of three timers in the Atmel '8535), you can produce dual pulse width modulated (PWM) signals, which are useful for controlling the speed of DC motors. In a practical circuit, you will need to interface the two PWM outputs of the BX-24 to a suitable transistor or H-bridge circuit in order to provide enough drive current to run the motors.

Working directly with the hardware registers of the BX-24 is not for the feint of heart, however. If you want to try this technique, first study the Atmel AT90S8535 data sheet and learn how the registers of the chip work. It's entirely possible to set the registers in a way that will crash the chip, rendering it inoperative (of course, you can always reset the BX-24 and try again with a new program).

A Sample BX-24 Program

Constructing a BX-24 program involves at least one subroutine, called Main, and one or more BasicX commands. In the following program example, the BX-24 flashes its red and green LEDs on and off several times each second.

LISTING 32.1

```
Sub Main()
' BX-24 LED demonstration.
Const GreenLED As Byte = 26
Const RedLED As Byte = 25
Const LEDon As Byte = 0
Const LEDoff As Byte = 1

Do
    '  Red pulse.
    Call PutPin(RedLED, LEDon)
    Call Delay(0.07)
    Call PutPin(RedLED, LEDoff)

    Call Delay(0.07)

    ' Green pulse.
```

```
        Call PutPin(GreenLED, LEDon)
        Call Delay(0.07)
        Call PutPin(GreenLED, LEDoff)

        Call Delay(0.07)
    Loop
    End Sub
```

Here's how the program works. The following commands,

```
Sub Main()
   . . .
End Sub
```

form the main subroutine that is automatically executed when the BX-24 is first turned on or when it is reset. You can have additional subroutines in the program, each with a different name, but at a minimum you need one subroutine called *Main* to get things started:

```
Const GreenLED As Byte = 26
Const RedLED As Byte = 25
Const LEDon As Byte = 0
Const LEDoff As Byte = 1
```

These lines define four constants, using the *Const* statement (similar to *Dim*). *Const* stands for "constant" and represents a variable that will never be changed again in the program. In this example, each *Const* statement defines three things:

- The name of the variable, such as *GreenLED* or *LEDon.*
- The type of variable (how many bits it requires). In all four instances the variables are of type *Byte* and each requires eight bits
- The value of each variable. For example, *GreenLED* is assigned the value 26; *LEDoff* is assigned the value 0.

All four constants are used elsewhere in the program, and they serve as a convenient way to change values should that ever be necessary. The statements,

```
Do
   . . .
Loop
```

set up an "infinite loop." That is, the loop repeats for as long as power is applied to the BX-24 (or until the chip is reset). Without the *Do...Loop* statements the commands in the program would execute just once. The loop provides a simple way to repeat the commands indefinitely:

```
' Red pulse.
Call PutPin(RedLED, LEDon)
Call Delay(0.07)
Call PutPin(RedLED, LEDoff)
Call Delay(0.07)
```

Each BasicX function, such as *PutPin,* is preceded by an optional *Call* statement. This tells the BasicX operating system to perform the named function. The *PutPin* function, called twice in this example, changes the state of a specified I/O line. Note the use of the constants. The syntax for *PutPin* is as follows:

```
PutPin (PinNumber; Value)
```

where *PinNumber* is the number of the pin you want to use (e.g., pin 25 for the red LED), and *Value* is either 1 for on (or logical HIGH) or 0 for off (or logical LOW).

The *Delay* function causes the BX-24 to pause a brief while, in this case 70 milliseconds. *Delay* is called twice, so there is a period of time between the on/off flashing of each LED:

```
' Green pulse.
Call PutPin(GreenLED, LEDon)
Call Delay(0.07)
Call PutPin(GreenLED, LEDoff)
Call Delay(0.07)
```

The process is repeated for the green LED.

Controlling RC Servos with the BX-24

You can easily control RC servos with the BX-24 using a few simple statements. While there is no built-in "servo command" as there is with the OOPic microcontroller (see Chapter 33), the procedure is nevertheless very easy to do in the BX-24. Here's a basic program that places a servo connected to pin 20 of the BX-24 at its approximate mid-point position. (I say "approximate" because the mechanics of RC servos can differ between makes, models, and even individual units):

```
Sub Main
Do
    Call PulseOut(20, 1.5E-3, 1)
    Call Delay(0.02)
Loop
End Sub
```

The program continuously runs because it's within an infinite *Do* loop. The *PulseOut* statement sends a short 1.5-millisecond (ms) HIGH pulse to pin 20. The *Delay* statement causes the BX-24 to wait 20 milliseconds before the loop is repeated all over again. With a delay of 20 milliseconds, the loop will repeat 50 times a second (50 * 20 milliseconds = 1000 milliseconds, or one second).

Note the optional use of scientific notation for the second parameter of *PulseOut*. Using the value 0.0015 would yield the same result. You should be aware that the BX-24 supports two versions of the *PulseOut* statement: a float version and an integer version:

■ The *float* version is used with floating-point numbers, that is, numbers that have a decimal point.
■ The *integer* version is used with integers, that is, whole numbers only.

The BX-24 compiler automatically determines which version to use based on the data format of the second parameter of the *PulseOut* statement. If you use

```
Call PulseOut(20, 20, 1)
```

it tells the BX-24 you want to send a pulse of 20 "units." A unit is 1.085 microseconds long; 20 units would produce a very short pulse of only 21.7 microseconds. To continue working in more convenient milliseconds, be sure to use the decimal point:

```
Call PulseOut(20, 0.020, 1)
```

This creates a pulse of 20 milliseconds in length.

Listing 32.2 shows a more elaborate servo control program and is based on an application note provided on the BasicX Web site. This program allows you to specify the position of the servo shaft as a value from 0 to 100, which makes it easier for you to use.

LISTING 32.2

```
Const ServoPin As Byte = 20
Const RefreshPeriod As Single = 0.02
Const NSteps As Integer = 100
Dim SetPosition As Byte
Dim Position As Single, PulseWidth As Single

Sub Main ()
' Moves a servo by sending a single pulse.
' Insert position as a value from 0 to 100
SetPosition = 50       ' move to mid-point

Position = CSng(SetPosition) / CSng(NSteps)
Do
      ' Translate position to pulse width, from 1.0 to 2.0 ms
      PulseWidth = 0.001 + (0.001 * Position)

      ' Generate a high-going pulse on the servo pin
      Call PulseOut(ServoPin, PulseWidth, 1)
      Call Delay(RefreshPeriod)
Loop
End Sub
```

The five lines at the beginning of the program set up all the variables that are used. The line

```
Const ServoPin As Byte = 20
```

creates a byte-sized constant and also defines the value of the constant as pin 20. Because it is a constant, the value assigned to *ServoPin* cannot be changed elsewhere in the program. Similarly, the lines

```
Const RefreshPeriod As Single = 0.02
Const NSteps As Integer = 100
```

create the constants *RefreshPeriod* and *NSteps*. *RefreshPeriod* is a single-precision floating-point number, meaning that it can accept numbers to the right of the decimal point. *Nsteps* is an integer and can accept values from -32768 to $+32767$.

The main body of the program begins with *Sub Main.* The statement

```
SetPosition = 50
```

sets the desired position of the servo relative to the total number of steps defined in *NSteps* (in the case of our example, 100). Therefore, a *SetPosition* of 50 will move the servo to its approximate midpoint. The line

```
Position = CSng(SetPosition) / CSng(NSteps)
```

produces a value from 0.0 to 1.0, depending on the number you used for *SetPosition.* With a value of 50, the *Position* variable will contain 0.5. The *Position* variable is then used within the *Do* loop that follows. Within this loop are the following statements:

```
PulseWidth = 0.001 + (0.001 * Position)
Call PulseOut(ServoPin, PulseWidth, 1)
Call Delay(RefreshPeriod)
```

The first statement sets the pulse width, which is between 1.0 and 2.0 milliseconds. The *PulseOut* statement sends the pulse through the indicated servo pin (the third parameter, 1, specifies that the pulse is positive-going, or HIGH). Finally, the *Delay* statement delays the BX-24 for the *RefreshPeriod,* in this case 20 milliseconds (0.02 seconds).

Reading Button Inputs and Controlling Outputs

A common robotics application is reading an input, such as a button, and controlling an output, such as an LED, motor, or other real-world device. Listing 32.3 shows some simple code that reads the value of a momentary push button switch connected to I/O pin 20. The switch is connected in a circuit, which is shown in Fig. 32.5, so when the switch is open, the BX-24 will register a 0 (LOW), and when it's closed the BX-24 will register a 1 (HIGH).

The instantaneous value of the switch is indicated in the LED. The LED will be off when the switch is open and on when it is closed.

LISTING 32.3

```
Sub Main()
Const InputPin As Byte = 20
Const LED As Byte = 26
Dim State as Byte
Sub Main()
Do
        ' Read I/O pin 20
        State = GetPin(InputPin)
        ' Copy it to the LED
        Call PutPin(LED, State)
```

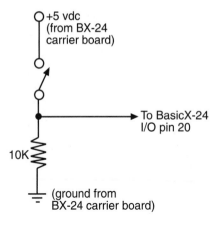

FIGURE 32.5 Wire the switch so it connects to the V+ (pin 21, *not* pin 24) of the BX-24. The resistors are added for safety.

```
Loop
End Sub
```

Now let's see how the program works. The lines,

```
Const InputPin As Byte = 20
Const LED As Byte = 26
Dim State as Byte
```

set the constant *InputPin* as I/O pin 20, and the constant LED as I/O pin 26. (Recall that one of the BX-24's on-board LEDs—the green one, by the way—is connected to I/O pin 26.) Finally, the variable *State* is defined as type *Byte:*

```
Do
        ' Read I/O pin 20
        State = GetPin(InputPin)
        ' Copy it to the LED
        Call PutPin(LED, State)
Loop
```

The *Do* loop repeats the program over and over. The *GetPin* statement gets the current value of pin 20, which will either be LOW (0) or HIGH (1). The companion *PutPin* statement merely copies the state of the input pin to the LED. If the switch is open, the LED is off; if it's closed, the LED is on.

Additional BX-24 Examples

So far we've just scratched the surface of the BX-24's capabilities. But fear not: throughout this book are several real-world examples of BX-24 being using in robotic applications. For instance, in Chapter 41 you'll learn how to use the BX-24 to interface to a sophisticated accelerometer sensor. In addition, you can find several application notes for the BX-24 (and its "sister" microcontrollers, such as the BX-01) on the BasicX Web page (www.basicx.com).

From Here

To learn more about...	*Read*
Stepper motors	Chapter 19, "Working with Stepper Motors"
How servo motors work	Chapter 20, "Working with Servo Motors"
Different approaches for adding brains to your robot	Chapter 28, "An Overview of Robot 'Brains'"
Connecting the OOPic microcontroller to sensors and other electronics	Chapter 29, "Interfacing with Computers and Microcontrollers"

33

USING THE OOPIC
MICROCONTROLLER

While the Basic Stamp described in Chapter 31 is a favorite among robot enthusiasts, it is not the only game in town. Hardware designers who know how to program their own microcontrollers can create a customized robot brain using state-of-the-art devices such as the PIC16CXXX family or the Atmel AVR family of eight-bit RISC-based controllers. The reality, however, is that the average robot hobbyist lacks the programming skill and development time to invest in custom microcontroller design.

Recognizing the large market for PIC alternatives, a number of companies have come out with Basic Stamp work-alikes. Some are pin-for-pin equivalents, and many cost less than the Stamp or offer incremental improvements. And a few have attempted to break the Basic Stamp mold completely by offering new and unique forms of programmable microcontrollers.

One fresh face in the crowd is the OOPic (pronounced "OO-pick"). The OOPic uses *object-oriented* programming rather than the "procedural" PBasic programming found in the Basic Stamp. The OOPic—which is an acronym for *O*bject-*O*riented *P*rogrammable *I*ntegrated *C*ircuit—is said to be the first programmable microcontroller that uses an object-oriented language. The language used by the OOPic is modeled after Microsoft's popular Visual Basic. And, no, you don't need Visual Basic on your computer to use the OOPic; the OOPic programming environment is completely stand-alone and available at no cost.

The OOPic, shown in Fig. 33.1, has built-in support for 31 input/output (I/O) lines. With few exceptions, any of the lines can serve as any kind of hardware interface. What enables them to do this is what the OOPic documentation calls "*hardware objects*," digital I/O lines that can be addressed individually or by nibble (4 bits), by byte (8 bits), or by word (16 bits). The OOPic also supports predefined objects that serve as analog-to-digital conversion inputs, serial inputs/outputs, pulse width modulation outputs, timers-counters, radio-controlled (R/C) servo controllers, and 4x4-matrix keypad inputs. The device can even be networked with other OOPics as well as with other components that support the Philips I2C network interface.

The OOPic comes with a 4K EEPROM for storing programs, but memory can be expanded to 32K, which will hold some 32,000 instructions. The EEPROM is "hot swappable," meaning that you can change EEPRPOM chips even while the OOPic is on and running. When a new EEPROM is inserted into the socket, the program stored in it is immediately started.

Additional connectors are provided on the OOPic for add-ins such as floating-point math; precision data acquisition; a combination DTMF, modem, musical-tone generator; a digital thermometer; and even a voice synthesizer (currently under development). The OOPic's hardware interface is an open system. The I2C interface specification, published by Philips, allows any IC that uses the I2C interface to "talk" to the OOPic.

While the hardware capabilities of the OOPic are attractive, its main benefit is what it offers robot hackers: Much of the core functionality required for robot control is already embedded in the chip. This feature will save you time writing and testing your robot con-

FIGURE 33.1   The OOPic supports 31 I/O lines and runs on 6–12 vdc power. Connectors are provided for the I/O lines, programming cable, memory sockets, and Philips I2C network.

trol programs. Instead of needing several dozen lines of code to set up and operate an RC servo, you need only about four lines when programming the OOPic.

A second important benefit of the OOPic is that its various hardware objects are multitasking, which means they run independently and concurrently of one another. For example, you might command a servo in your robot to go to a particular location. Just give the command in a single statement; your program is then free to activate other functions of your robot—such as move another servo, start the main drive motors, and so forth. Once started by your program, all of these functions are carried out autonomously by the objects embedded within the OOPic. This simplifies the task of programming and makes the OOPic capable of coordinating many hardware connections at the same time.

Fig. 33.2 shows a fire-fighting robot that uses several networked OOPics as its main processor. This two-wheeled robot hunts down small fires and literally snuffs them out with a high-powered propeller fan.

FIGURE 33.2 This fire-fighting robot, built by OOPic developer Scott Savage, uses three OOPics wired together in a network to control the machine's central command, sensors, and locomotion.

Objects and the OOPic

Mention the term *object-oriented programming* to most folks and they freeze in terror. Okay, maybe that's an exaggeration, but object-oriented programming seems like a black art to many, full of confusing words and complicated coding. Fortunately, the OOPic avoids the typical pitfalls of object-oriented programming. The OOPic chip supports an easy-to-use programming language modeled directly after Microsoft Visual Basic, so if you already know VB, you'll be right at home with the OOPic. Future versions of the OOPic software development platform will support C and Java syntax for those programmers who prefer these languages.

The OOPic VB-like language offers some 41 programming commands. That's not many commands actually, but it's important to remember that the OOPic doesn't derive its flexibility from the Basic commands. Rather, the bulk of the chip's functionality comes from its built-in 31 objects. Each of these objects has multiple *properties, methods,* and *events.* You manipulate the OOPic's hardware objects by working with these properties, methods, and events. The Basic commands are used for program flow.

Here's a sample OOPic program written in the chip's Basic language. I'll review what each line does after the code sample. This short program flashes a red LED on and off once a second. Fig. 33.3 shows how to connect the LED and a current-limiting resistor to I/O line 1 (pin 7 on the I/O connector) of the OOPic.

```
Dim RedLED As New oDio1

Sub Main()
RedLED.IOLine = 1
RedLED.Direction = cvOutput
Do
    RedLED.Value = OOPic.Hz1
Loop
End Sub
```

These lines comprise a complete, working program. Here's the program broken down:

```
Dim RedLED As New oDio1
```

The *Dim* statement creates a new *instance* of a particular kind of digital I/O object. This I/O object, referred to as *oDio1,* has already been defined within the OOPic. All of the behaviors of this object have been preprogrammed; your job is to select the behavior you want and activate the object. Note that all of the OOPic's object names start with a lowercase letter *O,* such as *oDio1, oServo,* and *oPWM.*

```
Sub Main()
    ...
End Sub
```

The main body of every OOPic program resides within a subroutine called *Main.* OOPic Basic permits you to add additional subroutines to your program, but every program must have a *Main* subroutine. As with Microsoft's Visual Basic, you refer to subroutines by name.

```
RedLED.IOLine = 1
RedLED.Direction = cvOutput
```

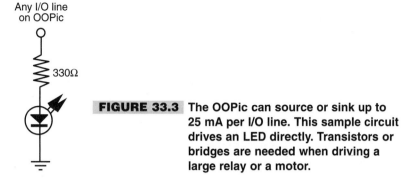

Any I/O line
on OOPic

330Ω

FIGURE 33.3 The OOPic can source or sink up to
25 mA per I/O line. This sample circuit
drives an LED directly. Transistors or
bridges are needed when driving a
large relay or a motor.

These two lines set up the I/O line connected to the *RedLED* object. In this case, we've
defined that the *RedLED* object is connected to I/O line 1 and that this object will serve as
an output (*cvOutput* is a predefined constant; you don't need to define its value ahead of
time). All digital I/O lines can be defined as either input or output. The OOPic does not
reserve certain lines as outputs and others as inputs.

```
Do
    RedLED.Value = OOPic.Hz1
Loop
```

The statement *RedLED.Value = OOPic.Hz1* makes the LED flash once a second. The *Do*
loop is used to keep the program running, so the LED continues to flash. Note the
OOPic.Hz1 value that is assigned to the *RedLED* object: *OOPic* is a built-in "system
object" that is always available to your programs. One property of the *OOPic* object is *Hz1*,
which is a one-bit value that can be used, for example, to change the state of an I/O line
(goes from HIGH to LOW) once a second. The following table describes other properties
of the OOPic system object you may find useful.

OOPIC PROPERTY	WHAT IT DOES
ExtVRef	Specifies the source of the voltage reference for the analog-toldigital module.
Hz1	1-bit value that cycles every 1 Hz.
Hz60	1-bit value that cycles every 60 Hz.
Node	Used when two or more OOPics "talk" to each other via the I2C network. A Node value of more than 0 is the OOPic's I2C network address.
Operate	Specifies the power mode of the OOPic.
Pause	Specifies if the program flow is suspended.
PullUp	Specifies the state of the internal pull-up resisters on I/O lines 8 –15.
Reset	Resets the OOPic.
StartStat	Indicates the cause of the last OOPic reset.

Using and Programming the OOPic

Other than a 6–12 *VDC* power source, you don't need any other components to begin using the OOPic. For adequate current handling when operating under battery power, I suggest that you use a set of eight alkaline AA batteries in a suitable holder. The OOPic Starter Package comes with a nine-volt transistor battery clip; you can use this clip with Radio Shack's part number 270-387 eight-cell AA battery holder. The holder has connectors for the transistor battery clip.

You can develop programs for the OOPic using a proprietary but free development software (see Fig. 33.4). The development software works under Windows 9x and NT, and it self-installs all the necessary system files.

To program the OOPic you connect a cable between the parallel port of your PC and the programming port of the OOPic. The programming cable is provided as part of the OOPic Starter Package or you can make your own by following the instructions provided on the OOPic home page (*www.oopic.com/*). Once you've written a program in the development software, it is compiled and downloaded through the programming cable. The OOPic is then ready to begin executing your program. Because the OOPic stores the downloaded program in nonvolatile EEPROM, the program will remain in the OOPic's memory until you erase it and replace it with another.

OOPic Objects That Are Ideal for Use in Robotics

Though the OOPic is meant as a general-purpose microcontroller, many of its objects are ideally suited for use with robotics. Of the built-in objects of the OOPic, the oA2D, oDiox, oKeypad, oPWM, oSerial, and oServo objects are probably the most useful for robotics work. In the following descriptions, the term *property* refers to the behavior of an object, such as reading or setting the current value of an I/O line.

ANALOG-TO-DIGITAL CONVERSION

The oA2D object converts a voltage that is present on an I/O line and compares it to a reference voltage. It then generates a digital value that represents the percentage of the volt-

FIGURE 33.4 Programs are written for the OOPic using a Windows-based software development platform. You open, save, debug, and compile your OOPic programs using pull-down menu commands.

age in relation to the reference voltage. The *Operate* property of the oA2D object initiates the conversion, and the *Value* property is updated with the result of the conversion. When the Operate value of the oA2D object is 1, the analog-to-digital conversion, along with the *Value* update, occurs repeatedly. Conversion ceases when the *Operate* property is changed to 0.

There are four physical analog-to-digital circuits implemented within the OOPic. They are available on I/O lines 1 through 4.

DIGITAL I/O

Several digital I/O objects are provided in 1-bit, 4-bit, 8-bit, or 16-bit blocks. In the case of the 1-bit I/O object (named oDio1), the *Value* property of the object represents the electrical state of a single I/O line. In the case of the remaining digital I/O objects, the *Value* property presents the binary value of all the lines of the group (4, 8, or 16, depending on the object used).

There are 31 physical 1-bit I/O lines implemented within the OOPic. The OOPic offers six physical 4-bit I/O groups, three 8-bit groups, and one 16-bit group.

R/C SERVO CONTROL

The oServo object outputs a servo control pulse on any IO line. The servo control pulse is tailored to control a standard radio-controlled (R/C) servo and is capable of generating a logical high-going pulse from 0 to 3 ms in duration in 1/36 ms increments.

A typical servo requires a five-volt pulse in the range of 1–2 ms in duration. This allows for a rotational range of 180°. The duration of the control pulse is determined by setting the *Value, Center,* and *InvertOut* properties of the oServer object. The *Value* property controls the position of the servo while the *Center* property adjusts the control pulse time to compensate for mechanical alignment. An *InvertOut* property is used to reverse the direction that the servo turns in response to the *Value* and *Center* properties. We will say more about servo control in a bit.

KEYPAD INPUT

The oKeypad object splits two sets of four I/O lines in order to read a standard 4x4-keypad matrix. The four row lines are individually and sequentially set low (0 volts) while the four column lines are used to read which switch within that row is pressed.

If any switch is pressed, the *Value* property of the oKeypad object is updated with the value of the switch. A *Received* property is used to indicate that at least one button of the keypad is pressed. When all the keys are released, the *Received* property is cleared to 0.

PULSE WIDTH MODULATION

The oPWM object provides a convenient pulse width modulated (PWM) output that is suitable for driving motors (through an appropriate external transistor output stage, of course). The oPWM object lets you specify the I/O line to use—up to two at a time for PWM output, the cycle frequency, and the pulse width.

ASYNCHRONOUS SERIAL PORT

The oSerial object transmits and receives data at a baud rate specified by the *Baud* property. The baud rate can be either 1200, 2400, or 9600 baud. The oSerial object is used to communicate with other serial devices, such as a PC or a serial LCD display.

Using the OOPic to Control a Servo Motor

Though R/C servo motors are intended to be used in model airplanes, boats, and cars, they are equally useful for robotics applications. Servo motors are inexpensive—basic models cost under $15 each—and they combine in one handy package a DC motor, a gearbox, and control electronics. The typical servo motor is designed to rotate 180° (or slightly more) in order to control the steering wheel on a model car or the flight control surfaces on an R/C airplane. For robotics, a servo can be connected to an armature to operate a gripper, to an arm or leg, and to just about anything else you can imagine.

SERVO MOTORS: IN REVIEW

Let's review the way servos operate so we can better understand how you can interface them to the OOPic. An R/C servo consists of a reversible DC motor. The high-speed output of the motor is geared down by a series of cascading reduction gears that can be made out of plastic, nylon, or metal (usually brass, but sometimes aluminum). The output shaft of the servo is connected to a potentiometer, which serves as the closed-loop feedback mechanism. A control circuit in the servo uses the potentiometer to accurately position the output shaft.

Servos use a single pulse width modulated (PWM) input signal that provides all the information needed to control the positioning of the output shaft. The pulse width varies from a nominal 1.25 milliseconds (ms) to roughly 1.75 ms, with 1.5 milliseconds representing the "center" (or neutral) position of the servo output shaft (note that servo specs vary; these are typical). Lengthening the pulse width causes the servo to rotate in one direction; shortening the pulse width causes the servo to rotate in the other direction. The position of the potentiometer acts to "null out" the input pulses, so when the output shaft reaches the correct location the motor stops.

R/C servos are engineered to accept a standard TTL-level signal, which typically comes from a receiver mounted inside a model car or plane. The OOPic can interface directly to an R/C servo and requires no external components such as power transistors.

CONTROLLING SERVOS VIA OOPIC CODE

You can theoretically control up to 31 servos with one OOPic—one servo per IO line. However, the more practical maximum is no more than 8 to 10 servos. The reason: Servos require a constant stream of pulses, or else they cannot accurately hold their position. The ideal pulse stream is at 30 to 60 Hz, which means that to operate properly each servo connected to the OOPic must be "updated" 30 to 60 times per second. The OOPic is engineered to provide pulses at 30-Hz intervals; with more than about eight servos the refresh rate is reduced to 15 Hz. While most servos will still function with this slow refresh rate, a kind of "throbbing" can occur if the motor is under load.

Some robotic projects call for controlling a half-dozen or more servos, such as the six-legged Hexapod II from Lynxmotion (which requires 12 servos working in tandem). However, the typical experimental robot uses only two or four servos. The OOPic is ideally suited for this task, and programming is easy. To operate a servo, you need only provide a few lines of setup code, then indicate the position of the servo using a positioning value from 0 to 63. This value corresponds to the 0–180° movement of the servo output shaft.

With 64 steps the OOPic is able to position a servo with 2.8° of accuracy. This assumes a maximum rotation of 180°, which not all servos are capable of. Note that if you need greater resolution than this you can make use of the OOPic's built-in pulse width modulation object, which can be programmed to provide your servos with far greater positional accuracy. However, for most applications, the OOPic's servo object provides adequate resolution and is easier to use.

Listing 33.1 shows a program written in the OOPic's native Basic syntax and demonstrates how to control an R/C servo using the oServo object. Fig. 33.5 shows how to connect the servo to the OOPic.

LISTING 33.1.

```
' OOPic servo demonstrator
' Uses a standard R/C servo

' This program cycles a servo, connected to IOLine 31,
'   for full rotation (0 to 180 degrees)

' Dimension needed objects
Dim S1 As New oServo
```

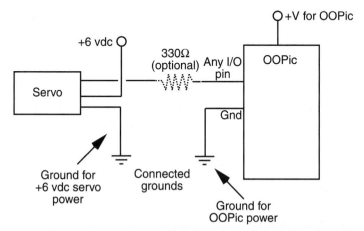

FIGURE 33.5 Follow this basic wiring diagram to connect a standard R/C servo to the OOPic. Most servos use consistent color coding for their wiring: black for ground, red for V+, and yellow or white for input (signal).

```
Dim x As New oByte
Dim i As New oNibble
'_____--
'First routine called when power is turned on
Sub Main()
Call Setup ' set up servo properties
For i = 1 to 5    ' repeat motions five times
     S1 = 0       ' set servo to 0 degrees, and wait a while
     Call longdelay
     S1 = 63      ' set servo to 180 degrees, and wait a while
     Call longdelay
Next i
End Sub
'_____--
' Delay loop routine
Sub longdelay()
For x = 1 To 200:Next x
End Sub
'_____--
' Setup routine
Sub Setup()
S1.Ioline = 31           ' Set servo to I/O line 31 (pin 26)
S1.Center = 31           ' Set center to 31 (experiment for best
                           results)
S1.Operate = cvTrue      ' Turn servo on
End Sub
```

POWERING THE SERVOS

Note that separate battery power supplies were used for the OOPic and the servo. Most hobby R/C servos are designed to be operated with 4.5 to 7.2 vdc. Connecting both OOPic and servo to a single 6-volt supply can cause the OOPic to reset itself. Most servos draw considerable current when turned on, and this current can cause the supply voltage of a 6-volt battery pack to sag below the 4.5-volt level required by the OOPic. When the voltage drops below 4.5 volts, the OOPic's built-in brownout circuit kicks in, which resets the processor. This repeats continuously, and the net effect is a nonfunctioning circuit.

One alternative is to power the whole shebang from a single 9- or 12-volt supply, but with higher voltage comes overpowered servos. Not all servos are built to handle the extra speed and heat caused by the higher voltage, and an early death for your servos could result. Therefore, it's best to use two different batteries. The OOPic is fine operating from a single 9-volt transistor battery. The servo runs from a set of four AA batteries.

HOW THE OOPIC SERVO CODE WORKS

The first three lines in Listing 33.1 "dimension" (create in memory) the objects used in the OOPic program. *S1* is the servo object; *x* and *I* are simple data objects that hold eight and four bits, respectively. The program itself begins with the *Main* subroutine, which is automatically run when the OOPic is first turned on or when it is reset. The first order of business is to call the *Setup* subroutine, located at the end of the program. In *Setup,* the program establishes that IO line 31 (pin 26 of the OOPic chip) is connected to the control input of the servo.

The servo is then centered using a value of 31 (half of 64, considering 0 as the first valid digit). You need to experiment to find the mechanical center of the servo you are using.

Each servo, particularly those that have different sizes and come from different manufacturers, can have a different mechanical center. Therefore, adjust this value up or down accordingly. Finally, the servo object is activated using the statement

```
S1.Operate = cvTrue
```

Notice the use of *properties* when working with the OOPic's objects. Properties are defined by specifying the name of the object, such as *S1* for servo 1, a period (known as the *member operator* in programming parlance), then the property name. So, *S1.Ioline* sets (or reads) the IO line property for the *S1* object. Similarly, *S1.Center* sets the center property, and *S1.Operate* turns the *S1* object on or off. Most OOPic properties are read and write, meaning that you can both set and read their value. A few are read-only or write-only.

Once you have set the servo up, you can manipulate it using the *S1.Value* property. In the demonstration program, the *Value* property is inferred because it is the so-called default property for servo objects. Therefore, it is only necessary to specify the name of the object and the value you want for it:

```
S1 = 0
```

This sets the servo all the way in one direction, and the following expression,

```
S1 = 63
```

sets the servo all the way in the other direction. Because the *Value* property is the default for the oServo object, the statement *S1 = 63* is the same as writing *S1.Value = 63*.

Note:

Exercise care when playing around with servos. Not all servos can travel a full 90° from center, especially if you have not properly set the mechanical center using the *S1.Center* property. For initial testing, use values slightly higher than 0 and slightly lower than 63 to represent the minimum and maximum servo movements, respectively. Otherwise, the OOPic may command the servo to move past an internal stop position, which can cause the gears to slip and grind. Left in this state the servo can be permanently damaged.

Operating Modified Servos

As designed, R/C servos are meant to travel in limited rotation, up to 90° to either side of some center point. But by modifying the internal construction of the servo, it's possible to make it turn freely in both directions and operate like a regular-geared DC motor. This modification is handy when you want to use servo motors for powering your robot across the floor.

The steps for modifying servos vary, but the general process is about the same:

1. Remove the case of the servo to expose the gear train, motor, and potentiometer. This is accomplished by removing the four screws on the back of the servo case and separating the top and bottom.

2. File or cut off the nub on the output gear that prevents full rotation. This typically requires removing one or more gears, so you should be careful not to misplace any parts. If necessary, make a drawing of the gear layout so you can replace things in their proper location!

3. Remove the potentiometer, and replace it with two 2.7K-ohm 1 percent (precision) resistors, wired as detailed in Chapter 20. This fools the servo into thinking it's always in the "center" position. Or relocate the potentiometer to the outside of the servo case, so you can make fine-tune adjustments of the center position. If needed, you can attach a new 5K- or 10K-ohm potentiometer to the circuit board outside the servo.

4. Reassemble the case.

See Chapter 20, "Working with Servo Motors," for a step-by-step tutorial on modifying commonly available servos for continuous rotation.

OOPIC CODE FOR MODIFIED SERVOS

Once modified, you can connect the servo to the OOPic just as you would an unmodified servo (see Fig. 33.5). Listing 33.2 shows how to use the OOPic with two modified servos acting as the drive motors for a two-wheeled robot. You can easily construct a demonstrator robot using LEGO parts, like the prototype shown in Fig. 33.6. I cemented two light-

FIGURE 33.6 You can construct a demonstrator for the OOPic two-wheel robot using LEGO bricks. The servos are glued to small LEGO parts to aid in mounting.

weight R/C airplane wheels to control horns (these come with the servos). I also cemented a 2x8 flat LEGO plate to the side of each servo to make it easier to snap the motors to the LEGO-made frame of the robot.

LISTING 33.2.

```
' OOPic two-motor (servo) robot demonstrator
' Requires the use of modified R/C servos (see text)

' This program cycles the robot through various movements,
'  including forward, backward, right spin, left spin,
'  and turns.

'  Dimension objects
Dim S1 As New oServo
Dim S2 As New oServo
Dim CenterPos as New oByte
Dim Button As New oDio1
Dim x as New oByte
Dim y as New oWord


'_____

Sub Main()
CenterPos = 31                    ' Set centering of servos
Call Setup
Do
      If Button = cvPressed Then
            ' Special program to calibrate servos
            S1 = CenterPos
            S2 = CenterPos
      Else
            ' Main program (IO line is held low)
            Call GoForward
            y = 200               ' Same as LongDelay
            Call Delay            ' Alternative to LongDelay

            Call HardRight
            Call LongDelay

            Call HardLeft
            Call LongDelay

            Call SoftRightForward
            Call ShortDelay

            Call SoftLeftForward
            Call ShortDelay

            Call GoReverse
            Call LongDelay
      End If
Loop
End Sub


'_____

' Set up IO lines and servos
Sub Setup()
Button.Ioline = 7                 ' Set IO Line 7 for function input
Button.Direction = cvInput        ' Make IO Line 7 input
S1.Ioline = 30                    ' Servo 1 on IO line 30
```

```
S1.Center = CenterPos          ' Set center of Servo 1
S1.Operate = cvTrue            ' Turn on Servo 1
S2.Ioline = 31                 ' Servo 2 on IO line 31
S2.Center = CenterPos          ' Set center of Servo 2
S2.Operate = cvTrue            ' Turn on Servo 2
S2.InvertOut = cvTrue          ' Reverse direction of Servo 2
End Sub


'_____--
' Short delay routine
Sub ShortDelay()
        For x = 1 To 80:Next x
End Sub


'_____--
' Long delay routine
Sub LongDelay()
        For x = 1 To 200:Next x
End Sub


'_____--
' Selectable delay routine
Sub Delay()
        For x = 1 To y:Next x
End Sub


'_____--
' Motion routines (forward, back, etc.)
' "Hard" turns spin robot in place
' "Soft" turns turn robot right or left in forward
' (or backward) motion
'_____--

Sub GoForward()
S1 = 0
S2 = 0
End Sub

Sub GoReverse()
S1 = 63
S2 = 63
End Sub

Sub HardRight()
S1 = 0
S2 = 63
End Sub

Sub HardLeft()
S1 = 63
S2 = 0
End Sub

Sub SoftRightBack()
S1 = CenterPos
S2 = 63
End Sub

Sub SoftRightForward()
S1 = 0
S2 = CenterPos
End Sub
```

```
Sub SoftLeftBack()
S1 = 63
S2 = CenterPos
End Sub

Sub SoftLeftForward()
S1 = CenterPos
S2 = 0
End Sub
```

We attached batteries and OOPic to the top of the robot using double-sided tape. Power to the OOPic is provided by a 9-volt battery; power to both servos is provided by a 6-volt pack of AAs. Note that I used a wire-wrap board as a terminal bus, and standard .100"-center connectors instead of hard-soldering any wiring to the various components. This makes it easier to test the robot and possibly add to it at a later date.

REVIEWING THE PROGRAM CODE

The program in Listing 33.2 is a modified version of the program in Listing 33.1. Its main difference, other than employing two oServo objects instead of one, is that the "center" position is used to turn the motor off. Values greater than this center position cause the servos to rotate in one direction; values less than the center position cause the servos to rotate in the opposite direction. The servos are made to turn one direction when their *Value* property is 0 and the other direction when their *Value* property is 63.

Note that in Listing 33.2 the "normal" direction of travel for servo 2 (S2) is reversed from S1, with the following statement:

```
S2.InvertOut = cvTrue
```

This is handy because in the two-wheeled robot the servos are mounted on opposite sides, and therefore one motor must operate in mirror image to the other. That is, one must turn clockwise while the other turns counterclockwise to move the robot forward or backward. Without the *InvertOut* property, you'd have to set the *Value* property of one servo to 0 and the other to 63 to maintain proper forward or backward motion.

Not shown in Listing 33.2 is a useful feature you may want to implement: values very close to the center position (+/- about five steps) will cause the servos to slow down by a proportional amount. For example, if the center position is 31, then a value of 32 for *S1* or *S2* may cause that servo to rotate clockwise very slowly. Higher values will modestly increase the speed in the same direction of travel. Conversely, a value of 30 for the *S1* or *S2* object may cause the servo to rotate counterclockwise very slowly. A value of 29 would make the motor go a little faster, and so on.

Listing 33.2 takes the robot through a series of patterned moves, including forward and backward movement, right and left spins, and turns. Delay routines allow you to specify how long each movement is to last. Vary the delay up or down to experiment with different motions. In the prototype for this book, the program in Listing 33.2 moves the robot back and forth about two feet. The program repeats itself until you reset the OOPic or disconnect the power.

The modified servos use an externally accessible trimmer potentiometer. The trimmer pot, which is attached to the case of the servo with a small piece of double-sided foam tape, serves to provide an accurate voltage divider by which the servos can be set to center, or neutral, position. The trimmer pots are set by temporarily taking IO line 7 high. This causes the program in Listing 33.2 to run an alternative routine in its *Main* loop, so you can set the *Center* property of both servos to a value of 31. The pots are then adjusted so that the motors just stop—this represents the center position. Using the potentiometer makes it *much* easier to calibrate the servos so they can be used with the program. Once calibrated, you can tie IO line 7 low again.

Using the OOPic to Control Stepper Motors

The OOPic is full of pleasant surprises, including the innate ability to control a standard four-phase unipolar stepper motor. Unlike R/C servos, however, the OOPic is not able to directly drive a stepper motor. For that you'll need an interface with a current and voltage rating for the stepper motor you are using. Chapter 19 provides additional information on using stepper motors.

Listing 33.3 shows a simple stepper motor driving program that uses a feature unique to the OOPic: virtual circuits. Instead of programming each of the four phases of a stepper with on/off values in code, this program uses two *processing* objects, oConverter and oCounter. Processing objects are used to construct virtual circuits, which are like real electronic circuits, only they are created solely using programming statements.

LISTING 33.3.

```
' OOPic stepper motor demonstrator
' Uses a standard four-phase unipolar stepper motor
' Operates motor in half-stepping mode

' Dimension objects
Dim Stepper as New oDio4      ' 4-bit IO for controlling stepper
Dim Driver as New oConverter
Dim Position as New oWord    ' 32-bit value for current position
Dim Mover as New oCounter


'_____--
Sub Main()
Call Setup
' The rest of your code here

' To reverse motor, use Mover.Direction = cvNegative
' or Mover.Direction = cvPositive
' To stop-and-hold motor, use Mover.Operate = cvFalse
' To restart motor, use Mover.Operate = cvTrue
' To stop and de-energize motor, use Driver.Blank = 1
End Sub


'_____--
' Set up stepper motor
Sub Setup()
Stepper.IOGroup = 1         ' Set stepper to use IO group 1 (pins 8-11)
Stepper.Nibble = 0          ' Picks lower 4 lines from IO group
Stepper.Direction = cvOutput    ' Make lines outputs
Driver.Output.Link(Stepper.Value) ' Set up virtual circuit
```

```
Driver.Input.Link(Position.Value)
Driver.Mode = cvPhase
Driver.Operate = cvTrue
Mover.ClockIn1.Link(OOPic.Hz60) ' Use OOPic 60 Hz object for stepping
Mover.Output.Link(Position.Value)
Mover.Operate = cvTrue       ' Enable counter
End Sub
```

The stepper motor program in Listing 33.3 demonstrates one of the uses for the oConverter numeric-conversion object. This program has the built-in "behavior" of being able to construct the proper phasing to control the forward and backward rotation of a four-phase unipolar stepper motor. The program also uses a counter object, which allows you to define the number of steps you wish to apply to the motor. Keep in mind that the oConverter object specifies an eight-phase cycle, which has the effect of moving the motor in half-step increments (this serves to improve the accuracy and torque of the motor). So, for example, if the motor is rated at 200 steps per revolution, it will require 400 pulses from the OOPic to turn it a full 360° degrees.

Experiment with the OOPic and you'll find it's a capable performer in the field of robotics. By using its objects judiciously, coupled with a liberal sprinkling of virtual circuits, you should be able to construct most any kind of robotic creature using a minimum number of external components.

From Here

To learn more about...	*Read*
Stepper motors	Chapter 19, "Working with Stepper Motors"
How servo motors work	Chapter 20, "Working with Servo Motors"
Different approaches for adding brains to your robot	Chapter 28, "An Overview of Robot 'Brains'"
Connecting the OOPic microcontroller to sensors and other electronics	Chapter 29, "Interfacing with Computers and Microcontrollers"

34

REMOTE CONTROL SYSTEMS

The most basic robot designs—just a step up from motorized toys—use a wired control box on which you flip switches to move the robot around the room or activate the motors in the robotic arm and hand. The wire link can be a nuisance and acts as a tether preventing your robot from freely navigating through the room. You can cut this physical umbilical cord and replace it with a fully electronic one by using a remote control receiver and transmitter.

This chapter details several popular ways to achieve links between you and your robot. You can use the remote controller to activate all of the robot's functions, or with a suitable on-board computer working as an electronic recorder, you can use the controller as a teaching pendant. You manually program the robot through a series of steps and routines, then play it back under the direction of the computer. Some remote control systems even let you connect your personal computer to your robot. You type on the keyboard, or use a joystick for control, and the invisible link does the rest.

Control Your Robot with an Atari-style Joystick

This easy project does not require you to assemble a remote control circuit out of ICs and components, and it is perfect when you want to control five or fewer functions. It uses a wired Atari 2600-style joystick. Though the Atari 2600 hasn't been sold for a long time, joysticks for it are still common finds at surplus stores. The joystick is designed to work with the Atari 2600 game machine and all computers that accept Atari-type (switch contact) joysticks. It is not intended for use with the Apple II or IBM PC, which use potentiometer-type joysticks. (If you're lucky, you may find wireless Atari-style joysticks. These come in two parts: a transmitter and receiver.)

Fig. 34.1 shows the functions of the pins on the joystick cable. To interface the joystick to your robot, wire the pins as shown in Fig. 34.2. You can interface to TTL or CMOS gates, but for any application where you want to drive heavy loads use relays or opto-isolators. An opto-isolator setup is shown in Fig. 34.2.

You can connect the joystick to discrete circuitry or to a computer or microprocessor. For example, you can use a Basic Stamp (see Chap. 31, "Using the Basic Stamp") to translate the joystick movements to motor control. Pushing the joystick forward ("Up" on the joystick) might makes both motors go forward. Pushing the joystick back (Down) might make both motors go backward, and so forth.

Build a Joystick "Teaching Pendant"

No doubt you've been to Disneyland or other theme parks that use robotic or "animatronic" performers. These on-stage automatons are operated via a sophisticated computerized system that plays back the audio portion of the program and controls every movement or every robot on the stage. Walt Disney was one of the early pioneers of this art and science. *Audio-animatronics,* the system his WED (Walt Elias Disney) Enterprises group developed, used audio tones on recorded tape as the control medium.

Animatronic shows are most commonly "acted out" by a human director who operates a joystick or other control in real time. As the sound portion of the program is played, the director moves the joystick to operate the various animatronic devices on stage. The movements of the joystick are recorded for later playback. This same concept is used in many kinds of manufacturing robots, whose actions are programmed not from the keyboard but from a "teaching pendant," a controller that records the actions of a human operator.

Using an ordinary joystick you can create your own teaching pendant for your robot (or animatron, if that's to your liking). For this next project, I'll use a common garden-variety IBM PC-style analog joystick, though you can apply the same techniques to any kind of joystick, analog or digital. IBM PC-style joysticks are inexpensive (mine cost $5) and

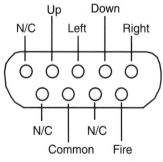

FIGURE 34.1 Pinout diagram for an Atari-style joystick.

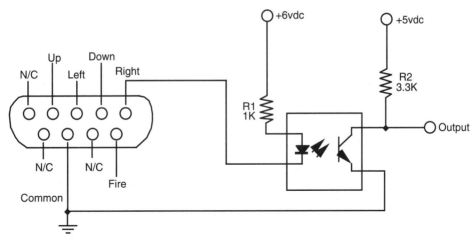

FIGURE 34.2 How to interface the output of the joystick to TTL or CMOS circuitry.

available everywhere. The joystick teaching pendant controls the motors of a two-wheel robot. You can record and play back up to 30 seconds of commands. You can also use the joystick teaching pendant in "free" (no record or playback) mode, where you control the robot by manually pushing the stick.

For the control electronics, we'll connect the joystick to a BasicX-24 microcontroller (see Chapter 32, "Using the BasicX Microcontroller") by way of a simple interface. The joystick interface, (see Fig. 34.3) shows how the joystick interface connects to the BX-24. The BX-24 in turn connects to whatever motor control electronics you are using. See Chapter 18, "Working with DC Motors," for more information on motor drive circuits for DC motors.

IBM PC-style joysticks contain analog potentiometers; the resistive value of these pots changes as you move the joystick around. We actually won't be using the analog nature of the joystick for this project, but you can add this feature in your own if you wish. For example, instead of controlling the power and direction of the motors, you could rig the joystick so that the more you push on the stick, the faster the motor goes.

Listing 34.1 provides the BX-24 code for the joystick teaching pendant.

Note:

This program requires the use of the SerialPort.Bas file, which is included with the BX-24 developer's kit (and available for download at the BasicX site). When creating the project file for the joystick teaching pendant, be sure to include SerialPort.Bas as well.

LISTING 34.1 JOYSTICK2.BAS

```
Private S1 As Byte
Private S2 As Byte
Private JoystickX As Integer
Private JoystickY As Integer
Private Steps As Integer
Private TempByte As Byte
```

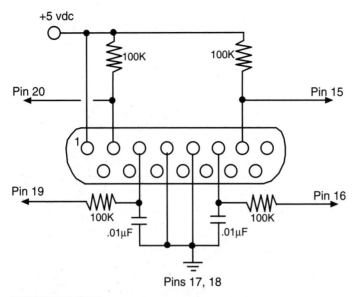

FIGURE 34.3 Connecting an IBM PC-style joystick to the BasicX-24 microcontroller.

```
Private Motors As Byte
Private MotorsStr As String * 10
Private TempStr As String * 20
Private RecordFlag As Boolean
Private PlayFlag As Boolean
Private Const GreenLED As Byte = 26
Private Const RedLED As Byte = 25
Private Const LEDon As Byte = 0
Private Const LEDoff As Byte = 1

Private Const MotLD As Byte = 9
Private Const MotLC As Byte = 10
Private Const MotRD As Byte = 11
Private Const MotRC As Byte = 12

Private Const Min As Integer = 450
Private Const Mid As Integer = 850
Private Const Max As Integer = 1200

Private Const CmdDelay As Integer = 256
Private Const MaxSteps As Integer = 60

Dim RecordArray (1 to MaxSteps) As Integer

        'DLDR four bit format (see text)
        'D=0, forward
        'D=1, reverse
        'L=0, left motor off
        'L=1, left motor on
        'R=0, right motor off
        'R=1, Right motor on

Sub Main()
        Dim Count As Integer
        Call PutPin (17, BxOutputLow)                          ' take low
```

```
Call PutPin (18, BxOutputLow)                        ' take low
Call OpenSerialPort(1, 19200)
RecordFlag = False
PlayFlag = False

Call PutPin (GreenLED, LedOff)
Call Sleep (1.5)
Do
        JoystickX =  GetPotValue (19)                ' X stick
        JoystickY = GetPotValue (16)                 ' Y stick
        Select Case JoyStickY
            Case Min to (Mid-51)                     'Forward
                Select Case JoystickX
                    Case Min to (Mid-51)
                        MotorsStr = "bx00000100"
                        Motors = bx00000100
                    Case (Mid-50) to (Mid+50)
                        MotorsStr = "bx00000101"
                        Motors = bx00000101
                    Case Mid+51 to Max
                        MotorsStr = "bx00000001"
                        Motors = bx00000001
                End Select
            Case (Mid-50) to (Mid+50)                ' Center
                Select Case JoystickX
                    Case Min to (Mid-51)
                        MotorsStr = "bx00001101"
                        Motors = bx00001101
                    Case (Mid-50) to (Mid+50)
                        MotorsStr = "bx00000000"
                        Motors = bx00000000
                    Case Mid+51 to Max
                        MotorsStr = "bx00000111"
                        Motors = bx00000111
                End Select
            Case Mid+51 to Max                       ' Reverse
                Select Case JoystickX
                    Case Min to (Mid-51)
                        MotorsStr = "bx00001100"
                        Motors = bx00001100
                    Case (Mid-50) to (Mid+50)
                        MotorsStr = "bx00001111"
                        Motors = bx00001111
                    Case Mid+51 to Max
                        MotorsStr = "bx00000011"
                        Motors = bx00000011
                End Select
        End Select
        Call SetMotors (Motors)
        Call PutStr (MotorsStr)
        Call Newline
        S1 = GetPin (20)
        S2 = GetPin (15)
        If S1 = 0 Then                  ' button 2, red led, pin 20
          RecordFlag = Not RecordFlag
          Call SetRedLed (RecordFlag)
          If RecordFlag = True Then
              Steps = 1
              TempStr = "***Recording On***"
              Call PutStr (TempStr)
              Call Newline
        End If
        If RecordFlag = False Then
```

```
                    TempStr = "***Recording Off***"
                    Call PutStr (TempStr)
                    Call Newline
                End If
            End If
            If S2 = 0 Then                         ' button 1, green led, pin 15
                PlayFlag = Not PlayFlag
                Call SetGreenLed (PlayFlag)
                If PlayFlag = True Then
                    TempStr = "***Playback On***"
                    Call PutStr (TempStr)
                    Call Newline
                End If
            End If
            If RecordFlag = True Then
                If Steps <= MaxSteps Then
                    RecordArray(Steps) = CInt(Motors)
                    Call PutI (RecordArray(Steps))
                    Call Newline
                    Steps = Steps + 1
                    End If
            End If
            If PlayFlag = True Then
                RecordFlag = False
                Call SetRedLed (RecordFlag)
                For Count = 1 to (Steps - 1)
                    Call PutI (RecordArray(Count))
                    TempByte = CByte(RecordArray (Count))
                    Call SetMotors (TempByte)
                    Call Newline
                    Call Sleep (CmdDelay)
                    If GetPin (15) = 0 Then
                        PlayFlag = False
                        Call PutPin (GreenLED, LedOff)
                        Exit For
                    End If
                        Next
                        PlayFlag = False
                        Call PutPin (GreenLED, LedOff)
                End If
                Call Sleep (CmdDelay)
        Loop
End Sub

Private Sub SetMotors (Motors As Byte)
Select Case Motors
        Case 0                   ' all stop
                Call PutPin (MotLD, 0)
                Call PutPin (MotLC, 0)
                Call PutPin (MotRD, 0)
                Call PutPin (MotRC, 0)
        Case 5                       ' forward
                Call PutPin (MotLD, 0)
                Call PutPin (MotLC, 1)
                Call PutPin (MotRD, 0)
                Call PutPin (MotRC, 1)
        Case 15                      ' reverse
                Call PutPin (MotLD, 1)
                Call PutPin (MotLC, 1)
                Call PutPin (MotRD, 1)
                Call PutPin (MotRC, 1)
        Case 1                         ' right
                Call PutPin (MotLD, 0)
```

```
              Call PutPin (MotLC, 0)
              Call PutPin (MotRD, 0)
              Call PutPin (MotRC, 1)
        Case 7                    ' hard right
              Call PutPin (MotLD, 0)
              Call PutPin (MotLC, 1)
              Call PutPin (MotRD, 1)
              Call PutPin (MotRC, 1)
        Case 4                    ' left
              Call PutPin (MotLD, 0)
              Call PutPin (MotLC, 1)
              Call PutPin (MotRD, 0)
              Call PutPin (MotRC, 0)
        Case 13                   ' hard left
               Call PutPin (MotLD, 1)
               Call PutPin (MotLC, 1)
               Call PutPin (MotRD, 0)
               Call PutPin (MotRC, 1)
        Case 12                   ' left reverse
               Call PutPin (MotLD, 1)
               Call PutPin (MotLC, 1)
               Call PutPin (MotRD, 0)
               Call PutPin (MotRC, 0)
        Case 3                    ' right reverse
              Call PutPin (MotLD, 0)
              Call PutPin (MotLC, 0)
              Call PutPin (MotRD, 1)
              Call PutPin (MotRC, 1)
End Select
End Sub

Private Sub SetRedLed (Flag As Boolean)
If Flag = True Then
     Call PutPin (RedLED, LedOn)
Else
     Call PutPin (RedLED, LedOff)
      RecordFlag = False
End If
End Sub

Private Sub SetGreenLed (Flag As Boolean)
If Flag = True Then
     Call PutPin (GreenLED, LedOn)
Else
     Call PutPin (GreenLED, LedOff)
      PlayFlag = False
End If
End Sub

Private Function GetPotValue(ByVal PinNumber As Byte) As Integer
    Const CapacitorDischargeTime As Integer = 4
    Call PutPin(PinNumber, bxOutputLow)
    Call Sleep(CapacitorDischargeTime)
    GetPotValue = RCtime(PinNumber, 0) ' Timeout returns 0.
End Function
```

USING THE JOYSTICK TEACHING PENDANT

In this section we briefly discuss how to use the joystick teaching pendant software, Joystick.bas. You'll want to read Chapter 32 to learn more about the BasicX-24 chip and how it's programmed. Follow these steps:

1. Insert the BasicX-24 chip into a suitable carrier.
2. Attach the serial programming cable between your PC and the BasicX-24 carrier.
3. Connect a joystick to the BasicX-24, as shown in Fig. 34.3. For best results, use ribbon cables that have the appropriate header connectors. I used a ribbon cable originally designed for use in PCs. It is outfitted with the proper DB-15 connector for the joystick on one end and a 16-pin dual-row male header on the other.
4. Apply power to the BX-24.
5. Create a new project (Joystick2.Bxp) and be sure to add the SerialPort.Bas file as one of its files.
6. Write the Joystick2.Bas program in Listing 34.1. When you are finished, be sure to save it (name it "Joystick2.Bas").
7. Set up the BasicX Development System main program for the proper download port and monitor port.
8. Compile and download the program. The program should automatically run after downloading is complete.

Test the program by pushing the joystick. For the purposes of verification and testing, the Joystick2.Bas program uses the BasicX debug window to display the binary value of the four motor control bits (only the last four bits are used). For example, when you push the joystick forward, the text *bx00000101* is shown in the debug window. The last four bits are *0101:*

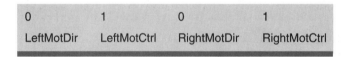

A value of 0 for LeftMotDir/RightMotDir means the motor is going forward (conversely, a value of 1 means the motor is going in reverse). A value of LeftMotCtrl/RightMotCtrl means that motor is activated. With the bits 0101, both motors are operating and are going forward. Note that the program samples the position of the joystick once every half-second.

RECORDING AND PLAYING BACK STEPS

Briefly depress button 1 (usually the "fire" button). The red LED on the BX-24 chip will light up to indicate that recording is on. In addition, a "Recording On" message is displayed in the debug window. The joystick is now in record mode, and the joystick positions are being stored in memory. Recording is simple in the Joystick2.Bas program: each half second the joystick position is stored in one element of a 60-element array. Since there are 60 elements and a new "snapshot" of the joystick controls is made every half second, this means there is a maximum of 30 seconds of recording.

You can revise the program to add longer programming time, but note that the BX-24 has 400 bytes of memory. The more elements there are, the more memory is consumed. As written, Joystick2.Bas consumes about 190 bytes of RAM, so there is room for expansion if you wish.

When you are done recording the steps you want, *briefly* depress button 1 again. The joystick will be taken out of record mode. You can play back your previously stored steps by *briefly* depressing button 2. While a previously stored set of steps is being played any joystick motions are ignored. If necessary you can abort play mode at any time by depressing button 2 again. When playback is complete, the program automatically goes back into "free-run" mode.

POSSIBLE ENHANCEMENTS

There are a number of enhancements you may wish to add. One is to increase the number of steps per second. This is done by decreasing the value as follows:

```
mdDelay As Integer = 256
```

The value 256 is approximately one half of a second, so 128 would be a quarter second, 64 would be an eighth of a second, and so forth. Be aware that the smaller the number is, the more steps are recorded per second, so the 60-element array will fill up faster.

Another enhancement is to add an additional joystick. The DB-15 connector is designed to accept signals from two joysticks at a time when used with a suitable joystick "Y" adapter (available at Radio Shack and elsewhere).

Commanding a Robot with Infrared Remote Control

I still remember the first television remote control I ever saw. The remote control, which looked like something from a 1950s sci-fi movie, was for a Zenith black-and-white TV, made circa 1962. The remote had just two functions: on/off and channel changing. To change the channel you had to keep depressing the channel button until the desired channel appeared (the channels changed up, from 2 through 13, then started over again). What was more amazing than the remote control itself was how it worked: by ultrasonic sound. Depressing one of the control buttons struck a hammer against a tuning fork. A microphone in the TV picked up the high-pitched *ping* and responded accordingly.

These days, remote control of TVs, VCRs, and other electronic devices is taken for granted. And instead of just two functions, the average remote handles dozens—more than the knobs and buttons on the TV or VCR itself. Except for some specialty remotes that use UHF radio signals, today's remote controls operate with infrared (IR) light. Pressing a button on the remote sends a specific signal pattern; the distinctive pattern is deciphered by the unit under control.

You can use the same remote controls to operate a mobile robot. A computer or microcontroller is used to decipher the signal patterns received from the remote via an infrared receiver. Because infrared receiver units are common finds in electronic and surplus stores (they're used heavily in TVs, VCRs, etc.), adapting a remote control for robotics use is actually fairly straightforward. It's mostly a matter of connecting the pieces together. With your infrared remote control you'll be able to command your robot in just about any way you wish—to start, stop, turn, whatever.

SYSTEM OVERVIEW

Here are the major components of the robot infrared remote control system:

- *Infrared remote.* Most any modern infrared remote control will work, but…the signal patterns they use vary considerably. You'll find it most convenient to use a "universal remote control" (about $10 at a department store). Specifically, you want the universal remote to support Sharp TVs and VCRs—99.99 percent do.
- *Infrared receiver module.* The receiver module contains an infrared light detector, along with various electronics to clean up, amplify, and demodulate the signal from the remote control. The remote sends a pattern of on/off flashes of light. These flashes are modulated at about 38–40 kHz in order to reduce interference from other light sources. The receiver strips out the modulation and provides just the on/off flashing patterns.
- *Computer or microcontroller.* You need some hardware to decode the light patterns, and a computer or microcontroller, running appropriate software, makes the job straightforward. For this project we'll use the BasicX-24 microcontroller, from NetMedia. The same programming techniques described here can be used with other microcontrollers and computers, but you'll need to adapt the program accordingly.

INTERFACING THE RECEIVER-DEMODULATOR

The first order of business is to interface the receiver-demodulator to the BX-24 chip. Most any receiver module for 38–42 kHz infrared operation should work well. I've speced out the Sharp GP1U57X because it's widely available, including at Radio Shack. You can also use units from LiteOn, Sony, Everlight, and others. (Hint: one good source for infrared receiver-demodulator units is salvaged VCRs. Carefully unsolder it from the circuit board in the VCR and use it in your robot remote control project.)

Fig. 34.4 shows the interface for the infrared receiver-demodulator. Note the electrolytic capacitor and pull-up resistor. These are required for proper operation. You'll find the circuit will work better if you solder the ground lead to the metal case of the receiver-demodulator (if it is so equipped). Keep lead lengths short.

PROGRAMMING THE BX-24

Listing 34.2 shows the demonstration program to be used with the BX-24 microcontroller. Please refer to Chapter 32, "Using the BasicX Microcontroller," for more information on how to use the BX-24, including how to compile, download, and execute programs. For now, I'll assume you're familiar with all these things and cut right to the chase.

To use the program, be sure to set your universal remote control to output Sharp TV infrared codes. The remote control will come with an instruction booklet on how to select the proper code setting. Look under the TV listing for Sharp and note which code number(s) to use. You may need to try several of them before you will have positive results. For reference, I used a General Electric Universal Remote, model RM94905, and selected Sharp TV code 112.

Important! To use this program you'll need to also include the SerialPort.Bas file as part of the project. This file comes with the BX-24 sample files, which are included with the Developer's Kit and are available for free download from the BasicX Web site (www.basicx.com/).

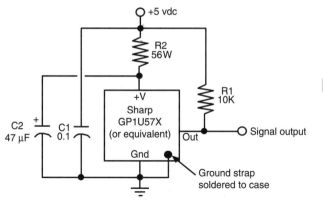

FIGURE 34.4 **The infrared receiver-demodulator requires relatively few external components to interface to the rest of the circuitry.**

LISTING 34.2 SHARPREMOTE.BAS.

```
Option Explicit

Private Const MaxPulses As Integer = 32
Private PulseTrain(1 To MaxPulses) As New UnsignedInteger
Private Const OverflowValue As Long = 65535

'_____
Sub Main()
    Dim I As Integer, Digit As Byte, Success As Boolean
    Dim BitValue As Integer, Value As Integer, Tx As String
    Dim Motors(1 to 9) as Byte, MotorVal As Byte

    ' define motor patterns
    'DLDR format
Motors(1) = bx00000100          ' forward left turn
Motors(2) = bx00000101          ' forward
Motors(3) = bx00000001          ' forward right turn
Motors(4) = bx00001101          ' spin left
Motors(5) = bx00000000          ' stop
Motors(6) = bx00000111          ' spin right
Motors(7) = bx00001100          ' reverse left turn
Motors(8) = bx00001111          ' reverse
Motors(9) = bx00000011          ' reverse right turn

Register.DDRA = bx00001111      ' PortA direction;
                                ' high nibble=input, low nibble=output

Call Initialize

MotorVal = 5
    Register.PORTA = (Register.PORTA AND bx11110000) OR Motors(CInt(MotorVal))
Do
        ' Record IR pulses on the input capture pin (pin 12 on BX-12)
        Call InputCapture(PulseTrain, MaxPulses, 0)

I = 2
        Digit = 1
        Value = 0
        Do
                Call TranslateSpace(PulseTrain(I), BitValue, Success)
                If (Success) Then
                    Value = Value * 2          ' Right shift
                    Value = Value + BitValue   ' Add LSBit
                Else
                    GoTo Continue
```

```
      End If

      I = I + 2                  ' Do even number elements only;
                                 ' spaces are in odd number elements
         Digit = Digit + 1
   Loop While I <= (MaxPulses - 1)

   ' Determine item selected
   ' Tx is for debug window display
   ' MotorVal is the value to use on PortA
   Select Case Value
         Case 16706
             Tx = "0"
             MotorVal = 0
         Case 16898
             Tx = "1"
             MotorVal = 1
         Case 16642
             Tx = "2"
             MotorVal = 2
         Case 17154
             Tx = "3"
             MotorVal = 3
         Case 16514
             Tx = "4"
             MotorVal = 4
         Case 17026
             Tx = "5"
             MotorVal = 5
         Case 16770
             Tx = "6"
           MotorVal = 6
         Case 17282
             Tx = "7"
             MotorVal = 7
         Case 16450
             Tx = "8"
             MotorVal = 8
         Case 16962
             Tx = "9"
           MotorVal = 9
         Case 16802
             Tx = "Power"
             MotorVal = 0
         Case 16930
             Tx = "Channel Up"
             MotorVal = 0
         Case 16674
           Tx = "Channel Down"
             MotorVal = 0
         Case 16546
             Tx = "Volume Up"
           MotorVal = 0
         Case 17058
             Tx = "Volume Down"
             MotorVal = 0
         Case Else
             Tx = "[other]"
   End Select

   ' Set PortA output to motor value (lower four bits only)
   Register.PORTA = (Register.PORTA AND bx11110000) OR _
             Motors(CInt(MotorVal))
   Call PutStr (Tx)                    ' display on debug window
```

```
Call NewLine
Call Delay(0.25)                      ' wait quarter of a second
Continue:
      Loop
End Sub


'_____
' Lifted from NetMedia BasicX code examples
Sub TranslateSpace(ByVal Space As UnsignedInteger, _
      ByRef BitValue As Integer, ByRef Success As Boolean)

   ' Translates the specified space into a binary digit.

   ' Each space must be within this range.
   Const MaxValue As Single = 1700.0E-6
   Const MinValue As Single = 300.0E-6

   ' This is the crossover point between binary 0 and 1.
   Const TripPoint As Single = 1000.0E-6
   Const UnitConversion As Single = 135.63368E-9 ' => 1.0 / 7372800.0
   Dim SpaceWidth As Single
   ' Convert to seconds.
   SpaceWidth = CSng(Space) * UnitConversion

   If (SpaceWidth < MinValue) Or (SpaceWidth > MaxValue) Then
      Success = False
      Exit Sub
   Else
      Success = True
   End If

   If (SpaceWidth > TripPoint) Then
      BitValue = 1
   Else
      BitValue = 0
   End If

End Sub


'_____
Sub Initialize()
   ' Wait for power to stabilize
   Call Delay(0.25)

' Used for serial port communications
Call OpenSerialPort(1, 19200)
End Sub
```

Of critical importance is the *Select Case* structure, which compares the values that are returned from the remote. These were the actual numeric values obtained using the Sharp TV code setting mentioned earlier. If your universal remote doesn't support these same values, you can easily determine the correct values to use for each button press on the remote through the code in Listing 34.3, RemoteTest.Bas. (Only the *Main* subroutine is shown; the other routines in SharpRemote.Bas, given in Listing 34.2, are used as is.)

LISTING 34.3 REMOTETEST.BAS.

```
Sub Main()
     Dim I As Integer, Digit As Byte, Success As Boolean
     Dim BitValue As Integer, Value As Integer, Tx As String
```

```
        Call Initialize
        Do
             Call InputCapture(PulseTrain, MaxPulses, 0)
             I = 2
             Digit = 1
             Value = 0
             Do
                Call TranslateSpace(PulseTrain(I), BitValue, Success)
                If (Success) Then
                    Call PutI(BitValue)
                        Value = Value * 2
                    Value = Value + BitValue
                Else
                    GoTo Continue
                End If
                I = I + 2
                Digit = Digit + 1
            Loop While I <= (MaxPulses - 1)
            Call PutByte (9)
            Call PutI (Value)
            Call Delay(0.5)
            Call NewLine
Continue:
        Loop
End Sub
```

When this program is run, pressing keys on the remote control should yield something like the following on the BasicX Development Program debug window:

```
100001000000010    16898
100000100000010    16642
100001100000010    17154
100000010000010    16514
100001010000010    17026
```

The first set of numbers is the signal pattern as on/off bits. The second set is the numeric equivalent of that pattern. If you see a string of 0s or nothing happens, either the circuit isn't working correctly or the remote is not generating the proper format of signal patterns.

CONTROLLING ROBOT MOTORS WITH THE SHARPREMOTE.BAS PROGRAM

The SharpRemote.Bas program assumes that you're driving the traditional two-motor robot, using DC motors (as opposed to stepper or modified servo motors). To operate a robot, connect a suitable motor driver circuit to pins 17 through 20 of the BX-24 chip. You can use most any motor driver that uses two bits per motor. One bit controls the motor direction (0 is "forward"; 1 is "backward"), and another bit controls the motor's power (0 is off; 1 is on). Chapter 18, "Working with DC Motors," presents a variety of motor drivers that you can use.

Note: Whatever motor drivers you use make sure that you provide adequate bypass filtering. This prevents excess noise from appearing on the incoming signal line from the IR receiver-demodulator. DC motors, particularly the inexpensive kind, generate copious noise from radio frequency (RF) interference as well as "hash" in the power supply lines of the circuit. If you fail to use adequate filtering unpredictable behavior will result. Your best bet is to use opto-coupling, with completely separate battery power supplies for the microcontroller and IR electronics on the one hand and the motors and motor driver on the other.

In SharpRemote.Bas, the following line,

```
Motors(1) = bx00000100
Motors(2) = bx00000101
[etc.]
```

set up the bit patterns to use for the four pins controlling the motors. Yes, the patterns show eight bits. We're only interested in the last four, so the first four are set to 0000. The bits are in "DLDR" format. That is, the left-most two bits control the left motor, and the right-most two bits control the right motor. The *D* represents direction; and *L* and *R* represent left and right, as you'd expect.

After the program has received a code pattern from the remote, it reconstructs that pattern as a 16-bit word. This word is then translated into a numeric equivalent, which is then used in the *Select Case* structure, as in the following example:

```
Select Case Value
        Case 16706
            Tx = "0"
            MotorVal = 5
        Case 16898
            Tx = "1"
            MotorVal = 1
 ....
```

The value *16706* represents the 0 button. When it's pressed, the program stores the string "0" (for display in the debug window) as well as the motor value, 5. Five is used as "stop," with both motors turning off (the numeric keypad on the remote forms a control diamond). The program interprets the value 16898 received from the remote as a 1 and sets the *MotorVal* to 1. The pattern for this value calls for the robot to turn left by turning off its right motor and turning on its left. Review SharpRemote.Bas for other variations, which are self-explanatory.

You will note that several of the buttons on the remote are not implemented and set the MotorVal to 5, or stop. You can add your own functionality to these buttons as you see fit. For example, pressing the *Volume Up/Down or Channel Up/Down* buttons could control the arm on your robot, if it's so equipped.

OPERATING THE ROBOT WITH THE REMOTE

Now that you have the remote control system working and you're done testing, it's time to play! Disconnect the BX-24 chip from its programming cable, set your robot on the ground, and apply all power. In the beginning, the robot should not move. Point the remote control at the infrared receiver-demodulator, and press the *2* button (forward). The robot should move forward. Press *5* to stop. Press other buttons to test out the other features.

GOING FURTHER

Perhaps the BX-24 is not your microcontroller of choice. Or maybe you don't want to use the signal patterns for Sharp TVs and VCRs. You can adapt the receiver-demodulator interface and the SharpRemote.Bas program for use with a wide variety of controllers, computers, and signal pattern formats. Of course, you'll need to revise the program as necessary, and determine the proper bit patterns to use.

You will probably find that the signal patterns used with a great many kinds of remote controls will be usable with the SharpRemote.Bas program. You merely need to test the remote, using the RemoteTest.Bas variation to determine the values to use for each button press. The program still works even if the signal from the remote contains more data bits than the 16 provided. For this application, it doesn't really matter that the last couple of bits are missing—we're not trying to control a TV or VCR but our own robot, and the code values for its control are up to us. The only requirement is that each button on the remote must produce a unique value. Things won't work if pressing 2 and 5 yield the same value.

Multifunction Encoder and Decoder Remote Control

Remote control devices, like those detailed in the previous section, typically use a special function integrated circuit called an *encoder* to generate unique sequences of digital pulses that can then be used to modulate an infrared light beam or radio signal. A matching *decoder,* on the receiving end, translates the digital pulses back into the original format.

For example, pressing the number 5 on the remote control might emit a sequence of 1s and 0s like this:

```
0111000100010111
```

The decoder receives this sequence and outputs a 5. In the previous section you used a BasicX-24 microcontroller as a kind of generic decoder for translating signals from a universal remote control. Another, less expensive, alternative is to use an encoder/decoder IC pair, which are available from a number of manufacturers such as National Semiconductor, Motorola, and Holtek. In this section, we'll examine a system that uses the popular Holtek HT-12E and HT-12D four-bit encoder/decoder. The HT-12 chips are available from a number of sources—you can also substitute most any encoder/decoder pair you wish to use. Most require minimal external parts and cost under $3 each, so you should feel free to experiment.

Fig. 34.5 shows the pinout diagram for the HT-12E encoder. Consult the data sheet for this chip (available at *www.holtek.com,* Web site for the maker of the chip) for a variety of circuit suggestions. The HT chips support up to 256 different addresses. To receive valid data, you must set the same address on both the encoder and the decoder. The data input is four bits (nibble), which you can connect to individual push-button switches. You can press multiple switches at a time for up to 16 different data values.

The output of the HT-12E gates an astable multivibrator circuit that oscillates at approximately 40 kHz. I have specified the use of two high-output infrared LEDs. The higher the output, the longer the range of your remote control system.

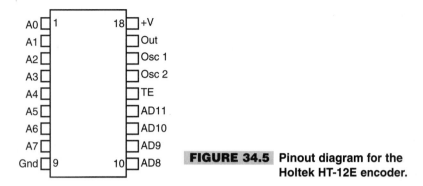

FIGURE 34.5 Pinout diagram for the Holtek HT-12E encoder.

Fig. 34.6 shows the pinout diagram for the HT-12D decoder. When used with infrared communications, the chip is typically connected to a receiver-demodulator that is "tuned" to the 40 kHz modulation from the encoder. When the 40 kHz signal is received, the modulation is stripped off, and only the digital signal generated by the HT-12E encoder remains. This signal is applied to the input of the HT-12D encoder. The Holtek Web page provides data sheets and circuit recommendations for the HT-12D chip.

There are five important outputs for the HT-12D: the four data lines (pins 10-13) and the valid data line (pin 17). The valid data line is normally low. When valid data is received, it will "wink" high then low again. At this point, you know the data on the data lines is good. The data lines are *latching,* which means their value remains until new data is received.

You can use the decoder with your robot in several ways. One way is to connect each of the output lines to a relay. This allows you to directly operate the motors of your robot. As detailed in Chapter 18, "Working with DC Motors," two relays could control the on/off operation of the motors; and two more relays could control the direction of the motors. Chapter 18 also shows you how to use solid-state circuitry and specialty motor driver ICs instead of relays.

Another way to use the decoder is to connect the four lines to a microcontroller, such as the Basic Stamp or the BasicX-24. In this way, you can send up to 16 different commands. Each command could be interpreted as a unique function for your robot.

Using Radio Control Instead of Infrared

If you need to control your robot over longer distances consider using radio signals instead of infrared. You can hack an old pair of walkie-talkies to serve as data transceivers, or even build your own AM or FM transmitter and receiver. But an easier (and probably more reliable) method is to use ready-made transmitter/receiver modules. Ming, Abacom, and several other companies make low-cost radio frequency modules that you can use to transmit and receive low-speed (less than 300 bits per second) digital signals. Fig. 34.7 shows transmitter/receiver boards from Ming. Attached to them are "daughter boards" outfitted with Holtek HT-12E and HT-12D encoder/decoder chips.

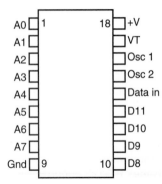

A0	1	+V
A1		VT
A2		Osc 1
A3		Osc 2
A4		Data in
A5		D11
A6		D10
A7		D9
Gnd	9 10	D8

FIGURE 34.6 Pinout diagram for the Holtek HT-12D decoder.

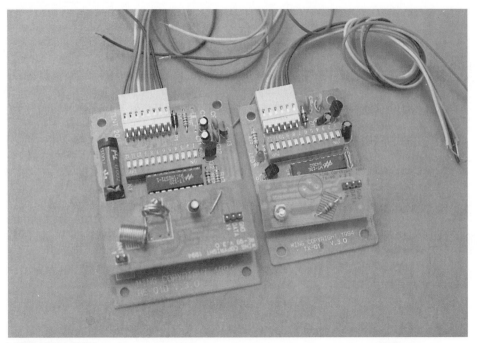

FIGURE 34.7 RF transmitter/receiver modules can be used to remotely control robots from a greater distance than with infrared systems.

The effective maximum range is from 20 to 100 feet, depending on whether you use an external antenna and if there are any obstructions between the transmitter and receiver. More expensive units have increased power outputs, with ranges exceeding one mile. You are not limited to using just encoder/decoders like the HT-12. You may wish to construct a remote control system using DTMF (dual-tone multifrequency) systems, the same technology found in Touch-Tone phones. Connect a DTMF encoder to the transmitter and a DTMF decoder to the receiver. Microcontrollers such as the Basic Stamp can be used as either a DTMF encoder or decoder, or you can use specialty ICs made for the job.

From Here

SENSORS AND NAVIGATION

35

ADDING THE SENSE OF TOUCH

Like the human hand, robotic grippers often need a sense of touch to determine if and when they have something in their grasp. Knowing when to close the gripper to take hold of an object is only part of the story, however. The amount of pressure exerted on the object is also important. Too little pressure and the object may slip out of grasp; too much pressure and the object may be damaged.

The human hand—indeed, nearly the entire body—has an immense network of complex nerve endings that serve to sense touch and pressure. Touch sensors in a robot gripper are much more crude, but for most hobby applications these sensors serve their purpose: to provide nominal feedback on the presence of an object and the pressure exerted on the object.

This chapter deals with the fundamental design approaches for several touch-sensing systems for use on robot grippers—or should the robot lack hands, else-where on the body of the robot. Modify these systems as necessary to match the specific gripper design you are using and the control electronics you are using to monitor the sense of touch.

Note that in this chapter I make the distinction between "touch" and collision. Touch is a proactive event, where you specifically wish the robot to determine its environment by making physical contact. Conversely, collision is a reactive event, where (in most cases) you wish the robot to stop what it's doing when a collision is detected and back away from the condition. Chapter 36, "Collision Avoidance and Detection," deals with the physical contact that results in collision.

Mechanical Switch

The lowly mechanical switch is the most common, and simple, form of tactile (touch) feedback. Most any momentary, spring-loaded switch will do. When the robot makes contact, the switch closes, completing a circuit (or in some cases, the switch opens, breaking the circuit). The switch may be directly connected to a motor or discrete circuit, or it may be connected to a computer or microcontroller, as shown in Fig. 35.1.

You can choose from a wide variety of switch styles when designing contact switches for tactile feedback. Leaf switches (sometimes referred to as Microswitch switches, after a popular brand name) come with levers of different lengths that enhance the sensitivity of the switch. You can also use miniature contact switches, like those used in keyboards and electronic devices, as touch sensors on your robot.

In all cases, mount the switch so it makes contact with whatever object you wish to sense. In the case of a robotic gripper, you can mount the switch in the hand or finger sections. In the case of "feelers" for a smaller handless robot, the switch can be mounted fore or aft. It makes contact with an object as it rolls along the ground. By changing the arrangement of the switch from vertical (see Fig. 35.2), you can have the "feeler" determine if it's reached the edge of a table or a stair landing.

Optical Sensors

Optical sensors use a narrow beam of light to detect when an object is within the grasping area of a gripper. Optical sensors provide the most rudimentary form of touch sensitivity and are often used with other touch sensors, such as mechanical switches.

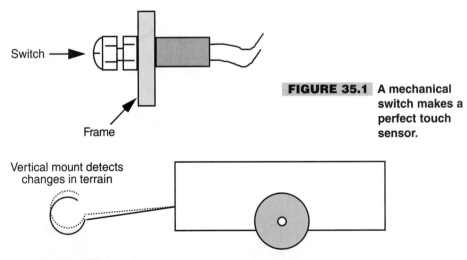

Switch →

Frame

FIGURE 35.1 A mechanical switch makes a perfect touch sensor.

Vertical mount detects changes in terrain

FIGURE 35.2 By orienting the switch to the vertical, your robot can detect changes in topography, such as when it's about to run off the edge of a table.

Building an optical sensor into a gripper is easy. Mount an infrared LED in one finger or pincher; mount an infrared-sensitive phototransistor in another finger or pincher (see Fig. 35.3). Where you place the LED and transistor along the length of the finger or pincher determines the grasping area.

Mounting the infrared pair on the tips of the fingers or pinchers provides for little grasping area because the robot is told that an object is within range when only a small portion of it can be grasped. In most gripper designs, two or more LEDs and phototransistors are placed along the length of the grippers or fingers to provide more positive control. Alternatively, you may wish to detect when an object is closest to the palm of the gripper. You'd mount the LED and phototransistor accordingly.

Fig. 35.4 shows the schematic diagram for a single LED-transistor pair. Adjust the value of R2 to increase or decrease the sensitivity of the phototransistor. You may need to place an infrared filter over the phototransistor to prevent it from triggering as a result of ambient light sources (some phototransistors have the filter built into them already). Use an LED-transistor pair equipped with a lens to provide additional rejection of ambient light and to increase sensitivity.

During normal operation, the transistor is on because it is receiving light from the LED. When an object breaks the light path, the transistor switches off. A control circuit connected to the conditioned transistor output detects the change and closes the gripper. In a practical application, using a computer as a controller, you'd write a short software program to control the actuation of the gripper.

Mechanical Pressure Sensors

An optical sensor is a go/no-go device that can detect only the presence of an object, not the amount of pressure on it. A pressure sensor detects the force exerted by the gripper on

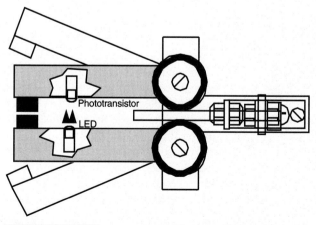

FIGURE 35.3 An infrared LED and phototransistor pair can be added to the fingers of a gripper to provide go/no-go grasp information.

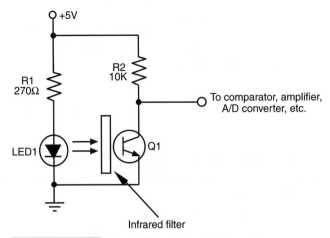

FIGURE 35.4 The basic electronic circuit for an infrared touch system. Note the infrared filter; it helps prevent the phototransistor from being activated by ambient light.

the object. The sensor is connected to a converter circuit, or in some cases a servo circuit, to control the amount of pressure applied to the object.

Pressure sensors are best used on grippers where you have incremental control over the position of the fingers or pinchers. A pressure sensor would be of little value when used with a gripper that's actuated by a solenoid. The solenoid is either pulled in or it isn't; there are no in-between states. Grippers actuated by motors are the best choices when you must regulate the amount of pressure exerted on the object.

CONDUCTIVE FOAM

You can make your own pressure sensor (or transducer) out of a piece of discarded conductive foam—the stuff used to package CMOS ICs. The foam is like a resistor. Attach two pieces of wire to either end of a one-inch square hunk and you get a resistance reading on your volt-ohm meter. Press down on the foam and the resistance lowers.

The foam comes in many thicknesses and densities. I've had the best luck with the semistiff foam that bounces back to shape quickly after it's squeezed. Very dense foams are not useful because they don't quickly spring back to shape. Save the foam from the various ICs you buy and test other types until you find the right stuff for you.

Here's how to make a "down-and-dirty" pressure sensor. Cut a piece of foam 1/4-inch wide by 1-inch long. Attach leads to it using 30-gauge wire-wrapping wire. Wrap the wire through the foam in several places to ensure a good connection, then apply a dab of solder to keep it in place. Use flexible household adhesive to cement the transducer onto the tips of the gripper fingers.

A better way is to make the sensor by sandwiching several pieces of material together, as depicted in Fig. 35.5. The conductive foam is placed between two thin sheets of copper or aluminum foil. A short piece of 30 AWG wire-wrapping wire is lightly soldered onto the foil (when using aluminum foil, the wire is wound around one end). Mylar plastic, like the kind used to make heavy-duty garbage bags, is glued on the outside of the sensor to provide

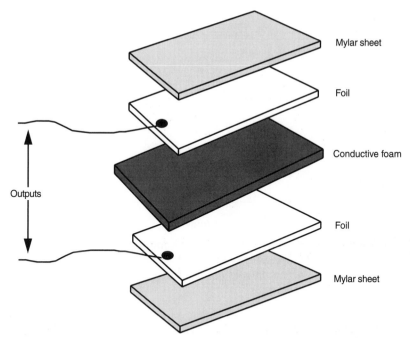

FIGURE 35.5 Construction detail for a pressure sensor using conductive foam. The leads are soldered or attached to foil (copper works best). Choose a foam that has a good "spring" to it.

electrical insulation. If the sensor is small and the sense of touch does not need to be too great, you can encase the foam and foil in heat-shrink tubing. There are many sizes and thicknesses of tubing; experiment with a few types until you find one that meets your requirements.

The output of the transducers changes abruptly when they are pressed in. The output may not return to its original resistance value (see Fig. 35.6). So in the control software, you should always reset the transducer just prior to grasping an object.

For example, the transducer may first register an output of 30K ohms (the exact value depends on the foam, the dimensions of the piece, and the distance between wire terminals). The software reads this value and uses it as the set point for normal (nongrasping) level to 30K. When an object is grasped, the output drops to 5K. The difference—25K—is the amount of pressure. Keep in mind that the resistance value is relative, and you must experiment to find out how much pressure is represented by each 1K of resistance change.

The transducer may not go back to 30K when the object is released. It may spring up to 40K or go only as far as 25K. The software uses this new value as the new set point for the next occasion when the gripper grasps an object.

STRAIN GAUGES

Obviously, the home-built pressure sensors described so far leave a lot to be desired in terms of accuracy. If you need greater accuracy, you should consider commercially available strain gauges. These work by registering the amount of strain (the same as pressure) exerted on various points along the surface of the gauge.

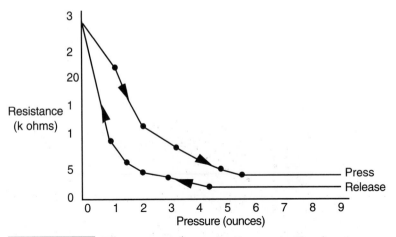

FIGURE 35.6 The response curve for the conductive foam pressure sensor. Note that the resistance varies depending on whether the foam is being pressed or released.

Strain gauges are somewhat pricey—about $10 and over in quantities of 5 or 10. The cost may be offset by the increased accuracy the gauges offer. You want a gauge that's as small as possible, preferably one mounted on a flexible membrane. See Appendix B, "Sources," for a list of companies offering such gauges."

CONVERTING PRESSURE DATA TO COMPUTER DATA

The output of both the homemade conductive foam pressure transducers and the strain gauges is analog—a resistance or voltage. Neither can be directly used by a computer, so the output of these devices must be converted into digital form first.

Both types of sensors are perfect for use with an analog-to-digital converter. You can use one ADC0808 chip (under $5) with up to eight sensors. You select which sensor output you want to convert into digital form. The converted output of the ADC0808 chip is an eight-bit word, which can be fed directly to a microprocessor or computer port. Fig. 35.7*a* shows a the basic wiring diagram for the ADC0808 chip, which can be used with conductive foam transducer; Fig. 35.7*b* shows how to connect a conductive foam transducer to one of the analog inputs of the ADC0808.

Notice the 10K resistor in Fig. 35.7, placed in series between the pressure sensor and ground. This converts the output of the sensor from resistance to voltage. You can change the value of this resistor to alter the sensitivity of the circuit. For more information on ADCs, see Chapter 29, "Interfacing with Computers and Microcontrollers."

Experimenting with Piezoelectric Touch Sensors

A new form of electricity was discovered just a little more than a century ago when the two scientists Pierre and Jacques Curie placed a weight on a certain crystal. The strain on the

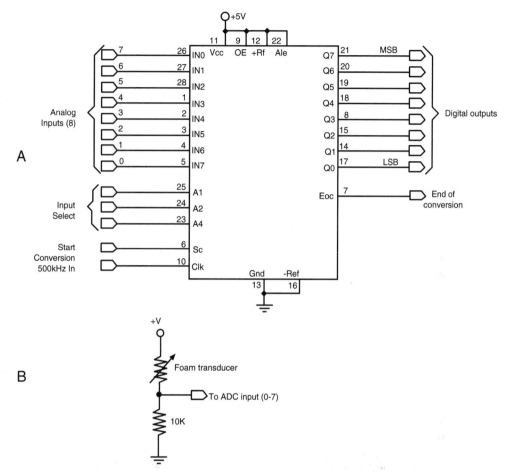

FIGURE 35.7 *a.* **The basic wiring diagram for converting pressure data into digital data, using an ADC0808 analog-to-digital converter (ADC) IC. You can connect up to eight pressure sensors to the one chip.** *b.* **How to interface a conductive foam sensor to the ADC0808 chip.**

crystal produced an odd form of electricity—significant amounts of it, in fact. The Curie brothers coined this new electricity "piezoelectricity"; *piezo* is derived from the Greek word meaning "press."

Later, the Curies discovered that the piezoelectric crystals used in their experiments underwent a physical transformation when voltage was applied to them. They also found that the piezoelectric phenomenon is a two-way street. Press the crystals and out comes a voltage; apply a voltage to the crystals and they respond by flexing and contracting.

All piezoelectric materials share a common molecular structure, in which all the movable electric dipoles (positive and negative ions) are oriented in one specific direction. Piezoelectricity occurs naturally in crystals that are highly symmetrical—quartz, Rochelle salt crystals, and tourmaline, for example. The alignment of electric dipoles in a crystal structure is similar to the alignment of magnetic dipoles in a magnetic material.

When the piezoelectric material is placed under an electric current, the physical distance between the dipoles change. This causes the material to contract in one dimension (or axis) and expand in the other. Conversely, placing the piezoelectric material under pressure (in a vise, for example) compresses the dipoles in one more axis. This causes the material to release an electric charge.

While natural crystals were the first piezoelectric materials used, synthetic materials have been developed that greatly demonstrate the piezo effect. A common human-made piezoelectric material is ferroelectric zirconium titanate ceramic, which is often found in piezo buzzers used in smoke alarms, wristwatches, and security systems. The zirconium titanate is evenly deposited on a metal disc. Electrical signals, applied through wires bonded to the surfaces of the disc and ceramic, cause the piezo material to vibrate at high frequencies (usually 4 kHz and above).

Piezo activity is not confined to brittle ceramics. PVDF, or polyvinylidene fluoride (used to make high-temperature PVDF plastic water pipes), is a semicrystalline polymer that lends itself to unusual piezoelectric applications. The plastic is pressed into thin, clear sheets and is given precise piezo properties during manufacture by—among other things—stretching the sheets and exposing them to intense electrical fields.

PVDF piezo film is currently used in many commercial products, including noninductive guitar pickups, microphones, even solid-state fans for computers and other electrical equipment. One PVDF film you can obtain and experiment with is *Kynar,* available directly from the manufacturer (see Measurement Specialists at *www.msiusa.com* for more information).

Whether you are experimenting with ceramic or flexible PVDF film, it's important to understand a few basic concepts about piezoelectric materials:

- Piezoelectric materials are voltage sensitive. The higher the voltage is, the more the piezoelectric material changes. Apply 1 volt to a ceramic disc and crystal movement will be slight. Apply 100 volts and the movement will be much greater.
- Piezoelectric materials act as capacitors. Piezo materials develop and retain an electrical charge.
- Piezoelectric materials are bipolar. Apply a positive voltage and the material expands in one axis. Apply a negative voltage and the material contracts in that axis.

EXPERIMENTING WITH CERAMIC DISCS

The ubiquitous ceramic disc is perhaps the easiest form of piezoelectric transducer to experiment with. A sample disc is shown in Fig. 35.8. The disc is made of nonferrous metal, and the ceramic-based piezo material is applied to one side. Most discs available for purchase have two leads already attached. The black lead is the "ground" of the disc and is directly attached to the metal itself.

You can use a ceramic disc as an audio transducer by applying an audio signal to it. Most piezo discs will emit sound in the 1K to 10K region, with a resonant frequency of between 3K and 4K. At this resonant frequency the output of the disc will be at its highest.

When the piezo material of the disc is under pressure—even a slight amount—the disc outputs a voltage proportional to the amount of pressure. This voltage is short lived: shortly after the initial change in pressure, the voltage output of the disc will return to 0. A negative voltage is created when the pressure is released (see the discussion of the bipolar nature of piezo materials earlier in the chapter).

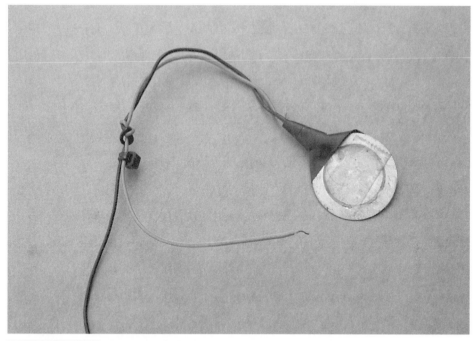

FIGURE 35.8 Piezo ceramic discs are ideally suited to be contact and pressure sensors for robotics.

You can easily interface piezo discs to a computer or microcontroller, either with or without an analog-to-digital converter. Chapter 36, "Collision Avoidance and Detection," discusses several interface approaches. See the section "Piezo Disc Touch Bar" in that chapter for more information.

EXPERIMENTING WITH KYNAR PIEZO FILM

Samples of Kynar piezoelectric film are available in a variety of shapes and sizes. The wafers, which are about the same thickness as the paper in this book, have two connection points, as illustrated in Fig. 35.9. Like ceramic discs, these two connection points are used to activate the film with an electrical signal or to relay pressure on the film as an electrical impulse.

You can perform basic experiments with the film using just an oscilloscope (preferred) or a high-impedance digital voltmeter. Connect the leads of the scope or meter to the tabs on the end of the film (the connection will be sloppy; later in this chapter we'll discuss ways to apply leads to Kynar film). Place the film on a table and tap on it. You'll see a fast voltage spike on the scope or an instantaneous rise in voltage on the meter. If the meter isn't auto-ranging and you are using the meter at a low setting, chances are that the voltage spike will exceed the selected range.

ATTACHING LEADS TO PIEZO FILM

Unlike piezoelectric ceramic discs, Kynar film doesn't usually come with preattached leads (although you can order samples with leads attached, but they are expensive). There

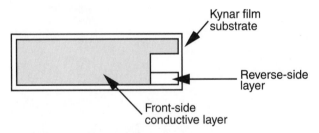

FIGURE 35.9 A close-up look at Kynar piezo film and its electrical contacts.

are a variety of ways to attach leads to Kynar film. Obviously, soldering the leads onto the film contact areas is out of the question. Acceptable methods include applying conductive ink or paint, self-adhesive copper-foil tape, small metal hardware, and even miniature rivets. In all instances, use small-gauge wire—22 AWG or smaller. I have had good results using 28 AWG and 30 AWG solid wire-wrapping wire. The following are the best methods:

- *Conductive ink or paint.* Conductive ink, such as GC Electronics' Nickel-Print paint, bonds thin wire leads directly to the contact points on Kynar film. Apply a small globule of paint to the contact point, and then slide the end of the wire in place. Wait several minutes for the paint to set before handling. Apply a strip of electrical tape to provide physical strength.
- *Self-adhesive copper-foil tape.* You can use copper-foil tape designed for repairing printed circuit boards to attach wires to Kynar film. The tape uses a conductive adhesive and can be applied quickly and simply. As with conductive inks and paints, apply a strip of electrical tape to the joint to give it physical strength.
- *Metal hardware.* Use small 2/56 or 4/40 nuts, washers, and bolts (available at hobby stores) to mechanically attach leads to the Kynar. Poke a small hole in the film, slip the bolt through, add the washer, and wrap the end of a wire around the bolt. Tighten with the nut.
- *Miniature rivets.* Homemade jewelry often uses miniature brass or stainless steel rivets. You can obtain the rivets and the proper riveting tool from many hobby and jewelry-making stores. To use them, pierce the film to make a small hole, wrap the end of the wire around the rivet post, and squeeze the riveting tool (you may need to use metal washers to keep the wire in place).

USING KYNAR AS A MECHANICAL TRANSDUCER

Fig. 35.10 shows a simple demonstrator circuit you can build that indicates each time a piece of Kynar film is struck. Tapping the film produces a voltage output, which is visually indicated when the LED flashes. The 4066 IC is an analog switch. When a voltage is applied to pin 3, the connection between pins 1 and 2 is completed and that finishes the electrical circuit to light the LED. For a robotic application, you can connect the output to a computer or microcontroller.

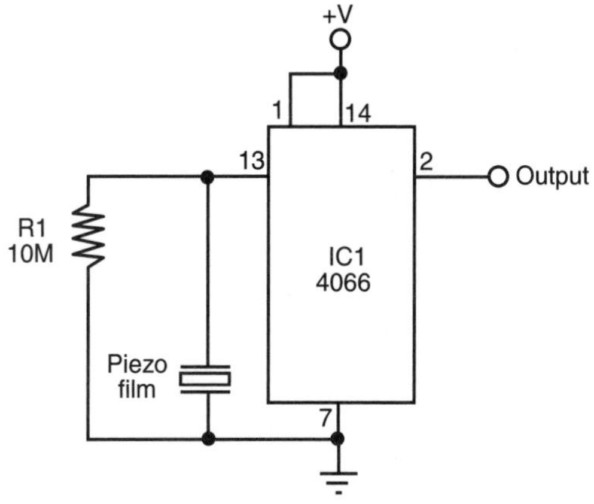

FIGURE 35.10 A strike/vibration indicator using Kynar piezo film.

CONSTRUCTING A KYNAR PIEZO BEND SENSOR

You can easily create a workable touch sensor by attaching one or two small Kynar transducers to a thick piece of plastic. The finished prototype sensor is depicted in Fig. 35.11. The plastic membrane could be mounted on the front of a robot, to detect touch contact, or even in the palm of the robot's hand. Any flexing of the membrane causes a voltage change at the output of one or both Kynar film pieces.

Other Types of "Touch" Sensors

The human body has many kinds of "touch receptors" embedded within the skin. Some receptors are sensitive to physical pressure, while others are sensitive to heat. You may wish to endow your robot with some additional touchlike sensors:

- *Heat sensors* can detect changes in the heat of objects within grasp. Heat sensors are available in many forms, including thermisters (resistors that change their value depending on temperature) and solid-state diodes that are specifically made to be ultrasensitive to changes in temperature. Chapter 39, "Fire Detection Systems," discusses using solid-state temperature sensors.
- *Air pressure sensors* can be used to detect physical contact. The sensor is connected to a flexible tube or bladder (like a balloon); pressure on the tube or bladder causes air to push into or out of the sensor, thereby triggering it. To be useful, the sensor should be sensitive down to about one pound per square inch, or less.
- *Resistive bend sensors,* originally designed for use with virtual reality gloves, vary their resistance depending on the degree of bending. Mount the sensor in a loop, and you can detect the change in resistance as the loop is deformed by the pressure of contact.
- *Microphones and other sound transducers* make effective touch sensors. You can use microphones, either standard or ultrasonic, to detect sounds that occur when objects touch. Mount the microphone element on the palm of the gripper or directly on one of the fingers or pinchers. Place a small piece of felt directly under the element, and

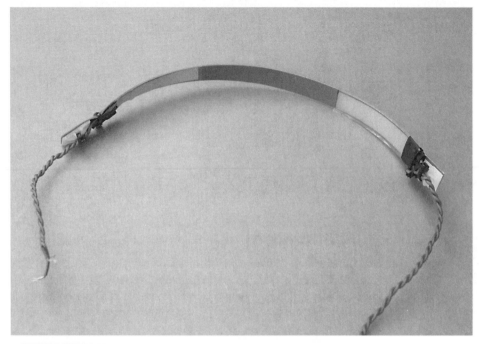

FIGURE 35.11 The prototype Kynar piezo bend sensor.

cement it in place using a household glue that sets hard. Run the leads of the microphone to the sound trigger circuit, which should be placed as close to the element as possible.

From Here

To learn more about...	*Read*
Designing and building robot hands	Chapter 27, "Experimenting with Gripper Designs"
Connecting sensors to computers and microcontrollers	Chapter 29, "Interfacing with Computers and Microcontrollers"
Collision detection systems	Chapter 36, "Collision Avoidance and Detection"
Building light sensors	Chapter 37, "Robotic Eyes"
Fire, heat, and smoke detection for robotics	Chapter 39, "Fire Detection Systems"

36

COLLISION AVOIDANCE AND DETECTION

You've spent hundreds of hours designing and building your latest robot creation. It's filled with complex little doodads and precision instrumentation. You bring it into your living room, fire it up, and step back. Promptly, the beautiful new robot smashes into the fireplace and scatters itself over the living room rug. You remembered things like motor speed controls, electronic eyes and ears, even a synthetic voice, but you forgot to provide your robot with the ability to look before it leaps.

Collision avoidance and detection systems take many forms, and all of the basic systems are easy to build and use. In this chapter, we present a number of passive and active detection systems you can use in your robots. Some of the systems are designed to detect objects close to the robot (called *near-object,* or *proximity, detection*), and some are designed to detect objects at distances of 10 feet or more (called *far-object detection*). All use sensors of some type, which detect everything from light and sound to the heat radiated by humans and animals.

Design Overview

Collision avoidance and collision detection are two similar but separate aspects of robot design. With *collision avoidance,* the robot uses noncontact techniques to determine the proximity and/or distance of objects around it. It then avoids any objects it detects.

Collision detection concerns what happens when the robot has already gone too far, and contact has been made with whatever foreign object was unlucky enough to be in the machine's path.

Collision avoidance can be further broken down into two subtypes: near-object detection and far-object detection. By its nature, all cases of collision detection involve making contact with nearby objects. All of these concepts are discussed in this chapter.

Note: In this book I make a distinction between a robot hitting something in its path ("collision") and sensing its environment tactilely by using grippers or feelers ("touch"). Both may involve the same kinds of sensors, but the goal of the sensing is different. Collision sensing is *reactive* with an emphasis on avoidance; tactile sensing is *active* with an emphasis on exploring. See Chapter 35, "Adding the Sense of Touch," for additional information on the sensors used for deliberate tactile feedback.

Additionally, robot builders commonly use certain object detection methods to navigate a robot from one spot to the next. Many of these techniques are introduced here because they are relevant to object detection, but we develop them more fully in Chapter 38, "Navigating through Space."

NEAR-OBJECT DETECTION

Near-object detection does just what its name implies: it senses objects that are close by, from perhaps just a breath away to as much as 8 or 10 feet. These are objects that a robot can consider to be in its immediate environment; objects it may have to deal with, and soon. These objects may be people, animals, furniture, or other robots. By detecting them, your robot can take appropriate action, which is defined by the program you give it. Your 'bot may be programmed to come up to people and ask them their name. Or it might be programmed to run away whenever it sees movement. In either case, it won't be able to accomplish either behavior unless it can detect objects in its immediate area.

There are two ways to effect near-object detection: proximity and distance:

■ *Proximity* sensors care only that some object is within a zone of relevance. That is, if an object is near enough in the physical scene the robot is looking at, the sensor detects it and triggers the appropriate circuit in the robot. Objects beyond the proximal range of a sensor are effectively ignored because they cannot be detected.

■ *Distance measurement* sensors determine the distance between the sensor and whatever object is within range. Distance measurement techniques vary; almost all have notable minimum and maximum ranges. Few yield accurate data if an object is smack-dab next to the robot. Likewise, objects just outside range can yield inaccurate results. Large objects far away may appear closer than they really are; very close small objects may appear abnormally larger than they really are, and so on.

Sensors have depth and breadth limitations: *depth* is the maximum distance an object can be from the robot and still be detected by the sensor. *Breadth* is the maximum height and width of the sensor detection area. Some sensors see in a relatively narrow zone, typically in a conical pattern, like the beam of a flashlight. Light sensors are a good example. Adding a lens in front of the sensor narrows the pattern even more. Other sensors have specific breadth patterns. The typical passive infrared sensor (the kind used on motion alarms) uses a Fresnel lens that expands the field of coverage on the top but collapses it on the bottom.

This makes the sensor better suited for detecting human motion instead of cats, dogs, and other furry creatures (humans being, on average, taller than furry creatures). The detector uses a pyroelectric element to sense changes in heat patterns in front of it.

FAR-OBJECT DETECTION

Far-object detection focuses on objects that are outside the robot's primary area of interest but still within a detection range. A wall 50 feet away is not of critical importance to a robot (conversely, the same wall one foot away is *very* important). Far-object detection is typically used for area and scene mapping to allow the robot to get a sense of its environment. Most hobby robots don't employ far-object detection because it requires fairly sophisticated sensors, such as narrow-beam radar or pulsed lasers.

The difference between near- and far-object detection is relative. As the designer, builder, and master of your robot, you get to decide the threshold between near and far objects. Perhaps your robot is small and travels fairly slowly. In that case, far objects are those 4 to 5 feet away; anything closer is considered "near." With such a robot, you can employ ordinary sonar distance systems for far-object detection, including area mapping.

In this chapter we'll concentrate on near-object detection methods since traditional far-object detection is beyond the reach and riches of most hobby robot makers (with the exception or sonar systems, which have a maximum range of about 30 feet). You may, if you wish, employ near-object techniques to detect objects that are far away relative to the world your robot lives in.

REMEMBERING THE KISS PRINCIPLE

Engineering texts like to tout the concept of KISS: "Keep It Simple, Stupid." If the admonition is intentionally insulting it is to remind all of us that usually the simple techniques are the best. Of course, "simplicity" is relative. An ant is simple compared to a human being, but so far no scientist has ever created the equivalent of a living ant (some cartoons have come close: who remembers Atom Ant?).

KISS certainly applies to using robotic sensors for object detection. We'd all like to put eyes on our robots to help them see the world the way we do. In fact, such eyes already exist in the form of CCD and CMOS video imagers. They're relatively cheap, too—less than $50 retail. What's missing in the case of vision systems are ways to use the wealth of information provided by the sensor. How do you make a robot differentiate between a can of Dr. Pepper and Mrs. Johnson's slobbering two-year-old —both of which are very wet when tipped over?

When you think about which object detection sensor or system to add to your robot, consider the system's relative complexity in relation to the rest of the project. If all your small 'bot needs is a bumper switch, then avoid going overboard with a $100 sonar system. Conversely, if the context of the robot merits it, don't *under* power your robot with inadequate sensors. Larger, heavier robots cry out for more effective object-detection systems—if for no other reason than to prevent injuring its master if your creation happens to run into you.

REDUNDANCY

Two heads are better than one? Maybe. One thing is for sure: two eyes are definitely better than one. The same goes for ears and many other kinds of sensors. This is *sensor redundancy* at work (having two eyes and ears also provides stereo vision or hearing, which aids in perception). Sensor redundancy—especially for object detection—is not intended primarily to compensate for system failure, the way NASA builds backups into its space projects in case some key system fails 25,000 miles up in space. Rather, sensor redundancy is meant as a way to "smooth out" and balance the results from sensors. If one sensor says an object is 10 feet away and another says it's a foot away, the robot's control computer knows something is amiss and can go about determining the truth.

With only one sensor the robot must blindly (excuse the pun) trust that the sensor data is reliable. This is not a good idea because even for the best sensors data is not 100 percent reliable. There are two kinds of redundancy:

- *Same-sensor redundancy* relies on two or more sensors of an identical type. Each sensor more or less sees the same scene. You can use sensor data in either (or both) of two ways: through *statistical analysis* or *interpolation* (my terms, for better or worse). With statistical analysis, the robot's control circuitry combines the input from the sensors and uses a statistical formula to whittle the data to a most likely result. For example, sensors with wildly disparate results may be rejected out of hand, and the values of the remaining sensors may be averaged out. With interpolation, the data of two or more sensors is combined and cross-correlated to provide a kind of 3-D representation, just like having two eyes and two ears adds depth to our visual and aural senses.
- *Complementary-sensor redundancy* relies on two or more sensors of different types. Since the sensors are fundamentally different—for example, they use completely different collection methods, among other differences—the data from the sensors is always interpolated. For instance, if a robot has both a sonar and an infrared distance-measuring system, it uses both because it understands that for some kinds of objects the data from the infrared system will be more reliable, and for other objects the data from the sonar system will be more reliable.

Budget and time constraints will likely be the limiting factors in whether you employ redundant sensor systems in your robots. So when combining sensors, do so logically: consider which sensors complement others well and if they can be reasonably added. For example, both sonar and infrared proximity sensors can use the same 40 kHz modulation system. If you have one, adding the other need not be difficult, expensive, or time-consuming.

Noncontact Near-Object Detection

Avoiding a collision is better than detecting it once it has happened. Short of building some elaborate radar distance measurement system, the ways for providing proximity detection to avoid collisions fall into two categories: light and sound. In the following section we'll take a closer look at several light- and sound-based techniques.

SIMPLE INFRARED LIGHT PROXIMITY SENSOR

Light may always travel in a straight line, but it bounces off nearly everything. You can use this to your advantage to build an infrared collision detection system. You can mount several infrared "bumper" sensors around the periphery of your robot. They can be linked together to tell the robot that "something is out there," or they can provide specific details about the outside environment to a computer or control circuit.

The basic infrared detector is shown in Fig. 36.1 (refer to the parts list in Table 36.1). This uses an infrared LED and infrared phototransistor. A suitable interface circuit is also shown. The output of the transistor can be connected to any number of control circuits. The comparator circuit for the whisker switches will work nicely and will provide a go/no-go output to a computer. Fig. 36.2 shows how the LED and phototransistor might be mounted around the base of the robot to detect an obstacle like a wall, chair, or person.

Sensitivity can be adjusted by changing the value of R2; reduce the value to increase sensitivity. An increase in sensitivity means that the robot will be able to detect objects farther way. A decrease in sensitivity means that the robot must be fairly close to the object before it is detected.

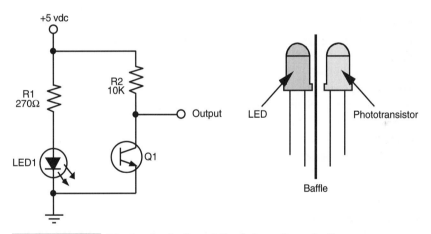

FIGURE 36.1 The basic design of the infrared proximity sensor.

TABLE 36.1 PARTS LIST FOR INFRARED PROXIMITY SWITCH	
R1	270-ohm resistor
R2	10K resistor
Q1	Infrared sensitive phototransistor
LED1	Infrared light-emitting diode
Misc.	Infrared filter for phototransistor (if needed)

All resistors have 5 or 10 percent tolerance, 1/4-watt; all capacitors have 10 percent tolerance, rated 35 volts or higher.

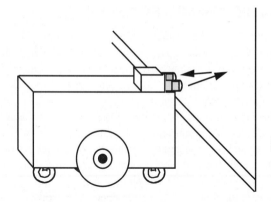

FIGURE 36.2 How the sensor is used to test proximity to a nearby object.

Bear in mind that all objects reflect light in different ways. You'll probably want to adjust the sensitivity so the robot behaves itself best in a room with white walls. But that sensitivity may not be as great when the robot comes to a dark brown couch or the coal gray suit of your boss.

The infrared phototransistor should be baffled—blocked—from both ambient room light as well as direct light from the LED. The positioning of the LED and phototransistor is very important, and you must take care to ensure that the two are properly aligned. You may wish to mount the LED-phototransistor pair in a small block of wood. Drill holes for the LED and phototransistor.

Or, if you prefer, you can buy the detector pair already made up and installed in a similar block. The device shown in Fig. 36.3 is a TIL139 (or equivalent) from Texas Instruments. This particular component was purchased at a surplus store for about $1.

PASSIVE INFRARED DETECTION

You can use commonly available passive infrared detection systems to detect the proximity of humans and animals. These systems, popular in both indoor and outdoor security systems, work by detecting the change in infrared thermal heat patterns in front of a sensor. This sensor uses a pair of pyroelectric elements that react to changes in temperature. Instantaneous differences in the output of the two elements are detected as movement, especially movement by a heat-bearing object, such as a human.

You can purchase pyroelectric sensors—commonly referred to as *PIR,* for passive infrared—new or salvage them from an existing motion detector. When salvaging from an existing detector, you can opt to unsolder the sensor itself and construct an amplification circuit around the removed sensor, or you can attempt to "tap" into the existing circuit of the detector to locate a suitable signal. Both methods are described next.

Using a new or removed-from-circuit detector   Using a new PIR sensor is by far the easiest approach since new PIR sensors will come with a data sheet from the manufacturer (or one will be readily available on the Internet). Some sensors—such as the Eltec 422—have built-in amplification, and you can connect them directly to a microcontroller or computer. Others require extra external circuitry, including amplification and signal filtering and conditioning.

FIGURE 36.3 The Texas Instruments TIL139 infrared emitter/detector sensor unit.
These types of units are often available on the surplus market.

If you prefer, you can attempt to salvage a PIR sensor from a discarded motion detector. Disassemble the motion detector, and carefully unsolder the sensor from its circuit board. The sensor will likely be securely soldered to the board so as to reduce the effects of vibration. Therefore, the unsoldered sensor will have fairly short connection leads. You'll want to resolder the sensor onto another board, being careful to avoid applying excessive heat.

Fig. 36.4 shows a typical three-lead PIR device. The pinouts are not industry standard, but the arrangement shown is common. Pin 1 connects to +V (often 5 volts); pin 2 is the output, and pin 3 is ground. Physically, PIR sensors look a lot like old-style transistors and come in metal cans with a dark rectangular window on top (see Fig. 36.5). Often, a tab or notch will be located near pin 1. As even "unamplified" PIR sensors include an internal FET transistor for amplification, the power connect and output of the sensor are commonly referred to by their common FET pinout names of "drain" and "source":

PIN	NAME
1	Drain
2	Source
3	Ground

If the sensor incorporates an internal output amplifier and signal conditioner, its output will be suitable for direct connection to a microcontroller or other logic input. A buffer circuit, like that shown in Fig. 36.6, is often recommended to increase input impedance. The

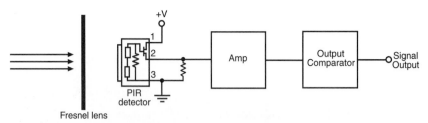

FIGURE 36.4 Most PIR sensors are large, transistorlike devices with a fairly common pinout arrangement. This is a block diagram of how the typical PIR sensor works.

FIGURE 36.5 The PIR sensor has an infrared window on the top to let in infrared heat radiated by objects. Movement of those objects is what the sensor is made to detect, not just the heat from an object.

circuit uses an op amp in unity gain configuration. If the sensor you are using lacks a preamplifier and signal condition, you can easily add your own with the basic circuit shown in Fig. 36.7.

With both the circuits shown in Figs. 36.6 and 36.7 the ideal interface to a robot computer or microcontroller is via an analog-to-digital converter (ADC). Many microcontrollers offer these onboard. If your control circuit lacks a built-in ADC, you can add one using one of the approaches outlined in Chapter 29, "Interfacing with Computers and Microcontrollers."

The output of the PIR sensor will be a voltage between ground and +V. For example's sake, let's assume the output will be the full 0–5 volts, though in practice the actual volt-

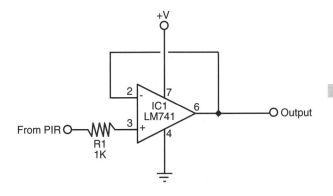

FIGURE 36.6 Use a buffer circuit between the output of the amplified PIR device and the microcontroller or other logic input.

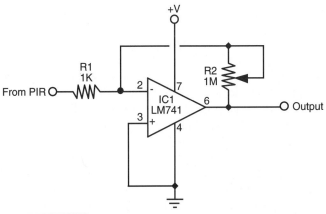

FIGURE 36.7 If the PIR sensor you are using lacks a built-in output amplifier, you can construct one using commonly available op amps.

age switch will be more restricted (e.g., 2.2 to 4.3 volts, depending on the circuitry you use). Assuming a 0–5 vdc output, with no movement detected, the output of the sensor will be 2.5 volts. As movement is detected, the output will swing first in one direction, then the other. It's important to keep this action in mind; it is caused by the nature of the pyroelectric element inside the sensor. It is also important to keep in mind that a heat source, even directly in front of the sensor, will not be detected if it doesn't move. For a PIR device to work the heat source must be in motion. When programming your computer or microcontroller, you can look for variances in the voltage that will indicate a rise or fall in the output of the sensor.

Hacking into a motion detector board Rather than unsolder the PIR sensor from a motion detector unit, you may be able to hack into the motion detector circuit board to find a suitable output signal. The advantage of this approach is that you don't have to build a new amplifier for the sensor. The disadvantage is that this can be hard to do depending on the make and model of the motion control unit that you use.

For best results, use a motion detector unit that is battery powered. This avoids any possibility that the circuit board in the unit also includes components for rectifying and reducing an incoming AC voltage. After disassembling the motion detector unit, connect +5 vdc power to the board. (Note: Some PIR boards operate on higher voltages, usually 9–12 volts. You may need to increase the supply voltage to properly operate the board.) Using a multitester or oscilloscope (the scope is the preferred method) *carefully* probe various points on the circuit board and observe the reading on the meter or scope. Wave your hand over the sensor and watch the meter or scope. If you're lucky, you'll find two kinds of useful signals:

- *Digital (on/off) output.* The output will normally be LOW and will go HIGH when movement is detected. After a brief period (less than one second), the output will go LOW again when movement is no longer detected. With this output you do not need to connect the sensor to an analog-to-digital converter.
- *Analog output.* The output, which will vary several volts, is the amplified output of the PIR sensor. With this output you will need to connect the sensor to an analog-to-digital converter (or an analog comparator).

You may also locate a timed output, where the output will stay HIGH for a period of time—up to several minutes—after movement is detected. This output is not as useful. Fig. 36.8 shows the innards of a hacked motion detector. In this model, I found a suitable analog located near a diode; I then soldered a wire to that diode. If the PIR board you are using operates with a 5 vdc supply, you can connect the wire you added directly to a microprocessor or microcontroller input. If the PIR board operates from a higher voltage, use a logic level translation circuit (see Appendix D for ideas), or connect the wire you added from the PIR board to the coil terminals of a 9 or 12 volt reed relay.

Use care when poking around inside the motion detector. In one unit I tried to hack I accidentally shorted out two pins of an IC, which promptly wrecked the device. Fortunately, I was still able to salvage the PIR sensor itself, so all was not lost.

Using a focusing lens   PIR sensors work by detecting electromagnetic radiation in the infrared region, especially about 5 to 15 micrometers (5000 to 15,000 nanometers). Infrared radiation in this part of the spectrum can be focused using optics for visible light. While you can use a PIR device without focusing, you'll find range and sensitivity are greatly enhanced when you use a lens. Most motion detectors use a specially designed Fresnel lens to focus infrared radiation. The lens, a piece of plastic with grooves, is made to gather more light at the top than at the bottom. With this geometry, when the sensor is mounted high and pointing down the motion detector is more sensitive to movement farther away than right underneath.

If you've gotten your PIR sensor by hacking a motion detector, you can use the same Fresnel lens for your robot. You may wish to invert the lens from its usual orientation (because your robot will likely be near the ground, looking up). Or, you can substitute an ordinary positive diopter lens and mount it in front of the PIR sensor. Chapter 37, "Robotic Eyes," has more information on the use and proper mounting of lenses. Note that, oddly enough, plastic lenses are probably a better choice than glass lenses. Several kinds of glass actively absorb infrared radiation, as do optical coatings applied to finer quality lenses. You may need to experiment with the lens material you use or else obtain a specialty lens (either regular or Fresnel) designed for use with PIR devices.

FIGURE 36.8 A hacked PIR detector, showing the DC-operated circuit board.

ULTRASONIC SOUND

Like light, sound has a tendency to travel in straight lines and bounce off any object in its path. You can use sound waves for many of the same things that light can be used for, including detecting objects. High-frequency sound beyond human hearing (*ultrasonic*) can be used to detect both proximity to objects as well as distance. In Chapter 38 you'll learn how to use an ultrasonic system to measure distance. The project in this section details a simple ultrasonic transmitter and receiver that you can use to detect proximity to nearby objects.

In operation, ultrasonic sound is transmitted from a transducer, reflected by a nearby object, then received by another transducer. The advantage of using sound is that it is not sensitive to objects of different color and light-reflective properties. Keep in mind, however, that some materials reflect sound better than others and that some even absorb sound completely. In the long run, however, proximity detection with sound is a more foolproof approach.

This system is adaptable for use with either a single transmitter/receiver pair or multiple pairs. Ultrasonic transmitter and receiver transducers are common finds in the surplus market and even new cost under $5 each (depending on make and model). Ultrasonic transducers are available from a number of retail and surplus outlets; see Appendix B, "Sources," for a more complete list of electronics suppliers.

You can also mount a single pair of transducers on a scanning platform (also called a turret or carousel), as shown in Fig. 36.9. The scanner can be operated using a standard RC servo (see Chapter 20, "Working with Servo Motors," for more information).

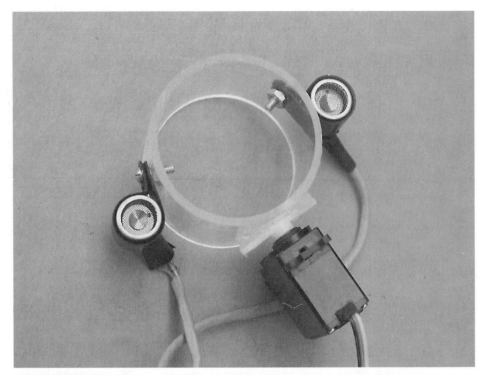

FIGURE 36.9 Ultrasonic sensors mounted on an RC servo scanner turret.

Figs. 36.10 and 36.11 show a basic circuit you can build that provides ultrasonic proximity detection and has two parts: a transmitter and a receiver (refer to the parts list in Tables 36.2 and 36.3). The transmitter circuit works as follows: a stream of 40 kHz pulses are produced by a 555 timer wired up as an astable multivibrator.

The receiving transducer is positioned two or more inches away from the transmitter transducer. For best results, you may wish to place a piece of foam between the two transducers to eliminate direct interference. The signal from the receiving transducer needs to be amplified; an op amp (such as an LM741, as shown in Fig. 36.11) is more than sufficient for the job. The amplified output of the receiver transducer is directly connected to another 741 op amp wired as a comparator. The ultrasonic receiver is sensitive only to sounds in about the 40 kHz range (± about 4 kHz).

The closer the ultrasonic sensor is to an object, the stronger the reflected sound will be. (Note, too, that the strength of the reflected signal will also vary depending on the material bouncing the sound.) The output of the comparator will change between LOW and HIGH as the sensor is moved closer to or farther away from an object.

Once you get the circuit debugged and working, adjust potentiometer R2, on the op amp, to vary the sensitivity of the circuit. You will find that, depending on the quality of the transducers you use, the range of this sensor is quite large. When the gain of the op amp is turned all the way up, the range may be as much as six to eight feet. (The op amp may ring, or oscillate, at very high gain levels, so use your logic probe to choose a sensitivity just below the ringing threshold.)

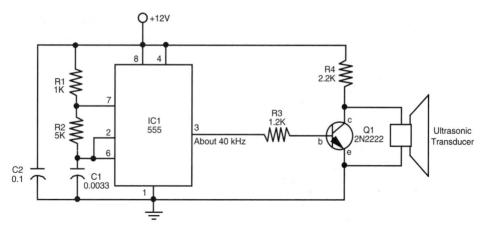

FIGURE 36.10 Schematic diagram for a basic ultrasonic proximity transmitter.

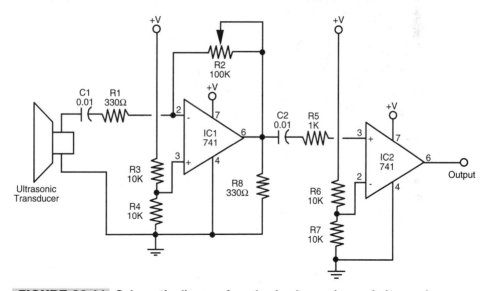

FIGURE 36.11 Schematic diagram for a basic ultrasonic proximity receiver.

Contact Detection

A sure way to detect objects is to make physical contact with them. Contact is perhaps the most common form of object detection and is often accomplished by using simple switches. In this section we'll review several contact methods, including "soft-contact" techniques where the robot can detect contact with an object using just a slight touch.

PHYSICAL CONTACT BUMPER SWITCH

An ordinary switch can be used to detect physical contact with an object. So-called "bumper switches" are spring-loaded push-button switches mounted on the frame of the

TABLE 36.2 PARTS LIST FOR ULTRAPROXIMITY TRANSMITTER.

IC1	555 Timer IC
Q1	2N2222 NPN transistor
R1	1K resistor
R2	5K resistor
R3	1.2K resistor
R4	2.2K resistor
C1	0.1 µF ceramic capacitor
C2	0.0033 µF monolithic, mica, or ceramic capacitor
TR1	Ultrasonic transmitter transducer (40 kHz nominal)

TABLE 36.3 PARTS LIST FOR ULTRASONIC PROXIMITY RECEIVER.

IC1,IC2	741 op amp IC
R1,R8	330 ohm resistor
R3,R4,R6,R7	10K resistor
R2	100K potentiometer
R5	1K resistor
C1,C2	0.01 µF ceramic capacitor
TR1	Ultrasonic receiver transducer (40kHz nominal)

All resistors have 5 or 10 percent tolerance, 1/4-watt; all capacitors have 10 percent tolerance, rated 35 volts or higher.

robot, as shown in Fig. 36.12. The plunger of the switch is pushed in whenever the robot collides with an object. Obviously, the plunger must extend farther than all other parts of the robot. You may need to mount the switch on a bracket to extend its reach.

The surface area of most push-button switches tends to be very small. You can enlarge the contact area by attaching a metal or plastic plate or a length of wire to the switch plunger. A piece of rigid 1/16-inch thick plastic or aluminum is a good choice for bumper plates. Glue the plate onto the plunger. Low-cost push-button switches are not known for their sensitivity. The robot may have to crash into an object with a fair amount of force before the switch makes positive contact, and for most applications that's obviously not desirable.

Leaf switches require only a small touch before they trigger. The plunger in a leaf switch (often referred to as a Microswitch, after the manufacturer that made them popular) is extra small and travels only a few fractions of an inch before its contacts close. A metal strip, or leaf, attached to the strip acts as a lever, further increasing sensitivity. You can mount a plastic or metal plate to the end of the leaf to increase surface area. If the leaf is wide enough, you can use miniature 4/40 or 3/38 hardware to mount the plate in place.

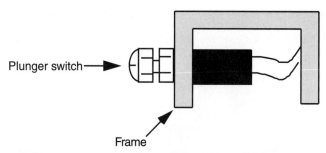

Plunger switch ➤

Frame

FIGURE 36.12 An SPST spring-loaded plunger switch mounted in the frame or body of the robot, used as a contact sensor. Experiment with different shapes and sizes for the plunger.

WHISKER

Many animal experts believe that a cat's whiskers are used to measure space. If the whiskers touch when a cat is trying to get through a hole, it knows there is not enough space for its body. We can apply a similar technique to our robot designs—whether or not kitty whiskers are actually used for this purpose.

You can use thin 20- to 25-gauge piano or stove wire for the whiskers of the robot. Attach the wires to the end of switches, or mount them in a receptacle so the wire is supported by a small rubber grommet.

By bending the whiskers, you can extend their usefulness and application. The commercially made robot shown in Fig. 36.13, the Movit WAO, has two whiskers that can be rotated in their switch sockets. When the whiskers are positioned so the loop is vertical they can detect changes in topography to watch for such things as the edge of a table, the corner of a rug, and so forth.

A more complex whisker setup is shown in Fig. 36.14. Two different lengths of whiskers are used for the two sides of the robots. The longer-length whiskers represent a space a few inches wider than the robot. If these whiskers are actuated by rubbing against an object but the short whiskers are not, then the robot understands that the pathway is clear to travel but space is tight.

The short whiskers are cut to represent the width of the robot. Should the short whiskers on only one side of the robot be triggered, then the robot will turn the opposite direction to avoid the obstacle. If both sides of short whiskers are activated, then the robot knows that it cannot fit through the passageway, and it either stops or turns around.

Before building bumper switches or whiskers into your robot, be aware that most electronic circuits will misbehave when they are triggered by a mechanical switch contact. The contact has a tendency to "bounce" as it closes and opens, so it needs to be conditioned. See the debouncing circuits in Chapter 29 for ways to clean up the contact closure so switches can directly drive your robot circuits.

PRESSURE PAD

In Chapter 35 you learned how to give the sense of touch to robot fingers and grippers. One of the materials used as a touch sensor was conductive foam, which is packaged with

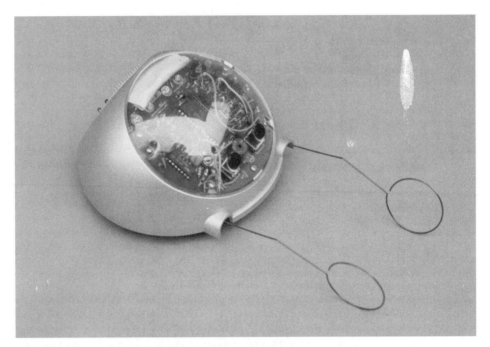

FIGURE 36.13 The Movit WAO robot (one of the older models, but the newer ones are similar). Its two tentacles, or whiskers, allow it to navigate a space.

most CMOS and microprocessor ICs. This foam is available in large sheets and is perfect for use as collision detection pressure pads. Radio Shack sells a nice five-inch square pad that's ideal for the job.

Attach wires to the pad as described in Chapter 35, and glue the pad to the frame or skin of your robot. Unlike fingertip touch, where the amount of pressure is important, the salient ingredient with a collision detector is that contact has been made with something. This makes the interface electronics that much easier to build.

Fig. 36.15 shows a suitable interface for use with the pad (refer to the parts list in Table 36.4). The pad is placed in series with a 3.3K resistor between ground and the positive supply voltage to form a voltage divider. When the pad is pressed down, the voltage at the output of the sensor will vary. The output of the sensor, which is the point between the pad and resistor, is applied to the inverting pin of a 339 comparator. (There are four separate comparators in the 339 package, so one chip can service four pressure pads.) When the voltage from the pad exceeds the reference voltage supplied to the comparator, the comparator changes states, thus indicating a collision.

The comparator output can be used to drive a motor direction control relay or can be tied directly to a microprocessor or computer port. Follow the interface guidelines provided in Chapter 29.

MULTIPLE BUMPER SWITCHES

What happens when you have many switches or proximity devices scattered around the periphery of your robot? You could connect the output of each switch to the computer, but

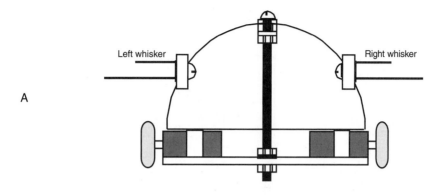

A

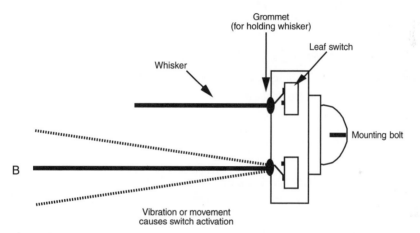

B

FIGURE 36.14 Adding whiskers to a robot. *a.* Whiskers attached to the dome of the Minibot (see Chap. 8); *b.* Construction detail of the whiskers and actuation switches.

that's a waste of interface ports. A better way to do it is to use a priority encoder or multiplexer. Both schemes allow you to connect several switches to a common control circuit. The robot's microprocessor or computer queries the control circuit instead of the individual switches or proximity devices.

Using a priority encoder The circuit in Fig. 36.16 uses a 74148 priority encoder IC. Switches are shown at the inputs of the chip. When a switch is closed, its binary equivalent appears at the A-B-C output pins. With a priority encoder, only the highest value switch is indicated at the output. In other words, if switch 4 and 7 are both closed, the output will only reflect the closure of pin 4.

Another method is shown in Fig. 36.17. Here, a 74150 multiplexer IC is used as a switch selector. To read whether a switch is or not, the computer or microprocessor applies a binary weighted number to the input select pins. The state of the desired input is shown in inverted form at the Out pin (pin 10). The advantage of the 74150 is that the state of any switch can be read at any time, even if several switches are closed.

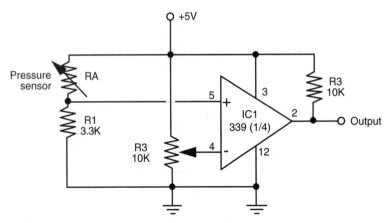

FIGURE 36.15 Converting the output of a conductive foam pressure sensor to an on/off type switch output.

TABLE 36.4	PARTS LIST FOR PRESSURE SENSOR BUMPER SWITCH.
IC1	LM339 Quad Comparator IC
R1	3.3K resistor
R2	10K potentiometer
R3	10K resistor
Misc	Conductive foam pressure transducer (see text)

Using a resistor ladder If the computer or microcontroller used in your robot has an analog-to-digital converter (ADC) or you don't mind adding one, you can use another technique for interfacing multiple switches: the resistor ladder. The concept is simple, as Fig. 36.18 shows. Each switch is connected to ground on one side and to V+ in series with a resistor on the other side. Multiple switches are connected in parallel to an ADC input, as depicted in the figure. The resistors form a voltage divider. Each resistor has a different value, so when a switch closes the voltage through that switch is uniquely different.

Note that because the resistors are in parallel, you can close more than one switch at one time. An "in-between" voltage will result. Feel free to experiment with the values of the resistors connected to each switch to obtain maximum flexibility.

"Soft Touch" and Compliant Collision Detection

The last nickname you'd want for your robot is "Bull in a China Closet," a not too flattering reference to your automaton's habit of crashing into and breaking everything.

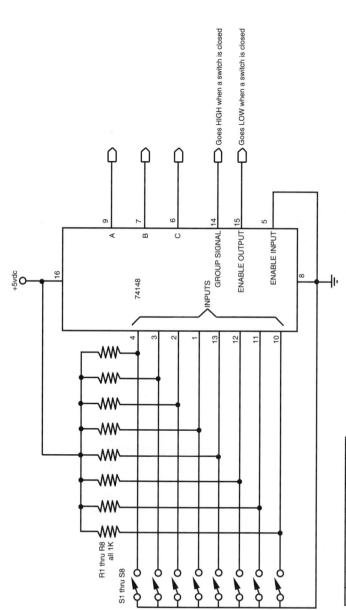

FIGURE 36.16 Multiple switch detection using the 74148 priority encoder IC. Basic wiring diagram.

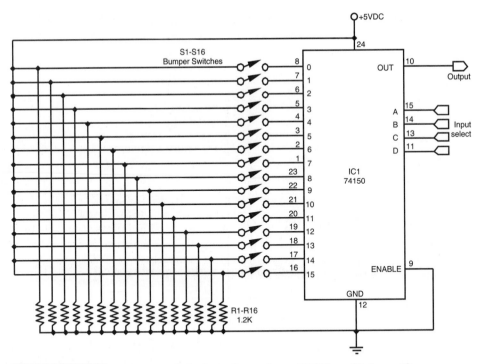

FIGURE 36.17 Multiple switch detection using a 74150 multiplexer IC.

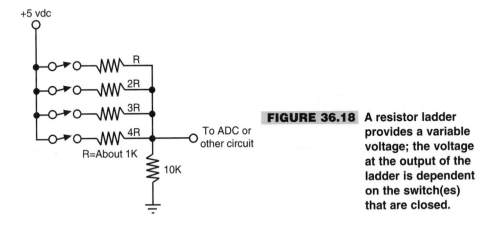

FIGURE 36.18 A resistor ladder provides a variable voltage; the voltage at the output of the ladder is dependent on the switch(es) that are closed.

Unfortunately, even the best behaved robots occasionally bump into obstacles, including walls, furniture, and the cat (your robot can probably survive a head-on collision with a solid wall, but the family feline . . . maybe not!).

Since it's impractical—not to mention darn near impossible—to always prevent your robot from colliding with objects, the next best thing is to make those collisions as "soft" as possible. This is done using so-called *soft touch* or *compliant* collision detection means. Several such approaches are outlined here. You can try some or all; mixing and matching sensors on one robot is not only encouraged, it's a good idea. As long as the sensor redundancy does not unduly affect the size, weight, or cost of the robot, having "backups" can make your robot a better behaved houseguest.

Laser Fiber "Whiskers"

You know about fiber optics: they're used to transmit hundreds of thousands of phone calls through a thin wire. They're also used to connect together high-end home entertainment gear and even to make "light sculpture" art. On their own, optical fibers offer a wealth of technical solutions, and when combined with a laser, optical fibers can do even more.

The unique "whiskers" project that follows makes use of a relatively underappreciated (and often undesirable) synergy between low-grade optical fibers and lasers. To fully understand what happens to laser light in an optical fiber, let's first take a look at how fiber optics work and then how the properties of laser light play a key role in making the fiber optic robowhiskers function.

FIBER OPTICS: AN INTRODUCTION

An optical fiber is to light what PVC pipe is to water. Though the fiber is a solid, it channels light from one end to the other. Even if the fiber is bent, the light follows the path, altering its course at the bend and traveling on. Because light acts as an information carrier, a strand of optical fiber no bigger than a human hair can carry the same amount of data as some 900 copper wires.

The idea for optical fibers is over 100 years old. British physicist John Tyndall once demonstrated how a bright beam of light was internally reflected through a stream of water flowing out of a tank. Serious research into light transmission through solid material started in 1934, when Bell Labs was issued a patent for the *light pipe.* In the 1950s, the American Optical Corporation developed glass fibers that transmitted light over short distances (a few yards). The technology of fiber optics really took off around 1970 when scientists at Corning Glass Works developed long-distance optical fibers.

Optical fibers are composed of two basic materials, as illustrated in Fig. 36.19: the core and the cladding. The *core* is a dense glass or plastic material that the light actually passes through as it travels the length of the fiber. The *cladding* is a less dense sheath, also of plastic or glass, that serves as a refracting medium. An optical fiber may or may not have an *outer jacket,* a plastic or rubber insulation used as protection.

Optical fibers transmit light by *total internal reflection* (TIR), as shown in Fig. 36.20. Imagine a ray of light entering the end of an optical fiber strand. If the fiber is perfectly straight, the light will pass through the medium just as it passes through a plate of glass. But if the fiber is bent slightly, the light will eventually strike the outside edge of the fiber. If the angle of incidence is great (more than the so-called critical angle), the light will be reflected internally and will continue its path through the fiber. But if the bend is large and the angle of incidence is small (less than the critical angle), the light will pass through the fiber and be lost.

Note the *cone* of *acceptance,* as shown in Fig. 36.20; the cone represents the degree to which the incoming light can be off axis and still make it into the fiber. The cone of

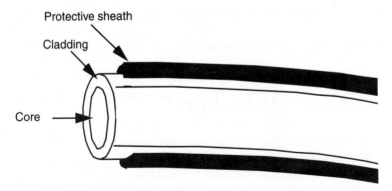

Protective sheath

Cladding

Core

FIGURE 36.19 The physical makeup of an optical fiber, consisting of core and cladding.

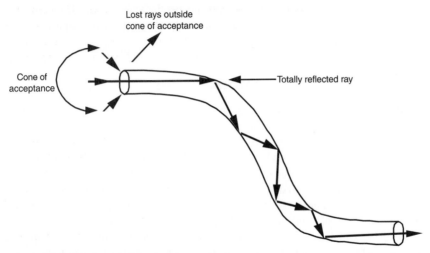

Lost rays outside cone of acceptance

Cone of acceptance

Totally reflected ray

FIGURE 36.20 Light travels through optical fibers due to a process called total internal reflection (TIR).

acceptance (usually 30°) of an optical fiber determines how far the light source can be from the optical axis and still manage to make it into the fiber. Though the cone of acceptance may be great, fiber optics perform best when the light source (and detector) are-aligned to the optical axis.

TYPES OF OPTICAL FIBERS

The classic optical fiber is made of glass, otherwise known as silica (which is plain ol' sand). Glass fibers tend to be expensive and are more brittle than stranded copper wire. But they are excellent conductors of light, especially light in the infrared region between 850 and 1300 nanometers (nm).

Less expensive optical fibers are made of plastic. Though light loss through plastic fibers is greater than through glass fibers, they are more durable. Plastic fibers are best used in communications experiments with near-infrared light sources—the 780 to 950 nm range. This nicely corresponds to the output wavelength and sensitivity of common infrared emitters and detectors.

Optical fiber bundles may be coherent or incoherent. These terms relate to the arrangement of the individual strands in the bundle. If the strands are arranged so that the fibers can transmit an image from one end to the other, they are said to be *coherent*. The vast majority of optical fibers are *incoherent:* an image or particular pattern of light is lost when it reaches the other end of the fiber.

The cladding used in optical fibers may be one of two types—step-index and graded-index. *Step-index* fibers provide a discrete boundary between more dense and less dense regions of core and cladding. They are the easiest to manufacture, but their design causes a loss of ray coherency when laser light passes through the fiber: that is, coherent light in, largely incoherent light out. The loss of coherency, which is due to light rays traveling slightly different paths through the fiber, reduces the efficiency of the laser beam. Still, it offers some very practical benefits, as you'll see later in this chapter.

There is no discrete refractive boundary in *graded-index* fibers. The core and cladding media slowly blend, like an exotic tropical drink. The grading acts to refract light evenly, at any angle of incidence. This preserves coherency and improves the efficiency of the fiber. As you might have guessed, graded-index optical fibers are the most expensive of the bunch.

WORKING WITH FIBER OPTICS

Optical fibers may be cut with wire cutters, nippers, or even a knife. But you must exercise care to avoid injuring yourself from shards of glass that may fly out when the fiber is cut (plastic fibers don't shatter when cut). Wear heavy cotton gloves and eye protection when working with optical fibers. Avoid working with fibers around food-serving or -preparation areas (that means stay out of the kitchen!). The bits of glass may inadvertently settle on food, plates, or eating utensils and could cause bodily harm.

One good way to cut glass fiber is to gently nick it with a sharp knife or razor, then snap it in two. Position the thumb and index finger of both hands as close to the nick as possible, then break the fiber with a swift downward motion (snapping upward increases the chance that glass shards will fly off in your direction).

BUILDING THE LASER-OPTIC WHISKER

Consider the arrangement in Fig. 36.21. A laser is pointed at one end of a stepped-index optical fiber. The fiber forms one or more loops around the front, side, or back of the robot. At the opposite end of the fiber is an ordinary phototransistor or photodiode (I'll just refer to it as a photodetector for now and not worry what kind it is). When the laser is turned on, the photodetector registers a certain voltage level from the laser light, say 2.5 volts. This is the *quiescent level.*

When one or more of the loops of the fiber are deformed—the robot has touched a person or thing, for instance—the laser light passing through the fiber is diverted in its path, and this changes the interference patterns at the photodetector end. The change in light level received by the photodetector does not span a very wide range, perhaps one volt total. But this one volt is enough to not only determine when the robot has touched an object but

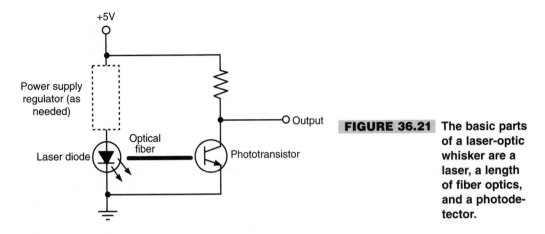

FIGURE 36.21 The basic parts of a laser-optic whisker are a laser, a length of fiber optics, and a photodetector.

the relative intensity of the collision. The more the robot has "connected" to some object, the more the fibers will deform and the greater the output change of the light as it reaches the photodetector.

The key benefit of the laser-optic whisker system is that a collision can be detected with just a *feather touch*. In fact, your robot may know when it's bumped into you before you do! Since contact with the robot is through a tiny piece of plastic, there's little chance the machine will damage or hurt anything it bumps into. The whiskers can protrude several inches from the body of the robot and omnidirectionally, if you desire. In this way it will sense contact from any direction.

Fig. 36.22 shows a prototype of this technique that consists of a hacked visible light penlight laser, several strands of cheap (*very* cheap) stepped-index optical fibers, and a set of three phototransistors. The optical fibers are tied together in a bundle using a small brass collar, electrical tape, and tie-wrap. This bundle is then inserted into the opening of the penlight laser and held in place with a sticky-back tie-wrap connector (available at Radio Shack and many other places).

On the opposite ends of the optical fibers are #18 crimp-type bullet connectors. These are designed to splice two #18 or #20 wires together, end to end. I (carefully) crimped them onto the ends of the fibers, so they act as plug-in connectors. As shown in Fig. 36.23, these ersatz connectors plug into makeshift "optical jacks," which are nothing more than 1/4-inch-diameter by 3/8-inch aluminum tubing. The tubing is glued over the ends of the phototransistors and the phototransistors are soldered near the edge of the protoboard.

Refer to Fig. 36.24 for a schematic wiring diagram of a power regulator for the penlight laser. Note the zener diode voltage regulator. The laser I used was powered by two AAA batteries, or roughly three volts. Diode lasers are sensitive to high input voltage, and many will burn out if fed a higher voltage than they are designed for. The penlight laser consumes less than about 30 mA. An alternative is to use three signal diodes (e.g. 1N4148) in series between the +V and the input of the laser to drop the 5 vdc voltage to about 2.7–3.0 volts. The diodes you use should be rated for 1/4-watt or higher.

INTERFACING THE PHOTODETECTORS

The output of a phototransistor is close to the full 0–5-volt range of the circuit's supply range. You'll want your robot to be able to determine the intensity changes as the whiskers

FIGURE 36.22 The prototype laser-optic sensor, showing the loose fibers (on the robot these fibers are neatly looped to create a kind of sensor antenna).

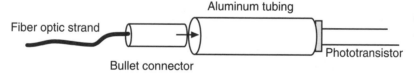

FIGURE 36.23 Use short lengths of aluminum tubing, available at hobby stores, and a crimp-on bullet connector to create "optical jacks" for the laser-optic whisker system.

bump against objects. If you're using a computer or microcontroller to operate your robot, this means you'll need to convert the analog signal produced by the detectors into a digital signal suitable for the brains on your 'bot. Several popular microcontrollers, such as the BasicX-24, OOPic, and 68HC11, have analog-to-digital converter (ADC) ports built in. If your computer or controller doesn't have ADC inputs, you can add an outboard ADC using an ADC0809 or similar chips. See Chapter 29 for more information on interfacing an analog signal to a digital input by way of an analog-to-digital converter.

CREATING THE WHISKER LOOPS

Okay, so the laser-optic whisker system may not use cat-type whiskers with ends that stick out. Still, the word *whisker* aptly describes the way the system works. If something even so much as brushes lightly against the whisker, the light reaching the photodetector will change, and your robot can react accordingly.

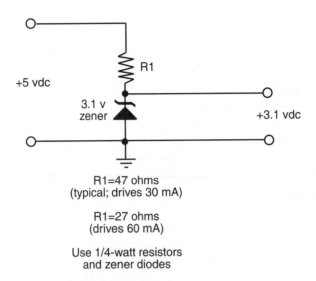

+5 vdc

R1

3.1 v
zener

+3.1 vdc

R1=47 ohms
(typical; drives 30 mA)

R1=27 ohms
(drives 60 mA)

Use 1/4-watt resistors
and zener diodes

FIGURE 36.24 **Most penlight lasers are designed
to operate with 3 vdc; use a zener
diode or voltage regulator to provide the proper voltage.**

The prototype system for this book used three "whiskers," all of which were formed into three small loops around the front and two sides of the test robot. The loops can he held in place with small screws, dabs of glue (don't use hot-melt glue!), or even LEGO parts should your robot be constructed with them. When forming the loops don't make them too tight. The more compliant the loops are, the more they will detect small amounts of pressure. If the loops are very tight, the fibers become rigid and not very compliant. This reduces the effectiveness of the whiskers.

At the same time, the loops should not be so loose that they tend to wobble or flap while the robot is in motion. Should this occur, the natural vibration and movement of the fiber will cause false readings. A loop diameter of from 4 to 6 inches should be sufficient given optical fiber pieces of average diameter and stiffness. Experiment with the optical fibers you obtain for the project. Your laser-optic whisker system does not need to use three separate fiber strands. One strand may be enough, especially if the robot is small. I elected to use three so the robot could independently determine in which direction (left, front, right) a collision or bump had occurred.

GETTING THE RIGHT KIND OF OPTICAL FIBER

Perhaps the hardest part of constructing this project is finding the right kind of optical fiber. You want to avoid any kind of graded-index fiber (described earlier) because these will not produce the internal interference patterns that the project depends on. In essence, what you want is the cheapest, lousiest fiber-optic strands you can find. The kind designed for "light fountain art" (popular in the early 1970s) is ideal. You *do not* want to use data communications-grade optical fiber.

Before you buy miles of optical fiber, test a two-foot strand with a suitable diode laser and phototransistor. Loop the fiber and tape it snugly to your desk or workbench. Connect the phototransistor to a sensitive volt-ohm meter or, better yet, an oscilloscope. Gently touch the fiber loops to deform them. You should observe a definite change of output in the phototransistor. If you do not, examine your setup to rule out a wiring error, and try again. Turn the laser off momentarily and observe the change in output.

WORKING WITH LASER DIODES

Penlight lasers can be easily hacked for a wide variety of interesting robot projects—the *soft-touch* fiber-optic whisker is just one of them. Penlight lasers use a semiconductor lasing element. While these elements are fairly hearty, they do require certain handling precautions. And even though they are small, they still emit laser light that can be potentially dangerous to your eyes. So keep the following points in mind:

- Always make sure the terminals of a laser diode are connected properly to the drive circuit.
- Never apply more than the rated voltage to the laser or it will burn up.
- Extend the same care to laser diodes that you do to any static-sensitive device. Wear an antistatic wrist strap while handling the bare laser element, and keep the device in a protective, antistatic bag until it's ready for use.
- Use only a grounded soldering pencil when attaching wires to the laser diode terminals. Limit soldering duration to less than five seconds per terminal.
- Never connect the probes of a volt-ohm meter across the terminals of a laser diode. The current from the internal battery of the meter may damage the laser.
- Use only batteries or *well-filtered* AC power supplies. Laser diodes are susceptible to voltage transients and can be ruined when powered by poorly filtered line-operated supplies.
- Take care not to short the terminals of the laser during operation.
- Avoid looking into the window of the laser while it is operating, even if you can't see any light coming out (is the diode the infrared type?).
- Unless otherwise specified by the manufacturer, clean the output window of the laser diode with a cotton swab dipped in ethanol. Alternatively, you can use optics-grade lens cleaning fluid.
- If you are using a laser from a laser penlight, bear in mind that the penlight casing acts as a heat sink. If you remove the laser from the penlight casing, be sure to attach the laser to a suitable heat sink to avoid possible damage. If you keep the laser in the casing, there is usually no need to add the heat sink—the casing should be enough.

Piezo Disc Touch Bar

The laser-optic whisker system described earlier is a great way to detect even your robot's minor collisions. But it may be overkill in some instances, providing too much sensitivity for a zippy little robot always on the go. The soft-touch collision sensor described in this section, which uses commonly available piezo ceramic discs, is a good alternative to the laser-optic whisker system for lower-sensitivity applications. This sensor is constructed with a half-round bar to increase the area of contact.

CONSTRUCTION OF THE PIEZO DISC TOUCH BAR

The main sensing elements of the piezo disc touch bar are two 1-inch-diameter bare piezo ceramic discs. These discs are available at Radio Shack and many surplus electronics stores; they typically cost under $1 or $2 each, and you can often find them for even less.

You attach the discs to a 6 1/2-inch long support bar, which you can make out of plastic, even a long LEGO Technic beam. As shown in Fig. 36.25, you glue the discs into place with 1/8-inch foam (available at most arts and craft stores) so it sticks to the ceramic surface of the disc and acts as a cushion. You then bend a length of 1/8-inch-diameter aluminum tubing (approximately 8–9 inches) into a half-circle; thread through two small grommets, as shown in Fig. 36.25; and glue the grommets to the support bar. You flatten the ends of the tubing and bend them at right angles to create a "foot"; the foot rests on the foam-padded surface of the discs. Fig. 36.26 shows a photograph of a finished piezo disc touch bar. The half-round tubing slopes downward slightly. This is intentional, so the robot can adequately sense objects directly in front of it near the ground.

To construct the piezo disc touch bar I used hot-melt glue to attach the discs and grommets to the support bar. You can use most any other adhesive or glue you wish, but be sure it provides a good, strong hold for the different materials used in this project (metal, plastic, and rubber).

CONSTRUCTING THE INTERFACE CIRCUIT

Piezo discs are curious creatures: when a voltage is applied to them, the crystalline ceramic on the surface of the disc vibrates. It is the nature of piezoelectricity to be both a *consumer* and a *producer* of electricity. When the disc is connected to an input, any physical tap or pressure on the disc will produce a voltage. The exact voltage is approximately proportional to the amount of force exerted on the disc: apply a little pressure or tap and you get a little voltage. Apply a heavier pressure or tap, and you get a bigger voltage.

The piezoelectric material on ceramic piezo discs is so efficient that even a moderately strong force on the disc will produce in excess of 5 or 10 volts. That's good in that it makes it easy to interface the discs to a circuit, since there is usually no need to amplify

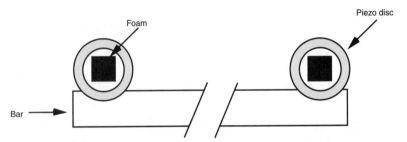

FIGURE 36.25 Glue the piezo discs to a piece of plastic; the plastic is a support bar for the discs that also makes it easier to mount the touch bar sensor onto your robot.

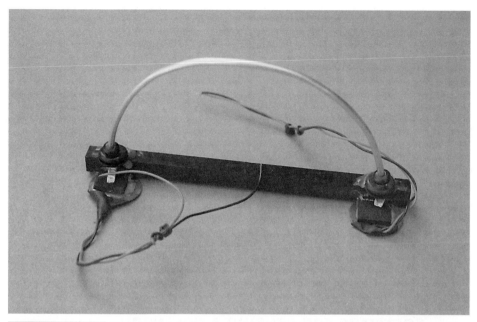

FIGURE 36.26 The finished prototype of the piezo disc touch bar. One variation is to mount the discs a little lower so the metal bar physically deforms the disc rather than pushes against its center.

the signal. But it's also bad in that the voltage from the disc can easily exceed the maximum inputs of the computer, microcontroller, or other electronic device you're interfacing with. (Pound on a piezo disc with a hammer, and, though it might be broken when you're done, it will also produce a thousand volts or more.)

To prevent damage to your support electronics, attach two 5.1-volt zener diodes as shown in Fig. 36.27, to each disc of the touch bar. The zener diodes limit the output of the disc to 5.1 volts, a safe enough level for most interface circuitry. For an extra measure of safety, use 4.7-volt zeners instead of 5.1 volt.

Note that piezoelectric discs also make great capacitors. This means that over time the disc will take a charge, and the charge will show up as a constantly changing voltage at the output of the disc. To prevent this, insert a resistor across the output of the disc and ground. In my experiments I found a resistor of about 82K eliminated the charge buildup without excessively diminishing the sensitivity of the disc. Experiment with the value of the resistor. A higher value will increase sensitivity, but it could cause an excessive charge buildup. A lower value will reduce the buildup but also reduce the sensitivity of the disc. It is also helpful to route the output of the disc to an op amp, preferably through a 100K or higher resistor.

MOUNTING THE TOUCH BAR

Once you have constructed the piezo disc touch bar and added the voltage-limiting circuitry, you can attach it to the body of the robot. The front of the robot is the likely choice, but you can add additional bars to the sides and rear to obtain a near 360° sensing pattern. The width of the bar makes it ideal for any robot that's between about 8 and 14 inches wide.

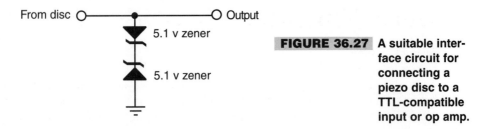

FIGURE 36.27 A suitable interface circuit for connecting a piezo disc to a TTL-compatible input or op amp.

Since the sensing element of the touch bar, the aluminum tube, has a half-round shape, the sensor is also suitable for mounting on a circular robot base. For added compliancy, you may wish to mount the bar using a thick foam pad, spring, or shock absorber (shocks made for model racing cars work well). If the bar is mounting directly to the robot the sensor exhibits relatively little compliancy.

You should mount the bar at a height that is consistent with the kinds of objects the robot is most likely to collide with. For a "wall-hugging" robot, for example, you may wish to mount the bar low and ensure that the half-round tube slopes downward. That way, the sensor is more likely to strike the baseboard at the bottom of the wall.

SOFTWARE FOR SENSING A COLLISION

Listing 36.1 shows a short sample program for reading the values provided by the piezo disc touch bar. The program is written for the BASIC Stamp II microcontroller and requires the addition of a one or more serial-output analog-to-digital converter chips (I used an ADC0831 for my prototype). You need only one ADC if it has multiple inputs; you'll need two ADCs if the chips have but a single input. See the comments in the program for hookup information.

LISTING 36.1.

```
' For the Basic Stamp II
' Uses an ADC081 serial ADC
Adress   var    byte     ' A-to-D result: one byte
CS       con    13       ' Chip select is pin 13
Adata    con    14       ' ADC data output is pin 14
CLK      con    15       ' Clock is pin 15
Vref     con    0        ' VRef

high Vref

high CS                          ' Deselect ADC to start

again:
    low CS                              ' Activate ADC
    shiftin AData,CLK,msbpost,[ADres\9] ' Shift in the data
    high CS                             ' Deactivate ADC
    debug ? Adres                       ' Display result
    pause 100                           ' Wait 1/10 second
goto again                              ' Repeat
```

Other Approaches for Soft-touch Sensors

There are several other approaches for using soft-touch sensors. For example, the resistive bend sensor changes its resistance the more it is curved or bent. Positioned in the front of your robot in a loop, the bend sensor could be used to detect the deflection caused by running into an object.

If you like the idea of piezoelectric elements but want a more localized touch sensor than the touch bar described in the previous section, you might try mounting piezoelectric material and discs on rubber or felt pads, or even to the "bubbles" of bubble pack shipping material, to create "fingers" for your 'bot.

From Here

To learn more about...	*Read*
Connecting analog and digital sensors to computers, microcontrollers, and other circuitry	Chapter 29, "Interfacing with Computers and Microcontrollers"
Building and using sensors for tactile feedback	Chapter 35, "Adding the Sense of Touch"
Giving your robots the gift of sight	Chapter 37, "Robotic Eyes"
Using sensors to provide navigation assistance to mobile robots	Chapter 38, "Navigating through Space"

37

ROBOTIC EYES

Giving sight to your robots is perhaps the kindest thing you can do for them. Robotic vision systems can be simple or complex to match your specific requirements and your itch to tinker. Rudimentary "Cyclops" vision systems are used to detect nothing more than the presence or absence of light. Aside from this rather mundane task, there are plenty of useful applications for an on/off light detector. More advanced vision systems decode relative intensities of light and can even make out patterns and crude shapes.

While the hardware for making robot "eyes" is rather simple, using the vision information they generate is not. Except for the one-cell light detector, vision systems must be interfaced to a computer to be useful. You can adapt the designs presented in this chapter to just about any computer using a microprocessor data bus or one or more parallel printer ports.

Simple Sensors for "Vision"

A number of simple electronic devices can be used as "eyes" for your robot. These include the following:

- *Photoresistors.* These are typically a cadmium-sulfide (CdS) cell (often referred to simply as a *photocell*). A CdS cell acts like a light-dependent resistor: the resistance of the cell varies depending on the intensity of the light striking it. When no light strikes the cell, the device exhibits very high resistance, typically in the high 100 kilohms, or even megohms. Light reduces the resistance, usually significantly (a few hundreds or thousands of ohms). CdS cells are very easy to interface to other electronics, but they are somewhat slow reacting and are unable to discern when light flashes more than 20 or 30 times per second. This trait actually comes in handy because it means CdS cells basically ignore the on/off flashes of AC-operated lights.

■ *Phototransistors.* These are very much like regular transistors with their metal or plastic tops removed. A glass or plastic cover protects the delicate transistor substrate inside. Unlike CdS cells, phototransistors are very quick acting and are able to sense tens of thousands of flashes of light per second. The output of a phototransistor is not "linear"; that is, there is a disproportionate change in the output of a phototransistor as more and more light strikes it. A phototransistor can become easily "swamped" with too much light. Even as more light shines on the device, the phototransistor is not able to detect any more change.

■ *Photodiodes.* These are the simpler diode versions of phototransistors. Like phototransistors, they are made with a glass or plastic cover to protect the semiconductor material inside them. And like phototransistors, photodiodes are very fast acting and can become "swamped" when exposed to a certain threshold of light. One common characteristic of most photodiodes is that their output is rather low, even when fully exposed to bright light. This means that to be effective the output of the photodiode must usually be connected to a small amplifier.

Photoresistors, photodiodes, and phototransistors are connected to other electronics in about the same way: you place a resistor between the device and either +V or ground. The point between the device and the resistor is the output, as shown in Fig. 37.1. With this arrangement, all three devices therefore output a varying voltage. The exact arrangement of the connection determines if the voltage output increases or decreases when more light strikes the sensor.

Light-sensitive devices differ in their spectral response, which is the span of the visible and near-infrared light region of the electromagnetic spectrum that they are most sensitive to. CdS cells exhibit a spectral response very close to that of the human eye, with the greatest degree of sensitivity in the green or yellow-green region (see Fig. 37.2). Both phototransistors and photodiodes have peak spectral responses in the infrared and near-infrared regions. In addition, some phototransistors and photodiodes incorporate optical filtration to decrease their sensitivity to the visible light spectrum. This filtration makes the sensors more sensitive to infrared and near-infrared light.

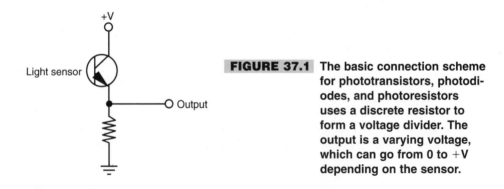

FIGURE 37.1 The basic connection scheme for phototransistors, photodiodes, and photoresistors uses a discrete resistor to form a voltage divider. The output is a varying voltage, which can go from 0 to +V depending on the sensor.

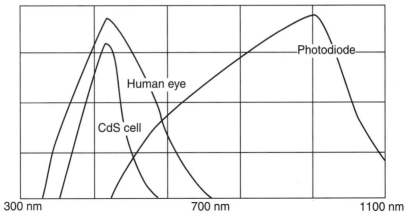

300 nm   700 nm   1100 nm

FIGURE 37.2 Light sensors vary in their sensitivity to different colors of the electromagnetic spectrum. The color sensitivity of CdS cells is very similar to that of the human eye.

One-Cell Cyclops

A single light-sensitive photocell is all your robot needs to sense the presence of light. The photocell is a variable resistor that works much like a potentiometer but has no control shaft. You vary its resistance by increasing or decreasing the light. Connect the photocell as shown in Fig. 37.3. Note that, as explained in the previous section, a resistor is placed in series with the photocell and that the output tap is between the cell and resistor. This converts the output of the photocell from resistance to voltage, the latter of which is easier to use in a practical circuit.

The value of the resistor is given at 3.3K ohms but is open to experimentation. You can vary the sensitivity of the cell by substituting a higher or lower value. For experimental purposes, connect a 1K resistor in series with a 50K pot (in place of the 3.3K ohm resistor) and try using the cell at various settings of the wiper. Test the cell output by connecting a volt-ohm meter to the ground and output terminals.

So far, you have a nice light-to-voltage sensor, and when you think about it there are numerous ways to interface this ultrasimple circuit to a robot. One way is to connect the output of the sensor to the input of a comparator. (The LM339 quad comparator IC is a good choice, but you can use just about any comparator.) The output of the comparator changes state when the voltage at its input goes beyond or below a certain "trip point."

In the circuit shown in Fig. 37.4 (refer to the parts list in Table 37.1), the comparator is hooked up so the noninverting input serves as a voltage reference. Adjust the potentiometer to set the *trip point*. To begin, set it midway, then adjust the trip point higher or lower as required. The output of the photocell circuit is connected to the inverting input of the comparator. When the voltage at this pin rises above or below the set point, the output of the comparator changes state.

One practical application of this circuit is to detect light levels that are higher than the ambient light in the room. Doing so enables your robot to ignore the background light level and respond only to the higher intensity light. To begin, set the trip point pot so the circuit

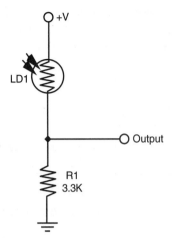

FIGURE 37.3 A one-cell robotic eye, using a CdS photocell as a light sensor.

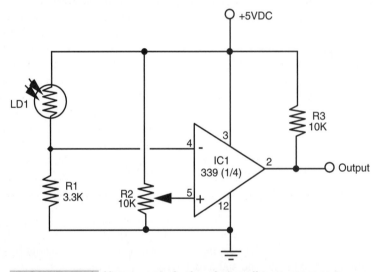

FIGURE 37.4 How to couple the photocell to a comparator.

TABLE 37.1 PARTS LIST FOR SINGLE-CELL ROBOTIC EYE	
IC1	LM339 Quad Comparator IC
R1	3.3K resistor
R2	10K potentiometer
R3	10K resistor
LD1	Photocell

just switches HIGH. Use a flashlight to focus a beam directly onto the photocell, and watch the output of the comparator change state. Another application is to use the photocell as a light detector—period. Set the pot all the way over so the comparator changes state just after light is applied to the surface of the cell.

Multiple-Cell Light Sensors

The human eye has millions of tiny light receptacles. Combined, these receptacles allow us to discern shapes, to actually "see" rather than just detect light levels. A crude but surprisingly useful implementation of human sight is given in Fig. 37.5 (refer to the parts list in Table 37.2). Here, eight photocells are connected to a 16-channel multiplexed analog-to-digital converter (ADC). The ADC, which has room for another eight cells, takes the analog voltages from the outputs of each photocell and one by one converts them into digital data. The eight-bit binary number presented at the output of the ADC represents any of 256 different light levels.

The converter is hooked up in such a way that the outputs of the photocells are converted sequentially, in a row and column pattern, following the suggested mounting scheme shown in Figs. 37.6 and 37.7. A computer hooked up to the A/D converter records the digital value of each cell and creates an *image matrix,* which can be used to discern crude shapes.

Each photocell is connected in series with a resistor, as with the one-cell eye presented earlier. Initially, use 2.2K resistors, but feel free to substitute higher or lower values to increase or decrease sensitivity. The photocells should be identical, and for the best results, they should be brand-new prime components. Before placing the cells in the circuit, test each one with a volt-ohm meter and a carefully controlled light source. Check the resistance of the photocell in complete darkness, then again with a light shining at it a specific distance away. Reject cells that do not fall within a 5 to 10 percent "pass" range. See Chapter 29, "Interfacing with Computers and Microcontrollers," for more information on using ADCs and connecting them to computer ports and microprocessors.

Note the short pulse that appears at pin 13, the End-of-Conversion (EOC) Output. This pin signals that the data at the output lines is valid. If you are using a computer or microcontroller, you can connect this pin to an interrupt line (if available). Using an interrupt line lets your computer do other things while it waits for the ADC to signal the end of a conversion. See Chapter 42, "Tips, Tricks, and Tidbits for the Robot Experimenter," for basic information on using the hardware interrupt port on computers and microcontrollers.

You can get by without using the EOC pin—the circuit is easier to implement without it—but you must set up a timing delay circuit or routine to do so. Simply wait long enough for the conversion to take place—a maximum of about 115 μs (microseconds)—then read the data. Even with a delay of 125 μs (to allow for settling, etc.), it takes no more than about 200 milliseconds to read the entire matrix of cells.

Eyes from Static CMOS Memory

Long before solid-state (CCD and CMOS) camcorders and digital cameras became common, robot experimenters used to play around with static CMOS RAM (random access memory), using modified chips as multicell eyes for their creations. Most all semiconductors are sensitive to light, even the memory matrix inside memory chips. By using static

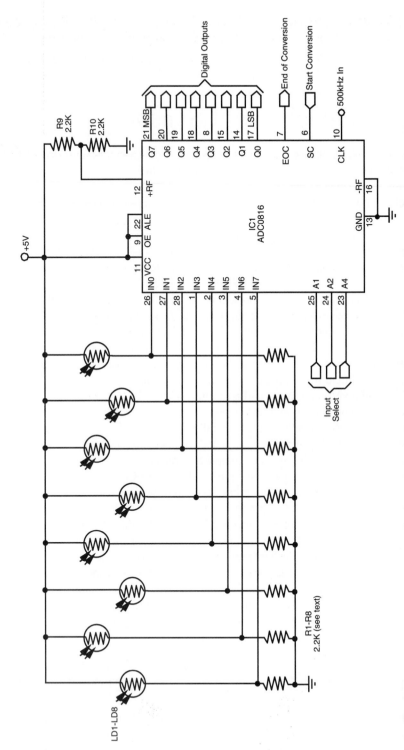

FIGURE 37.5

One way to make a robotic "eye." The circuit, as shown, consists of eight photocells connected to an ADC0816 8-bit, 16-input analog-to-digital converter IC. The output of each photocell is converted when selected at the Input Select lines. The ADC0816 can handle up to 16 inputs, so you can add another eight cells.

TABLE 37.2   PARTS LIST FOR MULTICELL ROBOTIC EYE.

IC1	ADC0816 8-bit analog-to-digital converter IC (okay to substitute another multiplexing ADC, such as the ADC0817, etc.)
R1–R8	2.2K resistor (adjust value to gain best response of photocells)
R9,R10	2.2K resistors
LD–L8	Photocell

All resistors have 5 or 10 percent tolerance, 1/4-watt.

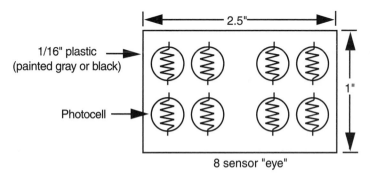

FIGURE 37.6 Mounting the photocells for an eight-cell eye.

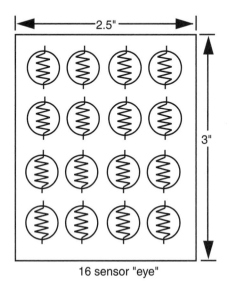

FIGURE 37.7 Mounting the photocells for a pair of four-cell eyes.

memory, you can keep the interface to the chip simple and straightforward. In fact, all you need to do is connect some wires from the chip to your computer or microcontroller.

You can sometimes find static CMOS memory chips that have already been modified for use as vision sensors. But if you cannot, you can make one yourself, using something like the 1K x 4-bit 2114L memory chip. You need to find the kind that come in a ceramic case, outfitted with a soldered metal lid. These were common in certain commercial, industrial, and military applications years ago, and they should still be available at the better electronics surplus stores. Aside from their use in ersatz robotic vision systems, these chips have such a low memory capacity (and are dog slow, to boot!) that they really aren't useful for much else. So it's a good idea to ask the store salespeople if they have any; this is the kind of stuff that collects dust in a back room.

Note:

> I used the 2114 because it was fairly common in its day and is still available in many surplus outlets. However, just about any RAM chip will work. Static RAM is preferable because it is easy to hook up, but you can also use dynamic RAM chips. Consult the RAM chip's data sheet for details on how to hook it up.)

You'll need to get the metal lid off in order to expose the semiconductor die inside the chip. The best way to do this is with a small butane torch. Secure the chip in a metal vise, and *carefully* apply even flame over the lid. After 5 to 10 seconds, the solder should melt. Quickly remove the flame, and slip the lid *all the way* off. Take care not to disturb the die or the connections inside the chip, or you'll ruin it. Fig. 37.8 shows a static RAM chip with its lid removed. *Do not* touch the removed metal lid, or the memory chip, until they have cooled off!

Fig. 37.9 is a hookup schematic for linking the 2114 with the Basic Stamp II, using all 10 address lines and all four output lines. (Each of the four output lines is connected to an LM308 precision op-amp, or requivalent.) Connect the 2114's write enable (*/WE*) line to pin 14 of the Stamp, and tie the chip select (/CS) low. The program in Listing 37.1 is a demonstrator of how to read a few of the pixels inside the chip. For each memory address you use you poll the four I/O lines. Though the listing shows only the use of the base address of 0000000000 as a demonstration, you can extrapolate the concept to access all of the memory locations in the chip. A logic 1 (HIGH) on an input line means high brightness on the chip.

LISTING 37.1.

```
A0      CON     4
A1      CON     5
A2      CON     6
A3      CON     7
A4      CON     8
A5      CON     9
A6      CON     10
A7      CON     11
A8      CON     12
A9      CON     13
WE      CON     14
```

FIGURE 37.8 A static RAM chip with its lid removed. Once the lid is off, the semiconductor die inside can see the light of day, and any other light for that matter.

```
Low WE                  ' set write enable to low to start
Low A0                  ' set all address lines to low (address 0000000000)
Low A1
Low A2
Low A3
Low A4
Low A5
Low A6
Low A7
Low A8
Low A9

' display data in debug window
Repeat_Loop:
      HIGH WE
      debug dec In0, tab, dec In1, tab, dec In2, tab, dec In3, cr
      Low WE
      Pause 150
      Goto Repeat_Loop:
```

Because the 2114 chip is designed for RAM applications and not vision, the organization of its memory cells is not ideal for rendering a visual scene. (It's basically four long rectangular blocks of 1K bit memory cells.) Each of the four outputs of any given address is located on separate portions of the semiconductor die. To assemble a usable picture, you will need to interlace the data from each of the four outputs.

Though the shape of the memory die in the 2114 makes for a poor video camera, it's ideal for such things as detecting patterns in reflected laser or focused infrared light. You could use

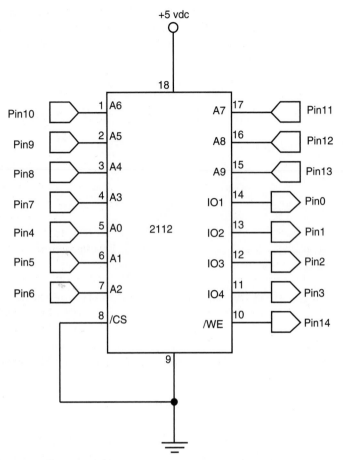

FIGURE 37.9 How to connect the 2114 static RAM chip to a Basic Stamp II. The hookup requires that quite a number of connections be made to the Stamp's pins.

the 2114 to make your own highly accurate triangulation distance measurement system, akin to the Sharp GP2D02 unit discussed in Chapter 38, "Navigating through Space."

While semiconductor vision systems like the 2114 are cheap and relatively easy to use, they suffer from rather poor sensitivity, so don't expect to be able to use them in low-light conditions! In fact, the scene scanned by the 2114 chip should be fairly well lit, and you should outfit the chip with a suitable lens. Either a single-element lens as described in the next section or a compound lens, like that used for security cameras, is ideal. The chip exhibits narrow dynamic range, which means the output line will tend to snap all the way HIGH with only a small change in brightness. This makes the 2114 and similar memory chips best suited for controlled lighting conditions where you want high contrast.

If your robot is used in low-light situations, I recommend a conventional CCD or CMOS imaging device. Black-and-white and color CCD cameras and imagers are now routinely available for under $100. With a digital output like that demonstrated with the 2114, you can easily connect the camera or imaging chip directly to a computer or microprocessor. Sadly, more elaborate vision systems are just a wee bit beyond the scope of this book (I had to draw the line somewhere). Fortunately, there are several ready-made products you can try that come complete with hookup diagrams and associated software. See Appendix B, "Sources," for additional information.

Using Lenses and Filters with Light-sensitive Sensors

Simple lenses and filters can be used to greatly enhance the sensitivity, directionality, and effectiveness of both single- and multicell-vision systems. By placing a lens over a small cluster of light cells, for example, you can concentrate room light to make the cells more sensitive to the movement of humans and other animate objects. The lens need not be complex; an ordinary 1/2- to 1-inch-diameter magnifying lens, purchased new or surplus, is all you need.

You can also use optical filters to enhance the operation of light cells. Optical filters work by allowing only certain wavelengths of light to pass through and blocking the others. CdS photocells tend to be sensitive to a wide range of visible and infrared light. You can readily accentuate the sensitivity of a certain color (and thereby de-accentuate other colors) just by putting a colored gel or other filter over the photocell.

In this section we'll briefly review the roles of lenses and filters, and how you can use them with light cells for your robots.

USING LENSES

Lenses are refractive media constructed so that light bends in a particular way. The two most important factors in selecting a lens for a given application are lens focal length and lens diameter:

■ *Lens focal length.* Simply stated, the focal length of a lens is the distance between the lens and the spot where rays are brought to a common point. (Actually, this is true of "positive" lenses only; "negative" lenses behave in an almost opposite way, as we will discuss later.)
■ *Lens diameter.* The diameter of the lens determines its light-gathering capability. The larger the lens is, the more light it collects.

There are six major types of lenses, shown in Fig. 37.10. Such combinations as plano-convex and bi-concave refer to each side of the lens. A plano-convex lens is flat on one side and curved outward on the other. A bi-concave lens curves inward on both sides. "Negative" and "positive" refer to the focal point of the lens, as determined by its design.

Lenses form two kinds of images: *real* and *virtual.* A real image is one that is focused to a point in front of the lens, such as the image of the sun focused to a small disc on a piece of paper. A virtual image is one that doesn't come to a discrete focus. You see a virtual image behind the lens, as when you are using a lens as a magnifying glass. Positive lenses, which magnify the size of an object, create both real and virtual images. Their focal

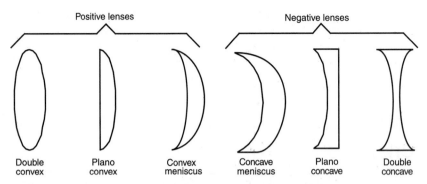

FIGURE 37.10 Lenses come in a variety of forms. Plano-convex and double-convex are among the most common.

length is stated as a positive number, such as +1 or +2.5. Negative lenses, which reduce the size of an object, create only virtual images. Their focal length is stated as a negative number.

Lenses are common finds in surplus stores, and you may not have precise control over what you get. For robotics vision applications, plano-convex or double-convex lenses of about 0.5 inch to 1.25 inch in diameter are ideal. Focal length should be fairly short—1 to 3 inches. When you are buying an assortment of lenses the diameter and focal length of each lens is usually provided, but if they are not, use a tape to measure the diameter of the lens and its focal length (see Fig. 37.11). Use any point source except the sun—focusing the light of the sun onto a small point can cause a fire! (As if you've never done this...)

To use the lens, position it over the light cell(s) using any convenient mounting technique. One approach is to glue the lens to a plastic or wood lens board. Or, if the lens is the correct diameter, you can mount it inside a short length of plastic PVC pipe; attach the other end of the pipe to the light cells. Be sure you block out stray light. You can use black construction paper to create light baffles. This will make the robot only "see" the light shining through the lens. If desired, attach a filter over the light cells. You can use a dab of glue to secure the filter in place.

Using Fig. 37.6 as a guide, you can create a kind of two-eyed robot by placing a lens over each group of four photocells. The lenses are mounted in front of the photocells, which are secured to a circuit board in two groups of four. For one project I used miniature photocells in TO-8 packages. These measure about 1/8 inch in diameter, and while they can be difficult to work with (avoid soldering them to your circuit board!) their small size is ideal for constructing compact multicell vision systems.

Each of the eight cells is connected to a separate input of an eight-input analog-to-digital converter (ADC) chip. By using an eight-input ADC, the values of all eight cells can be readily sensed without the need for separate ADC chips and extra wiring.

USING FILTERS

Filters accept light at certain wavelengths and block all others. A common filter used in robot design is intended to pass infrared radiation and block visible light. Such filters are commonly used in front of phototransistors and photodiodes to block out unwanted ambient (room) light. Only infrared light—from a laser diode, for instance—is allowed to pass

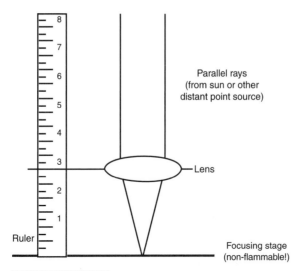

FIGURE 37.11 Use a bright light source, such as an incandescent lamp, and a tape to measure the focal point of a lens.

through and strike the sensor. Optical filters come in three general forms: colored gel, interference, and dichroic:

■ *Colored gel filters* are made by mixing dyes into a Mylar or plastic base. Good gel filters use dyes that are precisely controlled during manufacture to make filters that pass only certain colors. Depending on the dye used, the filter is capable of passing only a certain band of wavelengths. A good gel filter may have a bandpass region (the spectrum of light passed) of 40 to 60 nanometers (nm). Considering that the range of visible light ranges from about 400 to a little over 700 nm, an average bandpass region of 50 nm is roughly 15 percent of the visible light band. That equates to nearly one full color of the basic six-color rainbow.
■ *Interference filters* consist of several dielectric and sometimes metallic layers that each block a certain range of wavelengths. One layer may block light under 500 nm, and another layer may block light above 550 nm. The band between 500 and 550 nm is passed by the filter. Interference filters can be made to pass only a very small range of wavelengths.
■ *Dichroic filters* use organic dyes or chemicals to absorb light at certain wavelengths. Some filters are made from crystals that exhibit two or more different colors when viewed at different axes. Color control is maintained by cutting the crystal at a specific axis.

Introduction to Video Vision Systems

Single- and multicell-vision systems are useful for detecting the absence or presence of light, but they cannot make out the shapes of objects. This greatly limits the environment into which such a robot can be placed. By detecting the shape of an object, a robot might be able to make intelligent assumptions about its surroundings and perhaps be able to navigate those surroundings, recognize its "master," and more.

Even as recently as five years ago video vision was an expensive proposition for any robot experimenter. But the advent of inexpensive "pinhole" cameras—so called because they are used in place of the pinhole lens in the front door of a house or apartment—now makes the hardware for machine vision affordable.

A video system for robot vision need not be overly sophisticated. The resolution of the image can be as low as about 100 by 100 pixels (10,000 pixels total), though a resolution of no less than 300 by 200 pixels (60,000 pixels total) is preferred. The higher the resolution is, the better the image and therefore the greater the robot's ability to discern shapes. A color camera is not mandatory and, in some cases, makes it harder to write suitable video interpolating software.

Video systems that provide a digital output are generally easier to work with than those that provide only an analog video output. You can connect digital video systems directly to a PC, such as through a serial, parallel, or USB port. Analog video systems require that a video capture card, fast analog-to-digital converter, or other similar device be attached to the PC.

While the hardware for video vision is now affordable to most any robot builder, the job of translating a visual image a robot can use requires high-speed processing and complicated computer programming. Giving robots the ability to recognize shapes has proved to be a difficult task. Consider the static image of a doorway. Our brains easily comprehend the image, adapting to the angle at which we are viewing the doorway; the amount, direction, and contrast of the light falling on it; the size and kind of frame used in the doorway; whether the door is opened or closed; and hundreds or even thousands of other variations. Robot vision requires that each of these variables be analyzed, a job that requires computer power and programming complexity beyond the means of most robot experimenters.

Vision by Laser Light

Fortunately, there are some less complicated methods you can try as you experiment with robot vision. Here's one you might want to tackle, and it uses only about $30 worth of parts (minus the video camera). You need a simple penlight laser, a red filter, and a small piece of diffraction grating (available from Edmund Scientific Company and other sources for optical components; see Appendix B, "Sources," for additional information).

The system works on a principle similar to the three-beam focusing scheme used in CD players. In a CD player, laser light is broken into "sub-beams" by the use of a diffraction grating. A single, strong beam appears in the center, flanked by weaker beams on both sides, as shown in Fig. 37.12. The three-beam CD focusing system uses the two closest side beams, ignoring all the others.

The beam spacing increases as the distance from the lens to the surface of the disc increases. Similarly, the beam spacing decreases as the lens-to-CD distance decreases. A multicelled photodetector in the CD players integrates the light reflected by these beams and determines whether the lens should be moved closer to, or farther away, from the disc. For history buffs, the fundamental basis of this focusing technique is over a hundred years old and was pioneered by French physicist Jean Foucault.

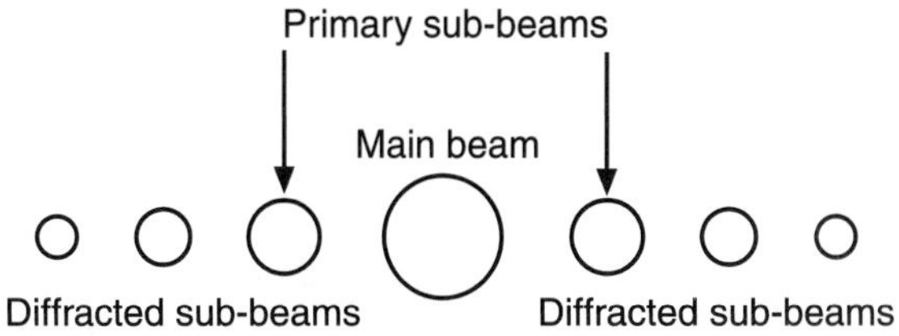

Primary sub-beams

Main beam

O O O ◯ ◯ ◯ O O

Diffracted sub-beams   Diffracted sub-beams

FIGURE 37.12 Most CD players use a diffraction grating to break up the
single laser beam into several
sub-beams. The sub-beams
are used to focus and track the
optical system.

CD players use a diffraction grating in which lines are scribed into a piece of plastic in only one plane. This causes the laser beam to break up into several beams along the same plane. With a diffraction grating that has lines scribed both vertically and horizontally, the laser beam is split up into multiple beams that form a "grid" when projected on a flat surface (see Figs. 37.13 and 37.14). The beams move closer together as the distance from the laser and surface is decreased; the beams move further apart as the distance from the laser and surface is increased.

As you can guess, when the beams are projected onto a three-dimensional scene, they form a kind of topographical map in which they appear closer or farther apart depending on the distance of the object from the laser.

The red filter placed in front of the camera lens filters out most of the light except for the red beams from the penlight laser. For best results, use a high-quality optical bandpass filter that accepts only the precise wavelength of the diode laser, typically 635 or 680 nanometers. Check the specifications of the laser you are using so you can get the correct filter. Meredith Instruments and Midwest Laser Products, among other sources, provide a variety of penlight lasers and optical filters you can use (see Appendix B).

The main benefit of the laser diffraction system is this: it's easier to write software that measures the distance between pixels than it is to write software that attempts to recognize shapes and patterns. For many machine vision applications, it is not as important for the robot to recognize the actual shape of an object as it is to navigate around or manipulate that shape. As an example, a robot may "see" a chair in its path, but there is little practical need for it to recognize the chair as an early-eighteenth-century Queen Anne-style two-seater settee. All it really needs to know is that something is there, and by moving left or right that object can be avoided.

Going Beyond Light-sensitive Vision

Sight provides a fast and efficient way for us to determine our surroundings. The eyes take in a wide field, and the brain processes what the eyes see to compose a "picture" of the immediate environment. Taking a cue from the special senses evolved by some animals, however, visual eyesight is not the only way to "see." For instance, bats use high-pitched

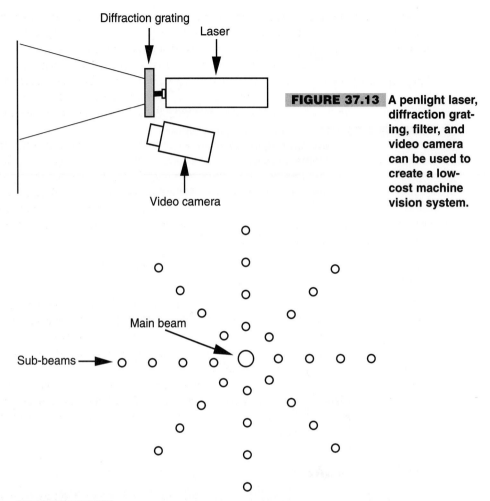

Diffraction grating

Laser

FIGURE 37.13 A penlight laser, diffraction grating, filter, and video camera can be used to create a low-cost machine vision system.

Video camera

Main beam

Sub-beams

FIGURE 37.14 When projected onto a flat surface, the beams from the diffracted laser light form a regular grid.

sound to quickly and efficiently navigate through dark caves. So accurate is their "sonar" that bats can sense tiny insects flying a dozen or more feet away.

Similarly, robots don't always need light-sensitive vision systems. You may want to consider using an alternative system, either instead of or in addition to light-sensitive vision. The following sections outline some affordable technologies you can easily use.

ULTRASONICS

Like a cave bat, your robot can use high-frequency sounds to navigate its surroundings. Ultrasonic transducers are common in Polaroid instant cameras, electronic tape-measuring devices, automotive backup alarms, and security systems. All work by sending out a high-frequency burst of sound, then measuring the amount of time it takes to receive the reflected sound.

Ultrasonic systems are designed to determine distance between the transducer and an object in front of it. More accurate versions can "map" an area to create a type of topographical image, showing the relative distances of several nearby objects along a kind of 3-D plane. Such ultrasonic systems are regularly used in the medical field. Some transducers are designed to be used in pairs—one transducer to emit a series of short ultrasonic bursts, another transducer to receive the sound. Other transducers, such as the kind used on Polaroid cameras and electronic tape-measuring devices, combine the transmitter and receiver into one unit.

An important aspect of ultrasonic imagery is that high sound frequencies disperse less readily than do low-frequency ones. That is, the sound wave produced by a high-frequency source spreads out much less broadly than the sound wave from a low-frequency source. This phenomenon improves the accuracy of ultrasonic systems. Both DigiKey and All Electronics, among others, have been known to carry new and surplus ultrasonic components suitable for robot experimenters. See Chapters 36 and 38 for more information on using ultrasonic sensors to guide your robots.

RADAR

Radar systems work on the same basic principle as ultrasonics, but instead of high-frequency sound they use a high-frequency radio wave. Most people know about the high-powered radar equipment used in aviation, but lower-powered versions are commonly used in security systems, automatic door openers, automotive backup alarms, and of course, speed-measuring devices used by the police.

Radar is less commonly found on robotics systems because it costs more than ultrasonics. But radar has the advantage that radar it is less affected by wind, temperature, and distance. For example, radar can be used up to several miles away; ultrasonics is useful only up to about 10 or 20 meters.

PASSIVE INFRARED

A favorite for security systems and automatic outdoor lighting, passive pyroelectric infrared (PIR) sensors detect the natural heat that all objects emit. This heat takes the form of infrared radiation—a form of light that is beyond the limits of human vision. The PIR system merely detects a rapid change in the heat reaching the sensor; such a change usually represents movement.

The typical PIR sensor is equipped with a Fresnel lens to focus infrared light from a fairly wide area onto the pea-sized surface of the detector. In a robotics vision application, you can replace the Fresnel lens with a telephoto lens arrangement that permits the detector to view only a small area at a time. Mounted onto a movable platform, the sensor could detect the instantaneous variations of infrared radiation of whatever objects are in front of the robot. See Chapter 36, "Collision Avoidance and Detection," for more information on the use of PIR sensors.

TACTILE FEEDBACK

Many robots can be effective navigators with little more than a switch or two to guide their way. Each switch on the robot is a kind of "touch sensor": when a switch is depressed, the robot knows it has touched some object in front of it. Based on this information, the robot can stop and negotiate a different path to its destination.

To be useful, the robot's touch sensors must be mounted where they will come into contact with the objects in their surroundings. For example, you can mount four switches along the bottom periphery of a square-shaped robot so contact with any object will trigger one of the switches. Mechanical switches are triggered only on physical contact; switches that use reflected infrared light or capacitance can be triggered by the proximity of objects. Noncontact switches are useful if the robot might be damaged by running into an object, or vice versa. See Chapter 35, "Adding the Sense of Touch," for more information on tactile sensors.

From Here

To learn more about...	*Read*
Using a brain with your robot	Chapter 28, "An Overview of Robot 'Brains'"
Connecting sensors to a robot computer or microcontroller	Chapter 29, "Interfacing with Computers and Microcontrollers"
Using touch to guide your robot	Chapter 35, "Adding the Sense of Touch"
Getting your robot from point A to point B	Chapter 38, "Navigating through Space"

NAVIGATING THROUGH SPACE

The projects and discussion in this chapter focus on navigating your robot through space—not the outer-space kind, but the space between two chairs in your living room, between your bedroom and the hall bathroom, or outside your home by the pool. Robots suddenly become useful once they can master their surroundings, and being able to wend their way through their surrounds is the first step toward that mastery.

The techniques used to provide such navigation are varied: path-track systems, infrared beacons, ultrasonic rangers, compass bearings, dead reckoning, and more.

A Game of Goals

A helpful way to look at robot navigation is to think of it as a game, like soccer. The aim of soccer is for the members of one team to kick the ball into a goal. That goal is guarded by a member of the other team, so it's not all that easy to get the ball into the goal. Similarly, for a robot a lot stands between it and its goal of getting from one place to another. Those obstacles include humans, chairs, cats, a puddle of water, an electrical cord—just about anything can prevent a robot from successfully traversing a room or yard.

To go from point A to point B, your robot will consider the following process (as shown in Fig. 38.1):

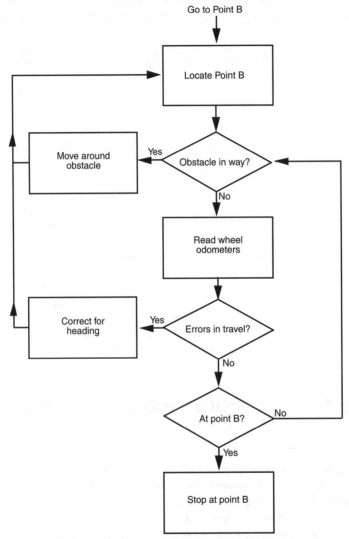

FIGURE 38.1 Navigation through open space requires that the robot be programmed not only to achieve the "goal" of a specific task but to self-correct for possible obstacles.

1. Retrieve instruction of goal: get to point B. This can come from an internal stimulus (battery is getting low; must get to power recharge station) or from a programmed or external command.
2. Determine where point B is in relation to current position (point A), and determine a path to point B. This requires obtaining the current position using known landmarks or references.
3. Avoid obstacles along the way. If an immovable obstacle is encountered, move around the obstacle and recalculate the path to get to point B.

4. Correct for errors in navigation ("in-path error correction") caused by such things as wheel slippage. This can be accomplished by periodically reassessing current position using known landmarks or references.

5. Optionally, time out (give up) if goal is not reached within a specific period of time or distance traveled.

Notice the intervening issues that can retard or inhibit the robot from reaching its goal. If there are any immovable obstacles in the way the robot must steer around them. This means its predefined path to get from point A to point B must be recalculated. Position and navigation errors are normal and are to be expected. You can reduce the effects of error by having the robot periodically reassess its position. This can be accomplished by using a number of referencing schemes, such as mapping, active beacons, or landmarks. More about these later in the chapter.

People don't like to admit failure, but a robot is just a machine and doesn't know (or care) that it failed to reach its intended destination. You should account for the possibility that the robot may never get to point B. This can be accomplished by using time-outs, which entails either determining the maximum reasonable time to accomplish the goal or, better yet, the maximum reasonable distance that should be traveled to reach the goal.

You can build other fail-safes into the system as well, including a program override if the robot can no longer reassess its current location using known landmarks or references. In such a scenario, this could mean its sensors have gone kaput or that the landmarks or references are no longer functioning or accurate. One course of action is to have the robot shut down and wait to be bailed out by its human master.

Following a Predefined Path: Line Tracing

Perhaps the simplest navigation system for mobile robots involves following some predefined path that's marked on the ground. The path can be a black or white line painted on a hard-surfaced floor, a wire buried beneath a carpet, a physical track, or any of several other methods. This type of robot navigation is used in some factories. The reflective tape method is preferred because the track can easily be changed without ripping up the floor.

You can readily incorporate a tape-track navigation system in your robot. The line-tracing feature can be the robot's only means of semi-intelligent action, or it can be just one part of a more sophisticated machine. You could, for example, use the tape to help guide a robot back to its battery charger nest.

With a line-tracing robot, you place a piece of white or reflective tape on the floor. For the best results, the floor should be hard, like wood, concrete, or linoleum, and not carpeted. One or more optical sensors are placed on the robot. These sensors incorporate an infrared LED and an infrared phototransistor. When the transistor turns on it sees the light from the LED reflected off the tape. Obviously, the darker the floor the better because the tape shows up against the background.

In a working robot, mount the LED and phototransistors in a suitable enclosure, as described more fully in Chapter 36, "Collision Avoidance and Detection." Or, use a commercially available LED-phototransistor pair (again, see Chapter 36). Mount the detectors on the bottom of the robot, as shown in Fig. 38.2, in which two detectors have been placed a little farther apart than the width of the tape. I used 1/4-inch art tape in the prototype for this book and placed the sensors 1/2 inch from one another.

Fig. 38.3 shows the basic sensor circuit and how the LED and phototransistor are wired. Feel free to experiment with the value of R2; it determines the sensitivity of the phototransistor. Fig. 38.4 shows the sensor and comparator circuit that forms the basis of the line-tracing system. Refer to this figure often because this circuit is used in many other applications.

You can use the schematics in Fig. 38.5 and Fig. 38.6 to build a complete line-tracing system (refer to the parts lists in Tables 38.1 and 38.2). You can build the circuit using just three IC packages: an LM339 quad comparator, a 7486 quad Exclusive OR gate, and a 7400 quad NAND gate. Before using the robot, block the phototransistors so they don't receive any light. Rotate the shaft of the set-point pots until the relays kick in, then back

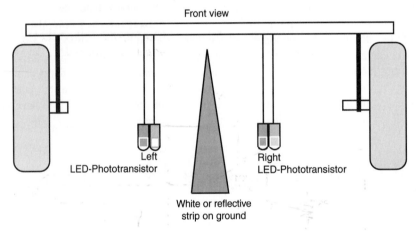

FIGURE 38.2 Placement of the left and right phototransistor-LED pair for the line-tracing robot.

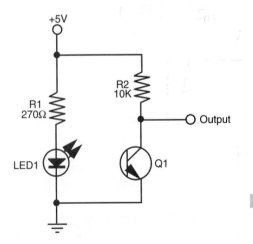

FIGURE 38.3 The basic LED-phototransistor wiring diagram.

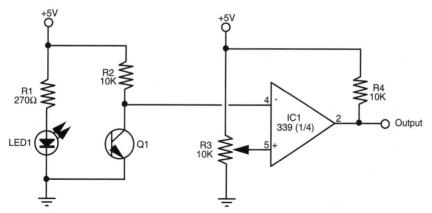

FIGURE 38.4 Connecting the LED and phototransistor to an LM339 quad comparator IC. The output of the comparator switches between HIGH and LOW depends on the amount of light falling on the phototransistor. Note the addition of the 10K "pullup" resistor on the output of the comparator. This is needed to assure proper HIGH/LOW action.

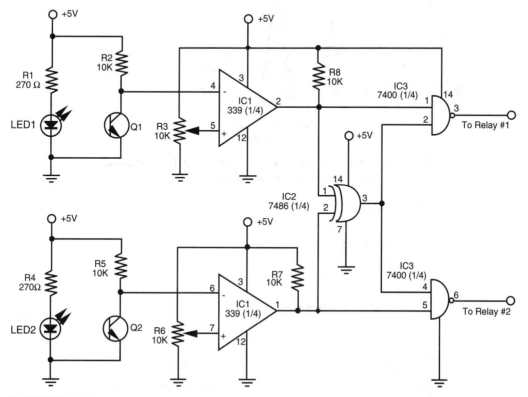

FIGURE 38.5 Wiring diagram for the line-tracing robot. The outputs of the 7400 are routed to the relays in Figure 38.6.

TABLE 38.1	PARTS LIST FOR LINE TRACER.
IC1	LM339 Quad Comparator IC
IC2	7486 Quad Exclusive OR Gate IC
IC3	7400 Quad NAND Gate IC
Q1,Q2	Infrared-sensitive phototransistors
R1,R4	270-ohm resistor
R2,R5, R7,R8	10K resistor
R3,R6	10K potentiometer
LED1,2	Infrared light-emitting diode
Misc.	Infrared filter for phototransistor (if needed)

All resistors are 5–10% tolerance.

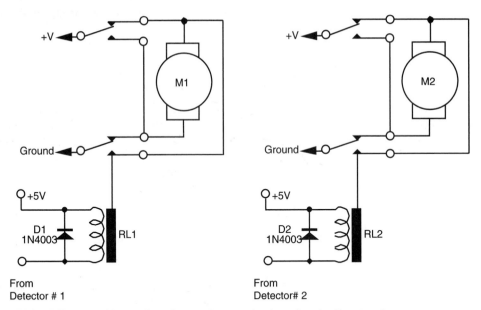

FIGURE 38.6 Motor direction and control relays for the line-tracing robot. You can substitute the relays for purely electronic control; refer to Chap. 18.

TABLE 38.2	PARTS LIST FOR RELAY CONTROL.
RL1,RL2	DPDT fast-acting relay; contacts rated 2 amps or more
D1, D2	1N4003 diodes

off again. You may have to experiment with the settings of the set point pots as you try out the system.

Depending on which motors you use and the switching speed of the relays, you may find your robot waddling its way down the track, overcorrecting for its errors every time. You can help minimize this by using faster-acting relays. Another approach is to vary the gap between the two sensors. By making it wide, the robot won't be turning back and forth as much to correct for small errors. I have also found that you can minimize this so-called *overshoot* effect by carefully adjusting the set-point pots.

You'll hardly ever see a railroad track with a turn tighter than about 8°. There is good reason for this. If the turn is made any tighter, the train cars can't stay on the track, and the whole thing derails. There is a similar limitation in line-tracing robots. The lines cannot be tighter than about 10° to 15°, depending on the robot's turning radius, or the thing can't act fast enough when it crosses over the line. The robot will skip the line and go off course.

The actual turn radius will depend entirely on the robot. If you need your robot to turn very tight, small corners, build it small. If your robot has a brain, whether it is a computer or central microprocessor, you can use it instead of the direct connection to the relays for motor control. The output of the comparators, when used with a +5 volt supply, is compatible with computer and microprocessor circuitry, as long as you follow the interface guidelines provided in Chapter 29. The two sensors require only two bits of an eight-bit port.

Wall Following

Robots that can follow walls are similar to those that can trace a line. Like the line, the wall is used to provide the robot with navigation orientation. One benefit of wall-following robots is that you can use them without having to paint any lines or lay down tape. Depending on the robot's design, the machine can even maneuver around small obstacles (doorstops, door frame molding, radiator pipes, etc.).

VARIATIONS OF WALL FOLLOWING

Wall following can be accomplished with any of four methods:

- *Contact.* The robot uses a mechanical switch, or a stiff wire that is connected to a switch, to sense contact with the wall, as shown in Fig. 38.7a. This is by far the simplest method, but the switch is prone to mechanical damage over time.
- *Noncontact, active sensor.* The robot uses active proximity sensors, such as infrared or ultrasonic, to determine its distance from the wall. No physical contact with the wall is needed. In a typical noncontact system, two sensors are used to judge when the robot is parallel to the wall (see Fig. 38.7b).
- *Noncontact, passive sensor.* The robot uses passive sensors, such as linear Hall effect switches, to judge distance from a specially prepared wall (Fig. 38.7c). In the case of Hall effect switches, you could outfit the baseboard or wall with an electrical wire through which a low-voltage alternating current is fed. When the robot is in the proximity of the switches the sensors will pick up the induced magnetic field provided by

the alternating current. Or, if the baseboard is metal the Hall effect sensor (when rigged with a small magnet on its opposite side) could detect proximity to a wall.

■ *"Soft-contact."* The robot uses mechanical means to detect contact with the wall, but the contact is "softened" by using pliable materials. For example, you can use a lightweight foam wheel as a "wall roller," as shown in Fig. 38.7*d*. The benefit of soft contact is that mechanical failure is reduced or eliminated because the contact with the wall is made through an elastic or pliable medium.

In all cases, upon encountering a wall the robot goes into a controlled program phase to follow the wall in order to get to its destination. In a simple contact system, the robot may back up a short moment after touching the wall, then swing in a long arc toward the wall again. This process is repeated, and the net effect is that the robot "follows the wall."

With the other methods, the preferred approach is for the robot to maintain proper distance from the wall. Only when proximity to the wall is lost does the robot go into a "find wall" mode. This entails arcing the robot toward the anticipated direction of the wall. When contact is made, the robot alters course slightly and starts a new arc. A typical pattern of movement is shown in Fig. 38.8.

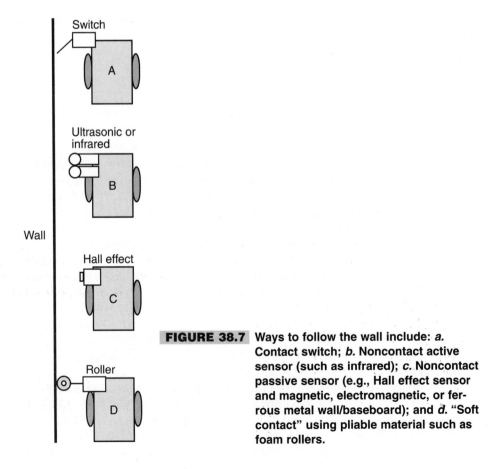

FIGURE 38.7 Ways to follow the wall include: *a.* Contact switch; *b.* Noncontact active sensor (such as infrared); *c.* Noncontact passive sensor (e.g., Hall effect sensor and magnetic, electromagnetic, or ferrous metal wall/baseboard); and *d.* "Soft contact" using pliable material such as foam rollers.

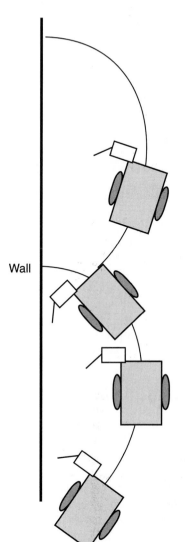

Wall

FIGURE 38.8 A wall-following robot that merely "feels" its way around the room might make wide, sweeping arcs. The arc movement is easily accomplished in a typical two-wheeled robot by running one motor slower than the other.

ULTRASONIC WALL FOLLOWING

A simple ultrasonic wall follower can use two ultrasonic transmitter/receiver pairs. Each transmitter and receiver is mounted several inches apart to avoid cross talk. Two transmitter/receiver pairs are used to help the robot travel parallel to the wall. Suitable ultrasonic transmitter and receiver circuits are detailed in Chapter 36, "Collision Avoidance and Detection."

Because the robot will likely be close to the wall (within a few inches), you will want to drive the transmitters at very low power and use only moderate amplification, if any, for the receiver. You can drive the transmitters at very low power by reducing the voltage to the transmitter.

SOFT-CONTACT FOLLOWING WITH FOAM WHEEL

Soft-contact wall following with a roller wheel offers you some interesting possibilities. In fact, you may be able to substantially simplify the sensors and control electronics by placing an idler roller made of soft foam as an outrigger to the robot and then having the robot constantly steer inward toward the wall. This can be done simply by running the inward wheel (the wheel on the side of the wall) a little slower than the other. The foam idler roller will prevent the robot from hitting the wall.

DEALING WITH DOORWAYS AND OBJECTS

Merely following a wall is, in essence, not that difficult. The task becomes more challenging when you want the robot to maneuver around obstacles or skip past doorways. This requires additional sensors, perhaps whiskers or other touch sensors in the forward portion of the robot. These are used to detect corners as well. This is especially important when you are constructing a robot that has a simple inward-arc behavior toward following walls. Without the ability to sense a wall straight ahead, the robot may become hopelessly trapped in a corner.

Open doorways that lead into other rooms can be sensed using a longer-range ultrasonic transducer. Here, the long-range ultrasonic detects that the robot is far from any wall and places the machine in a "go straight" mode. Ideally, the robot should time the duration of this mode to account for the maximum distance of an open doorway. If a wall is not detected within X seconds, the robot should go into a "look for wall" mode.

Odometry: The Art of Dead Reckoning

Hop into your car. Note the reading on the odometer. Now drive straight down the road for exactly one minute, paying no attention to the speedometer or anything else (of course, keep your eyes on the road!). Again note the reading on the odometer. The information on the odometer can be used to tell you where you are. Suppose it says one mile. You know that if you turn the car around exactly 180° and travel back one mile, at whatever speed, you'll reach home again.

This is the essence of odometry, reading the motion of a robot's wheels to determine how far it's gone. Odometry is perhaps the most common method for determining where a robot is at any given time. It's cheap and easy to implement and is fairly accurate over short distances. Odometry is similar to the "dead reckoning" navigation used by sea captains and pilots before the age of satellites, radar, and other electronic schemes. Hence, odometry is also referred to in robot literature as dead reckoning.

Unlike your car, robots don't have speedometers connected to their transmissions or front wheels to drive the odometer. Instead, a robot's "odometer" is typically devised using optical or magnetic sensors. Let's take a look at how each kind is used in a robot.

OPTICAL ENCODERS

You can use a small disc fashioned around the hub of a drive wheel, or even the shaft of a drive motor, as an optical shaft encoder (described in "Anatomy of a Shaft Encoder," in Chapter 18). The disc can be either the reflectance or the slotted type:

- With a *reflectance* disc, infrared light strikes the disc and is reflected back to a photodetector.
- With a *slotted* disc, infrared light is alternately blocked and passed and is picked up on the other side by a photodetector.

With either method, a pulse is generated each time the photodetector senses the light.

MAGNETIC ENCODERS

You can construct a magnetic encoder using a Hall effect switch (a semiconductor sensitive to magnetic fields) and one or more magnets. A pulse is generated each time a magnet passes by the Hall effect switch. A variation on the theme uses a metal gear and a special Hall effect sensor that is sensitive to the variations in the magnetic influence produced by the gear (see Fig. 38.9).

A bias magnet is placed behind the Hall effect sensor. A pulse is generated each time a tooth of the gear passes in front of the sensor. The technique provides more pulses on each revolution of the wheel or motor shaft, and without having to use separate magnets on the rim of the wheel or wheel shaft.

THE FUNCTION OF ENCODERS IN ODOMETRY

As the wheel or motor shaft turns, the encoder (optical or magnetic) produces a series of pulses relative to the distance the robot travels. Assume the wheel is 3 inches in diameter (9.42 inches in circumference), and the encoder wheel has 32 slots. Each pulse of the encoder represents 0.294 inches of travel (9.42/32). If the robot senses 10 pulses, it knows it has moved 2.94 inches.

If the robot uses the traditional two-wheel drive approach, you attach optical encoders to both wheels. This is necessary because the drive wheels of a robot are bound to turn at

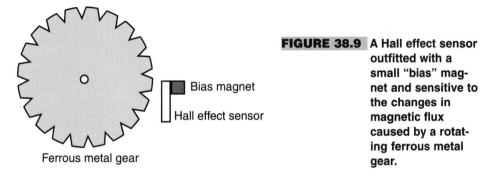

Bias magnet

Hall effect sensor

Ferrous metal gear

FIGURE 38.9 A Hall effect sensor outfitted with a small "bias" magnet and sensitive to the changes in magnetic flux caused by a rotating ferrous metal gear.

slightly different speeds over time. By integrating the results of both optical encoders, it's possible to determine where the robot really is as opposed to where it should be (see Fig. 38.10). As well, if one wheel rolls over a cord or other small lump, its rotation will be hindered. This can cause the robot to veer off course, possibly by as much as 3° to 5° or more. Again, the encoders will detect this change.

It's best to make odometry measurements using a microcontroller that is outfitted with a *pulse accumulator* or *counter* input. These kinds of inputs independently count the number of pulses received since the last time they were reset. To take an odometry reading, you clear the accumulator or counter and then start the motors. Your software need not monitor the accumulator or counter. Stop the motors, and then read the value in the accumulator or counter. Multiply the number of pulses by the known distance of travel for each pulse. (This will vary depending on the construction of your robot; consider the diameter of the wheels and the number of pulses of the encoder per revolution.)

If the number of pulses from both encoders is the same, you can assume that the robot traveled in a straight line, and you have only to multiple the number of pulses by the distance per pulse. For example, if there are 1055 pulses in the accumulator-counter, and if each pulse represents 0.294 inches of travel, then the robot has moved 310.17 inches straight forward.

ERRORS IN ODOMETRY

In a perfect world, robots would not need anything more than an odometer to determine exactly where they were at any given time. Unfortunately, robots live and work in a world that is far from perfect; as a result, their odometers are far from accurate. Over a 20- to 30-foot range, for example, it's not uncommon for the average odometer to misrepresent the position of the robot by as much as half a foot or more!

Why the discrepancy? First and foremost: wheels slip. As a wheel turns, it is bound to slip, especially if the surface is hard and smooth, like a kitchen floor. Wheels slip even

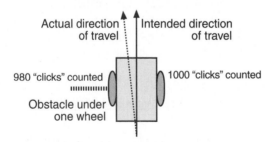

FIGURE 38.10 The relative number of "counts" from each encoder of the typical two-wheeled robot can be used to indicate deviation in travel. If an encoder shows that one wheel turned a fewer number of times than the other wheel, then it can be assumed the robot did not travel in a straight line.

more when they turn. The wheel encoder may register a certain number of pulses, but because of slip the actual distance of travel will be less. Certain robot drive designs are more prone to error than others. Robots with tracks are steered using slip—lots of it. The encoders will register pulses, but the robot will not actually be moving in proportion.

There are less subtle reasons for odometry error. If you're even a hundredth of an inch off when measuring the diameter of the wheel, the error will be compounded over long distances. If the robot is equipped with soft or pneumatic wheels, the weight of the robot can deform the wheels, thereby changing their effective diameter.

Because of odometry errors, it is necessary to combine it with other navigation techniques, such as active beacons, distance mapping, or landmark recognition. All three are detailed later in this chapter.

Compass Bearings

Besides the stars, the magnetic compass has served as humankind's principal navigation aid over long distances. You know how it works: a needle points to the magnetic north pole of the earth. Once you know which way is north, you can more easily reorient yourself in your travels.

Robots can use compasses as well, and a number of electronic and electromechanical compasses are available for use in hobby robots. One of the least expensive is the Dinsmore 1490, from Dinsmore Instrument Co. The 1490 looks like an overfed transistor, with 12 leads protruding from its underside. The leads are in four groups of three; each group represents a major compass heading: north, south, east, and west. The three leads in each group are for power, ground, and signal. A Dinsmore 1490, mounted on a circuit board, is shown in Fig. 38.11.

The 1490 provides eight directions of heading information (N, S, E, W, SE, SW, NE, NW) by measuring the earth's magnetic field. It does this by using miniature Hall effect sensors and a rotating compass needle (similar to ordinary compasses). The sensor is said to be internally designed to respond to directional changes much like a liquid-filled compass. It turns to the indicated direction from a 90° displacement in approximately 2.5 seconds. The manufacturer's specification sheet claims that the unit can operate with up to 12° of tilt with acceptable error, but it is important to note that any tilting from center will cause a corresponding loss in accuracy.

Fig. 38.12 shows the circuit diagram for the 1490, which uses four inputs to a computer or microcontroller. Note the use of pullup resistors. With this setup, your robot can determine its orientation with an accuracy of about 45° (less if the 1490 compass is tilted). Dinsmore also makes an analog-output compass that exhibits better accuracy.

Another option is the Vector 2X and 2XG. These units use magneto-inductive sensors for sensing magnetic fields. The Vector 2X/2XG provides either compass heading or uncalibrated magnetic field data. This information is output via a three-wire serial format and is compatible with Motorola SPI and National Semiconductor Microwire interface standards. Position data can be provided either 2.5 or 5 times per second.

Vector claims accuracy of ±2°. The 2X is meant to be used in level applications. The more pricey 2XG has a built-in gimbal mechanism that keeps the active magnetic-inductive element level, even when the rest of the unit is tilted. The gimbal allows tilt up to 12°.

FIGURE 38.11 The Dinsmore 1490 digital compass provides simple bearings for a robot. The sensor is accurate to about 45°.

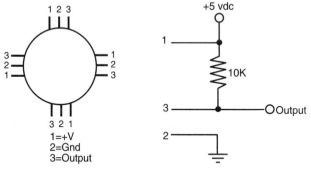

FIGURE 38.12 Circuit diagram for using the Dinsmore 1490 digital compass. When used with a +5 vdc supply, the four outputs can be connected directly to a microcontroller. One or two outputs can be activated at a time; if two are activated, the sensor is reading between the four compass points (e.g., N and W outputs denotes NW position).

Ultrasonic Distance Measurement

Police radar systems work by sending out a high-frequency radio beam that is reflected off nearby objects, such as your car as you are speeding down the road. The difference between the time when the transmit pulse is sent and when the echo is received denotes distance. Speed is calculated using the Doppler effect: the time between the sending pulse and echo increases or decreases proportionately depending on how fast you are going.

Radar systems are complex and expensive, and most require certification by a government authority, such as the Federal Communications Commission for devices used in the United States. There is another approach: you can use high-frequency sound instead to measure distance, and with the right circuitry you can even provide a rough indication of speed.

Ultrasonic ranging is, by now, an old science. Polaroid used it for years as an automatic focusing aid on their instant cameras. Other camera manufacturers have used a similar technique, though it is now more common to implement infrared ranging (covered later in the chapter). The Doppler effect that is caused when something moves toward or away from the ultrasonic unit is used in home burglar alarm systems. However, for robotics the more typical application of ultrasonic sound is either to detect proximity to an object (see "Ultrasonic Wall Following," earlier in the chapter) or to measure distance (also called ultrasonic ranging).

To measure distance, a short burst of ultrasonic sound—usually at a frequency of 40 kHz for most ultrasonic ranging systems—is sent out through a transducer (a specially built ultrasonic speaker). The sound bounces off an object, and the echo is received by another transducer (this one a specially built ultrasonic microphone). A circuit then computes the time it took between the transmit pulse and the echo and comes up with distance.

Certainly, the popularity of ultrasonics does not detract from its usefulness in robot design. The system presented here is suited for use with a computer or microcontroller. There are a variety of ways to implement ultrasonic ranging. One method is to use the ultrasonic transducer and driver board from an old Polaroid instant camera, such as the Polaroid Sun 660 or the Polaroid SX-70 One Step. However, the driver board used in these cameras may require some modification to allow more than one "ping" of ultrasonic sound without having to cycle the power to the board off, then back on. More about this in a bit. You can also purchase a new Polaroid ultrasonic transducer and driver board from a number of mail order sources, including on the Internet. Several of these outlets are listed in Appendix B, "Sources." These units are new, and most come with documentation, including hookup instructions for connecting them to popular microcontrollers, such as the Basic Stamp. Perhaps the most common Polaroid distance measuring kit is composed of the so-called 600 Series Instrument Grade transducer along with its associated Model 6500 Ranging Module.

The transducer, which is about the size of a silver dollar coin, acts as both ultrasonic transmitter and receiver. Because only a single transducer is used, the Polaroid system as described in this section cannot detect objects closer than about 1.3 feet. This is because of the amount of time required for the transducer to stop oscillating before it sets itself up to receive. The maximum distance of the sensor is about 35 feet when used indoors, and a little less when used outdoors, especially on a windy day. The system is powered by a single 6-vdc battery pack and can be interfaced to any computer or microcontroller.

FACTS AND FIGURES

First some statistics. At sea level, sound travels at a speed of about 1130 feet per second (about 344 meters per second) or 13,560 inches per second. This time varies depending on atmospheric conditions, including air pressure (which varies by altitude), temperature, and humidity. The time it takes for the echo to be received is in microseconds if the object is within a few inches or even a few feet of the robot. The short duration is really no problem, however, for fast-acting CMOS and TTL ICs. The overall time between transmit pulse and echo is divided by two to compensate for the round-trip travel time between the robot and the object.

Given a travel time of 13,560 inches per second for sound, it takes just 73.7 microseconds (0.0000737 seconds) for sound to travel one inch. With this figure in the back of our minds, let's consider how the Polaroid ranging system works. The Ranging Module is connected to a computer or microcontroller using only two wires: INIT (for INITiate) and ECHO. INIT is an output, and ECHO is an input. The Ranging Module contains other I/O connections, such as BLNK and BINH, but these are not strictly required when you are determining distance to a single object, and so they will not be discussed here.

To trigger the Ranging Module and have it send out a burst of ultrasonic sound, the computer or microcontroller brings the INIT line HIGH. The computer-microcontroller then waits for the ECHO line to change from LOW to HIGH. The time difference, in microseconds, is divided in two, and that gives you distance. To measure the time between the INIT pulse and the return ECHO, the computer or microcontroller uses a timer to precisely count the time interval.

Different timing-counting approaches are used depending on the computer or microcontroller you are using. For example, with the Basic Stamp or BasicX microcontrollers (see Chapters 31 and 32, respectively), you might use the RCTime function, which is normally used to time how long it takes for a capacitor to discharge. There is no capacitor to discharge in the Ranging Module, but the overall timing technique is still the same. With the OOPic microcontroller (see Chapter 33), you might use its oTimer object.

Let's suppose you're using the BasicX microcontroller. The short bit of code in Listing 38.1, which is taken from the BasicX application note on ultrasonic ranging, uses pins 15 and 16 of the chip to connect to the ECHO and INIT lines, respectively, of the Polaroid Ranging Module. The lines of the Ranging Module are connected as shown in Fig. 38.13.

Note that the BLNK and BINH lines are held LOW and that the power supply to the Polaroid Ranging Module *must not* come from the on-board regulator of the BasicX. The Polaroid Ranging Module needs a far more robust power supply that is capable of delivering an amp or two of current for a brief period of time. Four AA batteries connected in series will suffice. Connect the ground from the 6-vdc battery pack to the ground points of Polaroid Ranging Module and the BasicX.

LISTING 38.1.

```
' Connect pin 15 of BasicX to ECHO, pin 16 to INIT
Private Const EchoPin As Byte = 15
Private Const InitPin As Byte = 16
```

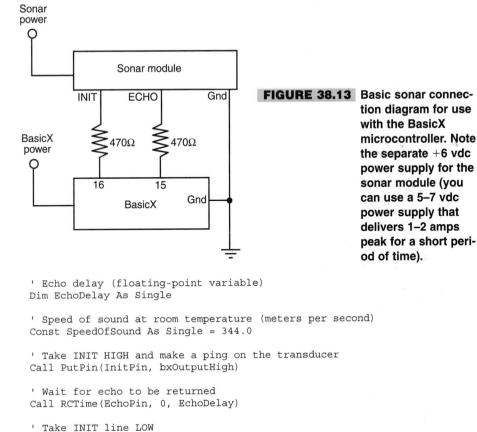

FIGURE 38.13 Basic sonar connection diagram for use with the BasicX microcontroller. Note the separate +6 vdc power supply for the sonar module (you can use a 5–7 vdc power supply that delivers 1–2 amps peak for a short period of time).

```
' Echo delay (floating-point variable)
Dim EchoDelay As Single

' Speed of sound at room temperature (meters per second)
Const SpeedOfSound As Single = 344.0

' Take INIT HIGH and make a ping on the transducer
Call PutPin(InitPin, bxOutputHigh)

' Wait for echo to be returned
Call RCTime(EchoPin, 0, EchoDelay)

' Take INIT line LOW
Call PutPin(InitPin, bxOutputLow)

' If no echo RCTime overflows and returns 0.0
If (EchoDelay = 0.0) Then
     Range = 11.0
Else
     Range = (EchoDelay / 2.0) * SpeedOfSound
End If
```

When INIT is taken HIGH, the Polaroid Ranging Module emits a short burst of ~50 kHz sound from the transducer. The module then waits for a period of 2.38 milliseconds for the transducer to stop ringing. This is the period of time it takes for the sonar ping to travel about 32 inches. Considering round-trip time, this equates to the 1.3-foot minimum imposed by the system. After this so-called "blanking" period, the Polaroid Ranging Module listens for the return ECHO. When an echo is detected, the ECHO line goes HIGH. Note that the module itself does not do any timing; this is the domain of the microcontroller that is connected to the module.

GUTTING A POLAROID SUN 660 CAMERA

Before moving on to the next subject of the chapter, it's worth noting that used Polaroid cameras are commonly available in thrift stores and on Internet auction sites such as eBay.

Even models with the built-in sonar ranging system are commonly available for under $10, and unless the camera has been damaged, they likely still work. (This is a testament to the excellent manufacturing quality of Polaroid cameras, despite their "snap-together" construction, as you'll see in a bit.)

Most of these cameras use a transducer and ranging module very similar to the units described already, though you will probably encounter a couple of variations. In addition, you must first disassemble the camera to extract the ultrasonic transducer and the ranging module board. This actually isn't as easy as it looks, because Polaroid was known for building their cameras with few, if any, screws. Instead, the cameras must be disassembled like a Chinese puzzle box: first this part, then that, then this one over here—all the while being careful you don't break anything important.

The methods for deconstructing the typical Polaroid camera are beyond the scope of this book, but here are a few tips. Bear in mind that when you dissect a Polaroid camera for its ultrasonic parts you render the camera completely inoperative:

- Start first by removing the film door and/or film rollers. This is typically accomplished by finding and removing the small hinges and pins that hold these parts onto the main body of the camera.
- Pry off the faceplate. Use a thin flat-bladed screwdriver and carefully look for the "snap" points. It's okay if you break off a little bit of plastic here and there.
- When you reach the innards of the camera, locate the small plastic pins that secure the transducer element. This is a delicate part of the camera, so very carefully pry the transducer loose. Under no circumstances should you disassemble the transducer or touch the gold-plated contact surface inside. Doing so will ruin the transducer (the gold plating will come right off in your hands).

Locate the ranging module and carefully pry it up (it'll likely be held down with two or more small plastic prongs). You can remove the wide multi-pin connector from the module, but keep the shielded wires to the ultrasonic transducer intact.

When you are done, you should have something that looks like the module and transducer in Fig. 38.14. This module came out of a Sun 660, and not all Polaroid modules look the same. Though similar in design, the pinouts of the multi-pin connector are different from those found in the Model 6500 Ranging Module described earlier in the chapter. Table 38.3 lists the pinouts for the Model 6500 Ranging Module as well as the ranging module from the Sun 660. Note that the Sun 660 module has an eight-pin connector; the Model 6500 has a nine-pin connector.

A disadvantage of using sonar ranging boards removed from Polaroid cameras is that in many cases, the board is not able to produce more than one "ping" of ultrasonic sound without recycling the power off, then back on. The Polaroid cameras from which the boards are taken are powered by a battery contained in the film pack. Between pictures, power to the electronics inside the camera is turned off, which resets the sonar ranging board.

There are a number of ways to modify the sonar board to permit it to ping more than once, and without recycling its power. I've found that one of the most effective—and easy—methods is to add a small single-pole, single-throw (SPST) five-volt relay between the sonar ranging module and the battery that powers it. The relay should be rated for at least one amp. The relay is controlled by the robot's microcontroller or microprocessor. A basic hook-up scheme for controlling a relay with a computer is shown in Chapter 18, "Working with DC Motors."

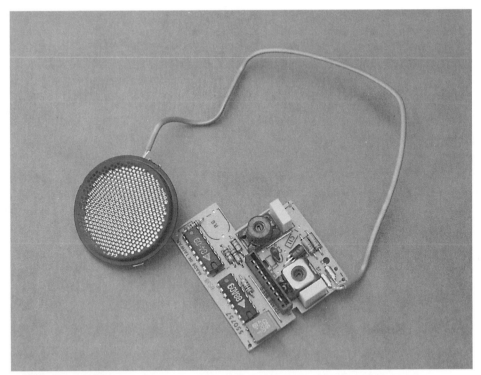

FIGURE 38.14 A ranging module pulled from a Polaroid Sun 660 camera. The camera was purchased for under $10 at a thrift store.

Infrared Distance Measurement

Ultrasonic sound is not the only method you can use to measure distance between your 'bot and some object. Infrared light can also be used. Unlike ultrasonic measurement, infrared distance sensors don't attempt to determine the time-of-flight for a light beam—it would be on the order of femto- and picoseconds for the distances we're interested in. Only the most costly electronic circuitry can handle these speeds.

Instead, infrared systems use a technical known as parallax, that is, the measurement of the angle of reflectance between a known light source and its return beam. Here's how the technique works: A beam of infrared light illuminates a scene. The beam reflects off an object in front of the sensor and bounces back into the sensor. The closer the object is, the greater the angle of displacement due to parallax. The reflected beam falls onto a linear array of very small photodetectors. This photodetector array is connected to circuitry that resolves the distance of the object. The circuitry can provide either a digital or an analog output. We'll cover both varieties here.

The premier maker of infrared distance measurement sensors for use in robotics is Japan-based Sharp. One of their infrared distance measurement sensors, the GP2D02, is shown in Fig. 38.15. Actually, Sharp doesn't make these sensors for the robotics industry; rather, they are principally intended for use in cars for proximity devices and copiers for

TABLE 38.3   PINOUTS FOR THE POLAROID MODEL 6500 RANGING MODULE AND "TAKE OUT" RANGING MODEL FROM THE POLAROID SUN 660 MODULE.

POLAROID MODULE 6500 RANGING MODULE	
PIN	FUNCTION
1	GND
2	BLNK
3	(not used; *do not* connect)
4	INIT
5	(not used; *do not* connect)
6	OSC
7	ECHO
8	BINH
9	V+
RANGING MODULE FROM SUN 660	
1	GND
2	BLNK
3	BINH
4	INIT
5	(not used; *do not* connect)
6	OSC
7	ECHO
8	V+

paper detection. Depending on the model, the sensors have a range of about 4 inches (10 cm) and 31.5 inches (80 cm).

We'll talk about three Sharp sensors, all of which have terribly nondescriptive names:

- *GP2D05*—Digital HIGH/LOW output registers whether an object is within a preset range.
- *GP2D02*—Digital serial output indicates range as an 8-bit value.
- *GP2D12*—Analog output indicates range as a voltage level.

(Also available is the GP2D15. It has a 3- to 30-cm range, and outputs a digital HIGH/LOW value depending on range to the detected object. It's not quite as useful for robotics work as the others, but works on the same principles.)

In all cases, the Sharp infrared sensors share better-than-average immunity to ambient light levels, so you can use them under a variety of lighting conditions (except perhaps

FIGURE 38.15 The Sharp infrared sensors are equipped with a focused infrared light source and a linear photodetector array. Distance can be determined by detecting where the reflected light touches the array.

very bright light outdoors). The sensors use a modulated—as opposed to a continuous—infrared beam that helps reject false triggering. It also makes the system accurate even if the detected object absorbs or scatters infrared light, such as heavy curtains or dark-colored fabrics.

USING THE GP2D05 INFRARED DISTANCE JUDGMENT SENSOR

The GP2D05 is a "distance judgment" sensor rather than a ranging sensor. It has a one-bit output that is either HIGH or LOW depending on whether an object has been detected within a threshold range. This range is set by adjusting a small potentiometer on the back of the sensor. Range is from 10 to 80 cm, depending on the adjustment of the pot. Fig. 38.16 shows a typical hookup diagram for the GP2DO5. To use the sensor, the *Vin* line is brought LOW for no more than 56 milliseconds (28 milliseconds is typical). If the *Vout* line goes HIGH after this period of time, it means that there was an object detected within the preset range of the sensor. If the line does not go high, it means no object was detected.

USING THE GP2D02 DIGITAL SERIAL OUTPUT INFRARED RANGING SENSOR

The GP2D02 digital serial output infrared sensor is probably the most commonly used of the Sharp units. Its output is an eight-bit serial digital train. The hookup diagram is

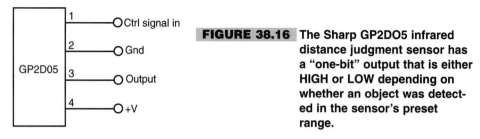

FIGURE 38.16 The Sharp GP2DO5 infrared distance judgment sensor has a "one-bit" output that is either HIGH or LOW depending on whether an object was detected in the sensor's preset range.

shown in Fig. 38.17. To use the GP2D02, you must send a clock signal to the sensor, then store each of the eight bits that are returned. Convert those eight bits into a value (from 0 to 255), and this is the range (in noncorrelated "units") from the sensor to the detected object.

Listing 38.2 demonstrates a simple Basic Stamp II program for use with the GP2D02 sensor. It displays the eight-bit result from the sensor in the debug window.

LISTING 38.2.

```
DataInput                  con     0
ClockOutput                con     1
StorageVariable            var     byte
RepeatLoop:
    LOW ClockOutput              ' activate detector
    Pause 70                     ' initial wait of 70 milliseconds
    Wait:
          If In0 = 0 Then Wait ' wait for output if needed
    ' shift in data
    SHIFTIN DataInput, ClockOutput, MSBPOST, [StorageVariable]
    HIGH ClockOutput     ' deactivate detector
    DEBUG dec StorageVariable, CR    ' display result
    PAUSE 1000           ' waits 1 sec; wait at least 2 ms before repeating
      GOTO RepeatLoop      ' repeat again
```

The eight-bit output value of the GP2D02 is not linear, which means that you can't expect a 1:1 ratio between the value you get and the distance separating the sensor and the detected object. For the value to be meaningful, you should conduct tests with objects placed set distances from the sensor (use a tape measure for accuracy). Note the values you get. The higher the value (say, 230 or 240) the closer the object is, and objects closer than 10 cm will yield unpredictable results. Values from 30 to 50 denote objects at the far end of the detection range, which is 80 cm.

The accuracy of the readings will depend greatly on the width of the target. You may wish to experiment by placing the sensor in front of a smooth white wall. Vary the distance between wall and sensor and note your results.

USING THE GP2D12 ANALOG OUTPUT INFRARED RANGING SENSOR

The GP2D12 is similar to the GP2D02 of the last section, except that it provides an analog output rather than a digital one. In some situations (and with some microcontrollers), an analog output is easier to deal with. This is the case if your microcontroller or computer has

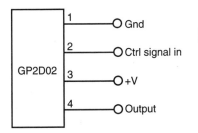

FIGURE 38.17 The Sharp GP2D02 sensor (pinout diagram shown here), which can be connected to a microcontroller or computer using a small diode (the diode drops the voltage to the sensor).

one or more analog-to-digital converter (ADC) inputs. Examples of microcontrollers with ADC inputs include the BasicX (see Chapter 32) and the OOPic (see Chapter 33).

Fig. 38.18 shows the connection of the GP2D12. When powered by +5 vdc, the GP2D12 outputs a voltage that is related to the distance between it and the detected object. The voltage span is approximately 0.6 volts to 3.1 volts. The lower the voltage, the farther away the object is, as shown in Fig. 38.19.

"Where Am I?": Sighting Landmarks

Explorers rely on landmarks to navigate wide-open areas. It might be an unusual outcropping of rocks or a bend in a river. Or the 7-Eleven down the street. In all cases, a landmark serves to give you general bearings. From these general bearings you can more readily navigate a given locale. Robots can use the same techniques, though rocks, rivers, and convenience stores are somewhat atypical as useful landmarks. Instead, robots can use such techniques as infrared beacons to determine their absolute position within a known area. The following sections describe some techniques you may wish to consider for your next robot project.

INFRARED BEACON

Unless you confine your robot to playing just within the laboratory, you'll probably want to provide it with a means to distinguish one room in your house from the next. This is particularly important if you've designed the robot with even a rudimentary form of object and area mapping. This mapping can be stored in the robot's memory and used to steer around objects and avoid walls.

For less than a week's worth of groceries, you can construct an infrared beacon system that your robot can use to determine when it has passed from one room to the next. You equip the robot with a receiver and place a transmitter in each room. The transmitters send out a unique code, which the robot interprets as a specific room. Once it has identified the room, it can retrieve the mapping information previously stored for it and use it to navigate through its surroundings.

The beacon system that follows is designed around a set of television and VCR remote control chips sold by Holtek. The chips are reasonably inexpensive but can be difficult to find. The chips used in this project are HT12D and HT12E; I bought mine at Jameco (*www.jameco.com,* but you should check the Internet for other sources as well).

You can, of course, use just about any wireless remote control system you desire. The only requirements are that you must be able to set up different codes for each transmitting

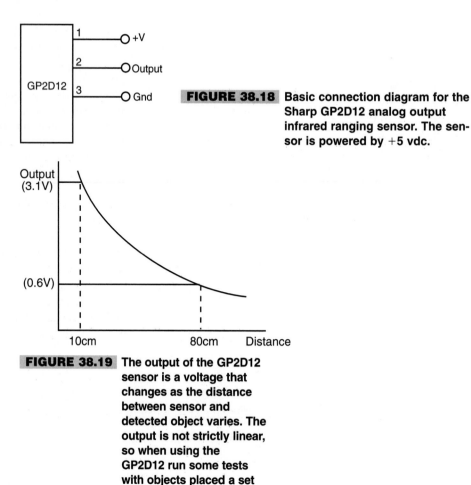

FIGURE 38.18 Basic connection diagram for the Sharp GP2D12 analog output infrared ranging sensor. The sensor is powered by +5 vdc.

FIGURE 38.19 The output of the GP2D12 sensor is a voltage that changes as the distance between sensor and detected object varies. The output is not strictly linear, so when using the GP2D12 run some tests with objects placed a set distance apart.

station and that the system must work with infrared light. You'll experience too much interference if you use radio control or ultrasonics.

You can connect the four-bit output of the HT-12D decoder IC to a microcontroller or computer. You will also want to connect the VD (valid data) line to a pin of your microcontroller or computer. When this line "winks" LOW, it means there is valid data on the four data lines. The value at the four data lines will coincide with the setting of the four-position DIP switch on each transmitter.

RADIO FREQUENCY IDENTIFICATION

Though a fairly old technology that dates back decades, radio frequency identification (RFID) uses small passive devices that radiate a digital signature when exposed to a radio frequency signal. RFID is found in products ranging from toys—most notably the *Star Wars* Episode 1 action figures—to trucking, farm animal inventories, automobile manufacturing, and more.

A transmitter/receiver, called the *interrogator* or *reader,* radiates a low- or medium-frequency carrier RF signal. If it is within range, a passive (unpowered) or active (powered) detector, called a *tag* or *transponder,* re-radiates (or "backscatters") the carrier frequency, along with a digital signature that uniquely identifies the device. RFID systems in use today operate on several common RF bands, including a low-speed 100–150 kHz band and a higher 13.5 MHz band.

The tag is composed of an antenna coil along with an integrated circuit. The radio signal provides power when used with passive tags, using well-known RF field induction principles. Inside the integrated circuit are decoding electronics and a small memory. A variety of data transmission schemes are used, including non-return-to-zero, frequency shift keying, and phase shift keying. Manufacturers of the RFID devices tend to favor one system over another for specific applications. Some data modulation schemes are better at long distances, for example.

Different RFID tags have different amounts of memory, but a common device might provide for 64 to 128 bits of data. This is more than enough to serve as room-by-room or locale-by-locale beacons. The advantage RFID has over infrared beacons (see earlier in this chapter) is that the coverage of the RF signal is naturally limited. While this limitation can certainly be a disadvantage, when properly deployed it can serve as a convenient way to differentiate between different areas of a house's robotic work space. The average working distance between interrogator and tag is several feet, though this varies greatly depending on the power output of the interrogator. Units with higher RF power can be used over longer distances. For room-by-room robotics use, however, we actually prefer a limited range, which also means a less expensive system.

While RFID systems are not complex, their cost is not quite in the super-affordable region (demonstration and developers' kits are available from some manufacturers for $100–$200, and this includes the reader and an assortment of tags). However, once implemented RFID is a low-maintenance, long-term solution for helping your robot know where it is.

LANDMARK RECOGNITION

As mentioned, humans navigate the real world by using landmarks: the red barn on the way to work signals you're getting close to your turnoff. Robots can use the same kind of visual cues to help them navigate a space. Landmarks can be "natural"—a support pillar in a warehouse for example—or they can be artificial, reflectors, posts, or bar codes positioned just for use by the robot. A key benefit of landmark recognition is that most systems are easy to install, cheap, and when done properly unmistakable from the robot's point of view.

Wide field bar code One technique to consider is the use of wide-field bar codes, which are commonly used in warehouses for quick and easy inventory. The bar code pattern is printed very large—perhaps as tall as two inches and as wide as a foot. A traditional laser bar code reader then scans the code. The large size of the bar code makes it possible to use the bar code reader even from a distance—10 to 20 feet or more.

You can adapt the same method to help your robot navigate from room to room, and even within a room. For each location you want to identify, print up a large bar code. Free and low-cost bar code printing software is available over the Internet and in several commercial packages. You can either make or purchase a wide-field bar code scanner and connect it to your robot's computer or microcontroller. As your robot roams about, the scanner can be constantly looking for bar codes. The laser light output from the scanner is very low and, if properly manufactured, is well within safe limits even if the beam should quickly scan past the eyes of people or animals).

Door frame "flags"   Yet another technique that merits consideration is the use of reflective tape placed around the frames of doors. Doorways are uniquely helpful in robot navigation because in the human world we tend to leave the space around them open and uncluttered. This allows us to enter and exit a room without tripping over something. It also typically means that line of sight of the door will not be blocked, creating a reliable landmark for a robot.

Imagine vertical strips of reflective tape on either side of the doorway. These strips could reflect the light from a scanning laser mounted on the robot, as shown in Fig. 38.20. The laser light would be reflected from the tape and received by a sensor on the robot. Since the speed of the laser scan is known, the timing between the return "pulses" of the reflected laser light would indicate the relative distance between the robot and the doorway. You could use additional tape strips to reduce the ambiguity that results when the robot approaches the doorway at an angle.

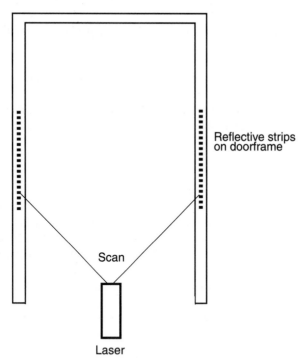

Reflective strips on doorframe

Scan

Laser

FIGURE 38.20   A scanning laser mounted on your robot can be used to detect the patterns of reflective tape located on or near doorways. Since the speed of the scan is known, electronics on your robot calculate distance (and position, given more strips) from robot to door.

Or consider using a CCD or CMOS camera. The robot could use several high-output infrared LEDs to illuminate the tape strips. Since the tape is much more reflective than the walls or door frame, it returns the most light. The CCD or CMOS camera is set with a high contrast ratio, so it effectively ignores anything but the bright tapes. Assuming the robot is positioned straight ahead of the door, the tapes will appear to be parallel. The distance between the tapes indicates the distance between the robot and the doorway. Should the robot be at an angle to the door, the tapes will not be parallel. Their angle, distance, and position can once again be interpolated to provide the robot's position relative to the door.

OTHER TECHNIQUES FOR "BEACONS" AND "LIGHTHOUSES"

There are scores of ways to relay position information to a robot. You've already seen two beacon-type systems: infrared and radio frequency. And there are plenty more. Sadly, there isn't enough space in this book to discuss them all, but the following sections outline some techniques you might want to consider. Many of these systems rely on line of sight between the beacon or lighthouse and the robot. If the line of sight is broken, the robot may very well get lost.

Three-point triangulation Traditional three-point triangulation is possible using either of two methods:

- *Active beacon.* A sensor array on the robot determines its location by integrating the relative brightness of the light from three active light sources.
- *Active robot.* The robot sends out a signal that is received by three sensors located around the room. The sensors integrate the robot's position, then relay this information back to the 'bot (via RF or an infrared radio link).

Coupled sonar and IR light This technique calculates time of flight using sound, and it offers excellent accuracy. You equip three active beacons with sonar transmitters and high-output infrared light-emitting diodes. You then connect the three beacons electrically so they will fire in sequence. When fired, both the sonar transmitter and IR LEDs emit a short 40 kHz signal. Because light travels much faster than sound, the robot will detect the IR signal first followed by the sound signal.

The difference in time between the reception of the IR and sound signals represents distance. Each beacon provides a "circle" path that the robot can be in. All three circles will intersect at only one spot in the room, and that will be the location of the robot. See Fig. 38.21 for a demonstration of how this works.

Exploring Other Position-referencing Systems

Over the years a number of worthwhile techniques have been developed to help robots know where they are. We've covered many of the most common techniques here. If your

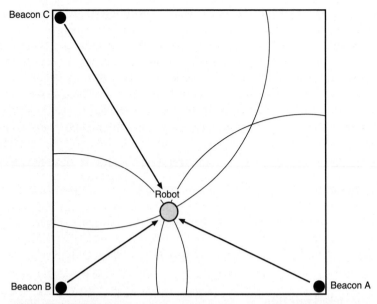

FIGURE 38.21 Three active beacons, connected to fire in sequence, provide both infrared and ultrasonic sound signals. Using a burst of infrared light, the robot times how long it takes for the sonar ping to reach it. Repeated three times—one for each beacon—the robot is able to make an accurate fix within the room.

budget and construction skill allow for it, however, you might want to consider any or all of the following.

GLOBAL POSITIONING SATELLITE

Hovering over the earth are some two dozen satellites that provide accurate world-positioning data to vehicles, ships, and aircraft. The satellite network, referred to as *global positioning system* (*GPS*), works by triangulation: the signals from three or more satellites are received and their timings are compared. The difference in the timings indicates the relative distances between the satellites and the receiver. This provides a "fix" by which the receiver can determine not only the latitude and longitude most anywhere on the earth, but also elevation.

GPS was primarily developed by the United States government for various defense systems, but it is also regularly used by private commerce and even consumers. Until recently, the signals received by a consumer-level GPS receiver have been intentionally "fuzzied" to decrease the accuracy of the device. (This is called "selective availability," imposed by the U.S. government for national security reasons.) Instead of the accuracies of a few feet or less that are possible with military-grade GPS receivers, consumer GPS receivers have had a nominal resolution of 100 meters, or about 325 feet. In practical use, with selective availability activated in the GPS satellites, the actual error is typically 50–100 feet. Selective availability has since been deactivated (but could be re-activated in the event of hostilities between the United States and another country), and the resolution of consumer GPS receivers can be under 20–25 feet.

Furthermore, a system called *differential GPS,* in which the satellite signals are correlated with a second known reference, demonstrably increases the resolution of GPS signals to less than five inches. When used outdoors (the signal from the satellites is too weak for indoor use) this can provide your robot with highly accurate positioning information, especially if your 'bot wanders hundreds of feet from its base station. Real-time differential GPS systems are still fairly costly, but their outputs can read into the robot's computer in real time. It takes from one to three minutes for the GSP system to "lock" onto the satellites overhead, however. Every time the lock is broken—the satellite signals are blocked or otherwise lost—it takes another one to three minutes to reestablish a fix.

If you're interested in experimenting with GPS, look for a receiver that has a NMEA-0183 or RS-232 compatible computer interface. A number of amateur radio sites on the Internet discuss how to use software to interpret the signals from a GPS receiver.

INERTIAL NAVIGATION

You can use the same physics that keep a bicycle upright when its wheels are in motion to provide motion data to a robot. Consider a bicycle wheel spinning in front of you while you hold the axle between your hands. Turn sideways and the wheel tilts. This is the gyroscopic effect in action; the angle of the wheel is directly proportion to the amount and time you are turning. Put a gyroscope in an airplane or ship and you can record even imperceptible changes in movement, assuming you are using a precision gyroscope.

Gyros are still used in airplanes today, even with radar, ground controllers, and radios to guide their way. While many modern aircraft have substituted mechanical gyros with completely electronic ones, the concept is the same. During flight, any changes in direction are recorded by the inertial guidance system in the plane (there are three gyros, for all three axes). At any time during the flight the course of the plane can be scrutinized by looking at the output of the gyroscopes.

Inertial guidance systems for planes, ships, missiles, and other such devices are far, far too expensive for robots. However, there are some low-cost gyros that provide modest accuracies. One reasonably affordable model is the Max Products MX-9100 micro piezo gyro, often used in model helicopters. The MX-9100 uses a piezoelectric transducer to sense motion. This motion is converted into a digital signal whose duty cycle changes in proportion to the rate of change in the gyro.

Laser- and fiber-optic-based gyroscopes offer another navigational possibility, though the price for ready-made systems is still out of the reach of most hobby robot enthusiasts. These devices use interferometry—the subtle changes in the measured wavelength of a light source that travels in opposite directions around the circumference of the gyroscope. The light is recombined onto a photosensor or a photocell array such as a CCD camera. In the traditional laser-based gyroscopes (e.g., the Honeywell "ring gyro"), the two light beams create a bull's-eye pattern that is analyzed by a computer. In simpler fiber-optic systems, the light beams are mixed and received by a single phototransistor. The wave patterns of the laser light produce sum and difference signals (heterodyning). The difference signals are well within audio frequency ranges, and these can be interpreted using a simple frequency-to-voltage converter. From there, relative motion can be ascertained.

You can also use accelerometers (similar to those described in detail in Chapter 41, "Experimenting with Tilt and Gravity Sensors") for inertial navigation. The nature of accelerometers, particularly the less-expensive piezoelectric variety, makes them difficult to employ in an inertial system. Accelerometers are sensitive to the earth's own gravity, and tilting on the part of the robot can introduce errors. By using multiple accelerometers—one to measure movement of the robot and one to determine tilt—it is generally possible to reduce (but perhaps not eliminate) these errors.

MAP MATCHING

Maps help us navigate strange towns and roads. By correlating what we see out the windshield with the street names on the map, we can readily determine where we are—or perhaps, just how lost we are! Likewise, given a map of its environment, a robot could use its various sensors to correlate its position with the information in a map. Map-based positioning, also know as *map matching,* uses a map you prepare for the robot or that the robot prepares for itself.

During navigation, the robot uses whatever sensors it has at its disposal (infrared, ultrasonic, vision, etc.) to visualize its environment. It checks the results of its sensors against the map previously stored in its memory. From there, it can compute its actual position and orientation. One common technique, developed by robot pioneer Hans Moravec, uses a "certainty grid" that consists of squares drawn inside the mapped environment (think of graph paper used in school). Objects, including obstacles, are placed within the squares. The robot can use this grid map to determine its location by attempting to match what it sees through its sensors with the patterns on the map.

Obviously, map matching requires that a map of the robot's environment be created first. Several consumer robots, like the Cyebot, are designed to do this mapping autonomously by "exploring" the environment over a period of time. Industrial robots typically require that the map be created using a CAD program and the structure and objects within it very accurately rendered. The introduction of new objects into the environment can drastically decrease the accuracy of the map matching, however. The robot may mistake a car for a foot stool, for example, and seriously misjudge its location.

From Here

To learn more about...	Read
Using computers and microcontrollers in your robots	Part 5—"Computers and Electronic Control"
Infrared and wireless communications techniques	Chapter 34, "Remote Control Systems"
Keeping your robot from crashing	Chapter 36, "Collision Avoidance and Detection" into things
Vision for your robot	Chapter 37, "Robotic Eyes"
Preventing your robot from falling over (or at least knowing when it happens)	Chapter 41, "Experimenting with Tilt and Gravity Sensors"

39

FIRE DETECTION SYSTEMS

Everyone complains that a robot is good for nothin'—except, perhaps, providing its master with a way to tinker with gadgets in the name of "science!" But here's one useful and potentially life-saving application you can give your robot in short order: fire and smoke detection. As this chapter will show, you can easily attach sensors to your robot to detect flames, heat, and smoke, making your robot a kind of mobile smoke detector.

Flame Detection

Flame detection requires little more than a sensor that detects infrared light and a circuit to trigger a motor, siren, computer, or other device when the sensor is activated. As it turns out, almost all phototransistors are specifically designed to be sensitive primarily to infrared or near-infrared light. You need only connect a few components to the phototransistor and you've made a complete flame detection circuit. Interestingly, the detector can "see" flames that we can't. Many gases, including hydrogen and propane, burn with little visible flame. The detector can spot them before you can, or before the flames light something on fire and smoke fills the room.

DETECTING THE INFRARED LIGHT FROM A FIRE

The simple circuit in Fig. 39.1 shows the most straightforward method for detecting flames. (See parts list in Table 39.1) You mask a phototransistor so it sees only infrared light by using an opaque infrared filter. Some phototransistors have the filter built in; with others, you'll have to add the filter yourself. If the phototransistor does not have an infrared-filtered lens, add one for the light to pass through. Place the transistor at the end of a small opaque tube, say one with a 1/4-inch or 1/2-inch I.D. (the black tubing for drip irrigation is a good choice). Glue the filter to the end of the tube. The idea is to block all light to the transistor except that which is passed through the filter.

In the circuit, when infrared light hits the phototransistor it triggers on. The brighter the infrared source is, the more voltage is applied to the inverting input of the comparator. If the input voltage exceeds the reference voltage applied to the noninverting input, the output of the comparator changes state.

Potentiometer R2 sets the sensitivity of the circuit. You'll want to turn the sensitivity down so ambient infrared light does not trigger the comparator. You'll find that the circuit does not work when the "background" light has excessive infrared content. You can't, for example, use the circuit outdoors or when the sensor is pointed directly at an incandescent light or the sun.

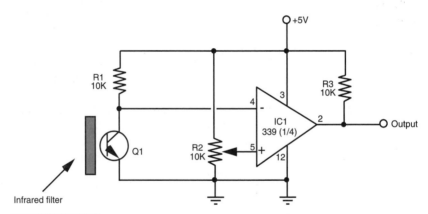

FIGURE 39.1 A flame detector built around the LM399 quad comparator IC.

TABLE 39.1 PARTS LIST FOR FLAME DETECTOR.	
IC1	LM339 Quad Comparator IC
R1	10K resistor
R2	10K potentiometer
R3	10K resistor
Q1	Infrared-sensitive phototransistor
Misc.	Infrared filter for phototransistor (if needed)

All capacitors have 10 percent tolerance unless noted; all resistors 1/4-watt.

Test the circuit by connecting an LED and 270-ohm resistor from the *Vout* terminal to ground. Point the sensor at a wall, and note the condition of the LED. Now, wave a match in front of the phototransistor. The LED should blink on and off. You'll notice that the circuit is sensitive to all sources of infrared light, which includes the sun, strong photolamps, and electric burners. If the circuit doesn't seem to be working quite right, look for hidden sources of infrared light. With the resistor values shown, the circuit is fairly sensitive; you can change them by adjusting the value of R1 and R2.

"WATCHING" FOR THE FLICKER OF FIRE

No doubt you've watched a fire at the beach or in a fireplace and noted that the flame changes color depending on the material being burned. Some materials burn yellow or orange, while others burn green or blue (indeed, this is how those specialty fireplace logs burn in different colors). Just like the color "signature" given off when different materials burn, the flames of the fire flicker at different but predictable rates.

You can use this so-called *flame modulation* in a robot fire detection system to determine what is a real fire and what is likely just sunlight streaming through a window or light from a nearby incandescent lamp. By detecting the rate of flicker from a fire and referencing it against known values, it is possible to greatly reduce false alarms. The technique is beyond the scope of this book, but you could design a simple flame-flicker system using an op amp, a fast analog-to-digital converter, and a computer or microcontroller. The analog-to-digital converter would translate the instantaneous brightness changes of the fire into digital signals.

The patterns made by those signals could then be referenced against those made by known sources of fire. The closer the patterns match, the greater the likelihood that there is a real fire. In a commercial product of this nature, it is more likely that the device would use more sophisticated digital signal processing.

Using a Pyroelectric Sensor to Detect Fire

A pyroelectric sensor is sensitive to the infrared radiation emitted by most fires. The most common use of pyroelectric infrared (or PIR) sensors is in burglar alarms and motion detectors. The sensor detects the change in ambient infrared radiation as a person (or animal or other heat-generating object) moves within the field of view of the sensor. The key ingredient here is *change:* a PIR sensor cannot detect heat per se but the changes in the heat within its field of view. In larger fires, the flickering flames create enough of a change to trigger the PIR detector.

Chapter 36, "Collision Avoidance and Detection," discusses how to use PIR sensors to detect the motion of people and animals around a robot. The same sensor, with little or no change, can be employed to detect fires. To be effective as a firefighter, you should ideally reduce the sensor's field of the view so the robot can detect smaller fires. The larger the field of view, the more the temperature and/or position of the heat source must change in order for the PIR sensor to detect it.

With a smaller field of view, the magnitude of change can be lower. However, with a small field of view, your robot will likely need to "sweep" the room, using a servo or stepper motor, in order to observe any possible fires. The sweeping must stop periodically so the robo can take a "room reading." Otherwise, the motion of the sensor could trigger false alarms.

Smoke Detection

"Where there's smoke, there's fire." Statistics show that the majority of fire deaths each year are caused not by burns but by smoke inhalation. For less than $15, you can add smoke detection to your robot's long list of capabilities and with a little bit of programming have it wander through the house checking each room for trouble. You'll probably want to keep it in the most "fire-prone" rooms, such as the basement, kitchen, laundry room, and robot lab.

You can build your own smoke detector using individually purchased components, but some items, such as the smoke detector cell, are hard to find. It's much easier to use a commercially available smoke detector and modify it for use with your robot. In fact, the process is so simple that you can add one to each of your robots. Tear the smoke detector apart and strip it down to the base circuit board.

Note:

> The active element used for detecting smoke—the radioactive substance Americium 241—has a half-life of approximately seven years. After about five to seven years, the effectiveness of the alarm is diminished, and you should replace it. Using a very old alarm will render your "Smokey the Robot" fairly ineffectual at detecting the smoke of fires.

HACKING A SMOKE ALARM

You can either buy a new smoke detector module for your robot or scavenge one from a commercial smoke alarm unit. The latter tends to be considerably cheaper—you can buy quality smoke alarms for as little as $7 to $10. In this section, I'll discuss hacking a commercial smoke alarm, specifically a Kidde model 0915K, so it can be directly connected to a robot's computer port or microcontroller. Of course, smoke alarms are not all designed the same, but the basic construction is similar to that described here. You should have relatively little trouble hacking most any smoke detector you happen to use.

However, you should limit your hacking attempts to those smoke alarms that use traditional 9-volt batteries. Certain smoke alarm models, particularly older ones, require you to use AC power or specialized batteries (such as 22-volt mercury cells). These are harder to salvage and, besides, their age makes them less suitable for sensitive smoke detection.

Start by checking the alarm for proper operation. If it doesn't have one already, insert a fresh battery into the battery compartment. *Put plugs in your ears* (or cover up the audio transducer hole on the alarm). Press the "Test" button on the alarm; if it is properly functioning the alarm should emit a loud, piercing tone. If everything checks okay, remove the battery, and disassemble the alarm. Less expensive models will not have screws but will likely use a "snap-on" construction. Use a small flat-headed screwdriver to unsnap the snaps.

Inside the smoke detector will be a circuit board, like the one in Fig. 39.2, that consists of the drive electronics and the smoke detector chamber.

Either mounted on the board or located elsewhere will be the piezo disc used to make the loud tone. Remove the circuit board, being careful you don't damage it. Examine the board for obvious "hack points," and note the wiring to the piezo disc. More than likely, there will be either two or three wires going to the disc:

- *Two wires to the piezo disc:* the wires will provide ground and +V power. This design is typical when you are using all-in-one piezo disc buzzers, in which the disc itself contains the electronics to produce the signal for audible tones.
- *Three wires to the piezo disc:* the wires will provide ground, +V power, and a signal that causes the disc to oscillate with an audible tone.

Using a volt-ohm meter or an oscilloscope, find the wire that serves as ground. (It is probably colored black or brown, but if no obvious color coding is used, examine the circuit board and determine where the wires are attached.) Connect the other test lead to the remaining wire. Or if the disc has three wires, connect the test lead to one of the remaining wires.

Replace the battery in the battery compartment, and depress the "Test" button on the alarm. Watch for a change in voltage. For a two-wire disc you should see the voltage change as the tone is produced. For a three-wire disc, try each wire to determine which produces the higher voltage; that is the one you wish to use. If you are using an oscilloscope, find the wire that produces a clean on/off pulse.

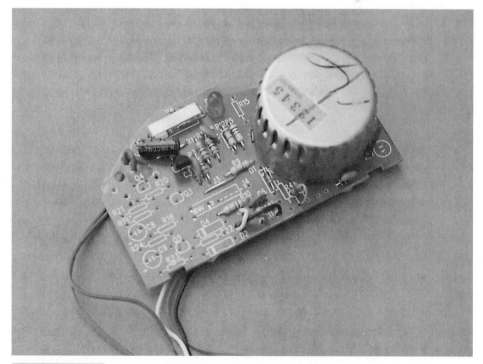

FIGURE 39.2 The guts of a smoke detector.

Once you have determined the functions of the wires to the piezo disc, clip off the disc and save it for some other project. Retest the alarm's circuit board to make sure you can still read the voltage changes with your volt-ohm meter or oscilloscope. Then clip off the wires to the battery compartment, noting their polarity. Connect the circuit to a +5 vdc power supply. Depress the "Test" button again. Ideally, the circuit will still function with the lower voltage. If it does not, you'll need to operate the smoke alarm circuit board with +9 vdc, which can complicate your robot's power supply and interfacing needs.

If you have an oscilloscope note the voltage. It should not be more than +5 volts. If it is, that means the circuit board contains circuitry for increasing the drive voltage to the piezo disc. You don't want this when you are interfacing the board to a computer port or microcontroller, so you'll need to limit the voltage by using a circuit such as that shown in Fig. 39.3. Here, the output of the smoke alarm circuit is clamped at no more than 5.1 volts, thanks to the 5.1-volt zener diode.

Because the output of the smoke alarm detector is often an oscillating signal, there is no effective way to measure the peak voltage by using a volt-ohm meter. The meter will only show an average of the voltage provided by the circuit. If you are limited to using only a volt-ohm meter for your testing, for safety's sake add the 5.1-volt zener circuit as shown in Fig. 39.4. While this may be unnecessary in some instances, it will help protect your digital interface from possible damage caused by over-voltage from the smoke alarm circuit board.

INTERFACING THE ALARM TO A COMPUTER

Assuming that the board works with the +5 vdc applied, your hacking is basically over, and you can proceed to interface the alarm with a computer port or microcontroller. By way of example, we'll assume that a simple microcontroller that periodically *polls* the input pin is connected to the smoke alarm circuit board. The program, checks the pin several times each second. When the pin goes HIGH, the smoke alarm has been triggered.

If your microcontroller supports interrupts, a better scheme is to connect the smoke alarm circuit board to an interrupt pin. Then write your software so that if the interrupt pin is triggered, a special "I smell smoke" routine is run. The benefit of an interrupt over polling is that the latter requires your program to constantly branch off to check the condition of the input pin. With an interrupt, your software program can effectively be ignorant of any smoke detector functionality. If and when the interrupt is triggered because the smoke alarm circuit was tripped, a special software routine takes over, commanding the robot to do something else. See Chapter 28 for more information on using interrupts in microcontrollers.

Rather than connect the output of the smoke alarm circuit board directly to the input pin, use a buffer to protect the microcontroller or computer against possible damage. You can construct a buffer using logic circuits (either TTL or CMOS) or with an op amp wired for unity-gain (with unity-gain, the op amp doesn't amplify anything). The buffer is optional, but I do recommend it. Note also that the smoke alarm circuit board derives its power from the robot's main +5 vdc power supply and not from the microcontroller.

Alternatively, you can use an opto-isolator. The opto-isolator bridges the gap between the detector and the robot. You do not need to condition the output of the opto-isolator if you are connecting it to a computer or microprocessor port or to a microcontroller. Several opto-isolator interfacing circuits are shown in Appendix D, "Interfacing Logic Families and ICs."

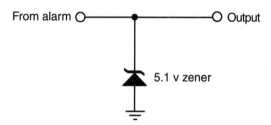

FIGURE 39.3 Use a 5.1 zener diode to ensure that the smoke alarm output does not drive the computer/micro-controller input above 5 vdc.

TESTING THE ALARM

Once the smoke alarm circuit board is connected to the microcontroller or computer port, test it and your software by triggering the "Test" button on the smoke alarm. The software should branch off to its "I smell smoke" subroutine. For a final test, light a match, and then blow it out. Wave the smoldering match near the smoke detector chamber. Again, the software runs the "I smell smoke" subroutine.

LIMITATIONS OF ROBOTS DETECTING SMOKE

You should be aware of certain limitations inherent in robot fire detectors. In the early stages of a fire, smoke tends to cling to the ceilings. That's why manufacturers recommend that you place smoke detectors on the ceiling rather than on the wall. Only when the fire gets going and smoke builds up, does it start to fill up the rest of the room.

Your robot is probably a rather short creature, and it might not detect smoke that confines itself only to the ceiling. This is not to say that the smoke detector mounted on even a one-foot high robot won't detect the smoke from a small fire; just don't count on it. Back up the robot smoke sensor with conventionally mounted smoke detection units, and *do not* rely only on the robot's smoke alarm.

DETECTING NOXIOUS FUMES

Smoke alarms detect the smoke from fires but not noxious fumes. Some fires emit very little smoke but plenty of toxic fumes, and these are left undetected by the traditional smoke alarm. Moreover, potentially deadly fumes can be produced in the absence of a fire. For example, a malfunctioning gas heater can generate poisonous carbon monoxide gas. This colorless, odorless gas can cause dizziness, headaches, sleepiness, and—if the concentration is high enough—even death.

Just as there are alarms for detecting smoke, so there are alarms for detecting noxious gasses, including carbon monoxide. Such gas alarms tend to be a little more expensive than smoke alarms, but they can be hacked in much the same way as a smoke alarm. Deduce the signal wires to the piezo disc and connect them (perhaps via a buffer and zener diode voltage clamp) to a computer port or microcontroller.

Combination units that include both a smoke and gas alarm are also available. You should determine if the all-in-one design will be useful for you. In some combination smoke-gas alarm units, there is no simple way to determine which has been detected. Ideally, you'll want your robot to determine the nature of the alarm, either smoke or gas (or perhaps both).

Heat Sensing

In a fire, smoke and flames are most often encountered before heat, which isn't felt until the fire is going strong. But what about before the fire gets started in the first place, such as when a kerosene heater is inadvertently left on or an iron has been tipped over and is melting the nylon clothes underneath?

If your robot is on wheels (or legs) and is wandering through the house, perhaps it'll be in the right place at the right time and sense these irregular situations. A fire is brewing, and before the house fills with smoke or flames the air gets a little warm. Equipped with a heat sensor, the robot can actually seek out warmer air, and if the air temperature gets too high it can sound an initial alarm.

Realistically, heat sensors provide the least protection against a fire. But heat sensors are easy to build, and, besides, when the robot isn't sniffing out fires it can be wandering through the house giving it an energy check or reporting on the outside temperature or…you get the idea.

Fig. 39.4 shows a basic but workable circuit centered around an LM355 temperature sensor. This device is relatively easy to find and costs under $1.50. The output of the device, when wired as shown, is a linear voltage. The voltage increases 10 mV for every rise in temperature of $1°$ Kelvin (K).

Degrees Kelvin uses the same scale as degrees Centigrade (C), except that the zero point is absolute zero—about $-273°$C. One degree Centigrade equals $1°$ Kelvin; only the start points differ. You can use this to your advantage because it lets you easily convert degrees Kelvin into degrees Centigrade. Actually, since your robot will be deciding when hot is *hot,* and doesn't care what temperature scale is used, conversion really isn't necessary.

You can test the circuit by connecting a volt-ohm meter to the ground and output terminals of the circuit. At room temperature, the output should be about 2.98 volts. You can

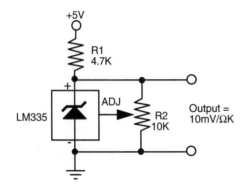

FIGURE 39.4 The basic wiring diagram for the LM355 temperature sensor.

TABLE 39.2	PARTS LIST FOR THE BASIC TEMPERATURE TRANSDUCER.
R1	4.7K resistor, 1 percent tolerance
R2	10K 10-turn precision potentiometer
D1	LM335 temperature sensor diode

All capacitors have 10 percent tolerance unless noted; all resistors 1/4-watt.

calculate the temperature if you get another reading by subtracting the voltage by 273 (ignore the decimal point but make sure there are two digits to the right of it, even if they are zeros). What's left is the temperature in degrees Centigrade. For example, if the reading is 3.10 volts, the temperature is 62°C (310 − 273 = 62). By the way, that's pretty hot! Time to turn on the air conditioner.

You can calibrate the circuit, if needed, by using an accurate bulb thermometer as a reference and adjusting R2 for the proper voltage. How do you know the "proper" voltage? If you know the temperature, you can determine what the output voltage should be by adding the temperature (in degrees C) to 273. If the temperature is 20°C, then the output voltage should be 2.93 volts. For more accuracy, float some ice in a glass of water for 15–20 minutes and stick the sensor in it (keep the leads of the testing apparatus dry). Wait 5 to 10 minutes for the sensor to settle and read the voltage. It should be exactly 2.73 volts.

The load presented at the outputs of the sensor circuit can throw off the reading. The schematic in Fig. 39.5 provides a buffer circuit so the load does not interfere with the output of the 355 temperature sensor. Note the use of the decoupling capacitors as recommended in the manufacturer's application notes. These aren't essential, but they are a good idea.

Fire Fighting

By attaching a small fire extinguisher to your robot, you can have the automaton put out the fires it detects. Obviously, you'll want to make sure that the fire detection scheme you've put into use is relatively free of false alarms and that it doesn't overreact in nonfire situations. Having your robot rush over to one of your guests and put out a cigarette he just lit is not only bad manners, it's potentially embarrassing.

It's a good idea to use a "clean" fire extinguishing agent for your fire-fighting 'bot. Halon is one of the best such agents, but, alas, the stuff is known to punch massive holes in the earth's ozone layer, and as a result it is no longer manufactured for general consumption. It's still legal to use, however, so if you have a working Halon fire extinguisher, you may wish to use it with your robot firefighter. You may also consider one of a number of Halon alternatives; select one that does not dispense a foam or powder. For example, any inert gas (helium, argon) and many noncombustible gasses (e.g., nitrogen) can be used to deplete a fire, and they will not leave a sediment on whatever they are sprayed on.

No matter what you use for the fire extinguisher, be sure to use caution as a guide when building any fire-fighting robot. Consider limiting your robot for experimental use, and test it only in well-ventilated rooms—or better yet—outside.

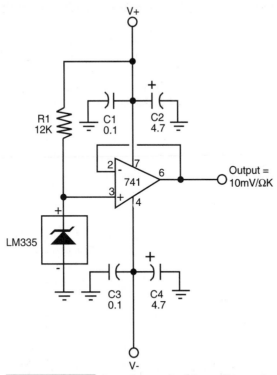

FIGURE 39.5 An enhanced wiring scheme for the LM355 temperature sensor. The load of the output is buffered and does not affect the reading from the LM355.

TABLE 39.3	PARTS LIST FOR THE BUFFERED TEMPERATURE TRANSDUCER.
R1	12K resistor, 1 percent tolerance
C1,C3	0.1 μF ceramic capacitor
C2,C4	4.7 μF tantalum capacitor
D1	LM335 temperature sensor diode

All capacitors have 10 percent tolerance unless noted; all resistors 1/4-watt.

The exact mounting and triggering scheme you use depends entirely on the design of the fire extinguisher. The bottle used in the prototype firebot is a Kidde Force-9, 2 1/2 pound Halon extinguisher. It has a diameter of about 3 1/4 inches. You can mount the extinguisher in the robot by using "plumber's tape," that flexible metallic strip used by plumbers to mount water and gas pipes. It has lots of holes already drilled into it for easy mounting. Use two strips to hold the bottle securely. Remember that a fully charged extinguisher is heavy—in this case over 3 pounds (2 1/2 pounds for the Halon chemical and about 1/2 pound for the bottle). If you add a fire extinguisher to your robot, you must relocate other components to evenly distribute the weight.

The extinguisher used in the prototype system for this book used a standard actuating valve. To release the fire retardant, you squeeze two levers together. Fig. 39.6 shows how to use a heavy-duty solenoid to remotely actuate the valve. You may be able to access the valve plunger itself (you may have to remove the levers to do so). Rig up a heavy-duty solenoid and lever system. A computer or control circuit activates the solenoid.

For best results, the valve should be opened and closed in quick bursts (200–300 milliseconds are about right). The body of the robot should also pivot back and forth so the extinguishing agent is spread evenly over the fire. Remember that to be effective, the extinguishing agent must be sprayed at the base of the fire, not at the flames. For most fires, this is not a problem because the typical robot stays close to the floor. If the fire is up high, the robot may not be able to effectively fight it.

You can test the fire extinguisher a few times before the bottle will need recharging. I was able to squeeze off several dozen short blasts before the built-in pressure gauge registered that I needed a new charge. For safety's sake, experiment with an extra extinguisher. Don't use your only extinguisher for your robot experiments; keep an extra handy in the unlikely event that you have to fight a fire yourself.

If the fire-fighting robot bug bites you hard, consider entering your machine in the annual Trinity College Fire Fighting Home Robot Contest (see *www.trincoll.edu/events/robot/* for additional information, including rules and a description of the event). This contest involves timing a robot as it goes from room to room in a houselike test field (the "house" and all its rooms are in a reduced scale). The object is to find the fire of a candle and snuff it out in the least amount of time. Separate competitions involving a junior division (high school and younger) and a senior division (everyone else) help to provide an even playing field for the contestants.

From Here

To learn more about...	*Read*
Connecting sensors to computers and microcontrollers	Chapter 29, "Interfacing with Computers and Microcontrollers"
Adding the sensation of "touch"	Chapter 35, "Adding the Sense of Touch"
Optical systems for detecting light	Chapter 37, "Robotic Eyes"
Enabling the robot to move around in a room or house	Chapter 38, "Navigating through Space"
Adding a siren or other warning device	Chapter 40, "Sound Output and Input"

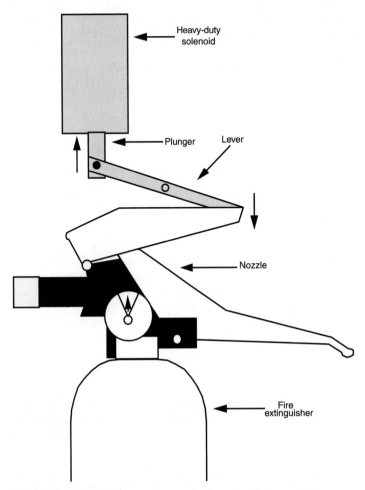

Heavy-duty
solenoid

Plunger

Lever

Nozzle

Fire
extinguisher

FIGURE 39.6 Using a heavy-duty solenoid to activate a
fire extinguisher.

40

SOUND OUTPUT AND INPUT

The robots of science fiction are seldom mute or deaf. They may speak pithy warnings— "Danger, Will Robinson, Danger"—or squeak out blips and beeps in some "advanced" language only other robots can understand. Voice and sound input and output make a robot more "humanlike," or at least more entertaining. What is a personal robot if not to entertain?

What's good for robots in novels and in the movies is good enough for us, so this chapter presents a number of useful projects for giving your mechanical creations the ability to make and hear noise. The projects include using recorded sound, generating warning sirens, recognizing and responding to your voice commands, and listening for sound events. Admittedly, this chapter only scratches the surface of what's possible today, especially with technologies like MP3 compressed digitized sound and ultracompact compact disc (and the ability to record them on a CD recorder connected to your computer). Alas, my publisher told me I had killed enough trees as it is and the book could not get any bigger, so this chapter must remain simply a primer on sound output and input.

Mechanically Recorded Sound Output

Before electronic doodads took over robotics there were mechanical solutions for just about everything. While they may not always have been as small as an electrical circuit, they were often easier to use. Case in point: you can use an ordinary cassette tape and playback mechanism to produce music, voice, or sound effects. Tape players and tape player mechanisms are common finds in the surplus market, and you can often find complete (and still working) portable cassette players-recorders at thrift stores. With just a few wires you can rig a cassette tape player in the robot and have the sound played back, on your command.

When looking for a cassette player try to find the kind shown in Fig. 40.1, which are solenoid controlled. These are handy for your robot designs because instead of pressing mechanical buttons, you can actuate solenoids by remote or computer control to play, fast-forward, or rewind the tape.

For most cassette decks you only need to provide power to operate the motor(s) and solenoids (if any) and a connection from the playback head to an amplifier. Since you are not using the deck for recording, you don't have to worry about the erase head, biasing the record head, and all that other stuff. If the deck already has a small preamplifier for the playback head, use it. It'll improve the sound quality. If not, you can use the tape head preamplifier shown in Fig. 40.2 (you can use a less expensive op amp than the one specified in the parts list in Table 40.1, but noise can be a problem). Place the preamplifier board as close to the cassette deck as possible to minimize stray pickup.

Electronically Recorded Sound Output

While mechanical sound playback systems are adequate, they lack the response and flexibility of a truly electronic approach. Fortunately, all-electronic reproduction of sound is fairly simple and inexpensive these days, in large part because of the wide availability of

FIGURE 40.1 A surplus cassette deck transport. This model is entirely solenoid driven and so is perfect for robotics.

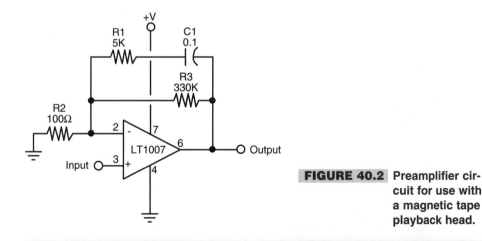

FIGURE 40.2 Preamplifier circuit for use with a magnetic tape playback head.

TABLE 40.1	PARTS LIST FOR CASSETTE TAPE HEAD PLAYER AMPLIFIER.
IC1	LT1007 low-noise operational amplifier (Linear Technology)
R1	330K-resistor
R2	4.9K resistor
R3	100–ohm resistor
C1	0. 1 μF ceramic capacitor

All resistors have 5 percent tolerance, 1/8- or 1/4-watt, metal film; all capacitors have 10 percent tolerance, rated 35 volts or higher.

custom-integrated circuits that are designed to record, store, and play back recorded sound. Most of these chips are made for commercial products such as microwave ovens, cellular phones, or car alarms.

In the following sections you'll learn about two approaches to electronically recorded sound output: hacking a sound recorder toy and using a special-purpose sound storage chip.

HACKING A TOY SOUND RECORDER

You can hack toy sound recorders, such as the Yak Bak, for use in your robot. These units, which can often be found at toy stores for under $10, contain a digital sound recording chip, microphone, amplifier, and speaker (and sometimes sound effects generator). To use them, you press the Record button and speak into the microphone. Then, stop recording and press the play button and the sound will play back until you make a new recording.

Fig. 40.3 shows a Yak Bak toy that was disassembled and hacked by soldering wires directly to the circuit board. The wires, which connect to a microcontroller or computer, are in lieu of pressing buttons on the toy to record and play back sounds. The buttons on most of these sound recorder toys are made of conductive rubber and are easily removed. To operate the unit via a microcontroller or computer, you bring the button inputs HIGH or LOW. (Which value you choose depends on the design of the circuit; you need to experiment to find out which to use.) Connect a 1K to 3K resistor between the microcontroller-computer port and the button input.

FIGURE 40.3 A hacked Yak Bak can be used to store and play short sound snips. You can record sounds for later playback, which can be via computer control. This model has two extra buttons for sound effects, which are also connected to the robot's microcontroller or computer.

Suppose you have a Yak Bak or similar toy connected to I/O pin 1 on a Basic Stamp II. Assume that on the toy you are using, bringing the button input HIGH triggers a previously recorded sound snip. The control program is as simple as this:

```
high 1
pause 10
low 1
```

The program starts by bringing the button input (the input of the toy connected to pin 1 of the Basic Stamp) HIGH. The *pause* statement waits 10 milliseconds and then places the button LOW again.

The built-in amplifier of these sound recorder/playback toys isn't very powerful. You may wish to connect the output of the toy to one of the audio output amplifiers described later in the chapter (see "Audio Amplifiers").

USING THE ISD FAMILY OF VOICE-SOUND RECORDER ICS

Toy sound recorders are limited to playing only a single sample. For truly creative robot yapping, you need a sound chip in which you can control the playback of any of several prerecorded snips. You can do this easily by using the family of sound storage and playback chips produced by Information Storage Devices (ISD). The company has made these "ChipCorder" ICs readily available to the electronics hobbyist and amateur robot builder.

You can purchase ISD sound recorder chips from a variety of sources, including Jameco and Digikey (see Appendix B). Prices for these chips vary depending on feature and recording time, but most cost under $15. While there are certainly other makers of sound storage/playback integrated circuits, the ISD chips are by far the most widely used and among the most affordable.

The ChipCorder products enjoy a rich assortment of data sheets and application notes, all of which are available from the ISD Web page at *www.isd.com.*

Sirens and Other Warning Sounds

If you use your robot as a security device or to detect intruders, fire, water, or whatever, then you probably want the machine to warn you of immediate or impending danger. The warbling siren shown in Fig. 40.4 will do the trick, assuming it's connected to a strong enough amplifier (refer to the parts list in Table 40.2). The circuit is constructed using two 555 timer chips (alternatively, you can combine the functions into the 556 dual timer chip). To change the speed and pitch of the siren, alter the values or R1and R4, respectively.

For maximum effectiveness, connect the output of the IC2 to a high-powered amplifier. You can get audio amplifiers with wattages of 8, 16, and even more volts in easy-to-build kit form. See Appendix B, "Sources," for a list of mail order companies that also sell electronic kits.

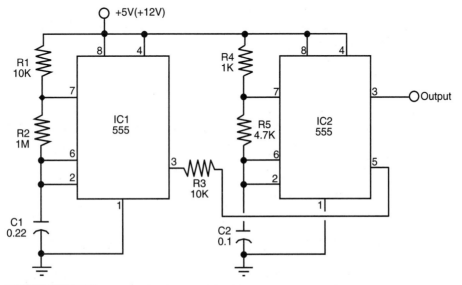

FIGURE 40.4 A warbler siren made from two 555 timer ICs.

TABLE 40.2 PARTS LIST FOR SIREN.	
IC1,IC2	555 Timer IC
R1	10K resistor
R2	1 megohm resistor
R3	10K resistor
R4	1K resistor
R5	4.7K resistor
C1	0.22 μF ceramic capacitor
C2	0.1 μF ceramic capacitor

All resistors have 5 or 10 percent tolerance, 1/4-watt; all capacitors have 10 percent tolerance, rated 35 volts or higher.

Sound Control

Unless you have all of the sound-making circuits in your robot hooked up to separate amplifiers and speakers (not a good idea), you'll need a way to select between the sounds. The circuit in Fig. 40.5 uses a 4051 CMOS analog switch and lets you choose from among eight different analog signal sources. You select input by providing a three-bit binary word to the select lines. You can load the selection via computer or set it manually with a switch. A binary-coded-decimal (BCD) thumbwheel switch is a good choice, or you can use a four-bank DIP switch. Table 40.3 shows the truth table for selecting any of the eight inputs.

You can route just about any of your sound projects through this chip, just as long as the outlet level doesn't exceed a few milliwatts. Do not pass amplified sound through the chip. Besides in all likelihood destroying the chip, it'll cause excessive cross talk between the channels. It's also important that each input signal not have a voltage swing that exceeds the supply voltage to the 4051.

Audio Amplifiers

Fig. 40.6 shows a rather straightforward 0.5-watt sound amplifier that uses the LM386 integrated amp. The sound output is minimal, but the chip is easy to get, cheap, and can be wired up quickly. It's perfect for experimenting with sound projects. The amplifier as shown has a gain of approximately 20, using minimal parts. You can increase the gain to about 200 by making a few wiring changes, as shown in Fig. 40.7. Either amplifier will drive a small (two- or three-inch) eight-ohm speaker. Refer to the parts lists in Tables 40.4 and 40.5 for these circuits.

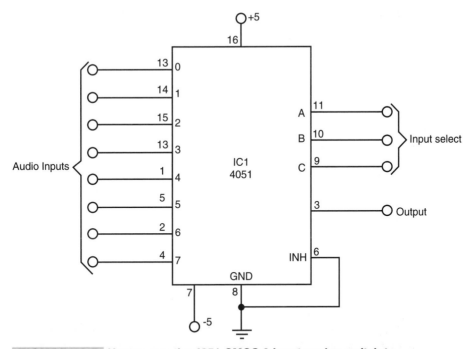

FIGURE 40.5 How to use the 4051 CMOS 8-input analog switch to control the output of the various sound-making circuits in your robot. You can choose the sound source you want routed to the output amplifier by selecting its input with the Input Select lines (they are binary weighted: A = 1, V =2, C = 4). For best results, the audio inputs should not already be amplified.

TABLE 40.3 4051 TRUTH TABLE.				
C	B	A	Selected Output	Pin
0	0	0	0	13
0	0	1	1	14
0	1	0	2	15
0	1	1	3	12
1	0	0	4	1
1	0	1	5	5
1	1	0	6	2
1	1	1	7	4

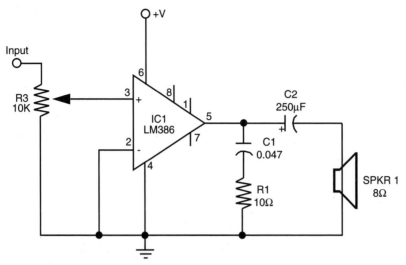

FIGURE 40.6 A simple gain-of-50 integrated amplifier, based on the popular LM386 audio amplifier IC.

TABLE 40.4 PARTS LIST FOR GAIN-200 AUDIO AMPLIFIER.	
IC1	LM386 Audio Amplifier IC
R1	10-ohm resistor
R2	10K potentiometer
C1	0.047 μF ceramic capacitor
C2	250 μF electrolytic capacitor
SPKR1	8-ohm miniature speaker

All resistors have 5 or 10 percent tolerance, 1/4-watt; all capacitors have 10 percent tolerance, rated 35 volts or higher.

Speech Recognition

Robots that listen to your voice commands and obey? Don't laugh; the technology is not only available, it's relatively inexpensive. Several companies, such as Sensory Inc. and Images Company, offer full-featured speech recognition systems for under $100. Both require you to "train" the system to recognize your voice patterns. Once trained, you simply repeat the command, and the system sets one or more of its outputs accordingly.

The Voice Direct, from Sensory Inc., is relatively easy to set up and use. The unit consists of a small double-sided circuit board that is ready to be connected to a microphone, speaker (for auditory confirmation), battery, and either relays or a microcontroller. The Voice Direct board recognizes up to 15 words or phrases and is said to have a 99 percent or better recognition accuracy. Phrases of up to 3.2 seconds can be stored, so you can tell you robot to "come here" or "stop, don't do that!" Fig. 40.8 shows Voice Direct module; the product comes with complete circuit and connection diagrams.

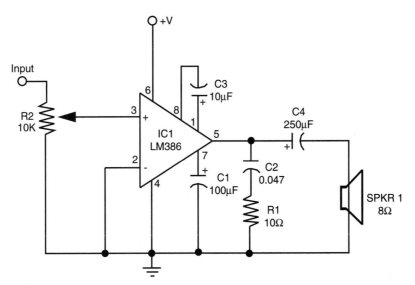

FIGURE 40.7 A simple gain-of-200 integrated amplifier.

TABLE 40.5 PARTS LIST FOR GAIN-OF-200 AUDIO AMPLIFIER.	
IC1	LM386 Audio Amplifier IC
R1	10-ohm resistor
R2	10K potentiometer
C1	100 μF electrolytic capacitor
C2	0.047 μF ceramic capacitor
C3	10 μF electrolytic capacitor
C4	250 μF electrolytic capacitor
SPKR1	8-ohm miniature speaker

All resistors have 5 or 10 percent tolerance, 1/4-watt; all capacitors have 10 percent tolerance, rated 35 volts or higher.

Keep the following in mind when using a voice recognition system:

■ You must be reasonably close to the microphone for the system to accurately understand your commands. The better the quality of the microphone you have, the better the accuracy of the recognition.

■ If you are using a voice recognition system on a mobile robot, you may wish to extend the microphone away from the robot so motor noise is reduced. For best results, you'll need to be fairly close to the robot and speak directly and clearly into the microphone.

■ Consider using a good-quality RF or infrared wireless microphone for your voice recognition system. The receiver of the wireless microphone is attached to your robot; you hold the microphone itself in your hand.

FIGURE 40.8 The Voice Direct voice recognition system from Sensory Inc.

Speech Synthesis

Not long ago, integrated circuits for the reproduction of human-sounding speech were fairly common. Several companies, including National Semiconductor, Votrax, Texas Instruments, and General Instrument offered ICs that were not only fairly easy to use, even on the hobbyist level, but surprisingly inexpensive. Voice-driven products using these chips included the Speak-and-Spell toys and voice synthesizers for the blind. In most cases, these ICs could create unlimited speech because they reproduced the fundamental sounds of speech (called phonemes).

With the proliferation of digitized recorded speech, however, unlimited speech synthesizers have become an exception instead of the rule. The companies that made stand-alone speech synthesizer chips either stopped manufacturing them or were themselves sold to other firms that no longer carry the old speech parts. (This was the case with General Instrument. and their once-popular SPO-256 speech synthesizer. General Instrument was sold to Microchip Technologies, makers of the PIC microcontroller.)

In addition, products such as the sound card for the IBM PC-compatible computers obviated the need for a separate, stand-alone speech circuit. Using only software and a sound card, it is possible to reproduce a male or female voice. In fact, Microsoft provides free speech-making tools for their Windows operating system, and you can use these tools under a variety of programming platforms, including Visual Basic. See the Microsoft site (*www.microsoft.com*) for more information.

Sure, unlimited speech synthesizer chips and chip sets are still available. But because they are not used as universally their price tends to be high. In the short sections that follow I'll review some of the unlimited speech synthesizers that are available in the hobbyist-amateur market as well as some of the alternatives available for providing a "synthesized" voice for your robotoid.

STAND-ALONE AND PC BOARD VOICE SYNTHESIZERS

If you were building robots in the late1970s and 1980s, odds are you either used, or wanted to use, a Votrax SC-01 or a General Instrument SPO-256 voice synthesis chip. Both were reasonably inexpensive (under $20) and could be connected to any computer. For several years, Radio Shack sold the SPO-256 and its companion text-to-speech converter IC as part of their regular inventory. Alas, these chips are no longer available except in occasional garage sales or surplus stores.

While they are not aimed at the hobbyist-amateur robotics market, such products as the RC8650 voice synthesizer chip set from RC Systems (*www.rcsys.com*) offer high-quality speech and fairly reasonable prices. The RC8650 costs under $50 in "low quantities," but as surface mount chips they can be hard for hobbyists to use. The company sells a demonstration kit, at about $175, that requires little or no electronic construction.

If you use an IBM PC-compatible motherboard for your robot's brain (see Chapter 28, "An Overview of Robot 'Brains'") you should consider adding a generic sound card to it and using text-to-speech software to program the card. As mentioned earlier, Microsoft provides a set of APIs (application programming interfaces) for speech output. Using the Internet, you can find other text-to-speech drivers for DOS and Linux as well.

VOICE SYNTHESIS TEXT-TO-SPEECH INTERNET SITES

If you plan to use recorded (mechanical or electronic) sound with your robot you may want to consider any of the several text-to-speech Internet sites, such as the Bell Labs TTS (text-to-speech) project at *http://www.bell-labs.com/projects/tts/*. At this Web site you type in the text you want to synthesize, and a sound file (*.wav, .au,* or *.aiff* format) is returned to you. Save the file and use it to produce a sound sample with a cassette tape or recording chip.

OF VODERS AND VOCODERS...AND ROBOTS

First, let's cover a little bit of the history and science behind the speech synthesizer. One of the earliest pioneers of the science of speech synthesis was Homer Dudley who, as a Bell Labs researcher during the 1930s and early 1940s, developed "parametric" methods for reproducing the human vocal tract. Among Dudley's accomplishments was the Voder, a mechanical speech-making machine that was controlled by a human operator. To make a word or sentence, the operator—who trained for about a year to become proficient on the machine—"played" the Voder using a small pianolike keyboard and foot pedals. The Voder was popular among the audiences who saw it in newsreel films as well as at the World's Fair. Though it had very little commercial value, the Voder was perhaps the world's first true speech synthesizer.

From the Voder came the Vocoder, an electronic device that literally took sound apart and put it back together again. The Vocoder combined noise and periodic pulses to produce a completely synthesized version of the human voice tract—all without human intervention. One of the aims of the Vocoder was to change the waveform of speech that was sent through the phone lines. Bell Labs hoped that it could not only reduce the bandwidth required to transmit speech but improve intelligibility as well. The Vocoder proved a huge success, and all-digital (and much improved) variations of it still exist. They are used in most all telephone systems, including cellular phones.

The Vocoder was also popular in radio, movies, and television as a way to produce eerie-sounding voices. Because the Voder-Vocoder model is based on the parameters of speech (pitch, modulation, noise, etc.) these parameters could be changed to electronically alter the sound of a voice. In operation, a human would speak normally into a microphone, which was connected to a Vocoder. The controls on the Vocoder were intentionally "mal-adjusted" to produce various effects such as monotones, warbles, and vibratos. Vocoder effects were used in the movie *Colossus: The Forbin Project,* in the 1970s TV series *Battlestar Galactica,* for background voice effects in *Star Wars,* and in many others.

A later version of a Bell Labs Vocoder synthesizer was programmed in the early 1960s to sing a song, "Bicycle Built for Two." Novelist Arthur C. Clarke saw the demonstration of the "singing computer" and used it in his book *2001: A Space Odyssey.* If you're familiar with the book or movie, you know that this is the song HAL the computer sings as he is being deactivated by astronaut Dave Bowman.

Vocoders are available today in both hardware and software form. Rock bands have long used analog vocoders (notice the lower case *v* to denote a generic Vocoder-like device) to create the "singing guitar" effect. You'll find vocoders of all sizes, shapes, and prices at most any well-stocked music store. All-software vocoders are available for use with Windows, DOS, and Macintosh. For example, the Prosoniq Orange Vocoder for use on Macs and PCs is designed to digitally manipulate any sound input, including mixing it with other tones and sounds to create unusual composite effects.

Where does all this lead us? For robotics, you can use vocoders to record voice samples and manipulate those voice samples so they don't sound like you, Uncle Bob, or Betty from next door (or whoever you used to record the voices). Instead, you can make your robot sound like the Colossus computer, a Cylon from *Battlestar Galactica*, a small child, a gnat, rustling leaves . . . even a robot!

The vocoder-processed voice samples can be stored using a cassette tape, a hacked digital recording toy, or a digital voice chip, such as the ISD series. These techniques were described earlier in this chapter. Under the control of bumper switches or a microprocessor or computer, your robot can then play back the appropriate vocoder-enriched voice clip.

Passive Sound Input Sensors

Next to sight, the most important human sense is hearing. And compared to sight, sound detection is far easier to implement in robots. Simple "ears" you can build in less than an hour let your robot listen to the world around it.

Sound detection allows your robot creation to respond to your commands, whether they take the form of a series of tones, an ultrasonic whistle, or a hand clap. It can also listen for the telltale sounds of intruders or search out the sounds in the room to look for and follow its master. The remainder of this chapter presents several ways to detect sound. Once detected, the sound can trigger a motor to motivate, a light to go lit, a buzzer to buzz, or a computer to compute.

MICROPHONE

Obviously, your robot needs a microphone (or mike) to pick up the sounds around it. The most sensitive type of microphone is the electret condenser, which is used in most higher-quality hi-fi mikes. The trouble with electret condenser elements, unlike crystal element mikes, is that they need electricity to operate. Supplying electricity to the microphone element really isn't a problem, however, because the voltage level is low—under 4 or 5 volts.

Most all electret condenser microphone elements come with a built-in field effect transistor (FET) amplifier stage. As a result, the sound is amplified before it is passed on to the main amplifier. Electret condenser elements are available from a number of sources, including Radio Shack, for under $3 or $4. You should buy the best one you can. A cheap microphone isn't sensitive enough.

The placement of the microphone is important. You should mount the mike element at a location on the robot where vibration from motors is minimal. Otherwise, the robot will do nothing but listen to itself. Depending on the application, such as listening for intruders, you might never be able to place the microphone far enough away from sound sources or make your robot quiet enough. You'll have to program the machine to stop, then listen.

AMPLIFIER INPUT STAGE

Use the circuit in Fig. 40.9 as an amplifier for the microphone (refer to parts list in Table 40.6). The circuit is designed around the common LM741 op amp, which is wired to operate from a single-ended power supply. Potentiometer R1 lets you adjust the gain of the op amp, and hence the sensitivity of the circuit to sound. After experimenting with the circuit and adjusting R1 for best sensitivity, you can substitute the potentiometer for a fixed-value resistor. Remove R1 from the circuit and check its resistance with a volt-ohm meter. Use the closest standard value of resistor.

By adding the optional circuit in Fig. 40.10, you can choose up to four gain levels via computer control. The resistors, R1 and R2 (you decide on their value based on the gain you wish), are connected to the inverting input of the op amp and the inputs of a 4066 CMOS 1-of-4 analog switch. Select the resistor value by placing a HIGH bit on the switch you want to activate. The manufacturer's specification sheets for this chip recommend that only one switch be closed at a time.

TONE DECODING DETECTION

The 741 op amp is sensitive to sound frequencies in a very wide band and can pick up everything that the microphone has to send it. You may wish to listen for sounds that occur only in a specific frequency range. You can easily add a 567 tone decoder IC to the amplifier input stage to look for these specific sounds.

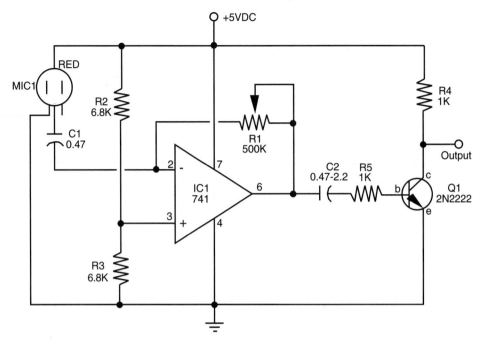

FIGURE 40.9 Sound detector amplifier. Adjust R1 to increase or decrease the sensitivity, or replace the potentiometer with the circuit that appears in Figure 40.10.

TABLE 40.6 PARTS LIST FOR SOUND DETECTOR.

IC1	LM741 Op Amp IC
Q1	2N2222 transistor
R1	500K potentiometer
R2,R3	6.8K resistor
R4,R5	1K resistor
C1,C2	0.47 μF ceramic capacitor
MIC1	Electret condenser microphone

The 567 is almost like a 555 in reverse. You select a resistor and capacitor to establish an operating frequency, called the *center frequency*. Additional components are used to establish a bandwidth—or the percentage variance that the decoder will accept as a desired frequency (the variance can be as high as 14 percent). Fig. 40.11 shows how to connect a 567 to listen to and trigger on about a 1 kHz tone (refer to the parts list in Table 40.7).

Before you get too excited about the 567 tone decoder, you should know about a few minor faults. The 567 has a tendency to trigger on harmonics of the desired frequency. You can limit this effect, if you need to, by adjusting the sensitivity of the input amplifier and decreasing the bandwidth of the chip.

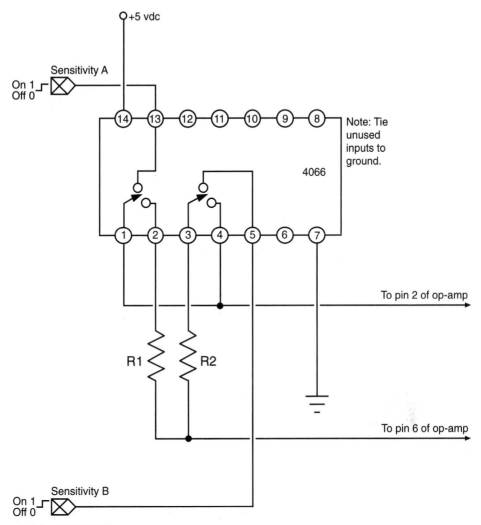

FIGURE 40.10 Remote control of the sensitivity of the sound amplifier circuit. Under computer control, the robot can select the best sensitivity for a given task.

Another minor problem is that the 567 requires at least eight wave fronts of the desired sound frequency before it triggers on it. This reduces false alarms, but it also makes detection of very-low-frequency sounds impractical. Though the 567 has a lower threshold of about 1 hertz, it is impractical for most uses at frequencies that low.

Yet another "problem" with the 567 is that it is officially "discontinued" by several of the manufacturers that used to make it. For the time being, however, you can still purchase 567 chips from most new and surplus retail and mail order electronics companies.

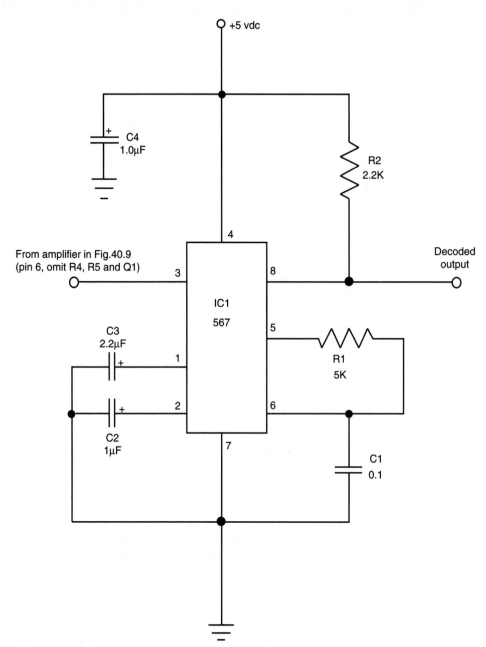

FIGURE 40.11 A 567 tone decoder IC, wired to detect tones at about 1 kHz.

TABLE 40.7	PARTS LIST FOR TONE DECODER.
IC1	567 Tone Decoder IC
R1	50K 3- to 15-turn precision potentiometer
R2	2.2K resistor
C1	0.1 μF ceramic capacitor
C2	2.2 μF tantalum or electrolytic capacitor
C3	1.0 μF tantalum or electrolytic capacitor
C4	1.0 μF electrolytic capacitor

Unless otherwise noted all resistors have 5 or 10 percent tolerance, 1/4-watt; all capacitors have 10 percent tolerance, rated 35 volts or higher.

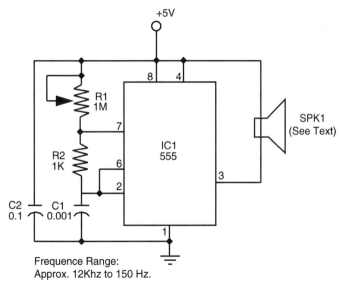

Frequence Range:
Approx. 12Khz to 150 Hz.

FIGURE 40.12 A variable frequency tone generator, built around the common 555 timer IC. The tone output spans the range of human hearing, and then some.

BUILDING A SOUND SOURCE

With the 567 decoder, you'll be able to control your robot using specific tones. With a tone generator, you'll be able to make those tones so you can signal your robot via simple sounds. Such a tone-generator sound source is shown in Fig. 40.12 (refer to the parts list in Table 40.8). The values shown in the circuit generate sounds in the 48-kHz to 144-Hz range. To extend the range higher or lower, substitute a higher or lower value for C1. Basic design formulas and tables for the 555 are provided in App. E.

For frequencies between about 5 kHz and 15 kHz, use a piezoelectric element as the sound source. Use a miniature speaker for frequencies under 5 kHz and an ultrasonic

TABLE 40.8 PARTS LIST FOR TONE GENERATOR.	
IC1	555 Timer IC
R1	1-megohm potentiometer
R2	1K resistor
C1	0.001 μF ceramic capacitor
C2	0.1 μF ceramic capacitor
SPKR1	4- or 8-ohm miniature speaker

All resistors have 5 or 10 percent tolerance, 1/4-watt; all capacitors have 10 percent tolerance, rated 35 volts or higher.

transducer for frequencies over 30 kHz. Cram all the components in a small box, stick a battery inside, and push the button to emit the tone. Be aware that the sound level from the speaker and especially the piezoelectric element can be quite high. *Do not* operate the tone generator close to your ears or anyone else's ears except your robot's.

From Here

To learn more about...	*Read*
Computer and microcontroller options for robotics	Chapter 28, "An Overview of Robot 'Brains'"
Interfacing sound inputs/outputs to a computer or microcontroller	Chapter 29, "Interfacing with Computers and Microcontrollers"
Sensors to prevent your robot from bumping into things	Chapter 36, "Collision Avoidance and Detection"
Eyes to go along with the ears of your robot	Chapter 37, "Robotic Eyes"

41

EXPERIMENTING WITH TILT AND GRAVITY SENSORS

Every schoolchild learns, the human body has five senses: sight, hearing, touch, smell, and taste. These are primary developed senses; yet the body is endowed with far more senses, including many we often take for granted. These more "primitive" human senses are typically termed "sixth senses"—a generic phrase for a sense that doesn't otherwise fit within the common five.

One of the most important "sixth senses" is the sense of balance. This sense is made possible by a complex network of nerves throughout the body, including those in the inner ear. The sense of balance helps us to stand upright and to sense when we're falling. When we're off balance, the body naturally attempts to reestablish an equilibrium. The sense of balance is one of the primary prerequisites for two-legged walking.

Our sense of balance combines information about both the body's angle and its motion. At least part of the sense of balance is derived from a sensation of gravity—the pull on our bodies from the earth's mass. Gravity is an extraordinarily strong physical force, but strangely enough it is not often used in hobby robotics because accurate sensors for measuring it have been prohibitively expensive.

But just consider the possibilities if a robot were given the ability to "feel" gravity. The same forces of gravity that help us to stay upright might provide a two-legged robot with the sensation that would keep it upright. Or a rolling robot—on wheels or tracks—might avoid tipping over and damaging something by determining if its angle is too steep. The sense of gravity might enable the robot to avoid traveling over that terrain, or it might tell the robot to shift some internal ballast weight (assuming it were so equipped) to change its center of balance.

In this chapter we'll explore various ways, from the simple to the not-so-simple, to endow your robot with a sense of balance so it can determine its motion and its physical orientation on this earth.

Sensors to Measure Tilt

One of the most common means for providing a robot with a sense of balance is to use a tilt sensor or tilt switch. The sensor or switch measures the relative angle of the robot with respect to the center of the earth. If the robot tips over, the angle of the sensor or switch changes, and this can be detected by electronics in the robot. Tilt sensors and switches come in various forms and packages, but the most common are the following:

■ Mercury-filled glass ampoules that form a simple on/off switch. When the tilt switch is in one position (say, horizontal), the liquid-mercury metal touches contacts inside the ampoule, and the switch is closed. But when the switch is rotated to vertical, the mercury no longer touches the contacts, and the switch is open. The major disadvantage of mercury tilt switches is the mercury itself, which is a *highly toxic* metal.

■ The ball-in-cage (see Fig. 41.1) is an all-mechanical switch popular in pinball machines and other devices where small changes in level are required. The switch is a square or round capsule with a metal ball inside. Inside the capsule are two or more electrical contacts. The weight of the ball makes it touch the electrical contacts, which forms a switch. The capsule may have multiple contacts so it can measure tilt in all directions.

■ Electronic spirit-level sensors use the common fluid bubble you see on ordinary levels at the hardware store plus some interfacing electronics. A spirit level is merely a glass tube filled, though not to capacity, with water or some other fluid. A bubble forms at the top of the tube since it isn't completely filled. When you tilt the tube gravity makes the bubble slosh back and forth. An optical sensor—an infrared LED and detector, for example–can be used to measure the relative size and position of the bubble.

■ Electrolytic tilt sensors are like mercury switches but more complex and a lot more costly. In an electrolytic tilt sensor a glass ampoule is filled with a special electrolyte liquid—that is, a liquid that conducts electricity but in very measured amounts. As the switch tilts, the electrolyte in the ampoule sloshes around, changing the conductivity between two (or more) metal contacts.

BUILDING A BALANCE SYSTEM WITH A MERCURY SWITCH

You can construct a simple but practical balance system for your robot using two small mercury switches. You want mercury switches that will open (or close) at fairly minor angles, perhaps 30–35° or so—just enough to signal to the robot that it is in danger of tipping over. You may have to purchase the switch with these specifications through a specialty industrial parts store, unless you're lucky enough to find one on the used or surplus market.

Mount the switch in an upright position. If the level of the robot becomes extreme the switch will trigger. You can directly interface the switch to an I/O (input/output) line on a

FIGURE 41.1 A typical ball-in-cage switch.

PC or microcontroller. However, you'll probably want to include a debounce circuit in line with the switch and I/O line since mercury switches can be fairly "noisy" electrically. A suitable debouncer circuit is shown in Chapter 29.

BUILDING A BALANCE SYSTEM WITH A BALL-IN-CAGE SWITCH

The four-conductor ball-in-cage switch is a rather common find in the surplus market, and it's very inexpensive. If the switch is tilted in any direction by more than about 25–30°, at least one of the four contacts in the switch will close, thus indicating that the robot is off level. You can use a debouncer circuit with the ball-in-cage tilt sensor.

Because the ball-in-cage sensor has four contacts (plus a center common), you can either provide independent outputs of the switch or a common output. With independent outputs, a PC or microcontroller on your robot can determine in which direction the robot is tilting (if two contacts are closed, then the ball is straddling two contacts at the same time). However, unless you come up with some fancy interface circuitry, you'll need to dedicate four I/O lines on the PC or microcontroller, one for each switch contact.

Conversely, with the common output approach you can wire all the outputs together in a serial chain. The switch will close if the ball touches any contact. This approach uses only one I/O line, but it deprives the robot of the ability to know in exactly which direction it is off level. A variation on this theme is to use resistors of specific values to form a voltage divider. When you connect the resistors to an I/O line capable of analog input (an analog-to-digital converter, for example), you can easily determine by the changing voltage at the input which contact switch has been closed.

Using an Accelerometer to Measure Tilt

One of the most accurate, yet surprisingly low-cost, methods for tilt measurement involves an accelerometer. Once the province only of high-tech aviation and automotive testing labs, accelerometers are quickly becoming common staples in consumer electronics. It's quite possible, for example, that your late-model car contains at least one accelerometer—if not as part of its collision safety system (such as an airbag), then perhaps as an integral part of its burglar alarm. Accelerometers are also increasingly used in high-end video game controllers, portable electric heaters, and in-home medical equipment.

New techniques for manufacturing accelerometers have made them more sensitive and accurate yet also less expensive. A device that might have cost upwards of $500 a few years ago sells in quantity to manufacturers for under $10 today. Fortunately, the same devices used in cars and other products is available to hobby robot builders, though the cost is a little higher because we're not buying 10,000 at a time!

In the following two sections, I'll show you how to construct highly accurate angle sensors using either of two affordable accelerometers from semiconductor maker Analog Devices. Both accelerometers are available through a number of retail outlets, and neither requires extensive external circuitry. While the text that follows is specific to the accelerometers from Analog Devices, you may substitute units from other sources after making the appropriate changes in the circuitry and computer interface software.

WHAT IS AN ACCELEROMETER?

The basic accelerometer is a device that measures *change* in speed. Put an accelerometer in your car, for example, and step on the gas. The device will measure the increase in speed. Most accelerometers only measure acceleration (or deceleration) and not constant speed or velocity. Such is the case with the accelerometers detailed here.

And though accelerometers are designed to measure changes in speed, many types of accelerometers—including the ones detailed in the following sections—are also sensitive to the constant pull of the earth's gravity. It is this latter capability that is of interest to us since it means you can use the accelerometer to measure the tilt, or "attitude," of your robot at any given time. This tilt is represented by a change in the gravitational forces acting on the sensor. The output of the accelerometer is either a linear AC or DC voltage or, more handily, a digital pulse that changes in response to the acceleration or gravity forces.

ADDITIONAL USES FOR ACCELEROMETERS

Before we go into the details of using accelerometers for tilt and angle measurement in robots, let's review the different robotics-based sensoric applications for these devices. Apart from sensing the angle of tilt, a gravity-sensitive accelerometer can also be used for the following tasks:

- *Shock and vibration.* If the robot bumps into something, the output of the accelerometer will "spike" instantaneously. Because the output of the accelerometer is proportional to the power of the impact, the harder the robot bumps into something, the larger the voltage spike. You can use this feature for collision detection obviously, but in ways that far exceed what is possible with simple bumper switches since an accelerometer is sensitive to shock from most any direction.
- *Motion detection.* An accelerometer can detect motion even if the robot's wheels aren't moving. This might be useful for robots that must travel over uneven or unpredictable terrain. Should the robot move (or stop moving) when it's not supposed to, this will show up as a change in speed and will therefore be sensed by the accelerometer.
- *Telerobotic Control.* You can use accelerometers mounted on your clothes to transmit your movements to a robot. For instance, accelerometers attached to your feet can detect the motion of your legs. This information could be transmitted (via radio or infrared link) to a legged robot, which could replicate those moves. Or you might construct an "air stick" wireless joystick, which would simply be a pipe with an accelerometer at the top or bottom and some kind of transmitter circuit. As you move the joystick your movements are sent to your robot, which acts in kind.

SINGLE- AND DUAL-AXIS SENSING

The basic accelerometer is single axis, meaning it can detect a change in acceleration (or gravity) in one axis only, as shown in Fig. 41.2. While this is moderately restrictive, you can still use such a device to create a capable and accurate tilt-and-motion sensor for your robot. The first accelerometer project we describe in this chapter uses such a single-axis device.

A dual-axis accelerometer detects changes in acceleration and gravity in both the X and Y planes (see Fig. 41.2). If the sensor is mounted vertically—so that the Y axis points straight up and down—the Y axis detects up and down changes, and the X axis will detect side-to-side motion. Conversely, if the sensor is mounted horizontally, the Y axis detects motion forward and backward, and the X axis detects motion from side to side.

Note:

To detect motion in 3D—forward/back, left/right, and up/down—you need both a single- and a dual-axis accelerometer mounted perpendicularly to one another.

THE ANALOG DEVICES' ADXL ACCELEROMETER FAMILY

Analog Devices is a semiconductor maker primarily for industrial- and military-grade operational amplifiers, digital-to-analog and analog-to-digital converters, and motion control

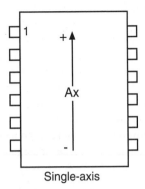 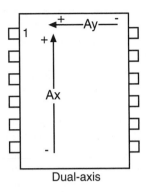

Single-axis   Dual-axis

FIGURE 41.2 *a.* A single-axis accelerometer; *b.* A dual-axis accelerometer.

products. One of their key product lines is accelerometers, and for them they use a patented fabrication process to create a series of near-microscopic mechanical beams. This "micro-machining" involves etching material out of a substrate. During acceleration, the beam is distended along its length. This distention changes the capacitance in nearby plates. The change in capacitance is correlated as acceleration and deceleration.

A thorough discussion of the theory of operation behind the ADXL family of accelerometers is beyond the scope of this chapter, but you can obtain much more on the subject directly from the manufacturer. Check The Analog Devices Website at www.analog.com.

In addition to the mechanical portions of the accelerometer, all the basic interface circuitry is part of the device. In fact, looking at one of the ADXL accelerometers you'd think it was just an integrated circuit of some kind. Because the basic circuitry is included as part of the accelerometer, only a minimum number of external parts are needed. In the first project that follows, only eight external parts are used, three of which are filtering/bypass capacitors to reduce the effects of electrical noise. The second project uses even fewer external parts.

Constructing a Single-axis Accelerometer Robotic Sensor

The single-axis accelerometer outlined here is based on the Analog Devices ADXL150JQC. The '150 combines a micro-machined mechanical accelerometer with on-board amplification electronics. The normal range of the '150 is plus or minus 50 g's (a *g* is the unit of measure of gravitational pull; one g is equal to the gravitational pull on the surface of the earth). This range is actually too high for most robotic applications, so we'll be scaling it down to about ±5 or 10 g's.

For reference, one *g* is the equivalent of a mass traveling at a speed of 32 feet per second, per second. That is, for every second of elapsed time, the speed of the mass increases (or decreases for deceleration) by 32 feet per second. However, note that objects falling to earth eventually reach a "terminal velocity" due to air friction and do not continue to accelerate.

The output signal of the '150 is a voltage. We'll be using the '150 in full DC mode, where the output will swing no less than about 2.5 volts as the sensor detects changes in acceleration. Depending on how you adjust the sensor board, you can get the '150 to deliver close to the full 0- to 5-volt output range. The broader the range of voltage, the more resolution the sensor will provide. At a scaling factor of ±5 g's, we'll be able to detect changes in 400 mV per *g*. With an 8-bit analog-to-digital converter (ADC) measuring a full 4000-mV (4-volt) range, this equates to a resolution of 15.625 mV per step, or roughly 0.06 g's. That's not bad; a 10- or 12-bit ADC will provide even greater resolution.

The one "gotcha" of the ADXL150 is that it is a 14-pin surface-mount component, which makes it more difficult to use in a homebrew circuit. There are a number of techniques you can use to make use of the '150, or most any surface mount part for that matter:

- Use an IC surface-mount carrier board. Carefully solder the surface-mount component onto the carrier board. You can then use the carrier in ordinary breadboards and prototyping boards. Because the leads of the ADXL150 are pre-tinned, you need only hold the part in place for the first "tack" of the soldering pencil. Repeat for the other leads. Be sure to buy a carrier board that has solder pads appropriately spaced for the '150. Most IC carrier boards are made for thinner packages; the ADXL150 is in a "fat" ceramic package and is wider than most ICs.
- Solder a short length of 30 AWG wire-wrap wire to each of the '150's leads. This is delicate work and requires expert soldering and a good eye (it was the technique I used for the prototype).
- Design and etch your own surface-mount board, custom-made for the ADXL150.
- Purchase the ADXL150 Evaluation Board, available directly from Analog Devices as well as several online and catalog merchants. This is perhaps the easiest method.

CONSTRUCTION DETAILS

Refer to Fig. 41.3 for a schematic showing how to use the ADXL150 as a general-purpose accelerometer suitable for measuring tilt, vibration, and, of course, acceleration. This circuit was adopted from the data sheet provided for the ADXL150, with the parts adjusted to the values appropriate for robotic endeavors.

The heart of the circuit is the ADXL150, which is powered by +5 vdc. Capacitors C1, C2, and C3 are for power supply bypass, and you should include them to reduce noise on the output. IC2 is an Analog Devices OP196 "rail-to-rail" operational amplifier. Though the circuit calls for the OP196 op amp, most any single-supply (V+ only, V- voltage not needed) rail-to-rail operational amplifier will probably work.

See Fig. 41.4 for a view of the prototype I built. Note the wire-wrap wires attached to the leads of the ADXL150. I mounted the '150 to a 14-pin wire-wrap socket using double-sided foam tape and attached the free ends of the wires to the pins of the socket. After inserting the socket into the prototyping board and soldering it in place, I clipped off the excess length of the socket pins since it was not needed. Note that in my prototype I soldered a wire to every pin on the '150 but this is not needed. Only pins 7, 8, 9, 10, and 14 need to be connected to anything. (Pin 9 is for the self-test function, which is not used in this project.)

After constructing the ADXL150 sensor board, connect the power leads to a suitable +5 vdc power supply, and connect the output to a fast-acting meter or oscilloscope. While you

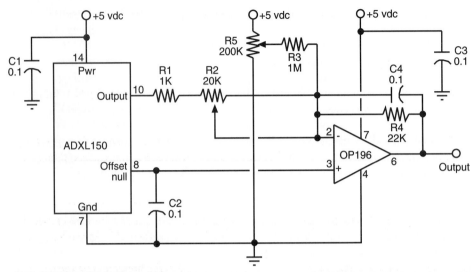

FIGURE 41.3 A basic schematic diagram for using the Analog Devices ADXL150 single-axis accelerometer.

FIGURE 41.4 My prototype of the ADXL150 accelerometer, which can be used for robotic tilt, motion, and vibration sensing.

slowly move the sensor board in different directions, adjust R2 and R4 for maximum voltage change.

ADJUSTING THE SENSOR

You will need to experiment with different settings to achieve the output you want for the application you have planned for the '150:

- As a level or tilt sensor, position the ADXL150 so it points arrow up (+1 g setting). Any tilt in any direction will then be registered as a negative-going voltage change. For this application, you want a low-g scale, so adjust R2 near its minimum and R4 near its maximum.
- As a movement sensor, position the ADXL150 so it points arrow forward (arrow to the front of the robot). For a slow robot, a rather low-g scale is likely the best choice, but adjust as you see fit. Some small robots turn and spin on their axis very quickly, producing momentary forces of 2 or 3 g's!
- As a shock or vibration sensor, position the ADXL150 in the horizontal or vertical position, as desired. Adjust the scale setting based on the sensitivity you need. If the robot is not supposed to be highly sensitive to minor bumps and grinds, for instance, set a high-g scale by increasing R2 and decreasing R4.

CONTROL INTERFACE AND SOFTWARE

Of course, the schematic in Fig. 41.3 still needs to be interfaced to a computer or microcontroller via an analog-to-digital converter to do your robot any good. Chapter 29, "Interfacing with Computers and Microcontrollers," discusses in more detail how to use ADCs, so we will dispense with that discussion for this project. If you plan to use your PC to interface to the ADXL150, for example, you need just a basic ADC chip, such as the ADC0804. You input a single analog voltage, and the output is converted to eight data bits, which you can connect to your PC via the parallel printer pot.

Conversely, you can use a microcontroller, such as the Parallax Basic Stamp II, which provides more than enough input/output lines from the ADC. Or perhaps an easier approach is to use a BasicX-24 microcontroller from NetMedia. As described in Chapter 32, the BasicX-24 (BX-24) is pin-for-pin compatible with the Basic Stamp II, but it includes an on-board analog-to-digital converter. This ADC is the "multiplexing" type, so you can use any (and all) of eight different data lines to read analog data. This feature of the BasicX-24 makes it particularly well suited for use with sensors such as the ADXL150 accelerometer, since there is no external ADC circuit.

Listing 41.1 shows a test program for the BasicX-24 microcontroller and how to use the ADXL150 as a tilt sensor. You connect the amplified output of the ADXL150 to pin 13 (I/O line 7) of the BasicX-24. The main body of the code—defined by the *Main()* subroutine—is an endless loop that constantly collects data from the accelerometer. A "software filter" is employed to average out the values of the '150. I've set the filter to average 254 samples of data from the accelerometer; you can select a lower value if you don't want to sample as many data points.

When running, the program changes the color of the LED built onto the BasicX-24 carrier board. With the '150 pointing "up" so the output is at its highest level, the green LED lights. As the '150 is tilted horizontally, the output decreases, and the red LED lights instead. You'll need to experiment with the "setpoint" (I used 825), depending on the actual values provided by your ADXL150 circuit.

LISTING 41.1

```
' ADXL150 test program
' For use with BasicX-24 microcontroller
' Output of ADXL150 is connected to pin 13 (IO line 7)
'   of the BasicX-24

Dim Voltage As Integer, BlinkTotal As Integer
Dim Total As Long
Const PinNumber As Byte = 13
Const GreenLED As Byte = 26
Const RedLED As Byte = 25
Const LEDon As Byte = 0
Const LEDoff As Byte = 1

Sub Main()
    Dim x as Byte
    Total = 0
    Do
        For x = 1 to 254                     ' adjust for "filter"
            Voltage = GetADC (PinNumber)
            Total = Total + CLng(Voltage)
        Next
        Total = Total \ 254                  ' adjust for "filter"
        BlinkTotal = CInt(Total)
        Call LEDs
        Call Delay(0.1)
    Loop
End Sub

Sub LEDs()
    If BlinkTotal > 825 Then                 ' adjust as needed
        Call PutPin(RedLED, LEDoff)
        Call PutPin(GreenLED, LEDon)
    Else
        Call PutPin(GreenLED, LEDoff)
        Call PutPin(RedLED, LEDon)
    End If
End Sub
```

Constructing a Dual-axis Accelerometer Robotic Sensor

The ADXL150 single-axis accelerometer, described in detail in the last section, has a close sibling: the ADXL250, which is like two accelerometers in one. The ADXL250 is a dual-axis device in which the axes are oriented at right angles to one another. When the accelerometer is positioned horizontally, it can therefore detect motion in 360° (it cannot detect up and down motion when in this position, however).

Using the ADXL250 is very similar to using the ADXL150—you just duplicate the interface electronics for the second axis. Refer to the data sheet for the ADXL250 on the pinout diagram for the device.

Analog Devices makes a less expensive line of accelerometers that is specifically designed for consumer products. Their ADXL202 is a dual-axis device with a ±2-g sensitivity (if you need more g's, check out the ADXL210, which is rated at ±10 g's). Besides being cheaper than the ADXL150/250, the '202 has a simplified output: instead of a linear voltage, the output is purely digital. As acceleration changes, the timing of the pulses at the output of the ADXL202 changes. This change can be readily determined by a PC or microcontroller, using simple software (see the example for the Basic Stamp later in the chapter). No op amp or scaling adjustment components are necessary, which makes the ADXL202 a breeze to use.

Like the ADXL150, the '202 is a surface-mount component. See the discussion in the section on the ADXL150 about alternative ways to interface the '202 electronics in your robot. By a long measure, the ready-made ADXL202 Evaluation Board is the easiest way to use this device. It comes on a small postage stamp carrier, which can be directly soldered to the Basic Stamp or other microcontroller.

WIRING DIAGRAM

The basic hookup diagram for the ADXL202 is shown in Fig. 41.5. Note that except for two filter capacitors and a single resistor, there are no external components. I have specified a rather low bandwidth of 10 Hz for the device. According to the ADXL202 data sheet, the value of C1 and C2 for this bandwidth should be 0.47 μF.

Resistor R1 sets the value of the timing pulse used for the output of the X and Y axes of the '202 chip. I have specified a modest timing pulse of 5 milliseconds; according to the data sheet this requires a nominal value of about 625K for R1 (I specified a reasonably close standard resistor value). Note that the exact timing of the pulse is not critical, as any variation will be accounted for in the software. You will want to select a higher or lower timing pulse based on the capabilities of the PC or microcontroller you are using and the resolution you desire.

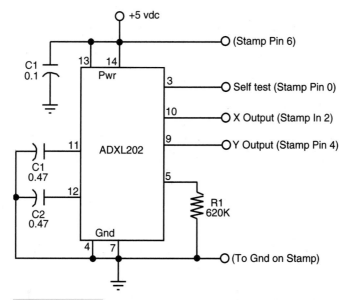

FIGURE 41.5 Hookup diagram for the Analog Devices ADXL202 dual-axis accelerometer chip, with connections shown for a Basic Stamp II microcontroller.

UNDERSTANDING THE OUTPUT OF THE ADXL202

The ADXL202 delivers a steady stream of digital pulses, or square waves. The timing of the pulses, defined as *T2*, is set by R1 (see the previous section). For our project, the pulses are 5 ms apart. Changes in acceleration change the width of each pulse (this is called *pulse width modulation,* or *PWM*). For the ADXL202, the width changes 12.5 percent for each g of acceleration—both positive and negative changes. Therefore, the width of these 5-ms pulses will change by 50 percent for the entire ±2-g range of the device. A zero g state is 50 percent duty cycle. The width of the pulses is defined as *T1*. Because the ADXL202 uses a pulse width modulated output, rather than a linear DC output, no analog-to-digital conversion is necessary.

ORIENTING THE ACCELEROMETER

Because the ADXL202 has two axes, it can detect acceleration and gravity changes in two axes at once. You can use the device in vertical or horizontal orientation. As a tilt sensor, orient the device horizontally; any tilt in any direction will therefore be sensed. In this position, the ADXL202 can also be used as a motion detector to determine the speed, direction, and possibly even the distance (given the resolution of the control circuitry you use) of that movement.

CONTROL INTERFACE AND SOFTWARE

The control interface for the ADXL202 is surprisingly simple. Fig. 41.6 shows the hookup diagram for connecting the ADXL202 surface-mount chip and evaluation board to a Basic Stamp II. In both cases, power for the '202 comes from one of the Stamp's I/O pins. As mentioned in an application note written by an Analog Devices engineer on the subject of interfacing the ADXL202 to a Basic Stamp, this isn't the overall best design choice, but for experimenting it's quick and simple.

Listing 41.2 shows a short program written in PBasic for the Basic Stamp II that allows continual reading of the two outputs of the ADXL202. The program works by first determining the period of the T2 basic pulse. It then uses the PULSIN command with both the *T1y* and *T1x* axis signals. PULSIN returns the length of the pulse; a longer pulse means higher *g;* a shorter pulse means lower g.

LISTING 41.2

```
Freq    Var Word
T1x Var Word
T1y Var Word
T2 Var Word

Low     0     ' self test, pin 0
Input   2     ' X accel, pin 2
Input   4     ' Y accel, pin 4
High    6     ' V+ pin 6

Count 4, 500, Freq
T2 = Freq * 2
```

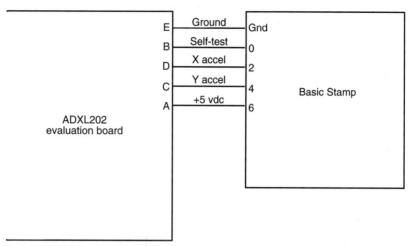

FIGURE 41.6 Connection diagram for hooking up an ADXL202EB (Evaluation Board) to a Basic Stamp II microcontroller.

```
Repeat_Loop:
    debug cls
    Pulsin 2,1,T1y
    T1y = 2 * T1y
    Pulsin 4,1,T1x
    T1x = 2 * T1x
    T1y = 8 * T1y / T2
    T1x = 8 * T1x / T2
    debug dec T1x, tab, dec T1y, tab, cr
    Pause 150
    Goto Repeat_Loop:
```

Because the Basic Stamp II has a clock frequency of 2 µs, the actual time of the *T1y* and *T1x* pulses are converted to microseconds with the lines,

```
T1y = 2 * T1y
T1x = 2 * T1x
```

In *T1y* and *T1x* are the pulse widths, in microseconds. These widths are then referenced to the T2 value previously obtained by the program with the following lines:

```
T1y = 8 * T1y / T2
T1x = 8 * T1x / T2
```

The typical results of this program are numbers like 200 and 170, for the *X* and *Y* axes, respectively. Note that even on a flat surface, the two outputs of the ADXL202 may not exactly match because of manufacturing tolerances.

The *Repeat_Loop* loop continually reads the outputs of the sensor. Without the *Pause* statement and *Debug* lines, the code loops very fast—just a few tenths of microseconds–which allows you to insert other programming for your robot. Note that once the loop has begun, the value of T2 is never read again (unless you restart the entire program). This is acceptable for low-accuracy applications like basic tilt sensing. But when higher accuracy is required, the timing of the T2 pulse train should be re-read every 5 or 10 minutes, and even more frequently if the robot will be subjected to sudden and sharp temperature changes. The output of the ADXL202 is sensitive to temperature, so changes in temperature will affect the timing of the T2 pulse.

As the program runs you will note that the value of the X and Y outputs will change ±50 to ±75 just by tilting the accelerometer on its sides. Sudden movement of the accelerometer will produce more drastic changes. Note the values you get and incorporate them into the accelerometer control software you devise for your robot.

ADDITIONAL USES

Though the ADXL202 accelerometer is ideally suited for use as a tilt sensor, it has other uses too. No additional hardware or even software is required to turn the sensor into a movement, vibration, and shock sensor. Assuming that the accelerometer is oriented so the robot travels in the chip's X-axis, then as the robot moves the '202 will register the change in acceleration.

Should the robot hit a wall or other obstacle, it will be sensed as a very high acceleration/deceleration spike. Your control software will need to loop through the code at a high enough rate to catch these momentary changes in output if you want your robot to react to shocks and vibrations. The *Repeat_ Loop* in the code in Listing 41.2 repeats often enough that your robot should detect most collisions with objects.

If you absolutely must detect *all* collisions you'll need to devise some kind of hardware interrupt that will trigger the microprocessor or microcontroller when the output of the '202 exceeds a certain threshold. Most hardware interrupts are not engineered to accept pulse width modulated signals, however, so additional external circuitry may be required. Another approach is to dedicate a fast-acting microcontroller just to the task of monitoring the output of the ADXL202. The low cost of microcontrollers these days makes such dedicated applications a reasonable alternative.

Alternatives to "Store-bought" Accelerometers

While factory-made accelerometers, such as the Analog Devices ADXL150 and ADXL202, are the most convenient for use with robotics, there are some low-cost alternatives you might want to experiment with. You can make your own homebrew accelerometer using a 50-cent piezo ceramic disc and a heavy steel ball or other weight. The homebrew piezo accelerometer isn't as accurate as the ADXL series or other factory-made accelerometers, but it'll do in a pinch and teach you about the physics of motion in the process.

The piezo disc accelerometer works by using a well-known behavior of piezoelectric material: it is both a *consumer* of energy and a *producer* of energy. Most applications of piezoelectric materials are in consumer products like speakers and beepers. Apply a voltage to the piezoelectric material, and it vibrates, producing a tone. Conversely, if you vibrate the piezoelectric material using some mechanical means, the output is an electrical signal.

Piezo discs are common finds in electronic and surplus stores. These units are typically used as the elements in low-cost speakers or tone-makers (like smoke alarms, car alarms, and what not). The typical piezo disc is about an inch in diameter and is made of brass or some other nonferrous metal. Deposited on one side of the disc is a ceramic material made of piezo crystals. It is these crystals that vibrate when a voltage is applied to the disc. Most piezo discs already have two wire leads conveniently soldered to them so they can be easily connected to the rest of your circuit.

CONSTRUCTING THE PIEZO DISC ACCELEROMETER

We'll be using the disc in electricity-producing mode, with the help of a steel ball or other heavy weight to provide mechanical energy. Place the ball or weight on the disc—ceramic side up—and tape the ball in place so it won't roll or fall off the disc. Connect the output of the disc to a fast-acting voltage meter or an oscilloscope. Lift the disc up and down rapidly, and you'll see the voltage output of the disc fluctuate, perhaps as much as a full volt or two. The faster you move the disc, the more the voltage will swing.

And, just as important, note that the polarity of the voltage changes depending on the direction of travel. The output of the disc might be in positive volts when moving up but negative volts when moving down.

To complete the construction, mount the disc either on a separate sensor board or on the robot itself. As an accelerometer that senses lateral motion, the disc can be mounted in a vertical position, though that will reduce its sensitivity since the ball or weight is being pulled off the disc by gravity. Be sure that the tape holding the ball is secure. You may wish to construct a more reliable captive mechanism, perhaps housing the disc and ball or weight in an enclosure. A 35-mm film can cut to size or a plastic "bug case" (like the kind used for prizes in bubble gum machines) are good options.

As with a factory-made accelerometer, you can use the piezo disc accelerometer for vibration and shock detection. Sudden jolts—like when the robot bumps into something—will translate into larger-than-normal variations in the output of the disc. When you connect this accelerometer to the brains of your robot, this information can be used to determine the machine's proper course of action.

Limitations of the Piezo Disc Accelerometer

While the piezo disc makes for a cheap and easy accelerometer, it's not without its limitations. Here are two you will need to consider:

- The disc will only measure changes in momentum since it is inherently an AC device. Once the momentum of the disc normalizes, the output voltage will fall back to its nominal state. Since gravity acts like a constant DC signal, this means you will not be able to use the piezo disc as a tilt sensor very easily.
- The output of the piezo disc can easily exceed the input voltage of your PC, microcontroller, or other interface electronics. Should the disc receive the blow of a sharp impact, the voltage output can easily exceed 20, 50, and even 100 volts. For this reason, you

must always place a zener diode to act as a voltage clamp, as shown in Fig. 29.14 of Chapter 29. Select a zener diode voltage that is compatible with the input voltage for the interface you are using. For example, if the interface voltage is 5 vdc, use a 5.1-volt zener.

■ The piezo disc is basically a capacitor so it stores a charge over time. You can reduce the effects of the capacitive charge by placing a 50K to 250K resistor across the output leads of the disc (this will help to "bleed off" the charge). You may also want to feed the output of the disc to an op amp.

From Here

To learn more about...	*Read*
Connecting hardware to a computer or microcontroller	Chapter 29, "Interfacing with Computers and Microcontrollers"
Using the Basic Stamp with an accelerometer	Chapter 31, "Using the Basic Stamp"
Using the BasicX microcontroller with an accelerometer	Chapter 32, "Using the BasicX Microcontroller"
Navigating through an environment	Chapter 38, "Navigating through Space"

42

TIPS, TRICKS, AND TIDBITS
FOR THE ROBOT EXPERIMENTER

Most every book has a "straggler" chapter that really doesn't fit with the rest. Well, this is the straggler chapter for *Robot Builder's Bonanza*. It contains various odds-and-ends discussions about robot building, including some of my own personal methodologies, rants, and observations.

But First...

All robots are different because their creators have different tasks in mind for their creations to accomplish. A robot designed to find empty soda cans is going to be radically different from one made to roam around a warehouse sniffing out the smoke and flames of a fire.

Consider that a true robot is a machine that not only *acts independently* within an environment but *reacts independently* of that environment. In describing what a robot is it's often easier to first consider what it *isn't:*

■ Your car is a machine, but it's *not* a robot. Unless you outfit it with special gizmos, it has no way of driving itself (okay, so "Q" can make a self-running car for James Bond). It requires you to control it, to steer the wheels and operate the gas and brake pedals, and to roll down the window to talk to the nice police officer.

■ Your refrigerator is a machine, but it's *not* a robot. It may have automatic circuitry that can react to an environment (increase the cold inside if it gets hot outside), but it cannot load or unload its own food, so it still needs you for its most basic function.

■ Your dishwasher is a machine, but it's *not* a robot. Like the refrigerator, the dishwasher is not self-loading, may not adjust itself in response to how dirty the dishes are, and cannot be reprogrammed to accommodate changes in the soap you use, nor can it detect that you've loaded it with $100-a-plate porcelain—so go easy on the rinse cycle, thank you very much.

Other machines around your home and office are the same. Consider your telephone answering machine, your copier, or even your personal computer. All need *you* to make them work and accomplish their basic tasks.

A real robot, on the other hand, doesn't need you to fulfill its chores. A robot is programmed ahead of time to perform some job, and it goes about doing it. Here, the distinction between a robot and an automatic machine becomes a little blurry because both can run almost indefinitely without human intervention (not counting wear and tear and the availability of power). However, most automatic machines lack the means to interact with their environment and to change that environment if necessary. This feature is often found in more complex robots.

Beyond this broad distinction, the semantics of what is and is not a robot isn't a major concern of this book. The main point is this: Once the robot is properly programmed, it should not need your assistance to complete its basic task(s), barring any unforeseen obstacles or a mechanical failure.

"What Does My Robot Do?": A Design Approach

Before you can build a robot you must decide what you want the robot to do. That seems obvious, but you'd be surprised how many first-time robot makers neglect this important step. By reducing the tasks to a simple list, you can more easily design the size, shape, and capabilities of your robot. Let's create an imaginary homebuilt robot named *RoBuddy,* for "Robotic Buddy," and go through the steps of planning its design. We'll start from the standpoint of the jobs it is meant to do. For the sake of simplicity, we'll design RoBuddy so that it's an "entertainment" 'bot—it's for fun and games and is not built for handling radioactive waste or picking up after your dog Spot.

I've found that one of the first things people ask me about my robots is, "So, what does it do?" That's not always an easy question to answer because the function of a robot can't always be summarized in a quick sentence. Yet most people don't have the patience to listen to a complex explanation. Such is the quandary of the robot builder!

AN ITINERARY OF FUNCTIONS

One of the best shortcuts to explaining what a robot can do is to simply give the darned thing a vacuum cleaner. That way, when you don't feel like repeating the whole litany of capabilities, you can merely say, "it cleans the floors." That's almost always guaranteed to elicit a positive response. So this is *Basic Requirement #1:* RoBuddy must be equipped with a vacuum cleaner. And since RoBuddy is designed to be self-powered from batteries, the vacuum cleaner needs to run under battery power too. Fortunately, auto parts stores carry a number of 12-volt portable vacuum cleaners from which you can choose.

Like the family dog that performs tricks for guests, a robot that mimics some activity amusing to humans is a great source of entertainment. One of the most useful—and effective—activities is pouring and serving drinks. That takes at least one arm and gripper, preferably two, and the arms must be strong and powerful enough to lift at least 12 ounces of beverage. We now have *Basic Requirement #2:* RoBuddy must be equipped with at least one appendage that has a gripper designed for drinking glasses and soda cans.

The RoBuddy must also have some kind of mobility so that at the very least it can move around and vacuum the floor. There are a number of ways to provide locomotion to a robot, and these were described in earlier chapters. But for the sake of description, let's assume we use the common two-wheel-drive approach, which consists of two motorized wheels counterbalanced by one or two nonpowered casters. That's *Basic Requirement #3:* RoBuddy must have two drive motors and two wheels for moving across the floor.

Since RoBuddy flits about your house all on its own accord, it has to be able to detect obstacles so it can avoid them. Obviously, then, the robot must be endowed with some kind of obstacle detection devices. We're up to *Basic Requirement #4:* RoBuddy must be equipped with passive and active sensors to detect and avoid objects in its path.

Serving drinks, vacuuming the floor, and avoiding obstacles requires an extensive degree of intelligence and is beyond the convenient capability of "hard-wired" discrete circuits consisting of some resistors, a few capacitors, and a handful of transistors. A better approach is to use a computer, which is capable of being programmed and reprogrammed at will. This computer is connected to the vacuum cleaner, arm and gripper, sensors, and drive motors. Finally, then, this is *Basic Requirement #5:* RoBuddy must be equipped with a computer to control the robot's actions.

These five basic requirements may or may not be important to you or applicable to all your robot creations. However, they give you an idea of how you should outline the functions of your robot and match them with a hardware requirement.

ADDITIONAL FEATURES

Depending on your time, budget, and construction skill, you may wish to endow your robot(s) with a number of other useful features, such as:

- *Sound output* perhaps combining speech, sound effects, and music.
- *Variable speed motors* so your robot can get from room to room in a hurry but slow down when it's around people, pets, and furniture.
- *Set-and-forget* motor control, so the "brains on board" that is controlling your 'bot needn't spend all its processing power just running the drive motors.
- *Distance sensors* for the drive motors so the robot knows how far it has traveled ("odometry").
- *Infrared and ultrasonic sensors* to keep the robot from hitting things.
- *Contact bumper switches* on the robot so it knows when it's hit something and to stop immediately.

- *LCD panels, indicator lights, or multidigit displays* to show current operating status.
- *Tilt switches, gyroscopes, or accelerometers* to indicate when the robot has fallen over, or is about to.
- *Voice input,* for voice command, voice recognition, and other neat-o things.
- *Teaching pendant and remote control* so you can move a joystick to control the drive motors and record basic movements.

Of course, we discussed all of these in previous chapters. Review the table of contents or index to locate the relevant text on these subjects.

Reality versus Fantasy

In building robots it's important to separate the reality from the fantasy. Fantasy is a *Star Wars* R2-D2 robot projecting a hologram of a beautiful princess. Reality is a homebrew robot that scares the dog as it rolls down the hallway—and probably hits the walls as it goes. Fantasy is a giant killer robot that walks on two legs and shoots a death ray from a visor in its head. Reality is foot-tall "trash can" robot that pours your houseguests a Diet Coke. Okay, so it spills a little every now and then . . . now you know why a robot equipped with a vacuum cleaner comes in handy!

It's easy to get caught up in the romance of designing and building a robot. But it's important to be wary of impossible plans. Don't attempt to give your robot features and capabilities that are beyond your technical expertise, budget, or both (and let's not also forget the limits of modern science). In attempting to do so, you run the risk of becoming frustrated with your inability to make the contraption work, and you miss out on an otherwise rewarding endeavor.

When designing your automaton, you may find it helpful to put the notes away and let them gel in your brain for a week. Quite often, when you review your original design, you will realize that some of the features and capabilities are mere wishful thinking, and beyond the scope of your time, finances, or skills. Make it a point to refine, alter, and adjust the design of the robot before, and even during, construction.

Understanding and Using Robot "Behaviors"

A current trend in the field of robot building is "behavior-based robotics," where you program a robot to act in some predictable way based on both internal programming and external input. For example, if the battery of your robot becomes weak, it can be programmed with a "find energy" behavior that will signal the robot to return to its battery charger. Behaviors are a convenient way to describe the core functionality of robots—a kind of "component" architecture to define what a robot will do given a certain set of conditions.

The concept of behavior-based robotics has been around since the 1980s and was developed as a way to simplify the brain-numbing computational requirements of artificial intelligence systems popular at the time. Behavior-based robotics is a favorite at the Massachusetts Institute of Technology, and Professor Rodney Brooks, a renowned leader in the field of robot intelligence, is one of its major proponents.

Since the introduction of behavior-based robotics, the idea has been discussed in countless books, papers, and magazine articles, and has even found its way into commercial products. The LEGO Mindstorms robots, which are based on original work done at MIT, use behavior principles. See Appendix A, "Further Reading," for books that contain useful information on behavior-based robotics.

WHEN A BEHAVIOR IS JUST A SIMPLE ACTION

Since the introduction of behavior-based robotics, numerous writers have applied the term *behavior* to cover a wide variety of things—to the point that *everything* a robot does becomes a "behavior." The result is that robot builders can become convinced their creations are really exhibiting human- or animal-like reactions, when all they are doing is carrying out basic instructions from a computer or simple electronic circuit. Delusions aside, this has the larger effect of distracting you from focusing on other useful approaches for dealing with robots.

Let me explain by way of an analogy. Suppose you see a magic show so many times that you end up believing the disappearing lady is really gone. Not so. It's an optical and psychological trick every time. Sometimes a robot displays a simple *action* as the result of rudimentary programming, and by calling everything it does a "behavior" we lose a clearer view of how the machine is really operating.

The following sections contain a brief discourse on behavior-based robotics, and my personal views on clarifying terms so we can get the most out of the behavior concept.

WALL FOLLOWING: A COMMON "BEHAVIOR?"

One common example of behavior-based robotics is the "wall follower," which is typically a robot that always turns in an arc, waiting to hit a wall. A sensor on the front of the robot detects the wall collision. When the sensor is triggered, the robot will back away from the wall, go forward a set amount, then repeat the whole process all over again.

This is a perfect example of how the term *behavior* has been misplaced: the *true* behavior of the robot is not to follow a *wall* but simply to turn in circles until it hits something. When a collision occurs, the robot moves to clear the obstacle, then continues to turn in a circle once again. In the absence of the wall—a reasonable change in environment—the robot would not exhibit its namesake "behavior." Or conversely, if there were additional objects in the room, the robot would treat them as "walls" too. In that case, the robot might be considered useless, misprogrammed, or worse.

If "wall following" is not a true behavior, then what is it? I won't presume to come up with an industry-standard term. The important thing to remember is that a *true behavior* is independent, or nearly so, of the robot's typical physical environment. That's *Rule Number One* to keep in mind.

Note that *environment* is not the same as a *condition*. A condition is a light shining on the robot that it might move toward or away from; an environment is a room or other area that may or may not have certain attributes. Conditions contribute to the function of the robot, just like batteries or other electric power contribute to the robot's ability to move its motors. Conversely, environments can be ever changing and in many ways unmanageable. Environments consist of physical parameters under which the robot may or may not operate at any given time.

Robotic behaviors are most useful when they encapsulate multiple variables, particularly those are in response to external input (senses). This is *Rule Number Two* of true behavior-based robotics. The more the robot is able to integrate and differentiate between different input (senses), irrespective of its environment—and still carry out its proper programming—the more it can demonstrate its true behaviors.

THE "WALT DISNEY EFFECT"

It is tempting to endow robots with human- or animal-like emotions and traits, such as hunger (battery power) or affinity/love (a beacon or an operator clicking a "clicker"). But in my opinion these aren't behaviors at all. They are anthropomorphic qualities that merely *appear* to result in a human-type response simply because we want them to.

In other words, it's completely made up. Imagine this in the extreme: Is a robot "suicidal" if it has a tendency to drive off the workbench and break as it hits the floor? Or is it that your workbench is too small and crowded, and your concrete floor is too hard? Emotions such as love are extremely complex; as a robot builder, it's easy to get confused about what your creation can really do and feel.

In his seminal book *Vehicles,* Valentino Braitenberg gives us a study of synthetic psychology on which fictional "vehicles" demonstrate certain behavioral traits. For example, Braitenberg's Vehicle 2 has two motors and two sensors (say, light sensors). By connecting the sensors to the motors in different ways the robot is said to exhibit "emotions," or at the least actions we humans may *interpret* as quasi-intelligent or human-like emotional responses. In one configuration, the robot may steer toward the light source, exhibiting "love." In another configuration, the robot may steer away, exhibiting "fear."

Obviously, the robot is feeling neither of these emotions, nor does Braitenberg suggest this. Instead, he gives us vehicles that are fictional representations of human-like traits. It's important not to get caught up in Disneyesque anthropomorphism. A good portion of behavior-based robotics centers around *human interpretation* of the robot's mechanical actions. We interpret those actions as intelligent, or even as cognition. This is valid up to a point, but consider that only we ourselves experience our own intelligence and cognition (that is, we are "self-aware"); a robot does not. Human-like machine intelligence and emotions are in the eye of a human beholder, not in the brain of the robot. This, however, may change in the future as new computing models are discovered, invented, and explored.

ROBOTIC FUNCTIONS AND ERROR CORRECTION

When creating behaviors for your robots, keep in mind the *function* that you wish to accomplish. Then, consider how that function is negatively affected by variables in the robot's likely environments. For practical reasons (budget, construction skill), you must consider at least some of the limitations of the robot's environment in order to make it reliably demonstrate a given behavior. A line-following robot, which is relatively easy to build and program, will not exhibit its line-following behavior without a line. By itself, such a robot would merely be demonstrating a simple action. But by adding *error correction*—to compensate for unknown or unexpected changes in environment—the line-following robot begins to demonstrate a useful behavior. This behavior extends beyond the robot's immediate environmental limits. The machine's ability to go into a secondary, error-correcting state to find a line to follow is part of what makes a valid line-following behavior—even more so if in the absence of a line to follow the robot can eventually make its own.

Error correction is *Rule Number Three* of behavior-based robotics. Without error correction, robots operating in restrictive environments are more likely to exhibit simple, even stupid, actions in response to a single stimulus. Consider the basic "wall-following" robot again: it *requires* a room with walls—and, at that, walls that are closer together than its turning radius. Outside or in a larger room, the robot "behaves" completely differently, yet its programming is *exactly* the same. The problem of the wall-following robot could be fixed either by adding error correction or by renaming the base behavior to more accurately describe what it physically is doing.

ANALYZING SENSOR DATA TO DEFINE BEHAVIORS

By definition, behavior-based robotics is reactive, so it requires some sort of external input by which a behavior can be triggered. Without input (a light sensor, ultrasonic detector, bumper switch, etc.) the robot merely plays out a preprogrammed set of moves—simple actions, like a player piano. More complex behaviors becomes possible if the following capabilities are added:

- The ability to analyze the data from an analog, as opposed to a digital, sensor. The output of an analog sensor provides more information than the simple on/off state of a digital sensor. Let's call this *sensor parametrics.*
- The ability to analyze the data from multiple sensors, either several sensors of the same type (a gang of light-sensitive resistors, for example) or sensors of different types (a light sensor and an ultrasonic sensor). This is commonly referred to as *sensor fusion.*

Let's consider sensor parametrics first. Suppose your robot has a temperature sensor connected to its onboard computer. Temperature sensors are analog devices; their output is proportional to the temperature. You use this feature to determine a set or range of preprogrammed actions, depending on the specified temperature. This set of actions constitutes a behavior or, if the actions are distinct at different temperatures, a variety of behaviors. Similarly, a photophilic robot that can discern the brightest light among many lights also exhibits sensor analysis from parametric data.

Sensor fusion analyzes the output of several sensors. Your robot initiates the appropriate behavioral response as a result. For example, your robot may be programmed to follow the brightest light but also detect obstructions in its path. When an obstruction is encountered, the robot is programmed to go around it, then continue—perhaps from a new direction—toward the light source.

Sensor fusion helps provide error correction and allows a robot to continue exhibiting its behavior (one might call it the robot's "prime directive") even in the face of unpredictable environmental variables. The variety, sophistication, and accuracy of the sensors determine how well the robot will perform in any given circumstance. Obviously, it's not practical—economic or otherwise—to ensure that your robot will work flawlessly under all environments and conditions. But the more you give your robots the ability to overcome common and reasonable environmental variables (such as socks on the floor), the better it will display the behavior you want.

THE ROLE OF SUBSUMPTION ARCHITECTURE

Subsumption architecture isn't an odd style of building. Rather, it's a technique devised by Dr. Brooks at MIT that has become a common approach for dealing with the complexities of sensor fusion and artificial machine intelligence. With subsumption, sensor inputs are prioritized. Higher-priority sensors override lower-priority ones. In the typical subsumption model, the robot may not even be aware that a low-priority sensor was triggered.

More complex hybrid systems may employ a form of simple subsumption along with more traditional artificial intelligence programming. The robot's computer may evaluate the relative merits of low-priority sensors and use this information to intuit a unique course of action, perhaps one in which direct programming for the combination of input variables does not yet exist. In some cases, the output of a low-priority sensor may moderate the interpretation of a high-priority one.

As an example, a fire-fighting robot may have both a smoke detector and a flame detector. The smoke detector is likely to sense smoke before any fire can be identified, since smoke so easily permeates a structure. Therefore, the smoke sensor will likely be given a lower priority to the flame detector, since it is so easily triggered. But consider that flames can exist without a destructive fire (e.g., a fireplace and candlelight, both of which do not emit much smoke under normal circumstances). Rather than have the robot totally ignore its other sensors when the high-priority flame detector is triggered, the robot instead integrates the output of both flame and smoke sensors to determine what is, and isn't, a fire that needs to be put out.

Multiple Robot Interaction

An exciting field of research is the interaction of several robots working together. Rather than build one big, powerful robot that does everything, multirobot scenarios combine the strengths of two or more smaller, simpler machines to achieve synergy: the whole is greater than the parts. Anyone who has seen the now-classic science fiction film *Silent Running* knows what three diminutive robots (named Huey, Luey, and Dewey, by the way) can do!

Robot "tag teams" are common in college and university robot labs, where groups of robot researchers compete with their robots as the players of a game (robo-soccer is popular). Each robot in the competition has a specific job, and the goal is to have them work together. There are three common types of robot-to-robot interaction:

■ *Peer-to-peer.* Each robot is considered equal, though each one may have a different job to do, based on predefined programming The workload may also be divided based on physical proximity to the work, and whether the other robots in the group are busy doing other things.

■ *Queen/drone.* One robot serves as the leader, and one or more additional robots serve as worker drones. Each drone takes it's work orders directly from the queen and may interact only peripherally with the other drone 'bots.

■ *Convoys.* Combining the first two types, the leader of the convoy is the "queen" robot, and the other robots act as peers among themselves. In convoy fashion, each robot may rely on the one just ahead for important information. This approach is useful when the "queen" is not capable, for computer processing reasons or otherwise, to control a large number of fairly mindless drones.

Why all the fuss with multiple robots? First and foremost because it's generally easier and cheaper to build many small and simple robots than a single big and complex one. Second, the mechanical failure of one robot can be compensated for by the remaining good robots. In many instances, the "queen" or leader robot is no different than the others, it just plays a coordinating role. In this way, should the leader 'bot go down for the count, any other robot can easily take its place. And third, work tends to get done faster with more hands helping.

Dealing with Failures

Few robots work perfectly when you flip the switch the first time. Failure is common in robot building and should be expected. As you learn from these failures you will build better robots. Failure can occur at the onset when you first try a new design, or it can occur at any point thereafter, as the robot breaks down for one reason or another.

MECHANICAL FAILURE

Mechanical problems are perhaps the most common failure. A design you developed just doesn't work well, usually because the materials or the joining methods you used were not strong enough. Avoid overbuilding your robots (that tends to make them too expensive and heavy), but at the same time strive to make them physically strong. Of course, "strong" is relative: a lightweight, scarab-sized robot needn't have the muscle to tote a two-year-old on a tricycle. At the very least, however, your robot construction should support its own weight, including batteries.

When possible, avoid "slap-together" construction, such as using electrical or duct tape. These methods are acceptable for quick prototypes but are unreliable for long-term testing. When gluing parts in your robot, select a glue that is suitable for the materials you are using. Epoxy and hot-melt glues are among the most permanent. You may also have luck with cyanoacrylate (CA) glues, though the bond may become brittle and weak over time (a few years or more, depending on humidity and stress).

Tip:

Use the "pull test" to determine if your robot construction methods are sound. Once you have attached something to your robot—using glue, nuts and bolts, or whatever—give it a healthy tug. If it comes off, the construction isn't good enough. Look for a better way.

ELECTRICAL FAILURE

Electronics can be touchy, not to mention extremely frustrating, when they don't work right. Circuits that functioned properly in a solderless breadboard may no longer work once you've soldered the components in a permanent circuit, and vice versa. There are many reasons for this, including mistakes in wiring, odd capacitive effects, even variations in tolerances due to heat transfer. Here are some pointers:

■ If a circuit doesn't work from the get-go, review your wiring and make necessary repairs.

■ If the circuit fails after some period of use, the cause may be a short circuit or broken wire, or it could be a burned-out component. Example: if your motors draw too much current from the drive circuitry you run the risk of permanently damaging some semiconductors.

Certain electronic circuit construction techniques are better suited for an active, mobile robot. Wire-wrap is a fast way to build circuits, but its construction can invite problems. The long wire-wrap pins can bend and short out against one another. Loose wires can come off. Parasitic signals and stray capacitance can cause "marginal" circuits to work, then not work, and then work again. For an active robot it may be better to use a soldered circuit board, perhaps even a printed circuit board of your design (see Chapter 6, "Electronic Construction Techniques," for more information).

Some electrical problems may be caused by errors in programming, weak batteries, or unreliable sensors. For example, it is not uncommon for sensors to occasionally yield totally wacky results. This can be caused by design flaws inherent in the sensor itself, spurious data (noise from a motor, for example), or corrupted or out-of-range data. Ideally, the programming of your robot should anticipate occasional bad sensor readings and basically ignore them. A perfectly acceptable approach is to throw out any sensor reading that is outside the statistical model you have decided on (e.g., a sonar ping that says an object is 1048 feet away; the average robotic sonar system has a maximum range of about 35 feet).

PROGRAMMING FAILURE

As more and more robots use computers and microcontrollers as their "brains," programming errors are fast becoming one of the most common causes of failure. There are three basic kinds of programming "bugs." In all cases, you must review the program, find the error, and fix it:

■ *Compile bug,* caused by bad syntax. You can instantly recognize these because the program compiler or downloader will flag these mistakes and refuse to continue. You must fix the problem before you can transfer the program to the robot's microcontroller or computer.

■ *Run-time bug,* caused by a disallowed condition. A run-time bug isn't caught by the compiler. It occurs when the microcontroller or computer attempts to run the program. An example of a common run-time bug is the use of an out-of-bounds element in an array (for instance, trying to assign a value to the thirty-first element in a 30-element array). Run-time bugs may also be caused by missing data, such as looking for data on the wrong input pin of a microcontroller.

■ *Logic bug,* caused by a program that simply doesn't work as anticipated. Logic bugs may be due to simple math errors (you meant to add, not subtract) or by mistakes in coding that cause a different behavior than you anticipated.

Task-Oriented Robot Control

As "workers," robots have a task to do. In many books on robotics theory and application, these tasks are considered "goals." Personally, I'm not big on the term *goal* because that suggests a human emotion involving desire. The robot you build will have no "desire" to fetch you a can of soda, but will merely do so because its programming tells it to. Instead, I prefer the term *task*—a defined job that the robot is expected to accomplish. A robot may be given multiple tasks at the same time, such as the following:

1. Get me a can of Dr. Pepper;
2. Avoid running into the wall while doing so;
3. Watch out for the cat and other ground-based obstacles;
4. And remember where you came from so you can bring the soda back to me.

These tasks form a hierarchy. Task 4 cannot be completed before task 1. Together, these two form the *primary directive tasks* (shades of *Star Trek* here—okay, I admit it: I'm a Trekker!). Tasks 2 and 3 may or may not occur; these are *error mode tasks.* Should they occur, they temporarily suspend the processing of the primary directive tasks.

PROGRAMMING FOR TASKS

From a programming standpoint, you can consider most any job you give a robot to look something like this:

```
Do Task X until
        on error Do Task Y
                repeat
                Task Y until no error
        resume Task X
Task X complete
```

X is the primary directive task, the thing the robot is expected to do. *Y* is a special function that gets the robot out of trouble should an error condition—of which there may be many—occurs. Most error modes will prevent the robot from accomplishing its primary directive task. Therefore, it is necessary to clear the error first before resuming the primary directive.

Note that it is entirely possible that the task will be completed without any kind of complication (no errors). In this case, the error condition is never raised, and the *Y* functionality is not activated. The robot programming is likewise written so that when the error condition is cleared, it can resume its prime directive task.

MULTITASKING ERROR MODES FOR OPTIMAL FLEXIBILITY

For a real-world robot, errors are just as important a consideration as tasks. Your robot programming must deal with problems, both anticipated (walls, chairs, cats) and unanticipated (water on the kitchen floor, no sodas in the fridge). The more your robot can recognize error modes, the better it can get itself out of trouble. And once out of an error mode, the robot can be reasonably expected to complete its task.

How you program various tasks in your robot is up to you and the capabilities of your robot software platform. If your software supports multitasking (BasicX, OOPic, LEGO Mindstorms, and others), then try to use this feature whenever possible. By dealing with tasks as discrete units, you can better add and subtract functionality simply by including or removing tasks in your program.

Equally important, you can make your robot automatically enter an error mode task without specifically waiting for it in code. In non-multitasking procedural programming, your code is required to repeatedly check (poll) sensors and other devices that warn the robot of an error mode. If an error mode is detected, the program temporarily branches to a portion of the code written to handle it. Once the error is cleared, the program can resume execution where it left off.

With a multitasking program, each task runs simultaneously. Tasks devoted to error modes can temporarily take over the processing focus to ensure that the error is fixed before continuing. The transfer of execution within the program is all done automatically. To ensure that this transfer occurs in a logical and orderly manner, the program should give priorities to certain tasks. Higher-priority tasks are able to take over ("subsume," a word now in common parlance) other running tasks when necessary. Once a high-priority task is completed, control can resume with the lower-priority activities, if that's desired.

GETTING A PROGRAM'S ATTENTION VIA HARDWARE

Even in systems that lack multitasking capability it's still possible to write a robot control program that doesn't include a repeating loop that constantly scans (polls) the condition of sensors and other input. Two common ways of dealing with unpredictable external events are using a timer (software) interrupt or a hardware (physical connection) interrupt.

Timer interrupt A timer built into the computer or microcontroller runs in the background. At predefined intervals—most commonly when the timer overflows its count—the timer grabs the attention of the microprocessor, which in turn temporarily suspends the main program. The microprocessor runs a special timer interrupt program, which in the case of a task-based robot would poll the various sensors and other input looking for possible error modes. (Think of the timer as a heart beat; at every beat the microprocessor pauses to do something special.)

If *no* error is found, the microprocessor resumes the main program. If an error is found, the microprocessor runs the relevant section in code that deals with the error. Timer interrupts can occur hundreds of times each second. That may seem like a lot in human terms, but it can be trivial to a microprocessor running at several million cycles per second.

Hardware interrupt   A hardware interrupt is a mechanism by which to immediately get the attention of the microprocessor. It is a physical connection on the microprocessor that can in turn be attached to some sensor or other input device on the robot. With a hardware interrupt the microprocessor can spend 100 percent of its time on the main program and temporarily suspend it if, and *only* if, the hardware interrupt is triggered.

Hardware interrupts are used extensively in most computers, and their benefits are well established. Your PC has several hardware interrupts. For example, the keyboard is connected to a hardware interrupt, so when you press a key the processor immediately fetches the data and makes it available to whatever program is currently running. The standard PC architecture has room for 16 hardware interrupts, even though the microprocessor uses just one interrupt pin. The one pin is *multiplexed* to make 16 separate inputs. You can do something similar in your own robot designs.

Glass half-empty, half-full   There are two basic ways to deal with error modes. One is to treat them as "exceptions" rather than the rule:

- In the exception model, the program assumes no error mode and only stops to execute some code when an error is explicitly encountered. This is the case with a hardware interrupt.
- In the opposite model, the program assumes the possibility of an error mode all the time and checks to see if its hunch is correct. This is the case with the timer interrupt.

The approach you use will depend on the hardware choices available to you. If you have both a timer and a hardware interrupt at your disposal, the hardware interrupt is probably the more straightforward method because it allows the microprocessor to be used more efficiently.

And Last...

Few other moments in life compare to the instant when you solder that last piece of wire, file down that last piece of metal, tighten that last bolt, and switch on your robot. Something *you* created comes to life, obeying your commands and following your preprogrammed instructions. This is the robot hobbyist's finest hour. It proves that the countless evenings and weekends in the workshop were worth it after all.

I started this book with a promise of adventure—to provide you with a treasure map of plans, diagrams, schematics, and projects for making your own robots. I hope you've followed along and built a few of the mechanisms and circuits that I described. Now, as you finish reading, you can make me a promise: improve on these ideas. Make them better. Use them in creative ways that no one has ever thought possible. Create that ultimate robot that everyone has dreamed about.

Your ideas, suggestions, and other comments are welcome. If you see a mistake in a circuit or mechanism, I'll make sure the next edition of this book is corrected. Write me, care of McGraw-Hill at the address in the front of the book, or visit me at *www.robotoid.com*. I can't always reply in a timely manner, but I assure you I'll consider your comments. If you have a unique robot design, why not share it with others? Send me details of your robots—finished or in process.

Now, stop reading—and do. Impress us all!

FURTHER READING

Interested in learning more about robotics? Sure you are! Here is a selected list of books than can enrich your understanding and enjoyment of all facets of robotics. These books are available at most better bookstores, as well as at many online bookstores, including Amazon (*www.amazon.com*), Barnes and Noble (*www.bn.com*), and Fatbrain (*www.fatbrain.com*).

This appendix also lists several magazines of interest to the robot experimenter. Both mailing and Internet addresses have been provided.

Contents

Hobby Robotics

LEGO Robotics and LEGO Building

Technical Robotics, Theory and Design

Artificial Intelligence and Behavior-Based Robotics

Mechanical Design

Electronic Components

Microcontroller/Microprocessor Programming

Electronics How-to and Theory

Power Supply Design and Construction

Motors and Motor Control

Lasers and Fiber Optics

Interfacing to IBM PC (and Compatibles)

Magazines

Hobby Robotics

Robots, Androids and Animatrons: 12 Incredible Projects You Can Build
John Iovine
McGraw-Hill
ISBN: 0070328048

The Personal Robot Navigator
Merl K. Miller, Nelson B. Winkless, Kent Phelps, Joseph H. Bosworth
A K Peters Ltd
ISBN: 188819300X
       (Contains CD-ROM of robot navigation simulator)

Applied Robotics
Edwin Wise
Howard W Sams & Co
ISBN: 0790611848

Muscle Wires Project Book
Roger G. Gilbertson
Mondo-Tronics
ISBN: 1879896141

Stiquito: Advanced Experiments with a Simple and Inexpensive Robot
James M. Conrad, Jonathan W. Mills
Institute of Electrical and Electronic Engineers
ISBN: 0818674083

Stiquito for Beginners: An Introduction to Robotics
James M. Conrad, Jonathan W. Mills
IEEE Computer Society Press
ISBN: 0818675144

LEGO Robotics and LEGO Building

Dave Baum's Definitive Guide to LEGO Mindstorms
Dave Baum
Apress
ISBN: 1893115097

Unofficial Guide to LEGO MINDSTORMS Robots
Jonathan B. Knudsen
O'Reilly & Associates
ISBN: 1565926927

Lego Crazy Action Contraptions
Dan Rathjen
Klutz, Inc
ISBN: 1570541574

Technical Robotics, Theory and Design

Mobile Robots: Inspiration to Implementation
Joseph L. Jones, Anita M. Flynn,
Bruce A. Seiger
A K Peters Ltd
ISBN: 1568810970

Sensors for Mobile Robots : Theory and Application
H. R. Everett
A K Peters Ltd
ISBN: 1568810482

Art Robotics: An Introduction to Engineering
Fred Martin
Prentice Hall
ISBN: 0805343369

Robot Evolution: The Development of Anthrobotics
Mark Rosheim
John Wiley & Sons
ISBN: 0471026220

Machines That Walk: The Adaptive Suspension Vehicle
Shin-Min Song
MIT Press
ISBN: 0262192748

Remote Control Robotics
Craig Sayers
Springer-Verlag
ISBN: 038798597

Artificial Vision for Mobile Robots: Stereo Vision and Multisensory Perception
Nicholas Ayache, Peter T. Sander
MIT Press
ISBN: 0262011247

Artificial Intelligence and Behavior-based Robotics

Robot: Mere Machine to Transcendent Mind
Hans Moravec
Oxford University Press
ISBN: 0195116305

Behavior-Based Robotics (Intelligent Robots and Autonomous Agents)
Ronald C. Arkin
MIT Press
ISBN: 0262011654

Artificial Intelligence and Mobile Robots: Case Studies of Successful Robot Systems
David Kortenkamp (editor), R. Peter Bonasso (editor), Robin Murphy (editor)
MIT Press
ISBN: 0262611376

Cambrian Intelligence: The Early History of the New AI
Rodney Allen Brooks
MIT Press
ISBN: 0262522632

Vehicles: Experiments in Synthetic Psychology
Valentino Braitenberg
MIT Press
ISBN: 0262521121

Intelligent Behavior in Animals and Robots
David McFarland, Thomas Bosser
MIT Press
ISBN: 0262132931

Mechanical Design

Mechanical Devices for the Electronics Experimenter
Britt Rorabaugh
Tab Books
ISBN: 0070535477

Mechanisms and Mechanical Devices Sourcebook, Second Edition
Nicholas P. Chironis, Neil Sclater
McGraw-Hill
ISBN: 0070113564

Home Machinist's Handbook
Doug Briney
Tab Books
ISBN: 0830615733

Electronic Components

Electronic Circuit Guidebook : Sensors
Joseph J. Carr
PROMPT Publications
ISBN: 0790610981

Electronic Circuit Guidebook (various volumes)
Joseph J. Carr
PROMPT Publications
  Volume 1: Sensors; ISBN: 0790610981
  Volume 2: IC Timers; ISBN: 0790611066
  Volume 3: Op Amps; ISBN: 0790611317

Build Your Own Low-cost Data Acquisition and Display Devices
Jeffrey Hirst Johnson
Tab Books
ISBN: 0830643486

Microcontroller/Microprocessor Programming

Programming and Customizing the Basic Stamp Computer
Scott Edwards
McGraw-Hill
ISBN: 0079136842

Microcontroller Projects with Basic Stamps
Al Williams
R&D Books
ISBN: 0879305878

The Basic Stamp 2—Tutorial and Applications
Peter H. Anderson (Author and Publisher)
ISBN: 0965335763

Programming and Customizing the Pic Microcontroller
Myke Predko
McGraw-Hill
ISBN: 007913646X

Easy Pic'N: A Beginner's Guide to Using Pic16/17 Microcontrollers
David Benson
Square One Electronics
ISBN: 0965416208

PIC'n Up Pace: An Intermediate Guide to Using PIC Microcontrollers
David Benson
Square One Electronics
ISBN: 0965416216

Design with Pic Microcontrollers
John B. Peatman
Prentice Hall
ISBN: 0137592590

Microcontroller Cookbook
Mike James
Butterworth-Heinemann
ISBN: 0750627018

Handbook of Microcontrollers
Myke Predko
McGraw-Hill
ISBN: 0079137164

Programming and Customizing the 8051 Microcontroller
Myke Predko
McGraw-Hill
ISBN: 0071341927

The 8051 Microcontroller
I. Scott MacKenzie
Prentice Hall
ISBN: 0137800088

The Microcontroller Idea Book
Jan Axelson
Lakeview Research
ISBN: 0965081907

Programming and Customizing the Hc11 Microcontroller
Thomas Fox
McGraw-Hill Professional Publishing
ISBN: 0071344063

AVR RISC Microcontroller Handbook
Claus Kuhnel
Newnes
ISBN: 0750699639

Electronics How-to and Theory

McGraw-Hill Benchtop Electronics Handbook
Victor Veley
McGraw-Hill
ISBN: 0070674965

The TAB Electronics Guide to Understanding Electricity and Electronics
G. Randy Slone
Tab Books
ISBN: 0070582165

Electronic Components: A Complete Reference for Project Builders
Delton T. Horn
Tab Books
ISBN: 0830633332

The Forrest Mims Engineer's Notebook
Forrest M. Mims, Harry L. Helms
LLH Technology Pub
ISBN: 1878707035

Engineer's Mini-Notebook (series)
Forrest M. Mims
Radio Shack

Logicworks 4: Interactive Circuit Design Software for Windows and Macintosh
Addison-Wesley
ISBN: 0201326825
(Book and CD-ROM; includes software)

Beginner's Guide to Reading Schematics
Robert J. Traister, Anna L. Lisk
Tab Books
ISBN: 0830676325

Printed Circuit Board Materials Handbook
Martin W. Jawitz (editor)
McGraw-Hill
ISBN: 0070324883

The Art of Electronics
Paul Horowitz, Winfield Hill
Cambridge University Press
ISBN: 0521370957

Student Manual for the Art of Electronics
Paul Horowitz, T. Hayes
Cambridge University Press
ISBN: 0521377099

Power Supply Design and Construction

DC Power Supplies
Joseph J. Carr
McGraw-Hill
ISBN: 007011496X

Power Supplies, Switching Regulators, Inverters, and Converters
Irving M. Gottlieb
Tab Books
ISBN: 0830644040

Motors and Motor Control: Electric Motors and Control Techniques, Second Edition
Irving M. Gottlieb
ISBN: 0070240124

Lasers and Fiber Optics

Lasers, Ray Guns, and Light Cannons: Projects from the Wizard's Workbench
Gordon McComb
McGraw-Hill
ISBN: 0070450358

Optoelectronics, Fiber Optics, and Laser Cookbook
Thomas Petruzzellis
McGraw-Hill
ISBN: 0070498407

Understanding Fiber Optics
Jeff Hecht
Prentice Hall
ISBN: 0139561455

Laser: Light of a Million Uses
Jeff Hecht, Dick Teresi
Dover
ISBN: 0486401936

Interfacing to IBM PC (and Compatibles)

Use of a PC Printer Port for Control & Data Acquisition
Peter H. Anderson (Author and Publisher)
ISBN: 0965335704

The Parallel Port Manual Vol. 2: Use of a PC Printer Port for Control and Data Acquisition
Peter H. Anderson (Author and Publisher)
ISBN: 0965335755

Programming the Parallel Port
Dhananjay V. Gadre
R&D Books
ISBN: 0879305134

Parallel Port Complete
Jan Axelson
Peer-to-Peer Communications
ISBN: 0965081915

Serial Port Complete
Jan Axelson
Lakeview Research
ISBN: 0965081923

Real-World Interfacing with Your PC
James Barbarello
PROMPT Publications
ISBN: 0790611457

Magazines

Robot Science and Technology
3875 Taylor Road, Suite 200
Loomis, CA 95650
www.robotmag.com

Poptronics
Gernsback Publications
500 Bi-County Blvd.
Farmingdale, NY 11735
www.poptronics.com

Nuts & Volts Magazine
430 Princeland Court
Corona, CA 91719
www.nutsvolts.com

Everyday Practical Electronics
Wimborne Publishing Ltd.
Allen House
East Borough, Wimborne
Dorset BH2 1PF
United Kingdom

Elektor
www.elektor-electronics.co.uk

SOURCES

Selected Specialty Parts and Sources

General Robotics Kits and Parts

Electronics/Mechanical Mail Order; New, Used, and Surplus

Microcontrollers, Single Board Computers, Programmers

Radio Control (RC) Retailers

Servo and Stepper Motors, Controllers

Ready-Made Personal and Educational Robots

Construction Kits, Toys, and Parts

Miscellaneous

Note:

Internet-based companies that do not provide a mailing address on their Web site are not listed. In addition, Internet-based companies hosted on a "free" Web hosting service (Tripod, Geocities, etc.) are also not listed because of fraud concerns.

The listing in this appendix is periodically updated at *www.robotoid.com.*

Selected Specialty Parts and Sources

BEAM Robots
  Solarbotics

Bend Sensor
  Images Company

Infrared Proximity/Distance Sensors
  Acroname
  HVW Technologies

Infrared Passive (PIR) Sensors
  Acroname
  Glolab

LCD Serial Controller
  Scott Edwards Electronics

Microcontroller Kits and Boards
  DonTronics
  microEngineering Labs
  Milford Instruments
  NetMedia
  Parallax, Inc.
  Savage Innovations
  Scott Edwards Electronics, Inc.

Motor Controllers ("Set and Forget")
  Solutions Cubed

Servo Motor Controller
  FerretTronis
  Lynxmotion
  Medonis Engineering
  Mister Computer
  Pontech
  Scott Edwards Electronics, Inc.

Shape Memory Alloy
  Mondo-Tronics

Sonar Sensors (Polaroid and others)
  Acroname

Speech Recognition
  Images Company

Surplus Mechanical Parts and Electronic Components
  All Electronics
  Alltronics

American Science & Surplus
B.G. Micro
C&H Sales
Halted Specialties Co.
Herbach & Rademan
Martin P. Jones & Assoc.

Wireless Transmitters (RF and Infrared)
  Abacom Technologies
  Glolab

General Robotics Kits and Parts

Acroname
P.O. Box 1894
Nederland, CO 80466
(303) 258-3161
www.acroname.com

Abacom Technologies
32 Blair Athol Crescent
Etobicoke, Ontario M9A 1X5
Canada
(416) 236-3858
www.abacom-tech.com

A.K. Peters, Ltd.
63 South Avenue
Natick, MA 01760
(508) 655-9933
www.akpeters.com

Amazon Electronics
Box 21
Columbiana, OH 44408
(888) 549-3749
www.electronics123.com

Design and Technology Index
40 Wellington Road
Orpington, Kent, BR5 4AQ
UK
+44 0 1689 876880
www.technologyindex.com

Images Company
39 Seneca Loop
Staten Island, NY 10314
(718) 698-8305
www.imagesco.com

Glolab Corp.
134 Van Voorhis
Wappingers Falls, NY 12590
www.glolab.com

HVW Technologies
Suite 473
300-8120 Beddington Blvd., SW
Calgary, Alberta T3K 2A8
Canada
(403) 730-8603
www.hvwtech.com

Hyperbot
905 South Springer Road
Los Altos, CA 94024-4833
(800) 865-7631
(415) 949-2566
www.hyperbot.com

Lynxmotion, Inc.
104 Partridge Road
Pekin, IL 61554-1403
(309) 382-1816
www.lynxmotion.com

Mekatronix
316 Northwest 17th Street, Suite A
Gainesville, FL 32603
www.mekatronix.com

Milford Instruments
120 High Street
South Milford, Leeds LS25 5AG
UK
+44 0 1977 683665
www.milinst.demon.co.uk

Mondo-Tronics, Inc
4286 Redwood Highway, #226
San Rafael, CA 94903
(415) 491-4600
www.robotstore.com

Mr. Robot
8822 Trevillian Road
Richmond, VA 23235
(804) 272-5752
www.mrrobot.com

Norland Research
8475 Lisa Lane
Las Vegas, NV 89113
(702) 263-7932
www.smallrobot.com

Personal Robot Technologies, Inc.
P.O. Box 612
Pittsfield, MA 01202
(800) 769-0418
www.smartrobots.com

RobotKitsDirect
17141 Kingview Avenue
Carson, CA 90746
(310) 515-6800 voice
www.owirobot.com

Sensory Inc
521 East Weddell Drive
Sunnyvale, CA 94089-2164
(408) 744-9000
www.sensoryinc.com

Solarbotics
179 Harvest Glen Way Northeast
Calgary, Alberta, T3K 3J4
Canada
(403) 818-3374
www.solarbotics.com

Technology Education Index
40 Wellington Road
Orpington, Kent, BR5 4AQ
UK
+44 0 1689 876880
www.technologyindex.com

Zagros Robotics
P.O. Box 460342
St. Louis, MO 63146-7342
(314) 176-1328
www.zagrosrobotics.com

Electronics/Mechanical Mail Order: New, Used, and Surplus

All Electronics
P.O. Box 567
Van Nuys, CA 91408-0567
(800) 826-5432
www.allelectronics.com

Alltech Electronics
2618 Temple Heights
Oceanside, CA 92056
(760) 724-2404
www.allelec.com

Alltronics
2300-D Zanker Road
San Jose, CA 95101-1114
(408) 943-9773
www.alltronics.com

American Science & Surplus
5316 North Milwaukee Avenue
Chicago, IL 60630
(847) 982-0870
www.sciplus.com

B.G. Micro
555 North 5th Street Suite #125
Garland, Texas 75040
(800) 276-2206
www.bgmicro.com

C&H Sales
2176 East Colorado Boulevard
Pasadena, CA 91107
(800) 325-9465
www.candhsales.com

DigiKey Corp.
701 Brooks Avenue South
Thief River Falls, MN 56701
(800) 344-4539
www.digikey.com

Edmund Scientific
101 East Gloucester Pike
Barrington, NJ 08007-1380
(800) 728-6999
www.edsci.com

Electro Mavin
2985 East Harcourt Street
Compton, CA 90221
(800) 421-2442
www.mavin.com

Electronic Goldmine
P.O. Box 5408
Scottsdale, AZ 85261
(480) 451-7454
www.goldmine-elec.com

Fair Radio Sales
1016 East Eureka Street
P.O. Box 1105
Lima, OH 45802
(419) 227-6573
www.fairradio.com

Gates Rubber Company
900 South Broadway
Denver, CO 80217-5887
(303) 744-1911
www.gates.com

Gateway Electronics
8123 Page Boulevard
St. Louis, MO 63130.
(314) 427-6116
www.gatewayelex.com

General Science & Engineering
P.O. Box 447
Rochester, NY 14603
(716) 338-7001
www.gse-science-eng.com

W. W. Grainger, Inc.
100 Grainger Parkway
Lake Forest, IL 60045-5201
www.grainger.com

Halted Specialties Co.
3500 Ryder Street
Santa Clara, CA 96051
(800) 442-5833
www.halted.com

Herbach and Rademan
16 Roland Avenue
Mt. Laurel, NJ 08054-1012
(800) 848-8001
www.herbach.com

Hi-Tech Sales, Inc.
134R Route 1 South
Newbury St.
Peabody, MA 01960
(978) 536-2000
www.bnfe.com

Hosfelt Electronics
2700 Sunset Boulevard
Steubenville, OH 43952
(888) 264-6464
www.hosfelt.com

Jameco
1355 Shoreway Road
Belmont, CA 94002
(800) 536-4316
www.jameco.com

JDR Microdevices
1850 South 10th Street
San Jose, CA 95112-4108
(800) 538-5000
www.jdr.com

Marlin P. Jones & Associates, Inc.
P.O. Box 12685
Lake Park, FL 33403-0685
(800) 652-6733
www.mpja.com

MCM Electronics
650 Congress Park Drive
Centerville, OH 45459
(800) 543-4330
www.mcmelectronics.com

McMaster-Carr
P.O. Box 740100
Atlanta, GA 30374-0100
(404) 346-7000
www.mcmaster.com

Mouser Electronics
958 North Main Street
Mansfield, TX 76063
(800) 346-6873
www.mouser.com

PIC Design
86 Benson Road
Middlebury, CT 06762
(800) 243-6125
www.pic-design.com

Scott Edwards Electronics Inc.
1939 South Frontage Road
Sierra Vista, AZ 85634
(520) 459-4802
www.seetron.com

Small Parts, Inc.
13980 Northwest 58th Court
P.O. Box 4650
Miami Lakes, FL 33014-0650
(800) 220-4242
www.smallparts.com

Surplus Traders
P.O. Box 276
Alburg, VT 05440 USA
(514) 739-9328
www.73.com

TimeLine, Inc.
2539 West 237 Street Building F
Torrance, CA 90505
(310) 784-5488
www.digisys.net/timeline/

Unicorn Electronics
1142 State Route 18
Aliquippa, PA 15001
(800) 824-3432
www.unicornelectronics.com

W.M. Berg, Inc.
499 Ocean Avenue
East Rockaway, NY 11518
(516) 599-5010
www.wmberg.com

Microcontrollers, Single-board Computers, Programmers

Boondog Automation
414 West 120th Street, Suite 207
New York, NY 10027
www.boondog.com/

DonTronics
P.O. Box 595
Tullamarine, 3043
Australia
(Check Web site for phone numbers)
www.dontronics.com

Gleason Research
P.O. Box 1494
Concord, MA 01742-1494
(978) 287-4170
www.gleasonresearch.com

Kanda Systems, Ltd.
Unit 17-18
Glanyrafon Enterprise Park
Aberystwyth, Credigion SY23 3JQ
UK
+44 0 1970 621030
www.kanda.com

microEngineering Labs, Inc.
Box 7532
Colorado Springs, CO 80933
(719) 520-5323
www.melabs.com

MicroMint, Inc.
902 Waterway Place
Longwood, FL 32750
(800) 635-3355
www.micromint.com

NetMedia (BasicX)
NetMedia Inc
10940 North Stallard Place
Tucson, AZ 85737
(520) 544-4567
www.basicx.com

Parallax, Inc.
3805 Atherton Road, Suite 102
Rocklin, CA 95765
(888) 512-1024
www.parallaxinc.com

Protean Logic
11170 Flatiron Drive
Lafayette, CO 80026
(303) 828-9156
www.protean-logic.com

Savage Innovations (OOPic)
2060 Sunlake Boulevard #1308
Huntsville, AL 35824
(603) 691-7688 (fax)
www.oopic.com

Technological Arts
26 Scollard Street
Toronto, Ontario
Canada M5R 1E9
(416) 963-8996
www.technologicalarts.com

Weeder Technologies
P.O. Box 2426
Fort Walton Beach, FL 32549
(850) 863-5723

Wilke Technology GmbH
Krefelder 147
D-52070 Aachen
Germany
+49 (241) 918 900
www.wilke-technology.com

Z-World
2900 Spafford Street
Davis, CA 95616
(530) 757-3737
www.zworld.com/

Radio Control (R/C) Retailers

Tower Hobbies
P.O. Box 9078
Champaign, IL 61826-9078
(800) 637-6050
(217) 398-3636
www.towerhobbies.com

Servo and Stepper Motors, Controllers

Effective Engineering
9932 Mesa Rim Road, Suite B
San Diego, CA 92121
(858) 450-1024
www.effecteng.com

FerretTronics
P.O. Box 89304
Tucson, AZ 85752-9304
www.FerretTronics.com

Hitec RCD Inc.
12115 Paine Street
Poway, CA 92064
www.hitecrcd.com

Medonis Engineering
P.O. Box 6521
Santa Rosa, CA 95406-0521
www.medonis.com

Mister Computer
P.O. Box 600824
San Diego, CA 92160
(619) 281-2091
www.mister-computer.com

Pontech
(877) 385-9286
www.pontech.com

Solutions Cubed
3029 Esplanade, Suite F
Chico, CA 95973
(530) 891-8045
www.solutions-cubed.com

Vantec
460 Casa Real Plaza
Nipomo, CA 93444
(888) 929-5055
www.vantec.com

Ready-Made Personal and Educational Robots

ActiveMedia Robotics
44-46 Concord Street
Peterborough, NH 03458
(603) 924-9100
www.activrobots.com

Advanced Design, Inc.
6052 North Oracle Road
Tucson, AZ 85704
(520) 575-0703
www.robix.com

Arrick Robotics
P.O. Box 1574
Hurst, TX, 76053
(817) 571-4528
www.robotics.com

General Robotics Corporation
1978 South Garrison Street, #6
Lakewood, CO 80227-2243
(800) 422-4265
www.edurobot.com

Newton Research Labs, Inc.
4140 Lind Avenue Southwest
Renton, WA 98055
(425) 251-9600
www.newtonlabs.com

Probotics, Inc
Suite 223
700 River Avenue
Pittsburgh, PA 15212
(888) 550-7658
www.personalrobots.com

Construction Kits, Toys, and Parts

Valient Technologies
(Inventa)
Valiant House
3 Grange Mills
Weir Road
London SW12 0NE
UK
+44 020 8673 2233
www.valiant-technology.com

Miscellaneous

Meredith Instruments
P.O. Box 1724
5420 West Camelback Rd., #4
Glendale, AZ 85301
(800) 722-0392
www.mi-lasers.com
Lasers

Midwest Laser Products
P.O. Box 262
Frankfort, IL 60423
(815) 464-0085
www.midwewst-laser.com
Lasers

Synergetics
Box 809
Thatcher, AZ 85552
(520) 428-4073
www.tinaja.com
Technical information

Techniks, Inc.
P.O. Box 463
Ringoes, NJ 08551
(908) 788-8249
www.techniks.com
Press-n-Peel transfer film

ROBOT INFORMATION ON THE INTERNET

Electronics Manufacturers

Shape Memory Alloy

Microcontroller Design

Robotics User Groups

General Robotics Information

Books, Literature, and Magazines

Surplus Resources

Commercial Robots

Video Cameras

Laser and Optical Components

LEGO Mindstorms Sources on the Web

Servo and Stepper Motor Information

Electronics Manufacturers

Analog Devices, Inc.
http://www.analog.com/

Atmel Corp.
http://www.atmel.com/

Dallas Semiconductor
http://www.dalsemi.com/

Infineon (Siemens)
http://www.infineon.com/

Maxim
http://www.maxim-ic.com/

Microchip Technology
http://www.microchip.com/

Motorola Microcontroller
http://www.mcu.motsps.com/

Precision Navigation
http://www.precisionnav.com/

Sharp Optoelectronics
http://www.sharp.co.jp/ecg/data.html

Xicor
http://www.xicor.com/

Shape Memory Alloy

(Portions of these pages are in Japanese)

http://www.toki.co.jp/BioMetal/index.html
www.toki.co.jp/MicroRobot/index.html

Microcontroller Design

Peter H. Anderson—Embedded Processor Control
http://www.phanderson.com/

"No-Parts" PIC Programmer
http://www.CovingtonInnovations.com/noppp/index.html

Iguana Labs
http://www.proaxis.com/~iguanalabs/tools.htm

LOSA—List of Stamp Applications
http://www.hth.com/losa/

Myke Predko's Microcontroller Reference
http://www.myke.com/

PICmicro Web Ring
http://members.tripod.com/~mdileo/pmring.html

Shaun's Basic Stamp II Page
http://www.geocities.com/SiliconValley/Orchard/6633/index.html

Robotics User Groups

Seattle Robotics Society
http://www.seattlerobotics.org/

Yahoo Robotics Clubs
http://clubs.yahoo.com/clubs/theroboticsclub
*http://search.clubs.yahoo.com/search/clubs?p=*robotics

The Robot Group
http://www.robotgroup.org/

Robot Builders
http://www.robotbuilders.com/

B-9 Builder's Club
http://members.xoom.com/b9club/index.htm

San Francisco Robotics Society
http://www.robots.org/

Nashua Robot Club
http://www.tiac.net/users/bigqueue/others/robot/homepage.htm

Mobile Robots Group
http://www.dai.ed.ac.uk/groups/mrg/MRG.html

Dallas Personal Robotics Group
http://www.dprg.org/

Portland Area Robotics Society
http://www.rdrop.com/~marvin/

General Robotics Information

Robotics Frequently Asked Questions
http://www.frc.ri.cmu.edu/robotics-faq/

Legged Robot Builder
http://joinme.net/robotwise/

Tomi Engdahl's Electronics Info Page
http://www.hut.fi/Misc/Electronics/

Boondog Automation Tutorials
http://www.boondog.com\tutorials\tutorials.htm

Find Chips Search
http://www.findchips.com/

Robotics Resources
http://www.eg3.com/ee/robotics.htm

Robotics Reference
http://members.tripod.com/RoBoJRR/reference.htm

Bomb Disposal Robot Resource List
http://www.mae.carleton.ca/~cenglish/bomb/bomb.html

Introduction to Robot Building
http://www.geckosystems.com/robotics/basic.html

Robot Building Information, Hints, and Tips
http://www.seattlerobotics.org/guide/extra_stuff.html

Suppliers for Robotics/control Models and Accessories
http://mag-nify.educ.monash.edu.au/measure/robotres.htm

Mobile Robot Navigation
http://rvl.www.ecn.purdue.edu/RVL/mobile-robot-nav/mobile-robot-nav.html

Robota Dolls
http://www-robotics.usc.edu/~billard/poupees.html

Basic Stamp, Microchip Pic, and 8051 Microcontroller Projects
http://www.rentron.com/

Hila Research QBasic
http://fox.nstn.ca/~hila/qbasic/qbasic.html

Dennis Clark's Robotics
http://www.verinet.com/~dlc/botlinks.htm

Dissecting a Polaroid Pronto One-step Sonar Camera
http://www.robotprojects.com/sonar/scd.htm

Polaroid Sonar Application Note
http://www.robotics.com/arobot/sonar.html

General Robot Info
http://www.employees.org:80/~dsavage/other/index.html

Standard Technologies of the Seattle Robotics Society
http://www.nwlink.com/~kevinro/guide/

Tech Wizards
http://www.hompro.com/techkids/

Android Workshop
http://www.tgn.net/~texpanda/library.htm

Hacking RAD Robot
http://www.netusa1.net/~carterb/radrobot.html

BEAM Robotics
http://nis-www.lanl.gov/robot/

Books, Literature, and Magazines

Robotics Book Reviews
http://www.weyrich.com/book_reviews/robotics_index.html

Robot Books
http://www.robotbooks.com/

Lindsay Publications
http://www.lindsaybks.com/

Circuit Cellar
http://www.circellar.com/

Midnight Engineering
http://www.midengr.com/

Robotics Bookstore
http://www.bectec.com/html/bookstore.html

Robohoo
http://www.robohoo.com/

Surplus Resources

Silicon Valley Surplus Sources
http://www.kce.com/junk.htm

Commercial Robots

Electrolux Vacuum Robot
http://www3.electrolux.se/robot/meny.html

Gecko Systems Carebot
http://www.geckosystems.com/

IS Robotics
http://www.isr.com/

Nomadic Technologies
http://www.robots.com/products.htm

Video Cameras

http://www.quickcam.com/
Logitech QuickCam

Laser and Optical Components

Ultrasonic Imaging Project
http://business.netcom.co.uk/iceni/usi_project/

Interfacing Polaroid Sonar Board
www.cs.umd.edu/users/musliner/sonar/

LEGO Mindstorms Sources on the Web

LEGO Mindstorms home page
http://www.legomindstorms.com/

LEGO Mindstorms Internals
http://www.crynwr.com/lego-robotics/

RCX Software Developer's Kit (from LEGO)
http://www.legomindstorms.com/sdk/index.html

RCX Internals
http://graphics.stanford.edu/~kekoa/rcx/

RCX Tools
http://graphics.stanford.edu/~kekoa/rcx/tools.html

Scout Internals (from LEGO)
http://www.legomindstorms.com/products/rds/hackers.asp

LEGO Dacta (educational arm of LEGO)
http://www.lego.com/dacta/

Pitsco (educational second sourcing for LEGO)
http://www.pitsco-legodacta.com/

LUGNET Newsgroups (technical LEGO discussion boards; robotics group is largest)
http://www.lugnet.com/

NQC (Not Quite C); (popular alternative programming environment for RCX)
http://www.enteract.com/~dbaum/nqc/

Gordon's Brick Programmer
http://www.umbra.demon.co.uk/gbp.html

LEGO on My Mind
http://homepages.svc.fcj.hvu.nl/brok//LEGOmind/

Mindstorms Add-Ons
http://www-control.eng.cam.ac.uk/sc10003/addon.html

MindStorms RCX Sensor Input
http://www.plazaearth.com/usr/gasperi/lego.htm

Servo and Stepper Motor Information

R/C Servo Fundamentals
http://www.seattlerobotics.org/guide/servos.html

Modifying R/C Servos to Full Rotation
http://www.seattlerobotics.org/guide/servohack.html

Definitive Guide to Stepper Motors
http://www.cs.uiowa.edu/~jones/step/index.html

Servo-Motor 101
http://www.repairfaq.org/filipg/RC/F_Servo101.html

Dual Axis Stepper Motor Controller
http://members.aol.com/drowesmi/dastep.html

Using the Allegro 5804 Stepping Motor Controller/Translator
http://www.phanderson.com/printer/5804.html

INTERFACING LOGIC FAMILIES

AND ICS

Most integrated circuits can be connected directly to one another, with no additional components. However, you should make some special design provisions when mixing CMOS and TTL logic families and when interfacing to or from mechanical switches, light-emitting diodes (LEDs), opto-isolators, relays, comparator ICs, and operational amplifiers (op amps). Refer to the following figures for more information on interfacing these components.

TTL to CMOS Level Translation

CMOS to TTL Level Translation

Op-Amp to CMOS and TTL Interfacing

Opt-Isolator Circuits

LED Drivers

CMOS and TTL to Relay and LED Drivers

TTL to CMOS Level Translation

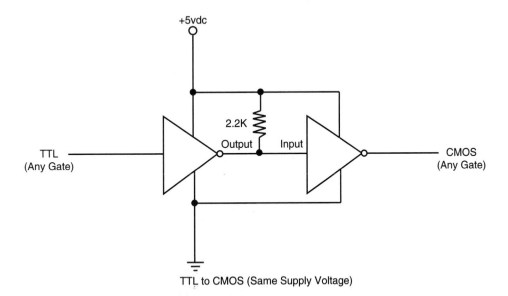

TTL to CMOS (Same Supply Voltage)

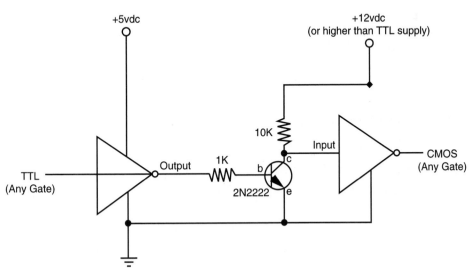

TTL to CMOS (Different Supply Voltage)

FIGURE GROUP D.1. **TTL to CMOS level translation.**

CMOS to TTL Level Translation

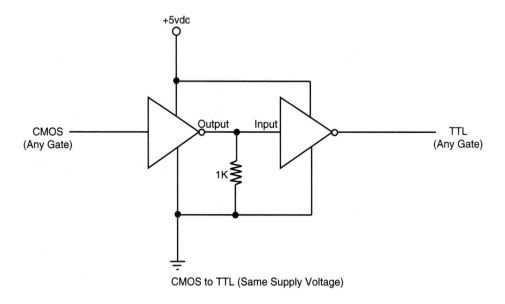

CMOS to TTL (Same Supply Voltage)

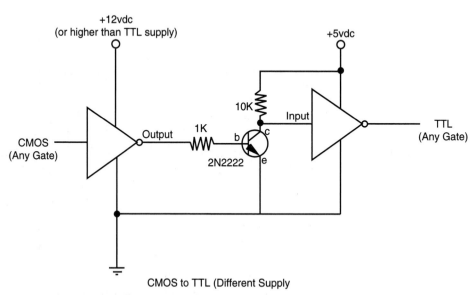

CMOS to TTL (Different Supply

FIGURE GROUP D.2.   **CMOS to TTL level translation.**

Op Amp to CMOS and TTL Interfacing

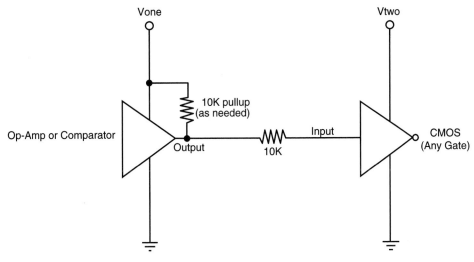

Op-Amp/Comparator to CMOS -- Vone = Vtwo

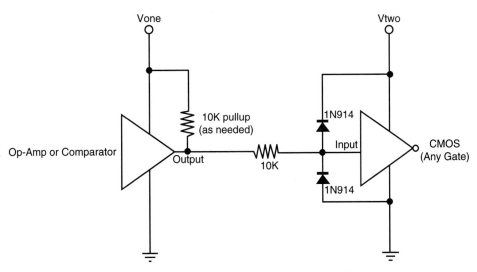

Op-Amp/Comparator to CMOS -- Vone <> Vtwo

FIGURE GROUP D.3.   **Op-amp—AMP to CMOS and TTL interfacing.**

Opto-Isolator Circuits

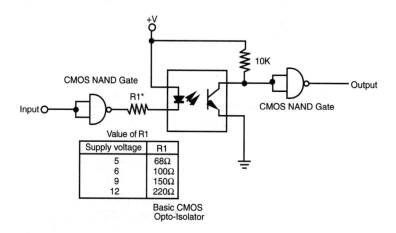

Value of R1

Supply voltage	R1
5	68Ω
6	100Ω
9	150Ω
12	220Ω

Basic CMOS
Opto-Isolator

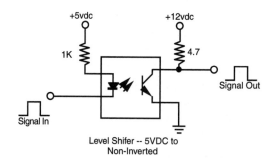

Level Shifer -- 5VDC to
Non-Inverted

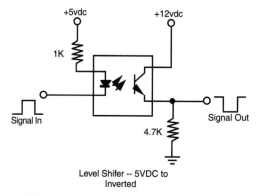

Level Shifer -- 5VDC to
Inverted

FIGURE GROUP D.4. **Opto-isolator interfacing.**

LED Drivers

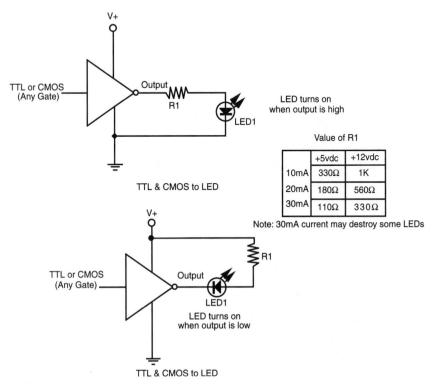

LED turns on
when output is high

TTL & CMOS to LED

Value of R1

	+5vdc	+12vdc
10mA	330Ω	1K
20mA	180Ω	560Ω
30mA	110Ω	330Ω

Note: 30mA current may destroy some LEDs

LED turns on
when output is low

TTL & CMOS to LED

FIGURE GROUP D.5.   **LED drivers.**

CMOS and TTL to Relay and LED Drivers

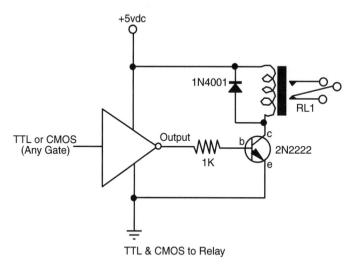

TTL & CMOS to Relay

FIGURE GROUP D.6. CMOS and TTL to RELAY and LED drivers.

APPENDIX E

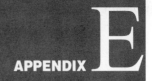

REFERENCE

Drill Bit and Bolt Sizing

Bolt	Size	Drill for hole	Drill # for NC tap
1/64		48	53
2/56		43	51
3/48	3/32″	38	47
4/40		33	43
5/40	1/8″	30	39
6/32		28	36
8/32		19	29
10/24	3/16″	10	25
12/24		7/32	16
1/4-20	1/4″	1/4	8
5/16-18	5/16″	5/16	F

FIGURE E.1

555 Circuit Reference

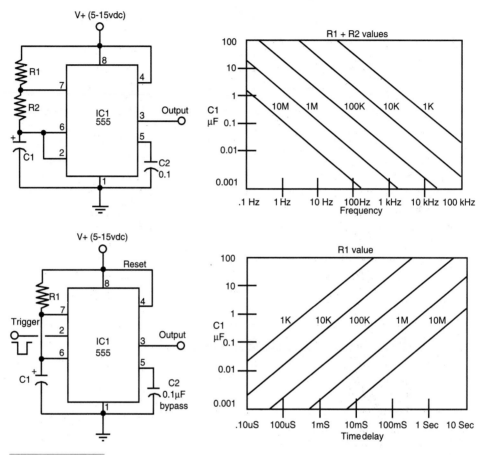

FIGURE E.2

567 Tone Decoder

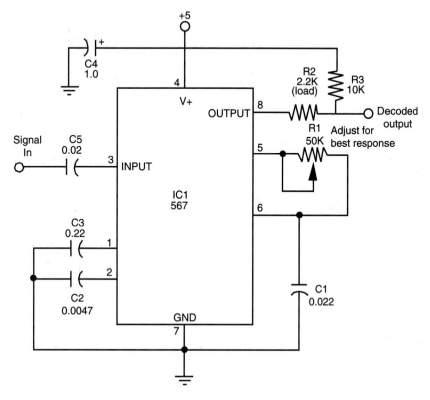

$$fo = \frac{1}{1.1R1C1}$$

$$\text{Bandwidth} = 1070 \sqrt{\frac{Vin}{foC2}} \text{ in \% of fo}$$

Components selected
for approx. 1kHz
response

FIGURE E.3

Wire Guage

TABLE E.1 SINGLE CONDUCTOR	
CONDUCTOR SIZE	**MAX. CURRENT CARRYING CAPACITY**
30 AWG	2
22 AWG	8
20 AWG	10
18 AWG	15
16 AWG	19
14 AWG	27
12 AWG	36

Source: Alpha Wire Company

INDEX

About the Author

Gordon McComb is an avid electronics hobbyist who has written for TAB Books for a number of years. He wrote the best-selling *Troubleshooting and Repairing VCRs* (now in its third edition), *Gordon McComb's Gadgeteer's Goldmine,* and *Lasers, Ray Guns, and Light Cannons.* He currently writes the widely read Robotics Workshop column for *Poptronics Magazine.*